Preventive and Predictive Genetics: Towards Personalised Medicine

Advances in Predictive, Preventive and Personalised Medicine

Volume 9

Series Editor:

Olga Golubnitschaja

Managing Editor:

Kristina Yeghiazaryan

For further volumes:
http://www.springer.com/series/10051

Godfrey Grech • Iris Grossman
Editors

Preventive and Predictive Genetics: Towards Personalised Medicine

Springer

Editors
Godfrey Grech
Pathology Department
University of Malta
Msida
Malta

Iris Grossman
IsraGene
Yakir
Israel

ISSN 2211-3495
ISBN 978-3-319-15343-8
DOI 10.1007/978-3-319-15344-5

ISSN 2211-3509 (electronic)
ISBN 978-3-319-15344-5 (eBook)

Library of Congress Control Number: 2015938453

Springer Cham Heidelberg New York Dordrecht London

Printed on acid-free paper

Springer is part of Springer Science+Business Media (www.springer.com)

What This Book Series is About . . .

Current Healthcare: What is Behind the Issue?

For many acute and chronic disorders, the current healthcare outcomes are considered as being inadequate: global figures cry for preventive measures and personalised treatments. In fact, severe chronic pathologies such as cardiovascular disorders, diabetes and cancer are treated after onset of the disease, frequently at near end-stages. Pessimistic prognosis considers pandemic scenario for type 2 diabetes mellitus, neurodegenerative disorders and some types of cancer over the next 10–20 years followed by the economic disaster of healthcare systems in a global scale.

Advanced Healthcare Tailored to the Person: What is Beyond the Issue?

Advanced healthcare promotes the paradigm change from delayed interventional to predictive medicine tailored to the person, from reactive to preventive medicine and from disease to wellness. The innovative Predictive, Preventive and Personalised Medicine (PPPM) is emerging as the focal point of efforts in healthcare aimed at curbing the prevalence of both communicable and non-communicable diseases such as diabetes, cardiovascular diseases, chronic respiratory diseases, cancer and dental pathologies. The cost-effective management of diseases and the critical role of PPPM in modernisation of healthcare have been acknowledged as priorities by global and regional organisations and health-related institutions such as the Organisation of United Nations, the European Union and the National Institutes of Health.

Why Integrative Medical Approach by PPPM as the Medicine of the Future?

PPPM is the new integrative concept in healthcare sector that enables to predict individual predisposition before onset of the disease, to provide targeted preventive measures and create personalised treatment algorithms tailored to the person.

The expected outcomes are conducive to more effective population screening, prevention early in childhood, identification of persons at-risk, stratification of patients for the optimal therapy planning, prediction and reduction of adverse drug-drug or drug-disease interactions relying on emerging technologies, such as pharmacogenetics, pathology-specific molecular patterns, sub/cellular imaging, disease modelling, individual patient profiles, etc. Integrative approach by PPPM is considered as the medicine of the future. Being at the forefront of the global efforts, the European Association for Predictive, Preventive and Personalised Medicine (EPMA, http://www.epmanet.eu/) promotes the integrative concept of PPPM among healthcare stakeholders, governmental institutions, educators, funding bodies, patient organisations and in the public domain.

Current Book Series, published by Springer in collaboration with EPMA, overview multidisciplinary aspects of advanced bio/medical approaches and innovative technologies. Integration of individual professional groups into the overall concept of PPPM is a particular advantage of this book series. Expert recommendations focus on the cost-effective management tailored to the person in health and disease. Innovative strategies are considered for standardisation of healthcare services. New guidelines are proposed for medical ethics, treatment of rare diseases, innovative approaches to early and predictive diagnostics, patient stratification and targeted prevention in healthy individuals, persons at-risk, individual patient groups, sub/populations, institutions, healthcare economy and marketing.

Prof. Dr. Olga Golubnitschaja

Book Series Editor

Dr. Golubnitschaja, Department of Radiology, Medical Faculty of the University in Bonn, Germany, has studied journalism, biotechnology and medicine and has been awarded fellowships for biomedical research in Paediatrics and Neuro-sciences (Medical Centres in Austria, Russia, UK, Germany, the Netherlands, and Switzerland). She is well-cited in the research fields of "gene hunting" and "subtractive hybridisation" applied to predictive prenatal and postnatal diagnostics published as

O. Labudova in years 1990–2000. Dr. Golubnitschaja is an expert in molecular diagnostics actively publishing in the fields of perinatal diagnostics, Down syndrome, diabetes mellitus, hyperhomocysteinemia, cardiovascular disease, neurodegenerative pathologies and cancer. She is the *cofounder* of the theory of multi-pathway organ-related blood fingerprintingwith specific molecular patterns at epi/genomic, transcriptional and post/translational levels and author of fundamental works in *integrative medicine.* Dr. Golubnitschaja holds appointments, at the rank of Professor, at several European Universities and in International Programmes for Personalised Medicine and is author of more than 300 international publications in the field. Awards: National and International Fellowship of the Alexander von Humboldt-Foundation; Highest Prize in Medicine and Eiselsberg-Prize in Austria; She is *Secretary-General* of the "European Association for Predictive, Preventive and Personalised Medicine" (EPMA in Brussels, www.epmanet.eu), Editor-in-Chief of *The EPMA-Journal* (BMC in London); Book Editor of *Predictive Diagnostics and Personalized Treatment: Dream or Reality,* Nova Science Publishers, New York 2009; Book Co-editor *Personalisierte Medizin,* Health Academy, Dresden 2010; Book Series Editor *Advances in Predictive, Preventive and Personalised Medicine,* Springer 2012; *European Representative* in the EDR-Network at the NIH/NCI, http://edrn.nci.nih.gov/; and *Advisor and Evaluator* of projects dedicated to personalisedmedicine at the EU-Commission in Brussels, NIH/NCI,Washington, DC, USA, and at Foundations and National Ministries of Health in several countries worldwide.

Preface

Traditionally, medical research comprised of the identification of the pathological causes of a disease, its epidemiology and empirical investigation of treatment response. Intensive genetic research, marked by the completion of the human genome project in 2003, heralded a new era in medical research. While epidemiology and gross pathology are still mainstay useful tools, genetics and genomics have gradually been shown to increase the resolution of drug response research, showing great potential in also informing and identifying the role of genes and their encoded products in the pathophysiology of diseases. This information is already being applied to effective early diagnosis, better risk assessment (prognosis), as well as targeted effective and safe treatment allocation (prediction and monitoring).

Genetic testing and genomics support personalised medicine by translating genome-based knowledge into clinical practice, offering enhanced benefit for patients and health-care systems at large. Current routine practice for diagnosing and treating patients is conducted by correlating parameters such as age, gender and weight with risks and expected treatment outcomes. In the new era of personalised medicine the healthcare provider is equipped with improved ability to prevent, diagnose, treat and predict outcomes on the basis of complex information sources, including genetic and genomic data. The support of regulatory bodies and policy makers internationally has been critical for the rapid translation of personalised medicines into the clinic. Notwithstanding, inequality in the utilisation of targeted therapies in different health care systems across the world exists, and ethical considerations, as well as economic cost-effectiveness analyses are in need to inform decision making. In addition to the benefits of pharmacogenomics in diagnosis and treatment, prevention of illness using genomic information is important to reduce the burden on the healthcare system, a methodology proven effective in many therapeutic areas, but, paradoxically still facing challenges in others. In current settings screening programmes (e.g. BRCA1/2 screening) address this by identifying susceptible families and preventive measures or ensuring appropriate treatment at the earliest stages of disease, hence increasing health management effectiveness.

The integration of pharmacogenomics into the various health care systems have been the responsibility of the respective national health authorities, which in turn follow recommendations by leading regulatory bodies such as the European

Medicines Agency and the US FDA. The dynamics and logistics of this integration therefore vary substantially across the globe. To this end, implementation of pharmacogenomics is an important component of PPPM (Predictive, Preventive and Personalised Medicine), which is the main focus of the European Association for Predictive, Preventive and Personalised Medicine (EPMA). Since 2009, EPMA (http://www.epmanet.eu/) embarked on various initiatives to promote PPPM including the launch of the EPMA Journal to ensure dissemination of current aspects of PPPM, and the organisation of the first World Congress in September 2011 bringing together participants from over 40 countries to discuss education, policy and implementation of PPPM. In addition, EPMA took the initiative to publish a series of books in advances in PPPM, including the present one entitled "**Preventive and Predictive Genetics: Towards Personalised Medicine**". The editors take this opportunity to thank all contributing authors and trust that the content meets the expectation of all readers.

The Editors

Godfrey Grech, PhD
and Iris Grossman, PhD

Acknowledgement of Reviewers

The Editors wish to acknowledge the reviewers for their generous contribution in assessing the Chapters providing constructive critique and recommendations. The reviewers provided the means to enhance the quality of the book and hence we sincerely are grateful.

Beena Koshy, Peter Shaw, Ann Daly, Anderson Wayne H, Dr Alex Gatt, Aruna Bansal, Michael Barnes, Godfrey Grech, Joseph Borg, George Patrinos, Christian Scerri.

Contents

Contributors

Patrick A. Akkari Cabernet Pharmaceuticals, Durham, NC, USA

James P. Bishop Teva Pharmaceuticals, Frazer, PA, USA

Anthonius de Boer Division of Pharmacoepidemiology and Clinical Pharmacology, Utrecht Institute for Pharmaceutical Sciences, Utrecht University, Utrecht, The Netherlands

Joseph Borg Department of Applied Biomedical Science, Faculty of Health Sciences, Mater Dei Hospital, Msida, Malta

Pablo Cacabelos Genomic Medicine, Camilo José Cela University, Madrid, Spain

EuroEspes Biomedical Research Center, Institute of Medical Science and Genomic Medicine, Corunna, Spain

Ramón Cacabelos Genomic Medicine, Camilo José Cela University, Madrid, Spain

EuroEspes Biomedical Research Center, Institute of Medical Science and Genomic Medicine, Corunna, Spain

Tomris Cesuroglu Department of Social Medicine, Maastricht University, Maastricht, Netherlands

Vasiliki Chondrou School of Health Sciences, Department of Pharmacy, University of Patras, Patras, GR, Greece

Bridget Ellul Faculty of Medicine and Surgery, University of Malta Medical School, Msida, Malta

Anthony G. Fenech Department of Clinical Pharmacology and Therapeutics, Faculty of Medicine and Surgery, University of Malta, Msida, Malta

Aiden Flynn Exploristics Ltd, Belfast, UK

Aikaterini Gravia School of Health Sciences, Department of Pharmacy, University of Patras, Patras, GR, Greece

Godfrey Grech Department of Pathology, Faculty of Medicine and Surgery, University of Malta, Msida, Malta

Iris Grossman IsraGene Ltd., Yakir, Israel

Personalized Medicine & Pharmacogenomics, Teva Pharmaceutical Industries, Netanya, Israel

Sonal B. Halburnt Cabernet Pharmaceuticals, Durham, NC, USA

Federico Innocenti UNC Institute for Pharmacogenomics and Individualized Therapy, University of North Carolina Eshelman School of Pharmacy, Chapel Hill, NC, USA

Theodora Katsila School of Health Sciences, Department of Pharmacy, University of Patras, Patras, GR, Greece

Craig Ledgerwood Exploristics Ltd, Belfast, UK

Marc Maliepaard Clinical Pharmacologist, Dutch Medicines Evaluation Board, Utrecht, The Netherlands

Felix J.M. van der Meer Department of Thrombosis and Hemostasis, Leiden University Medical Center, Leiden, The Netherlands

Janet Mifsud Department of Clinical Pharmacology and Therapeutics, University of Malta, Msida, Malta

Caroline O'Hare Exploristics Ltd, Belfast, UK

Jai N. Patel Levine Cancer Institute, Carolinas HealthCare System, Charlotte, NC

George P. Patrinos School of Health Sciences, Department of Pharmacy, University of Patras, Patras, GR, Greece

Munir Pirmohamed The Wolfson Centre for Personalised Medicine, Institute of Translational Medicine, University of Liverpool, Liverpool, UK

Michael A. Portelli Division of Respiratory Medicine, University of Nottingham, Queen's Medical Centre, Nottingham, UK

William K. Redekop Institute for Medical Technology Assessment, Erasmus University, Rotterdam, The Netherlands

Ian Sayers Division of Respiratory Medicine, University of Nottingham, Queen's Medical Centre, Nottingham, UK

Christian Scerri Department of Physiology and Biochemistry, Faculty of Medicine and Surgery, University of Malta, Msida, Malta

Jeanesse Scerri Department of Pathology, Mater Dei Hospital, Msida, Malta

Ron van Schaik Expert Center Pharmacogenetics, IFCC Reference Center for Pharmacogenetics, Dept. Clinical Chemistry, Erasmus University Medical Center, Rotterdam, The Netherlands

Tom Schalekamp Division of Pharmacoepidemiology and Clinical Pharmacology, Utrecht Institute for Pharmaceutical Sciences, Utrecht University, Utrecht, The Netherlands

Rianne M.F. van Schie Division of Pharmacoepidemiology and Clinical Pharmacology, Utrecht Institute for Pharmaceutical Sciences, Utrecht University, Utrecht, The Netherlands

Scott Sundseth Cabernet Pharmaceuticals, Durham, NC, USA

Clara Torrellas Genomic Medicine, Camilo José Cela University, Madrid, Spain

EuroEspes Biomedical Research Center, Institute of Medical Science and Genomic Medicine, Corunna, Spain

Richard Myles Turner The Wolfson Centre for Personalised Medicine, Institute of Translational Medicine, University of Liverpool, Liverpool, UK

Talitha I. Verhoef Division of Pharmacoepidemiology and Clinical Pharmacology, Utrecht Institute for Pharmaceutical Sciences, Utrecht University, Utrecht, The Netherlands

Christine M. Walko The DeBartolo Family Personalized Medicine Institute, Moffitt Cancer Center, Tampa, FL

Anke-Hilse Maitland-van der Zee Division of Pharmacoepidemiology and Clinical Pharmacology, Utrecht Institute for Pharmaceutical Sciences, Utrecht University, Utrecht, The Netherlands

About the Editor

Godfrey Grech, PhD

Senior Lecturer,
University of Malta

Dr. Godfrey Grech is Senior Lecturer at the University of Malta. Since 2006, Dr. Godfrey Grech is responsible for the coordination of an MSc in Biomedical Sciences and currently runs the Molecular Pathology laboratory at the University of Malta. Following his doctorate studies at Erasmus Medical Centre, Rotterdam, he published in high impact journals and presented in International conferences such as the American Society of Hematology, the EPMA Congress and the European Pathology Congress. In the last few years, Dr. Godfrey Grech was selected to lead the research arm of the National Transfusion Centre and was given the role of a Principle Investigator on the National Breast Cancer Research Project. In addition, he is highly recognised by the clinical sector and runs numerous projects with Mater Dei Hospital and was instrumental to set up infrastructures that allow the use of fresh surgical sections in molecular pathology research. Currently he runs various collaborative projects with International Institutions including Erasmus Medical Centre and the Molecular Medicine Institute in Leeds. The main research topic aims to identify biomarkers to classify breast cancer patients into a specific therapeutic group that shall benefit from activation of phosphatases as a main therapeutic option.

Dr. Godfrey Grech was nominated and is elected on international scientific committees including the International Scientific Council of the European Group for Molecular Pathology (EMP); Global Leader at the Genomic Medicine within the National Human Genome Research Institute (NHGRI) of the U.S. National Institutes of Health (NIH); member of the Pharmacogenomics Working Group of the Global Genomic Medicine Consortium (G2MC); Leader of the Cancer Position Paper at the European Association for Predictive, Preventive & Personalised Medicine (EPMA); and national contact point for the PharmacoGenetics for Every Nation Initiative (PGENI).

Iris Grossman, PhD

CEO and president,
IsraGene Ltd.

Vice President,
Global Head of Personalized Medicine and Pharmacogenomics, Teva Pharmaceuticals

Dr. Iris Grossman is CEO and president of the pharmacogenetics management consultancy IsraGene Ltd., as well as VP, global head of the Personalized Medicine and Pharmacogenomics (PMP) unit for Teva Global R&D. She has dedicated her research career, in both industry and academia, to the advancement of the field of personalised medicine. Dr. Grossman is currently charged with defining and implementing the global PMP strategy for Teva, a top-10 global pharmaceutical company, covering both discovery and development R&D programmes. Israel's leading financial magazine, Globes Magazine, selected Dr. Grossman as one of the country's top 40 professionals under 40 years of age in 2013.

This followed several years of spearheading pipeline pharmacogenetic programmes for industry and academia as director of pharmacogenetics at Cabernet Pharmaceuticals Inc. Dr. Grossman moved into consultancy having been responsible for running large-scale pharmacogenetic programmes at GlaxoSmithKline, with an emphasis on infectious and neurological diseases.

In academia, Dr. Grossman was a key member of Professor David Goldstein's team at the Center for Population Genomics and Pharmacogenetics, Institute for Genome Sciences and Policy, at Duke University. Dr. Grossman received her PhD from the Technion – Israel Institute of Technology, where her research project, conducted in collaboration with the Weizmann Institute for Science, investigated pharmacogenetic markers of multiple sclerosis treatment response.

Introduction

Godfrey Grech and Iris Grossman

Abstract Genetics evolved rapidly in the past decade, characterising genes that directly cause specific traits in monogenic diseases, as well as modifier genes that are associated with specific disorders but are not sufficient for causality, rather work in concert with additional gene and environmental factors to elicit the disease, mostly referred to as susceptibility genes. The science has also evolved from bench and lab discoveries to bedside implementations and further gathered momentum as knowledge acquired about allele and genotype frequencies in specific populations (via epidemiological studies) resulting in effective genetic disease prevention programmes. In parallel, numerous polymorphisms have by now been characterised and often formulated into drug labels, which play a role in the pharmacokinetics and pharmacodynamics of available therapeutics, predicting the efficacy of drugs in patients, and minimising the occurrence of drug adverse reactions.

The main two fronts of genetic contribution to personalised medicine address the preventive and the predictive aspects of medicine. Preventive genetics plays a major role in the characterisation of specific genetic disorders through populations-wide policies. Preventive programmes include population screening for carriers of rare, fully penetrant alleles that cause monogenic diseases, prenatal diagnosis of specific syndromes, and genotyping of susceptibility genes within families with high risk of developing a specific disease, providing the basis of Public Health Genetics. Carrier testing programmes go beyond the science and empirical testing, accompanied by premarital counselling aiming to provide necessary information to prevent the occurrence of disease, as well as support effective therapy and improved quality of life when such diseases are expressed. Preventing clinical manifestations by early

G. Grech (✉)
Department of Pathology, Faculty of Medicine and Surgery,
University of Malta, MSD2080 Msida, Malta
e-mail: godfrey.grech@um.edu.mt

I. Grossman
IsraGene, PO Box 339, Yakir, Israel
e-mail: irisg@isragene.com

Personalized Medicine & Pharmacogenomics, Teva Pharmaceutical Industries, Netanya, Israel

G. Grech, I. Grossman (eds.), *Preventive and Predictive Genetics: Towards Personalised Medicine,* Advances in Predictive, Preventive and Personalised Medicine 9,
DOI 10.1007/978-3-319-15344-5_1

diagnosis and intervention provides another level of preventive strategies that are not directed towards the prevention of the genetic defect, but avoiding emergence of symptoms by excluding exposure to exacerbating specific allergens/nutrients in disorders such as Glucose-6-Phosphate Dehydrogenase deficiency and phenylketonuria. Other strategies include the prevention of secondary complications of a disease, such as the use of antibiotics to prevent life-threatening bacterial infection in sickle cell disease. Overall, the multi-faceted approaches to newborn screening programmes are essential to prevent clinical manifestations or secondary complication of disease.

In complex diseases, genetic testing provides a means for measuring risk of manifesting disease given demographic and environmental risk factors. A "high risk" individual is then provided with recommendations for risk reduction interventions and/or access to screening and monitoring programmes at an earlier age, to prevent the occurrence or to delay onset of the disease. The efficiency of genetic testing to prevent disease depends on the weight of evidence, the predictive value of the variant(s) screened in the programme and the frequency of such alleles and genotypes in a specific population.

Another area of predictive genetics deals with the efficacy and toxicity of drugs in individuals. Pharmacogenomics is defined as the impact of the individual's genetic and genomic make-up on the body's response to drug therapy. This response depends on various factors additional to genotype and genomic expression, including environment, lifestyle and demographics (e.g. age and gender). Currently, the use of genetic information to treat patients is still in its early stages, with some clear success mostly in the oncology and infectious diseases therapeutic areas. Some successful examples include the targeting of tailored pharmaceuticals developed for the treatment of patients with a particular disease subtype or according to a specific genetic make-up pertaining to the drug's mode of action (e.g. zelboraf). In other examples, genetic information is being used to help determine the effective and safe dose of specific pharmaceuticals (e.g. warfarin). However, implementation of this pharmacogenetic knowledge to the clinic has proven to be challenging and to require collaboration between the various stakeholders throughout the discovery, development and validation stages so as to ensure the utility of actionable genetic testing in a cost-effective manner. Targeted therapy and reliable prediction of expected outcomes offer patients access to better healthcare management, by way of identifying the therapies effective for the relevant patient group, avoiding prescription of unnecessary treatment and reducing the likelihood of developing adverse drug reactions.

In accordance with the main themes that define preventive and predictive genetics and its utility and wide-spread adoption world-wide, the chapters of this book walk the reader through the principles of this discipline and the state of the art across key therapeutic areas. To this end, the second chapter (i.e. the one preceding the introduction) discusses the broad definition of public health and the specific role that genetic testing plays in advancing population-level health outcomes. Indeed, preventive genetics has demonstrated utility as a crucial component in the success of population-wide health policies that promote improved health outcomes. The identification of subjects at risk at the earliest age possible provides opportunities for tailoring actionable medical solutions when needed. Principles of preventive genetic programmes are outlined in details, and specific examples reviewed, including phenylketonuria, MCADD, homocystinuria, maple syrup urine disease, glutaric aciduria type 1, cystic fibrosis, haemoglobinopathies, hereditary haemochromatosis, familial hypercholesterolaemia, familial adenomatous polyposis and familial cancer predisposing syndromes. In addition, while complex traits are mostly yet to have been fully characterised in terms of the exact proportion explained by genetics on a population basis, characterisation of monogenic subtypes within complex diseases can be easily taken up into genetic testing programmes, as exemplified in detail by BRCA1 and BRCA2 for risk of breast and ovarian cancers, as well as other relevant cases. The chapter also touches on the practical, ethical and clinical aspects associated with biobanking of the required samples that facilitate the research, as well as application of screening genetics over time.

In the third chapter Bishop et al. describe the use of pharmacogenetics to the development of safer, more effective and differentiated therapies. This chapter describes the principles and requirements of an efficient and valuable pharmacogenetic strategy throughout the course of developing innovative medicines. This strategy combines a proven methodology with rigorous genetic science to create a "Pipeline Pharmacogenetic Program". By describing the pharma industry and the market forces shaping its drivers, pharmacogenetic applications are portrayed as aides to reduce attrition and enhance the scientific rigor, and over all benefit/risk profile of novel therapies. The authors review the scientific requirements, as well as sample collection and practical decision making perspectives that must be taken into account during R&D. Specific examples are shared throughout the drug development continuum and across a variety of therapeutic areas, including Alcohol Dependence, Oncology and Rare Diseases. Finally, unique features associated with the contemporenous development of drug and companion diagnostic are reviewed by way of describing an example dealing with Alzheimer's disease management.

The type of data and design of pharmacogenetic studies is a requirement to provide the necessary outcome and define actionable markers. In Chap. 4 Flynn et al. summarise the key statistical consideration required for successful and meaningful pharmacogenetic programmes. As a scientific discipline pharmacogenomics must demonstrate rigor of study design and significance of statistical findings, additional to biological and clinical relevance of variants identified. In addition, consideration of statistical factors unique to pharmacogenetics must be examined over the course of biomarker studies aspiring to implement prospective analyses. Furthermore, the nature of the biomarkers studied, whether predictive or prognostic, dictates a different suit of statistical considerations, as examplified by Flynn et al. This chapter provides the elements of good statistical practices in the pharmacogenomic space, spanning the entire field, from study design, source of variability, dimensionality, confirmation, model building, bioinformatics and ultimate development of diagnostics.

In Chap. 5 Mifsud et al. summarise decades of pharmacogenetic research dedicated to the various pharmacokinetics processes that drugs are subject to *in vivo* i.e.

absorption, distribution, metabolism and elimination. Examples span the earliest reports on succinylcholine from the 1950s through to the latest advances. In addition, authors provide a window into regulatory perspectives world-wide, which contribute to the adoption of existing knowledge, as well as dissemination of ADME genetics into novel drug development.

The critical role of genetics in predicting adverse drug reactions is described in Chap. 6 by Turner et al. Paradigm setting examples of the high predictive value of genetic variations for Immune-Mediated Adverse Drug Reaction are described in detail. However, the authors provide ample evidence that genetics can be a key determinant of adverse drug reactions associated with virtually any disease area and any drug mode of action, including analgesia, coagulation, cancer and cholesterol levels. These seminal examples have affected the medical profession in a profound fashion, ushering an industry of diagnostics that is widely accepted world-wide.

Chapter 7 builds on the insights revealed in the previous chapters and adds the first example of applied pharmacogenomics, describe the state of research and utility of pharmacogenomics in prescription of haemoglobinopathies therapeutics. Here, Gravias et al. describe the available treatment options and the genetic factors that have thus far been linked mostly to the β-globin gene cluster. These are believed to act by modulating HbF levels. The authors' analysis is concluded by the observation that the use of pharmacogenomics for haemoglobinopathies therapeutics are currently very limited, requiring larger studies in ethnically diverse patients groups.

Cacabelos et al. review the state of research and applicability of pharmacogenetics to neurodegenerative diseases in Chap. 8. The five categories of genetic variants associated with this field are defined as: (i) genes associated with disease pathogenesis (pathogenic genes); (ii) genes associated with the mechanism of action of drugs (mechanistic genes); (iii) genes associated with drug metabolism; (iv) genes associated with drug transporters; and (v) pleiotropic genes involved in multifaceted cascades and metabolic reactions. The role of each of these categories is then examined within the prototypic neurodegenerative diseases, i.e. Alzheimer's disease, Parkinson's disease, Amyotrophic Lateral Sclerosis (ALS), Multiple Sclerosis and Huntington's disease. The authors conclude by postulating future areas of focus for pharmacogenomic research, as well as overall policies associated with chronic, debilitating, late-onset diseases that affect the nervous system and expected to affect significant proportions of the aging world population.

The genetics and pharmacogenetics of asthma is reviewed in Chap. 9. Current approaches to asthma management call for clinical severity assessments, with regular re-evaluations of treatment, that is subsequently often redosed or switched. Specific pharmacogenetic considerations are provided which address the major drug classes in current use for asthma and/or chronic obstructive pulmonary disease (COPD), including leukotriene modifiers, glucocorticoids and anti-muscarinic agents. The evolving use of pharmacogenetic tools in novel drug development in respiratory indications in further details, providing great promise for patients.

Patel et al. provide an overview on the pharmacogenetics of antineoplastics in Chap. 10. In this context, the authors explain the unique attributes associated with germ-line versus somatic mutations, and the associated prognostic versus predictive value of the reported biomarkers within each category. Specifically, examples are shared which directly address clinical decision making for a variety of solid and liquid tumour types, including the use of 6-mercaptopurine (6-MP), irinotecan, tamoxifen, fluorouracil, crizotinib, imatinib, ATRA, vemurafenib. erlotinib, herceptin, panitumumab and cetuximab. With the reduction in sequencing technology cost it is expected that tumour profiling will be used for patient classification and drug development, as well as for identification of driver mutations that define causality, diagnosis, prognosis and life-saving treatment choices.

Chapter 11 entails the utility of pharmacogenomics to anticoagulant therapy. Here, van Schie et al. describe both the basic science and clinical evidence associated with multi-marker testing for coumarin anti-coagulant therapy. Furthermore, the authors expand the scope of their analysis to coumarin derivatives, clinical trials investigating the effectiveness of pre-treatment genotyping and the cost-effectiveness of pharmacogenetic-guided dosing. These are critical studies required for adoption of pharmacogenetics to common practice and serve as precedents for the entire field.

Having reviewed each therapeutic area separately, Chap. 12 serves to provide a common vision to lessons learned and remaining challenges associated with the practice of genomic medicine. Here, Grech et al. focus on the needs and recommendations to promote patient molecular classification; stratification of well-defined subgroups of predicted responders to specific therapies; the development of technologies and integrative information systems to provide the healthcare system with optimised and sustainable genetic testing protocols; the need of harmonised guidelines for the proper selection of patient groups for clinical trials; and advances in research to generate evidence based knowledge that can be smoothly translated for healthcare use. Key gaps in the uptake of Genomic Medicine in the Health Care System are attributed by the authors to insufficient education of the Healthcare professionals and lack of mechanisms for appropriate dissemination of genomic information within the healthcare system. Lastly, the research community is still lagging behind in providing real-world, validated evidence to the validity of pharmacogenomic findings.

The book is concluded by Ellul, summarising ethical considerations associated with pharmacogenomics. The chapter focuses on ethical issues affecting the individual patient through his or her experience undergoing pharmacogenetic testing for personalised treatment, enrolling in clinical trials, participating in genomic research or donating biological material for biobanking and research. Core concept leading the ethical discussion center around perceived and actual benefits and risks, and the relative relationship between the two. The ubiquitously acceptable tool of informed consent is presented, including its variable applications and its existing and evolving guiding rules. Aspects of discrimination, ethnicity, privacy, confidentiality and the responsibilities of each of the associated stakeholders is detailed as well, providing a rounded account of the complexities and opportunities associated with pharmacogenomic utilisation.

In summary, this book collates a comprehensive account of the state of the pharmacogenomic science and its application to the management of most common disease areas, as pertaining to available therapies as well as those in development for future, better patient use. The book provides an outstanding didactic content in both preventive and predictive genetics and intended for use in postgraduate courses in Molecular Biology and Genetics, Bioinformatics and other life sciences programmes that focus on applied Genetics for future medicine. The common theme concluded by each of the expert authors converges into the vision that targeted therapies will become mainstay across all disease areas. Further, with the exponential growth of omics Big Data, our ability to translate sequence variation into useful tools for drug development and utilisation will ensure a speedy, safe and efficaceuos drug development process for future generations.

Preventive and Predictive Genetics: A perspective

Godfrey Grech, Christian Scerri, Jeanesse Scerri and Tomris Cesuroglu

Abstract Public Health practices focus on the implementation of programmes for health improvement and disease prevention (Khoury et al., Am J Prev Med 40(4):486–493, 2011). Public health initiatives in diseases were initially targeted to prevent infectious diseases. Partly due to the availability of vaccines and antimicrobial therapy and partly due to better standard of living, the world is free of diseases such as small pox, almost free of polio and the prevalence of infections such as malaria and HIV is steadily on the decline. This has meant that the human race is living longer with the result that non-communicable diseases have become a global public health priority. Preventing non-communicable diseases is a more logical approach than treating them, even more so when modifiable, common lifestyle risk factors share a role in the onset and progression of the disease. Preventive genetics plays a crucial role in the identification of subjects at risk at a very early age, which would thus give public health officials the necessary time to take appropriate action.

Genetic tests can be classified into carrier, diagnostic and predictive testing. In carrier testing, the tests are directed towards the identification of carriers of autosomal recessive or X-linked genetic disorders to prevent disease. Preventive genetics can be defined as using genetics for the prevention of a future disease that has a genetic component either in the individual tested or in future offspring. Diagnostic testing is the process that identifies the current disease status of the subject and includes, among others, prenatal and newborn screening. The implementation of screening programmes allow the detection of genetic disorders at an early stage, so as to prevent these conditions or their serious consequences. Predictive testing

G. Grech (✉)
Department of Pathology, Faculty of Medicine and Surgery,
University of Malta, Msida MSD2080, Malta
e-mail: godfrey.grech@um.edu.mt

C. Scerri
Department of Physiology and Biochemistry, Faculty of Medicine and Surgery,
University of Malta, Msida MSD2080, Malta

J. Scerri
Department of Pathology, Mater Dei Hospital, Malta

T. Cesuroglu
Department of Social Medicine, Maastricht University, Maastricht, Netherlands

G. Grech, I. Grossman (eds.), *Preventive and Predictive Genetics: Towards Personalised Medicine,* Advances in Predictive, Preventive and Personalised Medicine 9,
DOI 10.1007/978-3-319-15344-5_2

determines whether a subject with a positive history but no symptoms of the disease, is at risk of developing the disorder at a future date. In this chapter, we will discuss the application of genetic screening tests to monogenic disorders and complex disorders with monogenic subsets, in view of the current practices. The multifactorial aetiology of complex disorders involves multiple gene effects and gene-lifestyle interactions that cannot be singled out to give a strong predictive value. However, a subset of the complex disorders are caused by highly penetrant genetic mutations. Hence, in this chapter we shall also address predisposing syndromes with high predictive value. In addition, the need of biobanks will be discussed.

Keywords Public health genomics · Preventive genetics · Predictive genetics · Screening programmes · Monogenic disorders · Monogenic syndromes · Biobanks

1 Screening Programmes

According to the UK National Health Services, the term *screening* signifies a public health service aimed at identifying individuals that, though apparently healthy, are at risk of or are already affected by a particular disease or its complications. This identification could be through a particular medical or biochemical test or a questionnaire. Once identified, the affected individuals could then be offered further information, complementary tests or treatment in a bid to reduce their risk of developing the disease or its complication. The use of genetic testing to improve healthcare, requires implementation of programmes as part of public health practice [1]. In their 2004 paper on the history of medical screening, Morabia and Zhang [2] identified the US army's 1917 screening programme, aimed to exclude individuals with clear psychological disorders from joining the army, as the first reported instance of a "screening programme." This programme consisted of the administration of psychological tests to officers, drafted and enlisted soldiers. Since this early example of a screening programme, other initiatives mostly directed towards the general public and aimed at the prevention or early treatment of important health conditions, have spread across the world.

There are three main types of organised screening programmes, namely population screening, newborn screening (NBS) and cascade screening. Population-based screening involves testing the majority of the population, which may either be defined as the whole population of a country or a specific population at risk (such as Ashkenazi Jews for Tay-Sachs disease, or women over a specific age for breast cancer screening). A highly targeted population for screening of genetic disorders is that of women at the prenatal or pre-conception stage, due to their high accessibility and ease of retraceability. Whereas pre-conception screening allows a wider choice of reproductive options than prenatal screening (i.e. opting to have no children, using a sperm donor, or pre-implantation genetic diagnosis (PGD) and selective implantation of embryos created through in-vitro fertilisation (IVF) in the former, against elective termination of pregnancy in the latter), antenatal groups are easier to target.

Meanwhile, NBS involves screening of all newborns in order to detect (generally and preferably) early-onset diseases before the occurrence of overt symptoms. Several genetic metabolic disorders, diagnosed through biochemical tests rather than genetic tests, constitute the core group of disorders for which established newborn screening programmes exist worldwide. These include phenylketonuria (PKU), for which the first NBS programme was set up in the early 1960s, maple syrup urine disease (MSUD) and congenital hypothyroidism (CH). Other specific disorders that are tested for within certain regions or ethnic groups include the haemoglobinopathies (by isoelectric focusing [IEF] and high-performance liquid chromatography [HPLC]) and Cystic Fibrosis (CF; by immunoreactive trypsinogen [IRT]). Cascade screening starts with an index case showing symptoms (the proband) and testing family members for mutations predisposing them to the same disease. Unlike the other forms of screening, cascade screening specifically targets individuals considered at high risk of acquiring the disease due to family history and is restricted to genetic (i.e. DNA) tests.

In the late 1960s, technological advances in medicine enabled the spread of screening in various fields of medicine but, at the same time, brought forward topics of controversy as well as ethical implications. Under this scenario, the World Health Organisation commissioned a report on screening from James M. G. Wilson and Gunner Jungner. In their 1968 report, Wilson and Jungner [3] recounted their pre-occupation that, while the "central idea of early disease detection and treatment is essentially simple", achieving its success of "bringing to treatment those with previously undetected disease", as opposed to "avoiding harm to those persons not in need of treatment", is not as easy as it might appear. In an attempt to simplify the process of screening, Wilson and Jungner proposed a set of criteria that has been adopted as the gold standard in the establishment of all screening programmes. According to these criteria, screening programmes should be considered for conditions fitting in within the following:

1. The condition should be an important health problem.
2. There should be a treatment for the condition.
3. Facilities for diagnosis and treatment should be available.
4. There should be a latent stage of the disease.
5. There should be a test or examination for the condition.
6. The test should be acceptable to the population.
7. The natural history of the disease should be adequately understood.
8. There should be an agreed policy on whom to treat.
9. The total cost of finding a case should be economically balanced in relation to medical expenditure as a whole.
10. Case-finding should be a continuous process, not just a "once and for all" project.

However, these criteria were targeted towards screening for diseases of significant burden in general and did not take into account certain aspects pertaining to genetic diseases, such as the serious debilitating nature of certain rare genetic conditions and the inheritable nature of these disorders. The accelerated rate of discovery of new disease-causing genes has opened up a whole new dimension of diagnosis by using genetic testing to detect diseases before the first clinical signs and/or symptoms appear, even before the disease starts its pathological course. Conversely, these advancements have also caused public health to lag behind in the introduction or expansion of genetic screening programmes, mostly because the decision-making process requires extensive risk-assessments and the implementation of pilot studies, as well as control and standardisation of such programmes [4]. This has resulted in different countries applying different criteria, most of which are based on the original ones by Wilson and Jungner but which take into account the mentioned additional factors, leading to a lack of standardisation or consensus. In the following sections, disorders for which screening programmes are either already in place or may be considered in the near future will be described, with the application of screening criteria to the decision-making process. In addition, the use of next-generation sequencing for newborn screening will be discussed as a future strategy in public health genomics.

2 Monogenic Disorders

Monogenic disorders are mainly rare disorders, caused by single-gene modifications that are present in all the cells of the body and following Mendelian modes of inheritance, i.e. dominant, recessive or X-linked [5]. Though the phenotype in monogenic disorders is almost dependent on a single genetic component, it is also influenced by the individual's genome (modifier genes), as well as environmental and lifestyle factors. Rare, or orphan, disorders are most commonly defined as disorders affecting 5/10,000 persons or less, but collectively they contribute to a significant degree of morbidity and mortality. Furthermore, eighty percent of rare disorders have genetic origins [6], and the advent of next-generation sequencing has brought about the accelerated discovery of new causative mutations [7]. There are also several monogenic disorders which are relatively common, either worldwide or in specific populations. Being easier to identify, either by biochemical or genetic tests, these disorders were the first genetic conditions for which population screening programmes were established.

2.1 Inborn Errors of Metabolism

An important subset of monogenic disorders consists of the inborn errors of metabolism. Neonatal screening for inborn errors of metabolism has been in place since the early 1960s, following the development of a fast and cheap blood test for phenylketonuria (PKU) [8]. After the success of the first newborn screening programme, maple syrup urine disease (MSUD) screening by the Guthrie ("heel prick") method was soon added. Up to the introduction of tandem mass spectrometry (MS/MS) technology in newborn screening, individual diseases where tested on a one test, one disorder system that is both expensive and relatively inefficient. The use of MS/MS has expanded the number of inborn errors of metabolism disorders that can be tested from a single dried blood spot, in one single analytical run. To date, MS/MS can identify 45 different disorders, including amino acid disorders, fatty acid oxidation disorders, and organic acidemias, from a single blood spot taken in the neonatal period [9]. Almost all of the national newborn screening programmes in the world include PKU, with medium-chain acyl-coA dehydrogenase deficiency (MCADD) found in a majority of programmes. Other disorders that are actively being added to the list are homocystinuria (HCU), maple syrup urine disease (MSUD) and glutaric aciduria type 1 (GA1).

2.1.1 Phenylketonuria (PKU)

PKU is an autosomal recessive disorder with a reported prevalence as high as 1 in 4500 in Ireland [10] to as low as less than 1 in 100,000 in Finland [11]. In its typical form, it is a result of a mutant phenylalanine hydroxylase (PAH) enzyme, which catalyses the conversion of phenylalanine to tyrosine. Reduced activity of PAH results in an accumulation of phenylalanine and its metabolite phenylpyruvate. This accumulation of phenylalanine leads to intellectual disability, seizures and other symptoms. The major problem is an inability of the brain to utilise other large neutral amino acids (LNAA), such as tryptophan and tyrosine, since the large quantities of phenylalanine block the blood-brain barrier's large neutral amino acid transporter (LNAAT). Thus, the brain is starved from amino acids that are essential for proper synthesis of neurotransmitters. Management of PKU consists in low phenylalanine diet and possibly the oral administration of tetrahydrobiopterin that is a cofactor in the oxidation process of phenylalanine. To be successful in achieving normal brain development, the diet needs to be initiated as early as possible and continued throughout life.

2.1.2 Medium-Chain Acyl-coA Dehydrogenase Deficiency (MCADD)

Medium-chain acyl-coA dehydrogenase deficiency (MCADD) is a disorder of fatty acid oxidation, in particular the conversion of medium chain fatty acids (MCFA) into acetyl-CoA. It is most prevalent in individuals of Northern European Caucasian descent, with reported worldwide prevalence of between 1:10000 and 1: 27000 [12]. The inability to convert MCFA into Acetyl-CoA results in a deficiency of fatty acid breakdown during periods of metabolic stress, such as a period of fasting or illness. The presenting symptoms include hypoglycaemic attacks, vomiting and encephalopathy that can lead to coma and sudden death. As studies have shown that over 25 % of undiagnosed children die during their first attack, with 16 % of the survivors having severe neurological deficiencies [13], early diagnosis and start of treatment is essential. The latter consists of a diet which is high in carbohydrates and low in fatty acids, in particular during periods of risk, and avoidance of fasting [14].

2.1.3 Homocystinuria (HCU)

Homocystinuria (HCU) is another rare, autosomal recessive condition, where the body is unable to convert homocysteine (derived from methionine) into cystathionine due to the reduced activity of the enzyme cystathionine beta-synthase. Thus, in a similar way to phenylketonuria, homocysteine and to an extent methionine, accumulate in the body while there is a deficiency of cystathionine. The reported worldwide prevalence is of 1 in 344,000, with certain countries such as Ireland (1: 65000) reporting comparatively very high rates [15]. The symptoms of HCU include developmental delay of the brain, behavioural changes such as mood and personality disturbances, dislocation of the lens of the eye, disproportionate length of limbs compared to the body, osteoporosis and higher risk of vascular thrombosis. Treatment consists of vitamin B6 (a cofactor of cystathionine beta-synthase) and a low methionine diet. In addition, the administration of oral betaine helps to convert homocysteine back into methionine and thus reduce the levels of homocysteine.

2.1.4 Maple Syrup Urine Disease (MSUD)

Maple syrup urine disease (MSUD) gets its name from the presence of a distinctive sweet smelling urine reminiscent of maple syrup. The condition can be the result of mutations in any one of the four genes that code for the 4 subunits ($E_1\alpha$, $E_1\beta$, E_2, and E_3) of the branched-chain alpha-keto acid dehydrogenase complex (BCKDC). The lack of BCKDC leads to a gradual increase of the branched-chain amino acids isoleucine, leucine and valine, resulting in a build-up of toxic ketoacids in the blood and consequently in the urine. Like most of the other inborn errors of metabolism, the condition is inherited in an autosomal recessive way. As the disorder can be due to the lack or reduced activity of any one of the 4 subunits, its clinical picture and prognosis is variable, but brain damage with mental problems as well as physical problems, such as lethargy, hypotonia, seizures, pancreatitis and ketoacidosis, are common events. The management of MSUD requires a life-long diet where the level of these essential branched chain amino acids is kept under strict control and at the barest minimum. If well managed from early years, those afflicted are usually able to live a normal life with little or no neurological damage. The prevalence of this disorder is reported to be approximately 1 in 200,000 [9], with increased incidence in persons of Amish, Mennonite, and Ashkenazi Jewish descent [16–18].

2.1.5 Glutaric Aciduria Type 1 (GA1)

Glutaric aciduria type 1 (GA1) is another inherited error of metabolism involving the reduced capability of the breakdown of amino acids. Its worldwide prevalence is reported as 1:100,000 [19]. GA1 is caused by mutations within the *GCDH* gene that encodes for the enzyme glutaryl-CoA dehydrogenase. The enzyme is required for the metabolism of the amino acids hydroxylysine, lysine and tryptophan. This reduction of effective enzyme activity results in a build-up of intermediate metabolites 3-hydroxyglutaric acid and glutaconic acid. Both these metabolites are toxic to the basal ganglia, with initial symptoms such as macrocephaly occurring at birth in some affected individuals, though it is not unusual for symptoms to become apparent during adolescence or early adulthood. Without treatment, the condition invariably leads to encephalitis-like crisis that leaves severe *sequelae* such as developmental delay, neurologic deterioration, and cerebral palsy. Encephalitis is usually triggered by an intercurrent illness such as infection, fever or prolonged fasting. Treatment consists of dietary manipulation to ensure a low level of lysine and tryptophan, avoidance of prolonged fasting (less than 4–6 h) as well as treatment with supplements such as riboflavin and L-carnitine.

The collective importance of the health burden constituted by inborn errors of metabolism, together with the cost-effective and acceptable patient testing and management strategies, make them ideal candidates for newborn testing programmes; in fact, several countries already have such programmes set-up, with a significant number offering extended screening through MS/MS (Table 3).

2.2 *Cystic Fibrosis*

Cystic fibrosis (CF) is the most common lethal autosomal recessive condition among Caucasians, with a prevalence of around 1/2500 live births [20] and a high mutation carrier rate at 1/25 [21]. It is caused by mutations in the *CFTR* gene encoding the cystic fibrosis transmembrane conductance regulator protein. The latter functions mainly as a chloride ion channel and is present in the epithelial cell membranes of several tissues having secretory functions, such as the lungs, sweat glands, gastrointestinal tract and pancreas [22]. Mutations in *CFTR* result in the loss or impaired function of the channel, leading to the pathological hallmarks of the disease. The main manifestations are in the lung and pancreas: 85–90 % of affected children develop pancreatic insufficiency by the first year of life, while pulmonary insufficiency causes more than 80 % of CF-related mortality. In the lungs, increased mucous viscosity inhibits its normal clearance by cilia and coughing; the excess mucous forms plaques with hypoxic spaces which can be colonised by opportunistic bacteria such as *Pseudomonas aeruginosa*. Chronic infections of the airways with inflammation and infiltration by polymorphonuclear cells are followed by bronchiectasis (irreversible dilation of the airways), hypoxaemia (low blood oxygen concentration) and hypercarbia (abnormally high levels of circulating carbon dioxide). Pancreatic insufficiency occurs due to intrapancreatic duct obstruction by viscous secretions, with autolysis and fat replacement; insulin insufficiency or carbohydrate intolerance are in fact commonly present in CF patients [23]. Most affected males are also infertile [24].

Fortunately, life expectancy for CF patients has increased to an average of 37 years through new management strategies, although no definitive cure is yet available, and new models predict newborns with CF today to have a life expectancy of around 50 years [25]. Since it is very important to diagnose the disease and provide treatment at an early stage, CF is one of the most important conditions included in newborn screening programmes (Table 3). The main testing strategy involves the measurement of immunoreactive trypsinogen (IRT) from heel-prick blood samples, followed by the sweat test and genetic mutation analyses [24]. More than 95 % of affected newborns have no recorded family history of CF [21], which provides an argument in favour of prenatal or (preferably) pre-conception screening for mutations to predict the risks of having an affected child. However, carrier screening for *CFTR* mutations has its disadvantages. Firstly, around 1000 mutations in the *CFTR* gene have been reported since its association with the disease; although in most populations, up to 90 % of affected individuals should have one of a few mutations included in a pan-ethnic gene panel, there may still be mutations which are not covered by the panel and this residual risk must be clearly stated both while obtaining informed consent and while reporting the results, to avoid instilling a false sense of security in mutation-negative patients [26, 27]. This might be overcome in the near future by high-throughput screening of the whole *CFTR* gene, but the latter in turn will identify sequence variations of undetermined significance and thus more research into each detectable variant will be required [21]. Secondly, not all mutations have been associated with classical CF; other *CFTR*-related conditions such as male infertility may constitute incidental findings and require careful genetic counselling [26]. Only the functional effects of a few mutations have been established to date, and associations between genotype and phenotype are weak, due to variability in the environmental and genetic background as well as phenotype heterogeneity in patients having the same mutations [23].

The recent discovery of mutation-specific orphan drugs for CF, most notably VX-770 for patients carrying the G551D mutation [28] and PTC124 for patients expressing premature stop codons [29], further highlights the importance of genetic testing for CF for early detection and treatment in target patients. Although these mutations constitute only a small fraction of CF patients (ca. 5 and 10 %, respectively) and, subsequently, treatment costs are still very high, research for new drugs is ongoing and will undoubtedly make population genetic screening more feasible in the near future.

2.3 The Haemoglobinopathies

Heamoglobinopathies comprise a heterogeneous group of autosomal recessive inherited disorders resulting in the production of abnormal or reduced synthesis of normal globin chains that constitute the building blocks of haemoglobin. Those conditions that completely or partially abolish the production of globin chains are known as thalassaemia, while the conditions that result in abnormal globin chain production constitute chain variants. The haemoglobin molecule is composed of two pairs of globin chains with each globin attached to a haem moiety. In adult haemoglobin, the two pairs of globin chains are known as alpha (α) and beta (β). The alpha chains are coded for by a pair of identical alpha genes (*HBA1 and HBA2*) on chromosome 16, thus an individual has four alleles. In contrast, the beta globin chains are encoded by a single gene (*HBB*) on chromosome 11 with an individual having two alleles. The worldwide distribution of haemoglobinopathies follows the worldwide distribution of endemic malaria [30]. As a result of human migratory patterns, haemoglobinopathies have become a major health issues in developed countries where the disease was not endemic. Since early identification and treatment of haemoglobinopathies improves morbidity and mortality rates, newborn, antenatal and/or cascade screening for these conditions has been established in various countries in Europe (Table 3) and around the world.

2.3.1 Haemoglobin Variants

In general, haemoglobin variants are due to mutations within the β or α globin genes, though foetal haemoglobin variants as a result of mutations in one of the two γ globin genes (*HBG1 and HBG2*) have also been reported. Individuals that carry only one mutated gene are known as carriers. Though considered as an autosomal recessive condition, it is more accurate to define a trait as autosomal codominant, as the globin chain variants are expressed and are present in the heamoglobin of 'carriers'. In contrast, carriers are usually considered healthy as the trait does not normally result in clinical consequences. Haemoglobin variant homozygotes or compound heterozygotes that result in significant functional alterations of the resultant haemoglobin would usually present with symptoms. The clinically relevant variants are easily identified through simple and relatively cheap electrophoretic techniques supplemented by HPLC methods to verify and quantify the variants. The worldwide clinically important variants include sickle cell haemoglobin, also known as HbS (β6 (A3) glutamic acid→valine), haemoglobin C (β6 (A3) glutamic acid→lysine) and haemoglobin E (β26 (B8) glutamic acid→lysine).

HbS is the most widespread haemoglobin variant, occurring mostly in persons of African origin but also found in persons of Mediterranean ethnicity and in the Indian subcontinent. The single point mutation substitutes a valine residue for glutamic acid in codon 6. This substitution induces the polymerisation of the deoxygenated haemoglobin variant through the formation of hydrophobic bonds between the inserted valine residues of adjacent haemoglobin molecules. This polymerisation results in deformation of the red blood cells that take the sickle cell shape form. Such deformed blood cells obstruct the microcirculation of sickle cell patients. Thus, any physiological stress that reduces oxygenation or increases oxygen requirements, results in the rapid polymerisation of HbS and the precipitation of sickle cell crises, with risks of strokes at a very young age. Treatment involves daily prophylactic antibiotics, transfusions and hydroxyurea treatment to reduce sickle cell crises.

In haemoglobin C, the glutamic acid to lysine change is in the same position to that of HbS but instead of polymerisation, the haemoglobin precipitates within the red cell and damages the membrane. Damaged red blood cells increase the viscosity of the blood, are haemolysed and sequestered both in the bone marrow and spleen. HbC disease is not as severe as HbS and might not be diagnosed before adulthood. Though mild, patients can still experience joint pains as well as gall stone problems. HbC is most common in persons of African ancestry.

Haemoglobin E is the most common beta globin variant in the Far East, where carrier frequency can reach 25 %, mainly in Thai and Chinese. The HbE mutation activates a cryptic splice site, leading to a slightly reduced rate of synthesis in addition to some instability. Thus in the homozygous state, HbE is considered a very mild haemoglobinopathy with mild anaemia that does not usually require treatment. In contrast, the condition is very severe in compound β^{E}/β^{Thal} heterozygotes.

2.3.2 α-Thalassaemia

As humans inherit two *HBA* genes from each parent and thus the normal genotype is αα/αα, the genetics of α-thalassaemia is somewhat complicated. Most of the alpha-thalassaemia abnormalities are the result of deletions of either one or two of the α-genes. Mutations that result in the inactivation of a single α-gene exist but are not common. The α-thalassaemias can be classified into two groups, α^{+}-thalassaemia and α^{0}-thalassaemia (Table 1).

The homozygous α^{0}-thalassaemia state is incompatible with life as no α-globin chains are produced from around the 6th week of foetal life, the α^{0}-thalassaemic foetus is severely anaemic, oedematous and has all the features of severe intrauterine hypoxia. The child is usually stillborn late in pregnancy and the pregnancy is usually complicated with toxaemia and difficulty during delivery, in part due to an enlarged placenta. The compound heterozygous state (- -/- α) is a condition known as HbH disease, which is characterised by anaemia and an enlarged spleen and might require lifelong treatment with transfusions and iron chelation. The α^{+}-thalassaemia and the heterozygous α^{0}-thalassaemia do not require treatment and can be considered as healthy carriers. α-thalassaemia can be identified during newborn screening through the identification of an abnormal haemoglobin made up of tetramers of γ globin chains (γ_4) called Hb Bart's and in adults through the identification of HbH (β_4).

2.3.3 β-Thalassaemia

Similar to α-thalassaemia, β-thalassaemia is prevalent in Mediterranean countries (with Cyprus having the highest prevalence), the Middle East, Central Asia, India, Southern China, and the Far East, as well as in countries along the north coast of Africa and in South America [31]. β-thalassaemia is mostly due to point mutations or small deletions, though rarely large deletions can also be the cause. Over 270 *HBB* gene mutations that give rise to a β-thalassaemia phenotype are listed in the database of human haemoglobin variants and thalassaemias (http://globin.bx.psu.edu/hbvar/menu.html). Each ethnic group has a small set of predominant mutations.

Table 1 The classification and genotypes of the α-thalassaemias

	Heterozygous state	Homozygous State
α^{+}-thalassaemia	−α/αα	−α/−α
α^{0}-thalassaemia	− −/αα	− −/− −

β-Thalassaemia can be classified into three groups, β-thalassaemia trait or carriers (β^A/β^{Thal}), β^+-thalassaemia (homozygous state for mild or moderate mutations or compound heterozygous for a moderate and severe mutation) and β^0-thalassaemia (homozygous state for severe mutations). Persons with the trait are considered healthy carriers. β^+- and β^0-thalassaemia patients, require regular transfusions, depending on the severity of the conditions, together with regular iron chelation. Newborn screening for β-thalassaemia is complicated by the fact that the majority of haemoglobin in a normal newborn is HbF and thus is not affected by the presence of the thalassaemia. The early identification of β-thalassaemia compound heterozygotes and homozygotes requires the establishment of antenatal screening programmes so as to identify β-thalassaemia carrier expectant mothers. Antenatal diagnosis offers the possibility of counselling and allows targeting of the baby once it is born to avoid complications of the disease.

2.4 *Hereditary Haemochromatosis*

Hereditary haemochromatosis (HH) is a rather common monogenic disorder, affecting 1 in 200 to 1 in 500 of Caucasians [32] and characterised by increased absorption of dietary iron, with subsequent progressive deposition in organs including the liver, pancreas and heart [33]. It is most prevalent in Northern European populations but is found worldwide [34]. The associated gene, *HFE*, is related to the human leukocyte antigen *(HLA)-A3* complex; two missense mutations, namely C282Y and H63D account for up to 95 % of probands. The HFE protein localises to the duodenal crypt cells, where dietary iron is absorbed, and negatively regulates absorption through association with cell-surface transferrin receptors. *HFE* mutations cause loss of HFE protein function, the C282Y mutation being more detrimental due to concurrent disruption of the association of HFE with β_2-microglobulin, which is important for proper function [33].

HH initially presents with relatively non-specific symptoms, such as fatigue, joint and abdominal pain and palpitations, leading to frequent misdiagnosis. Later complications of iron deposition include cirrhosis, diabetes mellitus (DM) cardiomyopathy and primary hepatocellular carcinoma, which are also relatively common primary disorders. Symptoms generally appear between 40 and 60 years, with later onset in females attributed to iron loss through menstrual cycles, pregnancy and lactation [33, 35]. Management of HH is by periodic venesection or phlebotomy to remove blood and, consequently, excess iron; this treatment is effective, safe and inexpensive [33].

Although inheritance of *HFE* mutations is simple Mendelian, incomplete penetrance makes predictive testing difficult and provides an argument against population screening, which may inevitably have psychosocial consequences. On the other hand, it meets most of the WHO criteria which justify population screening (Table 2). Cascade screening for *HFE* is in fact implemented in many countries throughout Europe (Table 3).

2.5 Familial Hypercholesterolaemia

Familial hypercholesterolaemia (FH) is also rather common, with an estimated prevalence of 1/500 (0.2%) in Caucasians. Some populations show yet higher frequencies attributed to founder effects [36]. FH is characterised by abnormally high plasma levels of low-density lipoproteins (LDL) and total cholesterol, with predisposition to early-onset coronary heart disease (CHD) due to the formation of atherosclerotic plaques. FH is mostly caused by mutations in the LDL receptor (*LDLR*; >1000 mutations identified), apolipoprotein B (*APOB*; 9 mutations identified) or proprotein convertase stabilisin/kexin type 9 (*PCSK9*) genes, with autosomal dominant inheritance [37]. The latter implies that the inheritance of one mutation from just one parent will result in the disease phenotype; in fact, heterozygous patients constitute the vast majority of cases. Also, couples where one partner is affected (with a heterozygous genotype) have a 50% chance of disease transmission to the offspring [38]. The high degree of risk among family members makes cascade screening for FH a valuable tool in the identification of affected individuals (Table 2); early detection of FH and subsequent treatment with statins significantly reduce morbidity and mortality from CHD associated with FH [38, 39].

Despite international efforts, an estimated 80% of FH patients are still not being diagnosed [40]. An important pitfall is the phenotypic and genotypic heterogeneity of the disease. Until recently, diagnosis was mostly made by means of LDL measurements, with specific cut-off points according to age and family history (lower cut-offs are used for those having an affected first-degree relative than those having an affected second-degree relative, for example) [41]. However, LDL and total cholesterol levels are highly variable in FH patients, even after adjustment for gender, age and body mass index (BMI), and may also overlap with those of the general population, resulting in reduced sensitivity and specificity [42].

The inclusion of mutation analysis has been proved to increase the specificity and sensitivity of the diagnosis of FH [36]. In fact, European criteria for FH developed by the Simon Broome Register Group (UK) and the Dutch Lipid Clinic Network, include the presence of a functional FH mutation, even in the absence of other criteria, as diagnostic of "definite" FH [43, 44]. This is due to the dominant nature of the genotype and the high penetrance of mutations, which is close to 100% [38]. Nonetheless, a wide genotypic variation is also observed: apart from populations with founder effects, which are characterised by a few mutations responsible for most cases, the situation is generally that of a large number of mutations giving rise to a highly heterogeneous population, such as in the UK, Italy and Germany [42]. Thus, each population must define the genotypic characteristics of its FH patients before a successful screening programme can be implemented, and even at this stage negative mutational analysis results do not necessarily exclude the presence of FH, since there might be mutations which are not included in the testing panel [38].

Table 2 describes how the Wilson and Jungner criteria apply to the monogenic disorders discussed above, for all of which some type of screening programme is currently implemented in most European countries (Table 3). It may be observed that the majority of the listed disorders do not satisfy the complete list of criteria,

Table 2 The Wilson and Jungner criteria as applied to a selected group of monogenic disorders

	Inborn errors of metabolism	Cystic fibrosis	Thalassaemia	Hereditary haemochro-matosis [5]	Familial hyper-choles-terolaemia
Important burden	Yes, collectively	Yes	Yes	Yes	Yes
Accepted & specific treatment	Yes, for most	Yes	Yes (blood transfusions)	Yes (phlebotomy)	Yes (statin administra-tion)
Facilities for diagnosis & treatment	Yes	Not yet evaluated	Yes	Not yet evaluated	Not yet evaluated
Early stage	Yes	Yes	Yes	Yes	Yes
Suitable test	Yes	Yes (in 90 % of cases)	Yes	Uncertain predictive value	Yes (dominant inheritance)
Test is Accept-able (Health benefits out-weigh psycho-social risks)	Uncertain	Uncertain (yes when limited to NBS by IRT?)	Yes	Uncertain	Uncertain
Known natural history	Yes	No	Yes	No	No
Agreement about when to treat	Yes (as early as possible)	No	Yes (CBC cut-offs)	No	No
Acceptable cost of care	Yes	Uncertain	Yes	Uncertain	Uncertain
Ongoing process	Yes, in dis-eases with established programmes (PKU, MSUD, etc.)	Yes	Yes	Cannot yet be addressed	Cannot yet be addressed

Table 3 Showing the number of European countries with established newborn, population-wide and cascade screening programmes for the conditions described in the text. Adapted from Javaher et al. (2010) [45].

Screening strategy	Inborn errors of metabolism	CF (*CFTR*)	Thalassaemia/ Haemoglobin-opathies	HH (*HFE*)	FH (*LDLR*)
NBS	26	15	8	–	–
Population-wide	–	7	9	–	–
Cascade	–	16	10	11	8

and that the most important criteria, which may have had the highest impact on the decision-making process, are the burden of the disease, the availability of a specific and acceptable treatment (or rather management, as no definitive treatment yet exists for any of the disorders), and the presence of an early stage during which the condition can be diagnosed (or predicted) and treatment initiated.

As can be observed in Table 3, all 26 European countries reviewed by Javaher et al. (2010) have taken up one form or another of newborn screening [45]. All states have NBS programmes for congenital hypothyroidism, and all except Malta for PKU.

It is evident that several factors played an important part in the decision-making process to set up the screening programmes, highlighting the importance of the previously described criteria. PKU can be defined as the classical type of disease ideal for NBS: although classified as a rare disease, it is of early onset (symptoms developing in the first few months of life), it is relatively easy to diagnose before symptoms develop and also easy to treat, and it has severe complications if left untreated. The same can be applied to congenital hypothyroidism, although it is estimated that only around 15 % of cases are genetic and 85 % are due to thyroid dysgenesis [46]; thus this disorder was not addressed in detail in this chapter. Newborn screening for other inborn errors of metabolism which are very rare has been made feasible through MS/MS, which is currently being applied in 11 European countries. It is important to note, however, that despite the ability of MS/MS to detect up to 45 disorders, most countries choose to report only a few of these disorders, mostly due to ethical and psychosocial reasons.

In the case of CF, NBS is generally by an established algorithm, involving a primary screening test for immunoreactive trypsinogen (IRT), followed by either repeated IRT testing or DNA testing. The second tier depends on whether repeat samples are taken from the newborns; where only one blood sample is available, a positive IRT is followed by DNA testing. The latter will defer in detection rate according to the number of mutations screened for, due to the large number of possible *CFTR* mutations causative of CF. Population-wide testing for carriers and couples at risk of having affected children are also dependent on mutation panels and thus include the inevitable degree of false-negative findings [47]. As has already been mentioned, psychosocial risks involved in genetic testing for *CFTR* mutations are also significant [11]. Cascade screening plays an important role since the mutations are narrowed down to those carried by the proband.

Currently, only cascade screening programmes are in place for HH and FH. This is due to the importance of family history in these disorders as well as the later onset, which makes screening targeted only at high-risk individuals more feasible. The screening process is generally initiated when the first case showing the symptoms and carrying a causative mutation, or proband, is found, and his/her relatives are followed-up to find whether they are carriers of the same mutation, in order to prevent the consequences of the disease.

The HH and FH cascade screening programmes provide valuable arguments in the implication of screening for monogenic subsets of common, multifactorial conditions. Being highly penetrant and autosomal dominant, *LDLR* mutations do not just confer a risk for FH but justify preventive measures, i.e. statin administration to prevent hypercholesterolaemia. Furthermore, FH in turn confers an increased risk of heart disease such as myocardial infarction. However, the low penetrance of HH mutations, which are very common especially in Northern Europe, makes screening for such mutations a risk status assessment, similar to that in multifactorial disorders. Thus, no population-wide screening programmes to detect *HFE* mutations are in place; rather, probands present with elevated iron levels in serum and their relatives are followed up to determine the risk of having inherited the same disorder. It is only at this level that genotyping of symptomless individuals take place [5].

An important observation to make is that there are significant discrepancies in the screening programmes present between European countries, and no consensus exists as yet as to which disorders should be screened for or, as in the case of MS/MS, reported [48].

Even more diverse are the European policies on population-based carrier screening programmes, targeting either pre-conception or prenatal individuals. In the case of such programmes, the epidemiology of the targeted disorders probably plays the most important part in their selection. A good model to illustrate this diversity are the population-wide carrier screening programmes for thalassaemia and other haemoglobinopathies established in several countries. Population-wide haemoglobinopathy screening is present in countries having higher prevalences, such as in the Mediterranean region, with ethnic-specific screening being preferred in countries such as Germany. Furthermore, in countries with high prevalence such as Cyprus, the population screening programme is run in parallel to a newborn screening programme, to ensure maximum coverage of affected individuals [45]. Increased public awareness and uptake of prenatal screening have largely contributed to the success of these programmes [49]. Other country specific differences involve pre-conception versus prenatal screening. While in Cyprus pre-conception screening of β-thalassaemia is carried out in all cases (followed by prenatal diagnosis in those cases that are at risk), in other countries such as the UK, screening is carried out prenatally.

3 The Way Forward—Public Health Genomics Perspectives on use of Next-Generation Sequencing for Newborn Screening

NBS started with the Guthrie test in the 1960s and rapidly expanded around the globe. It has been possible to test for a handful of disorders with this method. In the late 1990s, early 2000s, we have seen the introduction of MS/MS, which has been replacing the Guthrie test in many countries. With MS/MS, it is possible to expand the number of screened disorders extensively, without significantly increasing the overall costs.

In the transition from the Guthrie test to MS/MS, the major question was not the detectability of the disorders with this technology, but how to decide on which disorders to include in the NBS programmes. Although there has been a general consensus on screening criteria, such as the Wilson and Jungner criteria [3], various countries interpreted them differently or used modified criteria when selecting the disorders to be included in the NBS programmes. For example, Germany has been using three criteria [50], whereas the UK has been using 22 criteria, all of which must be fulfilled before a disorder is included in the NBS programme [51]. Additionally, various countries have different stakeholders and technology-push vs. market-pull dynamics for NBS, as well as different values, structures and processes in health technology assessment, all of which are rooted in differences in their health care systems [52]. These lead to different lists of disorders covered in NBS programmes, as seen in Table 3.

We are approaching another shift in NBS with the upcoming next-generation sequencing (NGS) technologies. Also termed as 'massively parallel sequencing' or 'second generation sequencing' [53], NGS has been reducing both the cost and time required to accomplish whole genome sequencing. With the rapidly decreasing cost of this technology, soon it will be possible to sequence the whole genome of an individual for less than 1000 US dollars. Once DNA sequencing technology is sufficiently robust and affordable, it will be possible for all babies to have their genomes sequenced at birth, replacing both newborn bloodspot screening and additional genetic tests required later in life [54]. This means that it will be possible to screen for a virtually unlimited number of disorders in NBS with almost no additional costs per disorder. However, it is not the technologic capacity to sequence entire genomes, but the analysis and interpretation of the generated data, that is the main bottleneck for the application of NGS for whole genome or exome sequencing [54].

The data generated from whole genome or exome sequencing in the newborn phase can be used not only to screen for monogenic disorders or monogenic subset of complex disorders, but also for pharmacogenetics (potential response to drugs), nutrigenetics (response to nutrients) and risk assessment programmes for major complex disorders, such as cardiovascular diseases or type 2 diabetes. Therefore, implementation of this technology will have an impact on the whole health care system, beyond the NBS programme.

There are several issues that need to be resolved when preparing for using NGS in the newborn phase. Some of these are presented very briefly below:

3.1 Human Resources

The knowledge, attitude and skills of health professionals to use genome-based knowledge are presently very limited. This shortcoming needs to be addressed effectively with professional education and training programmes. However, the effect of such training might be limited on health professionals that have already been practising in traditional ways for many years. The main target should be the under- and postgraduate education of health professionals.

Besides, to develop the tools for data analysis, bioinformaticians will be a crucial professional group required both in central levels and in local clinical services [54].

3.2 Informatics Capacity to Store and Process Data

Sequencing the whole genome or exome of large population groups will create massive amounts of data which need to be stored, retrieved and analysed. This requires data storage and processing infrastructures. Additionally, analytical tools to analyse and interpret the whole genome or exome must be developed.

3.3 Clinical Health Services

The systems must be in place to integrate the generated genome-based information in clinical health services. For example, in the first place, genome-based data can be used in the prescription of drugs which have significant pharmacogenetic interactions, such as warfarin. Nevertheless, the infrastructure to access such data and clinical work flows making use of the data must be in place, for NGS to be applicable to the healthcare setting.

3.4 Data

A recurring issue that arises when genome-based data are discussed is whether genetic data are different than other data concerning health. 'Genetic exceptionalism' claims that all genetic and genomic samples and data merit special protection, regardless of their medical sensitivity or predictive power. This claim has been rejected by various groups and reports, including the Public Health Genomics European Network [55], the Ickworth Group (an international group consisting of experts from multiple disciplines which came together in Ickworth, UK) [56] and others [57]. The main underlying idea is that genetic data deserve no separate status; they must satisfy equally high standards of data protection and confidentiality as other types of health data.

In the future paradigm of health care, the line dividing health care provision from health research will likely get thinner. In particular, in the context of the 'big data' approach, datasets created from the regular health care data and various other data collected from individuals are envisioned to be used by data mining to provide insights to health and diseases. Whole genome or exome sequence data of large populations will provide a great opportunity for such research and to discover the genetic basis of various diseases.

For this future vision, having longitudinal (long-data) and large sets of data (big-data) is required, but not enough. For the development of new forms of prevention, diagnosis and treatment of complex diseases, abundant and intricate health data must be combined with innovative analysis strategies in a cross-disciplinary environment [57]. However, recent developments in the EU legislation on data protection impede use of health care data for research, due to the restrictions proposed on the use of health data, even for research purposes [58–60]. This is an important regulatory bottleneck that needs to be overcome.

Genetic data hold a very promising future. However, precautions are required to protect the owner of the data, i.e. the individual. Regulatory mechanisms must be in place to prevent any discrimination that may arise due to genetic characteristics of the individual. GINA—Genetic Information Non-Discrimination Act in the USA, which came into force in 2009, is one of the major examples of a legislation that protects individuals from being discriminated by employers or insurance companies based on their genetic data.

3.5 *Ethical Issues*

An important issue that needs to be considered for whole genome or exome sequencing in the newborn phase is that the cost of sequencing the whole genome will soon practically be not more expensive than sequencing targeted genes. Therefore, sequencing the whole genome (or exome) at once seems to be the most practical solution. However, this brings ethical discussions on issues such as protection of the future autonomy of the screened infant, the right not to know, and the issues around incidental findings.

Incidental findings are any findings which are outside the scope of the clinical enquiry [54]. They are "the results of a deliberate search for pathogenic or likely pathogenic alterations in genes that are not apparently relevant to a diagnostic indication for which the sequencing test was ordered" [61]. In the context of whole genome or exome sequencing, the issues that may arise from incidental findings span multiple clinical, genetic and social dimensions, such as racial ancestry, misattributed parentage, consanguinity, disease susceptibility, and reproductive risks [54]. Several strategies are discussed to manage the issue of incidental findings, such as classification of genes and disorders according to net benefits [54] and providing lists of conditions that must be reported when found [61].

4 Complex Disorders

Common complex diseases can be defined as a group of disorders with similar symptoms but having a variable aetiology. Though the genetic component within complex diseases is inherited in a similar way to the inheritance in monogenic or Mendelian disorders, the main difference is that, while in Mendelian disorders a single gene variant is compulsory to reach the critical threshold in developing the disease, with modifier genes and environmental factors modifying the phenotype, in complex disorders both the threshold to develop the disease as well as its severity, are modified by complex aetiological factors. In between these two ends of the genetic spectrum, one finds a group of disorders in which alterations in susceptibility genes would bring one close to the required threshold for the condition, but other genetic, environmental and lifestyle factors are required for the disease to establish itself. This group can be considered as that of 'monogenic subtypes' and includes such genes as *BRCA1, BRCA2, MSH2* and *MLH1*.

Common complex disorders include cardiovascular disease (CVD), diabetes, cancer, dementia, auto-inflammatory and auto-immune diseases, amongst others. Preventive genetics in multifactorial disorders require an in-depth understanding of the genetic variability associated with the disorder, in order to define the complex disease into subtypes of variable aetiology. Genome-wide association studies (GWAS) have been used extensively to identify genetic variants associated with multifactorial diseases. Whereas current testing programmes include inherited monogenic diseases (Table 2), taken individually, the contributing genetic variants in multifactorial diseases still lack a high predictive value and hence genetic results are inconclusive and very difficult to interpret.

Of interest is the characterisation of monogenic subtypes within complex diseases that can be easily taken up into genetic testing programmes (Fig. 1). The term 'monogenic subtype' needs some clarification since the gene involved requires other contributing factors to cause the disease. The mutated gene in the monogenic subtypes within complex disorders, provides a cellular programme that is susceptible to the initiation of disease. The effect of the mutant will be exerted on tissues that normally express the gene. The resulting loss or gain of function will set a molecular threshold, requiring the contribution of additional factors, to initiate a specific disease. These susceptibility genes are masked by the presence of other variations that might occur during the disease state. Hence, low frequency susceptibility genes can only be identified through high penetrance in family studies.

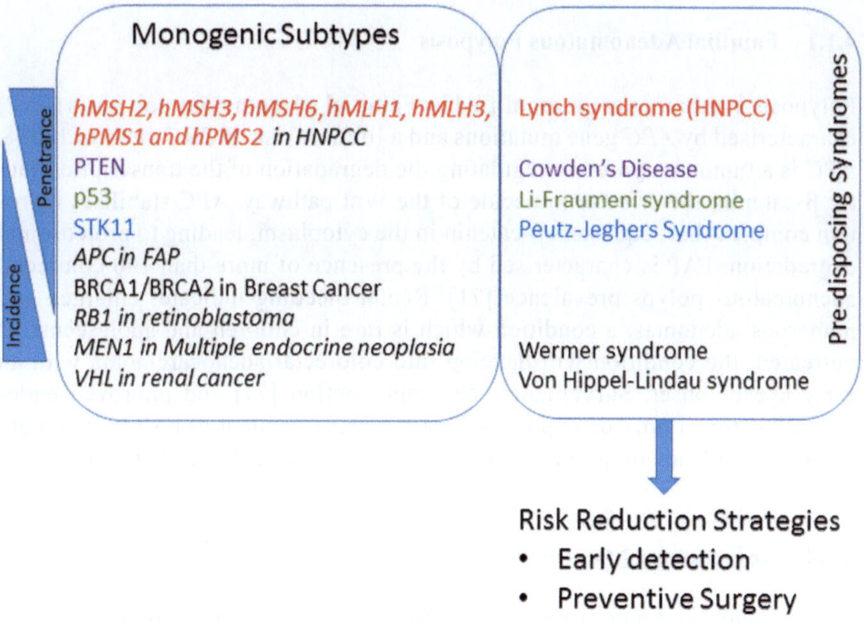

Fig. 1 Inherited susceptibility mutations in cancer. (Genes in red: mismatch repair genes; violet: phosphatases; green: apoptosis induction)

Major contributing genes in complex disorders have been identified due to their high penetrance, exemplified by *BRCA1* and *BRCA2* [62, 63] with increased risk of breast and ovarian cancers, *hMSH2* with increased risk of hereditary nonpolyposis colorectal cancer (HNPCC) [64], and α-synuclein in Parkinson disease [65].

The genetic basis of colorectal cancer (CRC) is well defined, and is used as a model to describe the molecular events that promote the progression of disease in cooperation with the defects in inherited susceptibility genes. The risk to develop CRC has been associated with inherited mutations in the mismatch repair genes [66]. Similarly, DNA repair genes predispose for breast cancer, including *BRCA1* and the Fanconi Anaemia (FA) genes *BRCA2 (FANCD1), FANCJ (BRIP1), FANCN (PALB2)* and *FANCO (RAD51C)* [67–69]. In addition, Fanconi genes are associated with other cancers, including acute myeloid leukaemia [70].

Inherited mutations in *PTEN, p53, RB1, MEN1* and *VHL* give rise to predisposing syndromes, namely Cowden's Disease [71], Li-Fraumeni syndrome [72], Retinoblastoma [73], Wermer syndrome [74], and von Hippel-Lindau syndrome [75], respectively. These inherited susceptibility genes have a low frequency with a high penetrance. Hence, these genes are candidates for a testing programme in specialised clinics, screening family members of patients, to identify risk and initiate preventive monitoring or discuss possible clinical solutions to reduce the risk significantly [76].

4.1 Familial Cancer Predisposing Syndromes

4.1.1 Familial Adenomatous Polyposis

Polyposis syndromes are exemplified by Familial adenomatous polyposis (FAP) characterised by *APC* gene mutations and a lifetime risk of CRC close to 100 %. APC is a tumour suppressor regulating the degradation of the transcription factor β-catenin, the effector molecule of the Wnt pathway. APC stabilises a protein complex that sequesters β-catenin in the cytoplasm, leading to proteosomal degradation. FAP is characterised by the presence of more than 100 colorectal adenomatous polyps prevalence [71]. Rectal bleeding indicates enlarged and numerous adenomas, a condition which is rare in children and adolescents. If untreated, the condition will develop into colorectal adenocarcinoma with an early age of onset. Surveillance, chemoprevention [77] and improved endoscopic treatment provide opportunities for better treatment of FAP [78] and decrease dependency on prophylactic cancer-preventive colorectal surgery [79].

4.1.2 Li-Fraumeni Syndrome

Li-Fraumeni Syndrome (LFS; OMIM 151623) is an autosomal dominant cancer predisposition syndrome associated with *p53* germline mutations [80]. Family studies show a high penetrance with early onset of sarcomas, breast cancer and other non-therapy-related neoplasms. The age of primary cancer onset ranges from 4 months to 49 years with a mean age of 25 years [81]. The variety of neoplasms and the early age of onset is attributed to mutant *p53*, a major gatekeeper of apoptosis in response to DNA damage. Of interest, mutations that result in truncation of the p53 protein are associated with higher cancer risk and earlier age of onset [82]. Defective p53 function results in the accumulation of mutations in proliferative tissues. The population frequency of germline *p53* mutations in Europe and United States is around 1:5000 individuals [83]. Eligibility for genetic screening is determined by established clinical criteria that classify individuals with LFS. The classic LFS classification scheme requires a proband with sarcoma diagnosed before the age of 45, a first-degree relative with any cancer before 45 years of age and another first- or second-degree relative with any cancer diagnosed at under 45 years or with a sarcoma at any age [84]. Other classification schemes were designed based on further family studies and the Chompret criteria [85] enhances the predictive value of the classic LFS classification, resulting in a testing sensitivity of 95% [81]. The Chompret criteria consider any proband with adrenocortical carcinoma at any age of onset eligible to *p53* mutation analysis, irrespective of family history. In addition, the criteria include other proband neoplasms such as breast cancer and brain tumour diagnosed at an early age of 36 years. Presymptomatic testing for germline *p53* mutations predicts the susceptibility to various neoplasms, imposing ethical issues due to lack of complete clinical surveillance, preventive measures and treatment recommendations. The high penetrance of germline *p53* mutations in familial breast cancer patients predicts an average age of onset of 31 years [86], and hence provides eligibility for breast cancer screening in p53-mutant women who are in their mid-20s, followed by implementation of risk reduction strategies.

4.1.3 Cowden Syndrome

Cowden Syndrome (CS; MIM 158350) is a rare autosomal dominant cancer predisposition syndrome with a prevalence of 1 in 200,000 [87] and an age-related penetrance of around 80% [88]. The susceptibility gene in CS is *PTEN*, predisposing individuals to breast, endometrial and thyroid cancer [89]. PTEN is a ubiquitously expressed phosphatase involved in the attenuation of the PI3K pathway, hence acting as a proliferation suppressor [90]. The clinical symptoms of CS include multi-organ hamartomatous polyps in the majority of the affected subjects [91]. In addition to Cowden Syndrome in adults, germline *PTEN* mutations result in Bannayan-Riley-Ruvalcaba syndrome (BRRS; MIM 153480) in children [92], collectively known as the *PTEN* hamartoma tumour syndrome (PHTS). The highest age-adjusted standardised incidence ratio of germline *PTEN* mutants occurs in thyroid cancer, followed by endometrial, kidney, breast and colorectal cancer and melanoma (Table 4). Of interest, promoter mutations in the *PTEN* gene were associated with

Table 4 Age-adjusted estimated lifetime risk for cancers in some examples of cancer-predisposing syndromes

Syndrome	Mutant genes	Tumour	Estimated lifetime risk (70 years) (%)
Cowden syndrome Tan et al. 2012 [93]	*PTEN*	Breast	85.2 (71.4–99.1%)
		Thyroid	35.2 (19.7–50.7%)
		Endometrium	28.2 (17.1–39.3%)
		Colorectal	9.0 (3.8–14.1%)
		Kidney	33.6 (10.4–56.9%)
		Melanoma	6 (1.6–9.4%)
Lynch syndrome Bonadona et al. 2011 [95]	*MLH1*	Colorectal	41 (25–75%)
		Endometrium	54 (20–80%)
		Ovarian	20 (1–65%)
	MSH2	Colorectal	48 (30–77%)
		Endometrium	21 (8–77%)
		Ovarian	24 (3–52%)
	MSH6	Colorectal	12 (8–22%)
		Endometrium	16 (8–32%)
		Ovarian	1 (0–3%)
Peutz-Jeghers syndrome Hearle et al. 2006 [105]	*STK11/LKB1*	Gastrointestinal	57 (39–76%)
		Breast	45 (27–68%)
		Gynaecological	18 (9–34%)
		Pancreas	11 (5–24%)
		Lungs	17 (8–36%)

breast cancer, while nonsense mutants were associated with colorectal cancers [93]. The age of onset of any of the core cancers within the family will establish the age for clinical observations, taken as 5 years less than the age of the youngest family member affected.

4.1.4 Lynch Syndrome

Lynch Syndrome (MIM 120435) is an autosomal dominant cancer predisposing syndrome caused by germline mutations in the mismatch repair (MMR) genes, including *MLH1, MSH2, MSH6 and PMS2* [94]. Lynch Syndrome is also known as hereditary nonpolyposis colorectal cancer (HNPCC) syndrome and confers a high risk to early-onset colorectal and endometrial cancer [95]. In contrast to polyposis syndromes, HNPCC lacks characteristic diagnostic features and clinics depend on standardised criteria, such as the Amsterdam criteria, to establish diagnosis. Eligibility for carrier screening requires a family history characterised by immunohistochemistry-verified CRC cases in at least three relatives, including a first-degree relative, the presence of the disease in at least two successive generations, and one of the relatives diagnosed with CRC with an age of onset less than 50 years, with absence of polyposis syndrome [96]. In addition, the inheritance of syndrome-associated cancers within the family and testing for microsatellite instability are used to predict patients with HNPCC [96]. Lynch syndrome accounts for 2–5 % of the total CRC cases, as reflected in the high estimated lifetime risk of *MLH1* and *MSH2* germline mutation carriers (Table 4). Of interest, 10–15 % of CRCs have microsatellite instability, a useful marker of mismatch repair genes loss of function [97]. The majority of cases with no familial predisposition are caused by sporadic hypermethylation of the *MLH1* gene promoter, supporting the high risk of inherited *MLH1* mutants in the development of carcinoma. Lynch syndrome patients are also at risk of developing other cancers, including stomach, ovarian, renal, pancreatic, small intestinal, brain and skin tumours; it can thus be included in the subset of cancer-predisposing syndromes [98].

4.1.5 Peutz-Jeghers Syndrome

Peutz-Jeghers syndrome (PJS; MIM 175200) is an autosomal dominant condition arising from germline mutations in the serine/threonine kinase gene (*STK11/ LKB1*) [99]. Since PJS condition is rare, estimates of frequency within populations vary significantly. Patients present with mucocutaneous pigmentation and gastrointestinal polyposis [100] and have an increased risk of gastrointestinal cancers (as in most polyposis syndromes), breast ovarian, uterine, cervical, lung and testicular cancers [101, 102] (Table 4). Loss of heterozygosity (LOH) of the normal *STK11* allele in 70 % of PJS patients, followed by the progression of hamartomas to adenocarcinomas, suggest that *LKB1* is a tumour suppressor gene [103]. Interestingly, somatic mutations in *LKB1* are common in lung cancer [104], supporting the involvement of inherited *LKB1* mutants in the progression of disease.

4.2 Tools in Predictive (& Preventive) Genetics

Unlike in monogenic disorders, screening for complex diseases presents a greater challenge since the aim is that of risk assessment rather than presymptomatic diagnosis. Furthermore, the concept of personalised (genomic) medicine seems to conflict with the aim of public health, that is, *to improve health from a population perspective* [1]. However, predictive genetics reduces the burden imposed on the health care system by admission of fewer patients with advanced stages of disease; thus, its importance is being increasingly recognised. Family history and personal genomics (the assessment of individual genetic variations at multiple loci) are two important predictive genetics tools. These have a recognised value in the diagnosis of monogenic conditions (as observed in cascade screening programmes such as for HH); however, their value in the management of risk of complex diseases remains to be established [106]. Although assessment of family history provides information on the inheritance of a phenotype [107], the likelihood of an individual carrying a mutant susceptibility gene to develop cancer depends on other genetic variations (modifier genes), as well as on dietary, lifestyle and environmental factors influencing the age of onset and severity of the disease. The routine use of family history or personal genomics alone as a measure of risk of common complex conditions does not generate sufficient evidence in the primary care setting; however, when used together and in conjunction with evidence-based medicine (EBM-the application of population-derived data), they comprise a very useful tool in the improvement of disease prevention [106]. Public health genomics shall play an essential role in designing more effective genetic screening programmes, by applying data derived from personal genomics to public health. This becomes especially important with the advent of cost-effective services for whole genome sequencing or microarrays detecting large panels of mutations, which may undoubtedly lead to over-diagnosis.

Cascade screening for genetic predisposition to cancer is a form of "systematic predictive testing", where asymptomatic individuals at an increased risk due to their genotypic inheritance may benefit from surveillance programmes to ensure early detection of tumours.

Colorectal cancer is a major contributor to cancer morbidity and mortality. 15 % of CRCs are familial with 2–5 % caused by HNPCC and less than 1 % associated with polyposis syndromes (FAP and PJS). Following diagnosis of Lynch disease, family members benefit from clinical screening by colonoscopy [108]. Recommendations include offering a colonoscopy every 1 or 2 years starting at the age of 25 years or at 5 years before the youngest diagnosed member of the family, whichever is the earlier [109]. In the case of FAP families, individuals with an APC mutation should strongly consider a prophylactic colectomy before the age of 25 years. Referral of genetic testing for *APC* mutations will provide confirmation of familial adenomatous polyposis (FAP) or, in the case of family members not exhibiting adenomatous polyps, will evaluate relatives at risk and initiation of surveillance programmes. Surveillance should be offered for *APC* mutant members that defer surgery [110]. For patients diagnosed with PJS, risk reduction strategies include upper and lower endoscopy, breast examination, endoscopic ultrasound and CA19–9 tumour marker testing for pancreatic tumour surveillance and ultrasound, cervical cytology and CA125 testing tumour marker testing for ovarian cancer [111].

Familial cases account for 10% of breast cancer in Western countries. Susceptibility genes are inherited in an autosomal dominant pattern, but with limited penetrance. High risk families testing positive for the *BRCA1* and *BRCA2* mutant genes are associated with a four or more times higher incidence of the disease in close relatives. The use of surveillance programmes and/or risk reduction strategies in healthy *BRCA1* and *BRCA2* mutation carriers are instrumental to ensure a positive impact of genetic screening on the health care system. Risk-reducing mastectomy (RRM) supresses breast cancer development by 90% [112]. Inherited mutations in *p53* (Li-Fraumeni) and *PTEN* (Cowden syndrome) have a high incidence of breast cancer development, but the syndromes are very rare. Testing for mutations within these genes requires the use of further diagnostic criteria as detailed below.

Risk assessment tools for complex diseases take into consideration various contributing factors, including the incidence of disease in first-degree relatives; previous diagnostic test results; monitoring results (if any); and presence of genetic mutations/polymorphisms associated with the disease of interest. In the case of breast cancer, risk assessment includes age at first live birth, the use of hormone replacement therapy and other risk factors specifically associated with the disease. The breast cancer risk assessment tool (GAIL model) is used to measure contribution of variants to the calculated risk [113]. The selection of the genetic contribution to multifactorial diseases is not a simple task. Also, the interrogation of genetic variation through screening programmes or referral for testing depends on the minor allele frequency within a population and also on the penetrance within a family having a history of the phenotype. Less penetrant genes with higher prevalence are more significant from a public health point of view. For instance, *the factor V Leiden mutation*, with increased risk of thrombotic events [114], is integrated in the testing regime of health care genetic clinics.

Most of the common chronic disorders, such as asthma, cancer, cardiovascular diseases, obesity, diabetes, hypertension, psychiatric disorders, arthritis, Parkinson's and Alzheimer's, have a complex aetiology and pathophysiology. The current view is that these disorders are due to numerous small, additive genetic defects compounded with environmental and lifestyle causes. Considering that these conditions constitute a major health and economic burden and are the cause of substantial morbidity and mortality, a concerted action is required to elucidate the pathophysiology and thus open the road for successful treatments and preventive strategies. One major drawback in the determination of aetiological factors that singly confer a small increase in risk is the need of a large number of affected and unaffected individuals ('cases' and 'controls'), so as to achieve statistically significant results. It is very difficult for single clinical and research centres to obtain such large numbers in a relatively short time and with reasonable budgets. For this reason, biobanks and their related databanks have become an important tool for the elucidation of the pathophysiology of these disorders.

5 Biobanks and Preventive Genetics

The Organisation for Economic Co-operation and Development (OECD) definition of a Biobank is "A collection of biological material and the associated data and information stored in an organised system, for a population or a large subset of a population." This definition brings about the need to consider a number of terms, basically "biological material", "associated data and information," "stored in an organised system," and "population or a large subset of a population."

It is current practice, if not a legal requisite, for hospitals and laboratories to collect and store whole organs and tissues that have been excised for diagnostic or therapeutic aims. In addition to these archival banks, since the early 70s specialised collections have been initiated, targeting cells and their products including DNA, RNA and proteins. Collections of biological material, without any associated data, have very limited usefulness as biobanks. The collection of data, including medical histories, lifestyle, social and environmental information, increases the research value of these biobanks as the samples can now be separated into different case and control categories. The next important step in the establishing of a useful biobank is the establishment of associated databases that results in an easier and more efficient search and classification tool as compared to pen and paper processes. Finally, to be considered as a biobank, the samples have to be collected either from the whole population or from a subset of the population that has a particular disorder. In each case, the information collected can either be retrospective, transvers or prospective, depending on the final aim and use of the biobank.

5.1 *Future Biobanks*

The networking of biobanks from different countries or centres, in particular those where both the samples and data have been collected with standardised protocols, has the potential of becoming an ideal platform to collect the necessary data of thousands of individuals with the same medical condition. This has been the main impetus behind various international initiatives in forming large, virtual biobanks through the networking of individual databases corresponding to individual biobanks. Apart from data on demographic, environmental risk factors, health, lifestyle, nutrition and socioeconomic variables, these biobanks might also hold 'omics' (genomics, transcriptomics, proteomics and metabolomics) data. These data and related bioinformatics software offer advanced possibilities in understanding the disease pathophysiology, thus paving the way for therapeutic and prevention programmes for a number of chronic diseases.

6 Conclusion and Recommendations

Consensus is being sought as to which genetic diseases should be included in population screening programmes and new criteria are being defined to achieve this aim [115]. These emerging criteria include important psychosocial aspects of screening, which become especially important when the test is only an assessment of risk rather than a definitive prediction, as well as ethical procedures such as informed consent [5]. Probably, the most effective way of reaching a suitable conclusion is to organise the established and emerging criteria into a whole process of policy-making, involving a thorough assessment of public health requirements, evaluation of the involved tests and interventions, and the actual development of the policy and implementation of the screening programme (Fig. 2). It is only through rigorous planning and organisation that a screening programme may be truly efficacious and cost-effective. In addition, the decision to integrate genetic testing for the identity of carriers at pre-conception, prenatal or cascade screening is important for autosomal recessive and low penetrance disorders. Furthermore, adverse results in both types of conditions lead to decision-making which involves many psychological and ethical issues [49, 116, 117]. This is especially because, in screening for complex or untreatable conditions, the benefits conferred do not necessarily involve treatment/management, but other life-plans such as reproductive or lifestyle choices.

A global effort should be made to standardise the design of such screening programmes, since integration of preventive and predictive genetics into the diverse health care systems would always remain under the responsibility of national/regional health authorities. Hence, guidelines should be set-up and implemented by leading regulatory bodies, such as the European Medicines Agency and the US FDA, under the recommendation of global experts in the field. Finally, standardisation of screening programmes may be also achieved by setting up regional centres of expertise, for example across the European Union. Such centres would provide standardisation and cost-effectiveness by carrying out tests for rare genetic conditions for all the participating countries.

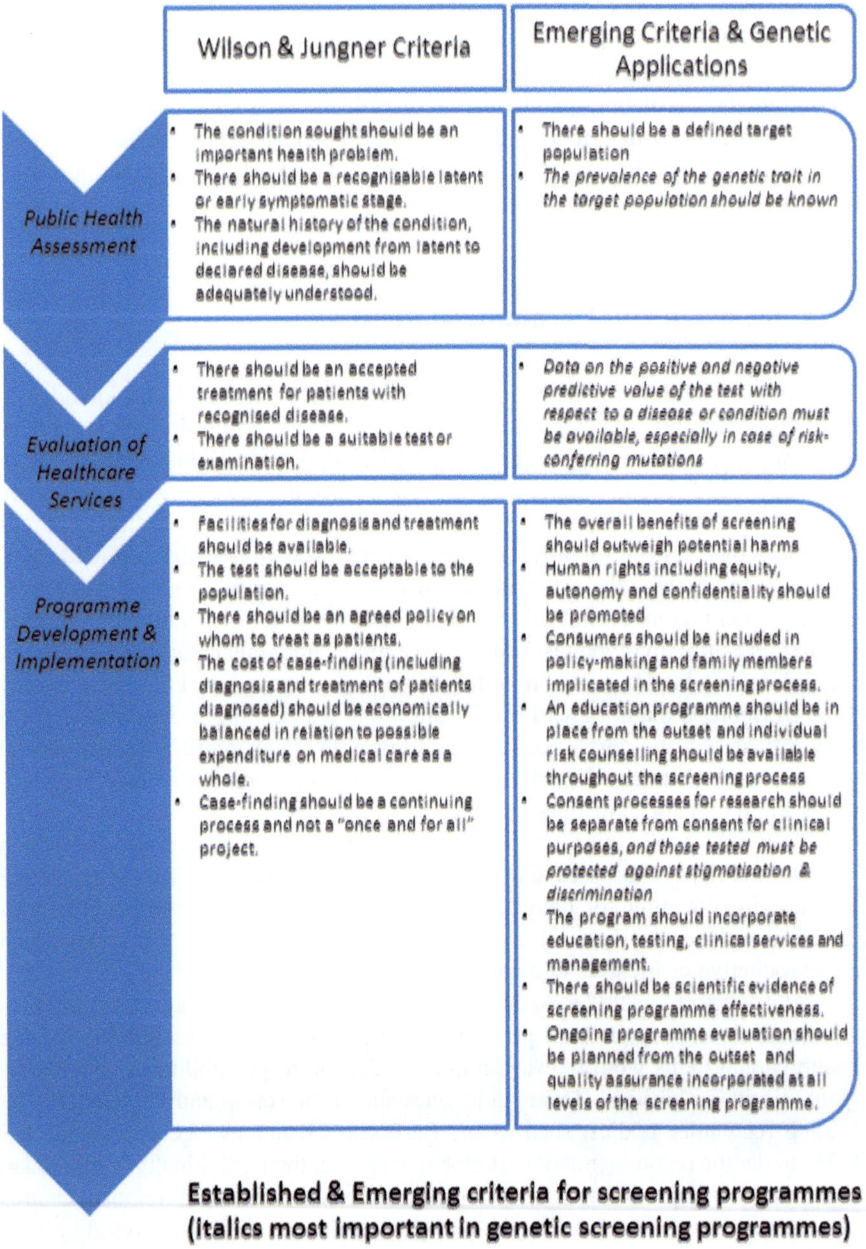

Fig. 2 Established [3] & emerging [5, 115] criteria for screening programmes, with criteria which specifically apply to genetic screening programmes in italics

References

1. Khoury MJ, Bowen MS, Burke W, Coates RJ, Dowling NF, Evans JP, Reyes M, St Pierre J (2011) Current priorities for public health practice in addressing the role of human genomics in improving population health. Am J Prev Med 40(4):486–493. doi:10.1016/j.amepre.2010.12.009
2. Morabia A, Zhang FF (2004) History of medical screening: from concepts to action. Postgrad Med J 80(946):463–469. doi:10.1136/pgmj.2003.018226
3. Wilson JM, Jungner YG (1968) Principles and practice of mass screening for disease. Bol Oficina Sanit Panam 65 (4):281–393
4. Andermann A, Blancquaert I, Beauchamp S, Déry V (2008) Revisiting Wilson and Jungner in the genomic age: a review of screening criteria over the past 40 years. Bull World Health Organ 86(4):241–320
5. Becker F, van El CG, Ibarreta D, Zika E, Hogarth S, Borry P, Cambon-Thomsen A, Cassiman JJ, Evers-Kiebooms G, Hodgson S, Janssens AC, Kaariainen H, Krawczak M, Kristoffersson U, Lubinski J, Patch C, Penchaszadeh VB, Read A, Rogowski W, Sequeiros J, Tranebjaerg L, van Langen IM, Wallace H, Zimmern R, Schmidtke J, Cornel MC (2011) Genetic testing and common disorders in a public health framework: how to assess relevance and possibilities. Eur J Hum Genet 19(Suppl 1):S6–S44. doi:10.1038/ejhg.2010.249
6. IRDiRC (2012) International rare diseases research. http://www.irdirc.org. Accessed 22 Apr 2014
7. Gilissen C, Hoischen A, Brunner HG, Veltman JA (2011) Unlocking Mendelian disease using exome sequencing. Genome Biol 12(9). doi:10.1186/gb-2011-12-9-228
8. Guthrie R, Susi A (1963) A simple phenylalanine method for detecting Phenylketonuria in large populations of newborn infants. Pediatrics 32:338–343
9. Ozben T (2013) Expanded newborn screening and confirmatory follow-up testing for inborn errors of metabolism detected by tandem mass spectrometry. Clin Chem Lab Med 51(1):157–176. doi:10.1515/cclm-2012-0472
10. O'Neill CA, Eisensmith RC, Croke DT, Naughten ER, Cahalane SF, Woo SL (1994) Molecular analysis of PKU in Ireland. Acta Paediatr Suppl 407:43–44
11. Guldberg P, Henriksen KF, Sipila I, Guttler F, de la Chapelle A (1995) Phenylketonuria in a low incidence population: molecular characterisation of mutations in Finland. J Med Genet 32(12):976–978
12. Oerton J, Khalid JM, Besley G, Dalton RN, Downing M, Green A, Henderson M, Krywawych S, Leonard J, Andresen BS, Dezateux C (2011) Newborn screening for medium chain acyl-CoA dehydrogenase deficiency in England: prevalence, predictive value and test validity based on 1.5 million screened babies. J Med Screen 18(4):173–181. doi:10.1258/jms.2011.011086
13. Wilcken B, Leung KC, Hammond J, Kamath R, Leonard JV (1993) Pregnancy and fetal long-chain 3-hydroxyacyl coenzyme A dehydrogenase deficiency. Lancet 341(8842):407–408. doi:10.1016/0140-6736(93)92993-4
14. Leonard JV, Dezateux C (2009) Newborn screening for medium chain acyl CoA dehydrogenase deficiency. Arch Dis Child 94(3):235–238. doi:10.1136/adc.2007.134957
15. Yap S, Naughten E (1998) Homocystinuria due to cystathionine beta-synthase deficiency in Ireland: 25 years' experience of a newborn screened and treated population with reference to clinical outcome and biochemical control. J Inherit Metab Dis 21(7):738–747
16. Carleton SM, Peck DS, Grasela J, Dietiker KL, Phillips CL (2010) DNA carrier testing and newborn screening for maple syrup urine disease in Old Order Mennonite communities. Genet Test Mol Biomarkers 14(2):205–208. doi:10.1089/gtmb.2009.0107
17. Edelmann L, Wasserstein MP, Kornreich R, Sansaricq C, Snyderman SE, Diaz GA (2001) Maple syrup urine disease: identification and carrier-frequency determination of a novel founder mutation in the Ashkenazi Jewish population. Am J Hum Genet 69(4):863–868. doi:10.1086/323677
18. Puffenberger EG (2003) Genetic heritage of the Old Order Mennonites of southeastern Pennsylvania. Am J Med Genet C Semin Med Genet 121C(1):18–31. doi:10.1002/ajmg.c.20003

19. Kolker S, Christensen E, Leonard JV, Greenberg CR, Burlina AB, Burlina AP, Dixon M, Duran M, Goodman SI, Koeller DM, Muller E, Naughten ER, Neumaier-Probst E, Okun JG, Kyllerman M, Surtees RA, Wilcken B, Hoffmann GF, Burgard P (2007) Guideline for the diagnosis and management of glutaryl-CoA dehydrogenase deficiency (glutaric aciduria type I). J Inherit Metab Dis 30(1):5–22. doi:10.1007/s10545-006-0451-4
20. Collins FS (1992) Cystic fibrosis: molecular biology and therapeutic implications. Science 256(5058):774–779. doi:10.1126/science.256.5058.774
21. Massie J, Delatycki MB (2013) Cystic fibrosis carrier screening. Paediatric Respir Rev 14(4):270–275. doi:10.1016/j.prrv.2012.12.002
22. Barrett PM, Alagely A, Topol EJ (2012) Cystic fibrosis in an era of genomically guided therapy. Human Mol Genet 21(R1):R66–R71. doi:10.1093/hmg/dds345
23. O'Sullivan BP, Freedman SD (2009) Cystic fibrosis. The Lancet 373(9678):1891–1904. doi:10.1016/s0140-6736(09)60327-5
24. Hurt K, Bilton D (2012) Cystic fibrosis. Medicine 40(5):273–276. doi:10.1016/j.mpmed.2012.02.006
25. Dodge JA, Lewis PA, Stanton M, Wilsher J (2007) Cystic fibrosis mortality and survival in the UK: 1947–2003. Eur Respir J 29(3):522–526. doi:10.1183/09031936.00099506
26. Grody WW, Cutting GR, Klinger KW, Richards CS, Watson MS, Desnick RJ, Subcommittee on Cystic Fibrosis screening AoGSCAACoMG (2001) Laboratory standards and guidelines for population-based cystic fibrosis carrier screening. Genet Med 3(2):149–154. doi:10.109700125817-200103000-00010
27. Watson MS, Cutting GR, Desnick RJ, Driscoll DA, Klinger K, Mennuti M, Palomaki GE, Popovich BW, Pratt VM, Rohlfs EM, Strom CM, Richards CS, Witt DR, Grody WW (2004) Cystic fibrosis population carrier screening: 2004 revision of American College of Medical Genetics mutation panel. Genet Med 6(5):387–391. doi:10.1097/01.gim.0000139506.11694.7c
28. Accurso FJ, Rowe SM, Clancy JP, Boyle MP, Dunitz JM, Durie PR, Sagel SD, Hornick DB, Konstan MW, Donaldson SH, Moss RB, Pilewski JM, Rubenstein RC, Uluer AZ, Aitken ML, Freedman SD, Rose LM, Mayer-Hamblett N, Dong Q, Zha J, Stone AJ, Olson ER, Ordonez CL, Campbell PW, Ashlock MA, Ramsey BW (2010) Effect of VX-770 in persons with cystic fibrosis and the G551D-CFTR mutation. N Engl J Med 363(21):1991–2003. doi:10.1056/NEJMoa0909825
29. Sermet-Gaudelus I, Boeck KD, Casimir GJ, Vermeulen F, Leal T, Mogenet A, Roussel D, Fritsch J, Hanssens L, Hirawat S, Miller NL, Constantine S, Reha A, Ajayi T, Elfring GL, Miller LL (2010) Ataluren (PTC124) induces cystic fibrosis transmembrane conductance regulator protein expression and activity in children with nonsense mutation cystic fibrosis. Am J Respir Crit Care Med 182(10):1262–1272. doi:10.1164/rccm.201001-0137OC
30. Roberts DJ, Williams TN (2003) Haemoglobinopathies and resistance to malaria. Redox Rep 8(5):304–310. doi:10.1179/135100003225002998
31. Galanello R, Origa R (2010) Beta-thalassemia. Orphanet J Rare Dis 5:11. doi:10.1186/1750-1172-5-11
32. Limdi JK, Crampton JR (2004) Hereditary haemochromatosis. QJM 97(6):315–324. doi:10.1093/qjmed/hch065
33. Hanson EH, Imperatore G, Burke W (2001) HFE gene and hereditary hemochromatosis: A HuGE review. Am J Epidemiol 154(3):193–206. doi:10.1093/aje/154.3.193
34. Merryweather-Clarke AT, Pointon JJ, Jouanolle AM, Rochette J, Robson KJH (2000) Geography of HFE C282Y and H63D Mutations. Genetic Test 4(2):183–198. doi:10.1089/10906570050114902
35. Niederau C, Fischer R, Pürschel A, Stremmel W, Häussinger D, Strohmeyer G (1996) Long-term survival in patients with hereditary hemochromatosis. Gastroenterology 110(4):1107–1119
36. Austin MA, Hutter CM, Zimmern RL, Humphries SE (2004) Familial hypercholesterolemia and coronary heart disease: A HuGE association review. Am J Epidemiol 160(5):421–429. doi:10.1093/aje/kwh237

37. Varret M, Abifadel M, Rabès JP, Boileau C (2008) Genetic heterogeneity of autosomal dominant hypercholesterolemia. Clin Genet 73(1):1–13. doi:10.1111/j.1399-0004.2007.00915.x
38. Ned RM, Sijbrands EJG (2011) Cascade screening for familial hypercholesterolemia (FH). PLoS Curr 3:RRN1238. doi:10.1371/currents.RRN1238
39. Neil A, Cooper J, Betteridge J, Capps N, McDowell I, Durrington P, Seed M, Humphries SE (2008) Reductions in all-cause, cancer, and coronary mortality in statin-treated patients with heterozygous familial hypercholesterolaemia: a prospective registry study. Eur Heart J 29(21):2625–2633. doi:10.1093/eurheartj/ehn422
40. Williams RR, Hamilton-Craig I, Kostner GM, Hegele RA, Hayden MR, Pimstone SN, Faergeman O, Schuster H, Steinhagen-Thiessen E, Beisiegel U, Keller C, Czeizel AE, Leitersdore E, Kastelein JC, Defesche JJP, Ose L, Leren TP, Seftel HC, Raal FJ, Marais AD, Eriksson M, Keller U, Miserez AR, Jeck T, Betterridge DJ, Humphries SE, Day INM, Kwiterovich PO, Lees RS, Stein E, Illingworth R, Kane J, Boulyjenkov V (1996) MED-PED: An integrated genetic strategy for preventing early deaths. In: Dr KBP, Dr VB, Dr YC (eds) Genetic approaches to noncommunicable diseases. Springer, Berlin, pp 35–45
41. Williams RR, Hunt SC, Schumacher MC, Hegele RA, Leppert MF, Ludwig EH, Hopkins PN (1993) Diagnosing heterozygous familial hypercholesterolemia using new practical criteria validated by molecular genetics. Am J Cardiol 72(2):171–176
42. Bertolini S, Cantafora A, Averna M, Cortese C, Motti C, Martini S, Pes G, Postiglione A, Stefanutti C, Blotta I, Pisciotta L, Rolleri M, Langheim S, Ghisellini M, Rabbone I, Calandra S (2000) Clinical expression of familial hypercholesterolemia in clusters of mutations of the LDL receptor gene that cause a receptor-defective or receptor-negative phenotype. Arterioscler Thromb Vasc Biol 20(9):e41–e52. doi:10.1161/01.ATV.20.9.e41
43. Scientific Steering Committee on behalf of the Simon Broome Register Group (1991) Risk of fatal coronary heart disease in familial hypercholesterolaemia. BMJ 303(6807):893–896
44. World Health O (1999) Familial hypercholesterolaemia: report of a second WHO consultation. WHO, Geneva
45. Javaher P, Nyoungui E, Kaariainen H, Kristoffersson U, Nippert I, Sequeiros J, Schmidtke J (2010) Genetic screening in Europe. Public Health Gen 13(7–8):524–537. doi:10.1159/000294998
46. Rastogi MV, LaFranchi SH (2010) Congenital hypothyroidism. Orphanet J Rare Dis 5:17. doi:10.1186/1750–1172-5-17
47. Comeau AM, Accurso FJ, White TB, Campbell PW 3rd, Hoffman G, Parad RB, Wilfond BS, Rosenfeld M, Sontag MK, Massie J, Farrell PM, O'Sullivan BP (2007) Guidelines for implementation of cystic fibrosis newborn screening programs: Cystic Fibrosis Foundation workshop report. Pediatrics 119(2):e495–518. doi:10.1542/peds.2006-1993
48. Loeber JG, Burgard P, Cornel MC, Rigter T, Weinreich SS, Rupp K, Hoffmann GF, Vittozzi L (2012) Newborn screening programmes in Europe; arguments and efforts regarding harmonization. Part 1. From blood spot to screening result. J Inherit Metab Dis 35(4):603–611. doi:10.1007/s10545-012-9483-0
49. Godard B, ten Kate L, Evers-Kiebooms G, Aymé S (2003) Population genetic screening programmes: principles, techniques, practices, and policies. Eur J Hum Genet 11(S2):S49–S87. doi:10.1038/sj.ejhg.5201113
50. Gemeinsamer Bundesausschuss (G-BA) Mandate of the Federal Joint Committee. (2014) http://www.english.g-ba.de/special-topics/prevention/mandate/. Accessed 05 June 2014
51. UK National Screening Committee (2013) Programme appraisal criteria—Criteria for appraising the viability, effectiveness and appropriateness of a screening programme. http://www.screening.nhs.uk/criteria. Accessed 05 Jun 2014
52. Sanderman LK (2011) Newborn screening in Europe: lessons learned from tandem mass spectrometry in Germany and the UK. (Master of Science) Maastricht University, Maastricht
53. Stratton MR, Campbell PJ, Futreal PA (2009) The cancer genome. Nature 458(7239):719–724
54. Wright C, Burton H, Hall A, Moorthie S, Pokorska-Bocci A, Sagoo G, Sanderson S, Skinner R (2011) Next steps in the sequence. The implications of whole genome sequencing for health in the UK. PHG Foundation, Cambridge

55. Schröder P, Zimmern R (2007) PHGEN Working Group 2– The status of genetic information and genetic testing. Consequences for Public Health Genomics and the Public Health Genomics European Network (PHGEN). http://ec.europa.eu/health/screening_genetics/genomics/index_en.htm, Accessed 05 Jun 2014
56. Burke W, Burton H, Hall AE, Karmali M, Khoury MJ, Knoppers B, Meslin EM, Stanley F, Wright CF, Zimmern RL (2010) Extending the reach of public health genomics: what should be the agenda for public health in an era of genome-based and "personalized" medicine? Genet Med 12(12):785–791
57. EAPM (2013) Innovation and patient access to personalised medicine—Report from Irish Presidency Conference March 20th/21st 2013. http://euapm.eu/wp-content/uploads/2012/07/EAPM-REPORT-on-Innovation-and-Patient-Access-to-Personalised-Medicine.pdf. Accessed 05 Jun 2014
58. Wellcome Trust (2013) Impact of the draft European Data Protection Regulation and proposed amendments from the rapporteur of the LIBE committee on scientific research. http://www.wellcome.ac.uk/stellent/groups/corporatesite/@policy_communications/documents/web_document/wtvm054713.pdf. Accessed 16 Nov 2013
59. Ploem MC, Essink-Bot ML, Stronks K (2013) Proposed EU data protection regulation is a threat to medical research. BMJ 346:f3534. doi:10.1136/bmj.f3534
60. Brice P (2013) EU legal amendments threaten genomic medicine and research. PHG Foundation. http://www.phgfoundation.org/news/14842/. Accessed 16 Nov 2013
61. Green RC, Berg JS, Grody WW, Kalia SS, Korf BR, Martin CL, McGuire AL, Nussbaum RL, O'Daniel JM, Ormond KE, Rehm HL, Watson MS, Williams MS, Biesecker LG (2013) ACMG recommendations for reporting of incidental findings in clinical exome and genome sequencing. Genet Med 15(7):565–574. doi:10.1038/gim.2013.73
62. Easton DF, Ford D, Bishop DT (1995) Breast and ovarian cancer incidence in BRCA1-mutation carriers. Breast Cancer Linkage Consortium. Am J Hum Genet 56(1):265–271
63. Ford D, Easton DF, Stratton M, Narod S, Goldgar D, Devilee P, Bishop DT, Weber B, Lenoir G, Chang-Claude J, Sobol H, Teare MD, Struewing J, Arason A, Scherneck S, Peto J, Rebbeck TR, Tonin P, Neuhausen S, Barkardottir R, Eyfjord J, Lynch H, Ponder BA, Gayther SA, Zelada-Hedman M, the Breast Cancer Linkage Consortium (1998) Genetic heterogeneity and penetrance analysis of the BRCA1 and BRCA2 genes in breast cancer families. The Breast Cancer Linkage Consortium. Am J Hum Genet 62(3):676–689. doi:10.1086/301749
64. Lynch HT, Smyrk T, Lynch JF (1998) Molecular genetics and clinical-pathology features of hereditary nonpolyposis colorectal carcinoma (Lynch syndrome): historical journey from pedigree anecdote to molecular genetic confirmation. Oncology 55(2):103–108
65. Polymeropoulos MH, Lavedan C, Leroy E, Ide SE, Dehejia A, Dutra A, Pike B, Root H, Rubenstein J, Boyer R, Stenroos ES, Chandrasekharappa S, Athanassiadou A, Papapetropoulos T, Johnson WG, Lazzarini AM, Duvoisin RC, Di Iorio G, Golbe LI, Nussbaum RL (1997) Mutation in the alpha-synuclein gene identified in families with Parkinson's disease. Science 276(5321):2045–2047
66. Vogelstein B, Kinzler KW (2004) Cancer genes and the pathways they control. Nat Med 10(8):789–799. doi:10.1038/nm1087
67. Rahman N, Seal S, Thompson D, Kelly P, Renwick A, Elliott A, Reid S, Spanova K, Barfoot R, Chagtai T, Jayatilake H, McGuffog L, Hanks S, Evans DG, Eccles D, Easton DF, Stratton MR (2007) PALB2, which encodes a BRCA2-interacting protein, is a breast cancer susceptibility gene. Nat Genet 39(2):165–167. doi:10.1038/ng1959
68. Seal S, Thompson D, Renwick A, Elliott A, Kelly P, Barfoot R, Chagtai T, Jayatilake H, Ahmed M, Spanova K, North B, McGuffog L, Evans DG, Eccles D, Easton DF, Stratton MR, Rahman N (2006) Truncating mutations in the Fanconi anemia J gene BRIP1 are low-penetrance breast cancer susceptibility alleles. Nat Genet 38(11):1239–1241. doi:10.1038/ng1902
69. Somyajit K, Subramanya S, Nagaraju G (2010) RAD51C: a novel cancer susceptibility gene is linked to Fanconi anemia and breast cancer. Carcinogenesis 31(12):2031–2038. doi:10.1093/carcin/bgq210

70. Lensch MW, Tischkowitz M, Christianson TA, Reifsteck CA, Speckhart SA, Jakobs PM, O'Dwyer ME, Olson SB, Le Beau MM, Hodgson SV, Mathew CG, Larson RA, Bagby GC, Jr. (2003) Acquired FANCA dysfunction and cytogenetic instability in adult acute myelogenous leukemia. Blood 102(1):7–16. doi:10.1182/blood-2002–09-2781
71. Garber JE, Offit K (2005) Hereditary cancer predisposition syndromes. J Clin Oncol 23(2):276–292. doi:10.1200/JCO.2005.10.042
72. McBride KA, Ballinger ML, Killick E, Kirk J, Tattersall MH, Eeles RA, Thomas DM, Mitchell G (2014) Li-Fraumeni syndrome: cancer risk assessment and clinical management. Nat Rev Clin Oncol. doi:10.1038/nrclinonc.2014.41
73. Francke U (1976) Retinoblastoma and chromosome 13. Cytogenet Cell Genet 16(1–5):131–134
74. Chandrasekharappa SC, Guru SC, Manickam P, Olufemi SE, Collins FS, Emmert-Buck MR, Debelenko LV, Zhuang Z, Lubensky IA, Liotta LA, Crabtree JS, Wang Y, Roe BA, Weisemann J, Boguski MS, Agarwal SK, Kester MB, Kim YS, Heppner C, Dong Q, Spiegel AM, Burns AL, Marx SJ (1997) Positional cloning of the gene for multiple endocrine neoplasia-type 1. Science 276(5311):404–407
75. Gnarra JR, Duan DR, Weng Y, Humphrey JS, Chen DY, Lee S, Pause A, Dudley CF, Latif F, Kuzmin I, Schmidt L, Duh FM, Stackhouse T, Chen F, Kishida T, Wei MH, Lerman MI, Zbar B, Klausner RD, Linehan WM (1996) Molecular cloning of the von Hippel-Lindau tumor suppressor gene and its role in renal carcinoma. Biochim Biophys Acta 1242(3):201–210
76. Hirshfield KM, Rebbeck TR, Levine AJ (2010) Germline mutations and polymorphisms in the origins of cancers in women. J Oncol 2010:1–11. doi:10.1155/2010/297671
77. Samoha S, Arber N (2005) Cyclooxygenase-2 inhibition prevents colorectal cancer: from the bench to the bed side. Oncology 69(Suppl 1):33–37. doi:10.1159/000086630
78. Lynch PM (2007) Prevention of colorectal cancer in high-risk populations: the increasing role for endoscopy and chemoprevention in FAP and HNPCC. Digestion 76(1):68–76. doi:10.1159/000108395
79. Rozen P, Samuel Z, Rabau M, Goldman G, Shomrat R, Legum C, Orr-Urtreger A (2001) Familial adenomatous polyposis at the Tel Aviv Medical Center: demographic and clinical features. Fam Cancer 1(2):75–82
80. Malkin D, Li FP, Strong LC, Fraumeni JF, Jr., Nelson CE, Kim DH, Kassel J, Gryka MA, Bischoff FZ, Tainsky MA, Friend SH (1990) Germ line p53 mutations in a familial syndrome of breast cancer, sarcomas, and other neoplasms. Science 250(4985):1233–1238
81. Gonzalez KD, Noltner KA, Buzin CH, Gu D, Wen-Fong CY, Nguyen VQ, Han JH, Lowstuter K, Longmate J, Sommer SS, Weitzel JN (2009) Beyond Li Fraumeni Syndrome: clinical characteristics of families with p53 germline mutations. J Clin Oncol 27(8):1250–1256. doi:10.1200/jco.2008.16.6959
82. Olivier M, Goldgar DE, Sodha N, Ohgaki H, Kleihues P, Hainaut P, Eeles RA (2003) Li-Fraumeni and related syndromes: correlation between tumor type, family structure, and TP53 genotype. Cancer Res 63(20):6643–6650
83. Lalloo F, Varley J, Ellis D, Moran A, O'Dair L, Pharoah P, Evans DG (2003) Prediction of pathogenic mutations in patients with early-onset breast cancer by family history. Lancet 361(9363):1101–1102. doi:10.1016/S0140-6736(03)12856-5
84. Li FP, Fraumeni JF, Jr., Mulvihill JJ, Blattner WA, Dreyfus MG, Tucker MA, Miller RW (1988) A cancer family syndrome in twenty-four kindreds. Cancer Res 48(18):5358–5362
85. Chompret A, Abel A, Stoppa-Lyonnet D, Brugieres L, Pages S, Feunteun J, Bonaiti-Pellie C (2001) Sensitivity and predictive value of criteria for p53 germline mutation screening. J Med Genet 38(1):43–47
86. Hwang SJ, Lozano G, Amos CI, Strong LC (2003) Germline p53 mutations in a cohort with childhood sarcoma: sex differences in cancer risk. Am J Hum Genet 72(4):975–983. doi:10.1086/374567
87. Farooq A, Walker LJ, Bowling J, Audisio RA (2010) Cowden syndrome. Cancer Treat Rev 36(8):577–583. doi:10.1016/j.ctrv.2010.04.002
88. Hobert JA, Eng C (2009) PTEN hamartoma tumor syndrome: an overview. Genet Med 11(10):687–694. doi:10.1097/GIM.0b013e3181ac9aea

89. Liaw D, Marsh DJ, Li J, Dahia PL, Wang SI, Zheng Z, Bose S, Call KM, Tsou HC, Peacocke M, Eng C, Parsons R (1997) Germline mutations of the PTEN gene in Cowden disease, an inherited breast and thyroid cancer syndrome. Nat Genet 16(1):64–67. doi:10.1038/ng0597-64
90. Li J, Yen C, Liaw D, Podsypanina K, Bose S, Wang SI, Puc J, Miliaresis C, Rodgers L, McCombie R, Bigner SH, Giovanella BC, Ittmann M, Tycko B, Hibshoosh H, Wigler MH, Parsons R (1997) PTEN, a putative protein tyrosine phosphatase gene mutated in human brain, breast, and prostate cancer. Science 275(5308):1943–1947
91. Scheper MA, Nikitakis NG, Sarlani E, Sauk JJ, Meiller TF (2006) Cowden syndrome: report of a case with immunohistochemical analysis and review of the literature. Oral Surg Oral Med Oral Pathol Oral Radiol Endod 101(5):625–631. doi:10.1016/j.tripleo.2005.06.026
92. Gorlin RJ, Cohen MM, Jr., Condon LM, Burke BA (1992) Bannayan-Riley-Ruvalcaba syndrome. Am J Med Genet 44(3):307–314. doi:10.1002/ajmg.1320440309
93. Tan MH, Mester JL, Ngeow J, Rybicki LA, Orloff MS, Eng C (2012) Lifetime cancer risks in individuals with germline PTEN mutations. Clin Cancer Res 18(2):400–407. doi:10.1158/1078-0432.CCR-11-2283
94. Vasen HF, Boland CR (2005) Progress in genetic testing, classification, and identification of Lynch syndrome. JAMA 293(16):2028–2030. doi:10.1001/jama.293.16.2028
95. Bonadona V, Bonaiti B, Olschwang S, Grandjouan S, Huiart L, Longy M, Guimbaud R, Buecher B, Bignon YJ, Caron O, Colas C, Nogues C, Lejeune-Dumoulin S, Olivier-Faivre L, Polycarpe-Osaer F, Nguyen TD, Desseigne F, Saurin JC, Berthet P, Leroux D, Duffour J, Manouvrier S, Frebourg T, Sobol H, Lasset C, Bonaiti-Pellie C (2011) Cancer risks associated with germline mutations in MLH1, MSH2, and MSH6 genes in Lynch syndrome. JAMA 305(22):2304–2310. doi:10.1001/jama.2011.743
96. Vasen HF, Watson P, Mecklin JP, Lynch HT (1999) New clinical criteria for hereditary nonpolyposis colorectal cancer (HNPCC, Lynch syndrome) proposed by the International Collaborative group on HNPCC. Gastroenterology 116(6):1453–1456
97. Boland CR, Goel A (2010) Microsatellite instability in colorectal cancer. Gastroenterology 138 (6):2073–2087 e2073. doi:10.1053/j.gastro.2009.12.064
98. Aarnio M, Sankila R, Pukkala E, Salovaara R, Aaltonen LA, de la Chapelle A, Peltomaki P, Mecklin JP, Jarvinen HJ (1999) Cancer risk in mutation carriers of DNA-mismatch-repair genes. Int J Cancer 81(2):214–218. doi:10.1002/(SICI)1097-0215(19990412)81:2<214::AID-IJC8>3.0.CO;2-L
99. Jenne DE, Reimann H, Nezu J, Friedel W, Loff S, Jeschke R, Muller O, Back W, Zimmer M (1998) Peutz-Jeghers syndrome is caused by mutations in a novel serine threonine kinase. Nat Genet 18(1):38–43. doi:10.1038/ng0198-38
100. Tomlinson IP, Houlston RS (1997) Peutz-Jeghers syndrome. J Med Genet 34(12):1007–1011
101. Giardiello FM, Brensinger JD, Tersmette AC, Goodman SN, Petersen GM, Booker SV, Cruz-Correa M, Offerhaus JA (2000) Very high risk of cancer in familial Peutz-Jeghers syndrome. Gastroenterology 119(6):1447–1453. doi:10.1053/gast.2000.20228
102. Fearon ER, Vogelstein B (1990) A genetic model for colorectal tumorigenesis. Cell 61(5):759–767. doi:10.1016/0092-8674(90)90186-I
103. Gruber SB, Entius MM, Petersen GM, Laken SJ, Longo PA, Boyer R, Levin AM, Mujumdar UJ, Trent JM, Kinzler KW, Vogelstein B, Hamilton SR, Polymeropoulos MH, Offerhaus GJ, Giardiello FM (1998) Pathogenesis of adenocarcinoma in Peutz-Jeghers syndrome. Cancer Res 58(23):5267–5270
104. Sanchez-Cespedes M, Parrella P, Esteller M, Nomoto S, Trink B, Engles JM, Westra WH, Herman JG, Sidransky D (2002) Inactivation of LKB1/STK11 is a common event in adenocarcinomas of the lung. Cancer Res 62(13):3659–3662
105. Hearle N, Schumacher V, Menko FH, Olschwang S, Boardman LA, Gille JJ, Keller JJ, Westerman AM, Scott RJ, Lim W, Trimbath JD, Giardiello FM, Gruber SB, Offerhaus GJ, de Rooij FW, Wilson JH, Hansmann A, Moslein G, Royer-Pokora B, Vogel T, Phillips RK, Spigelman AD, Houlston RS (2006) Frequency and spectrum of cancers in the Peutz-Jeghers syndrome. Clin Cancer Res 12(10):3209–3215. doi:10.1158/1078-0432.CCR-06-0083

106. Khoury MJ, Feero WG, Valdez R (2010) Family history and personal genomics as tools for improving health in an era of evidence-based medicine. Am J Prev Med 39(2):184–188. doi:10.1016/j.amepre.2010.03.019
107. Valdez R, Yoon PW, Qureshi N, Green RF, Khoury MJ (2010) Family history in public health practice: a genomic tool for disease prevention and health promotion. Annu Rev Public Health 31:69–87. doi:10.1146/annurev.publhealth.012809.103621
108. Jarvinen HJ, Mecklin JP, Sistonen P (1995) Screening reduces colorectal cancer rate in families with hereditary nonpolyposis colorectal cancer. Gastroenterology 108(5):1405–1411. doi:10.1016/0016-5085(95)90688-6
109. Burke W, Daly M, Garber J, Botkin J, Kahn MJ, Lynch P, McTiernan A, Offit K, Perlman J, Petersen G, Thomson E, Varricchio C (1997) Recommendations for follow-up care of individuals with an inherited predisposition to cancer. II. BRCA1 and BRCA2. Cancer Genetics Studies Consortium. JAMA 277(12):997–1003
110. De Cosse JJ, Bulow S, Neale K, Jarvinen H, Alm T, Hultcrantz R, Moesgaard F, Costello C (1992) Rectal cancer risk in patients treated for familial adenomatous polyposis. The Leeds Castle Polyposis Group. Br J Surg 79(12):1372–1375
111. Dunlop MG (2002) Guidance on gastrointestinal surveillance for hereditary non-polyposis colorectal cancer, familial adenomatous polypolis, juvenile polyposis, and Peutz-Jeghers syndrome. Gut 51(Suppl 5):V21–27
112. van Sprundel TC, Schmidt MK, Rookus MA, Brohet R, van Asperen CJ, Rutgers EJ, Van't Veer LJ, Tollenaar RA (2005) Risk reduction of contralateral breast cancer and survival after contralateral prophylactic mastectomy in BRCA1 or BRCA2 mutation carriers. Br J Cancer 93(3):287–292. doi:10.1038/sj.bjc.6602703
113. Wacholder S, Hartge P, Prentice R, Garcia-Closas M, Feigelson HS, Diver WR, Thun MJ, Cox DG, Hankinson SE, Kraft P, Rosner B, Berg CD, Brinton LA, Lissowska J, Sherman ME, Chlebowski R, Kooperberg C, Jackson RD, Buckman DW, Hui P, Pfeiffer R, Jacobs KB, Thomas GD, Hoover RN, Gail MH, Chanock SJ, Hunter DJ (2010) Performance of common genetic variants in breast-cancer risk models. N Engl J Med 362(11):986–993. doi:10.1056/NEJMoa0907727
114. Rosendaal FR, Koster T, Vandenbroucke JP, Reitsma PH (1995) High risk of thrombosis in patients homozygous for factor V Leiden (activated protein C resistance). Blood 85(6):1504–1508
115. Andermann A, Blancquaert I, Beauchamp S, Costea I (2011) Guiding policy decisions for genetic screening: developing a systematic and transparent approach. Public Health Gen 14(1):9–16. doi:10.1159/000272898
116. Nussbaum RL, McInnes RR, Willard HF (2007) Thompson & Thompson Genetics in Medicine. Elsevier Health Sciences, Makati City
117. Health Council of the Netherlands (2008) Screening: between hope and hype. http://www.emgo.nl/files/Bonn%202010%20genetic%20screening%20criteria%20in%20the%20age%20of%20genomics.ppt. Accessed 20 Sep 2014

Roadmap to Drug Development Enabled by Pharmacogenetics

James P. Bishop, Sonal B. Halburnt, Patrick A. Akkari, Scott Sundseth and Iris Grossman

Abstract The primary goal of the pharmaceutical industry is to develop safe and effective medications. As the industry matures and the existing arsenal of marketed therapeutics grows, novel drugs must exhibit greater efficacy and safety to achieve registration and favorable reimbursement. Furthermore, gaining market-share has become extremely competitive, in terms of both meaningful clinical effects and tolerated safety profiles. As a result, the pharmaceutical industry has experienced a steady decline in productivity in recent decades. However, the achievement of regulatory approvals for targeted therapeutics may reverse this drop in productivity. The convergence of high-throughput genetic analysis technologies and the exponentially expanding biological and genomic knowledgebase have provided many clear examples that genetic variation can affect both disease risk and drug response. Therefore, evaluation of genetic variation in clinical trial populations should be considered essential and routine from the earliest phases of drug development. Pharmacogenetics (PGx) in particular has gained considerable attention from drug developers, regulators and payers over the past decade as a means to achieving safer, efficacious and more cost-effective drugs. While PGx science has great potential to impact positively the success of developing a new medicine, the integration of PGx into the decision making processes of the drug development pipeline has been difficult. The goal of this chapter is to describe the principles and requirements of an efficient

J. P. Bishop (✉)
Teva Pharmaceuticals, Frazer, PA, USA
e-mail: james.bishop02@tevapharm.com

S. B. Halburnt · P. A. Akkari · S. Sundseth
Cabernet Pharmaceuticals, Durham, NC, USA
e-mail: sonal.halburnt@cabernetpharma.com

P. A. Akkari
e-mail: pakkari@cabernetpharma.com

S. Sundseth
e-mail: sundseth@cabernetpharma.com

I. Grossman
IsraGene Ltd., Yakir, Israel
e-mail: IrisG@IsraGene.com

Personalized Medicine & Pharmacogenomics, Teva Pharmaceutical Industries, Netanya, Israel

G. Grech, I. Grossman (eds.), *Preventive and Predictive Genetics: Towards Personalised Medicine,* Advances in Predictive, Preventive and Personalised Medicine 9,
DOI 10.1007/978-3-319-15344-5_3

and valuable PGx strategy that makes use of every opportunity during the course of developing innovative medicines. This strategy combines a proven methodology with rigorous genetic science to create a "*Pipeline Pharmacogenetic Program*".

Keywords Novel drugs · Companion diagnostics · Pipeline pharmacogenetics · Clinical trials · Drug development · Project management methodology

1 Pharmacogenetics in Today's Market-Place

Consumer demand for customized products and services is well established and evident in mainstream retail markets as well as emerging technologies. Gone are the days of "one-size fits all" and if a product or service is mass produced, then the available combinations, flavors and add-ons are so numerous that most consumer experiences can be, or at least feel, truly personalized. Similar pressures exist in healthcare markets. In fact, personalized medicine has the potential to benefit the consumer more than most retail products. The complexities of health and disease, underlined by each patient's specific environmental and genetic factors, call for a truly personalized approach given the suboptimal performance of standard therapy (most drugs exhibit response rates lower than 60%) [1]. Recognizing this growing need for individualized healthcare, many USA healthcare providers and hospitals offer services through "Centers of Personalized Medicine", like the Duke University Health System, MD Anderson Cancer Center, and Cleveland Clinic. Similarly, clinical pharmacologists and medical laboratories throughout Europe provide Personalized Medicine services, including the Karolinska Medical System and the Erasmus University-Rotterdam. Furthermore, international collaborative networks on personalized medicine are quickly forming to enhance knowledge acquisition and leverage capabilities. The Personalized Medicine Coalition, for instance, consists of over 200 academic, industry, healthcare provider and payer groups "seeking to advance the understanding and adoption of personalized medicine concepts and products for the benefit of patients" [2]. The European Commission is dedicating considerable investment in Horizon 2020 for innovation across European member countries with an emphasis on personalized medicine and systems medicine [3]. A key element of Personalized Medicine concepts and products has been and will likely continue to be in the area of PGx. There are currently 128 FDA-approved drugs that contain pharmacogenomic information in their label [4]. Indeed, regulatory agencies promote using genetic information in the drug development process in order to improve safety and efficacy by using pharmacogenomics information to decrease adverse events and to identify non-responders [5]. The general public also appears to have considerable interest and willingness in PGx testing to predict side effects, guide dosing and assist with drug selection [6]. The growing genetic testing market, estimated at $ 5.9 billion in 2016, and numerous direct-to-consumer and physician-provided genetic test companies are evidence of the economic forces driving the industry [7, 8]. The need for tailored medicines and the favorable regulatory environment to facilitate their development is driving increased availability of genetic testing services, thus creating market forces that reduce the cost of acquiring individual genetic information. For example, currently the cost for whole genome sequencing is $ 5800 per sample [9]. It is therefore an unquestionable fact that Personalized Medicine has arrived and its utilization and effects on healthcare is growing. For instance, in a recent McKinsey report the authors indicate that already over a third of the drugs currently in clinical studies are associated with a companion biomarker, indicating that newly approved drugs in the coming years will increasingly be dominated by targeted therapeutics [10]. However, the process of investigating, validation and qualifying companion PGx tests is challenging. It requires early investments in scientific infrastructure, and it hinges on clear *a priori* commercial and regulatory strategies to ensure the timely and cost-effective launch of the two end-products (i.e. the drug (Rx) and the diagnostic (Dx)). In addition to the principles of an efficient and valuable PGx strategy, we outline below the requirements, advantages and challenges associated with integrating PGx investigations into the drug research and development (R&D) pipeline. The information reported is based on our deep expertise in "Pipeline Pharmacogenetics" acquired over cumulative decades of application across diverse therapeutic areas and several global pharmaceutical companies.

2 Pharmacogenetics-by-Design: the R&D Environment

The application of PGx to currently marketed drugs as a method to predict safety and efficacy is of significant value to patients, physicians, regulators, payers and industry (some examples include warfarin, abacavir and multiple oncologic agents). The inability to predict the risk of adverse drug reactions (ADR) each time a patient is exposed to a new medicine continues to dramatically affect patients' morbidity and mortality. For example, in two separate studies, researchers reported that ADRs are estimated to be the 7th most common cause of death in a 2001 Swedish population based study [11] and that the incidence of serious ADRs was estimated to be 6.7 % of hospitalized patients in the US [12]. In addition, most medicines display significant inter-individual variability in efficacy, but the current clinical practice approach addresses this problem by passive and reactive empirical methodology: treatment is administered according to standard protocols and outcome is assessed during later visits to determine efficacy. This "trial and error" practice is usually followed by either dose adjustments or triage onto other medicines if the patient poorly responds or fails treatment. This practice also results in prolonged procedures, including delay of efficacious treatment (sometimes over the course of months and years in the case of immunomodulatory treatments), risk of exposure to unnecessary drugs (which are always associated with a host of side effects), protracted suffering of patients and their caregivers, and, finally, additional cost to payers.

The development of PGx tests for registered medicines aims to identify optimum benefit-risk ratios and allow prospective testing prior to administration of drug. The availability of PGx testing also permits a differentiation strategy that guides the pharma industry to develop medicines tailored specifically for non-responder populations, thus addressing true unmet medical needs. Regulatory approved PGx safety tests prospectively predict who is at risk of considerable harm and provide high value as a warning to healthcare providers and patients regarding drugs about to be launched or currently on the market. Equally important, predicting specifically who is going to be at risk of ADRs and excluding them from treatment prevents valuable medicines from being withdrawn from the marketplace. The PGx test thus serves to identify those patients who should be administered a drug and expect meaningful efficacy and safety.

For formularies such as Australia's pharmaceutical benefits scheme (PBS) as well as commercial payers such as those in the United States who are looking at drugs and their value to the public to whom they are responsible, PGx testing allows identification of subpopulations for whom there is an unmet need and greatest benefit [13, 14]. However, little progress has been made on the pharmacoeconomics of the prospective use of genomic biomarkers in the prediction of the benefit-risk ratio for patients. Notwithstanding, there are many examples of genetic variation being significantly associated with ADRs and effective as Dx in clinical practice, such as hepatotoxicity and hypersensitivity reactions [15]. Case studies teach us that genetic variation in the drug target (e.g., receptor) and signal transduction pathway of the majority of drugs accounts for much of the variability in response to medicines [13]. Variation in genes associated with immunological reactions and pathways can also be implicated in drug safety, most notably the MHC/HLA system. One such example is in the use of a PGx test for HLA-B*5701 prior to the administration of abacavir has resulted in the complete mitigation of cases of serious hypersensitivity reaction to the drug. Subjects who are HLA-B*5701 negative almost never develop immunologically-confirmed hypersensitivity reaction upon secondary administration of abacavir, on the other hand, HLA-B*5701 positive subjects (5 % of the Caucasian population) have a 70 % chance of developing a serious hypersensitivity reaction leading to hospitalization and possibly death if untreated [16].

The FDA and EMEA in addition to other regulatory agencies around the world, have experience with PGx integration into drug development. The FDA has substantial experience with how PGx may be used and there are now several FDA-approved drugs with PGx information in their labeling [5]. This illustrates that there is now a clear expectation that PGx data would be available on safety and efficacy and the FDA has published guidance on this [5]. The FDA has now seen PGx used where variability in response or exposure is observed, where adverse events are a concern, where drug dosage adjustment based on genotype is suggested and where known polymorphism at the target and or signal transduction pathway is evaluated. ADME gene variation involved in the metabolism of molecules has also been seen by the regulators and several molecules approved have ADME genotyping recommendations in the label (aripiprazole and CYP2D6 metabolizer status is an example).

3 The Roadmap to "Pipeline Pharmacogenetics"

3.1 Scientific Rationale

Pharmacogenetics, like any discipline employed for the purpose of improving the way drugs are designed and developed, is first and foremost a science. It is critical that during the course of PGx application this perspective remains the leading principle during the selection of methodologies, technologies and analysis procedures. This is particularly true given the exponential growth seen in recent years in scientists' capability to sequence genomes, analyze Big Data and integrate complex phenotypic and medical information into clinically meaningful health management decisions. Still, one may ask—what is the advantage of embarking on the PGx process at early development phases, given limited power considerations associated with the size of these studies (often only a few hundred patients are collectively exposed to an investigational drug leading up to Phase III of its development)? After all, one could argue that postponing the investment would enable focusing efforts on drugs only after demonstrating favorable proof-of-concept (PoC) results and passing the initial safety hurdles. The counter argument lies in the very premise of the concept of "Pipeline PGx", and is well supported by positive, as well as negative, examples: the initial clinical development phases stand to benefit the most from the PGx methodology. PGx-enabled PoC design can maximize efficacy signals and exclude safety outliers so as to shift the overall benefit/risk ratio, resulting in increased probability of technical success early on for the entire program. *Post hoc* attempts to rescue development programs incur costs and waste valuable time depriving patients of effective treatments. History has repeatedly shown that only pre-emptive and systematic application of available scientific understanding of the mode-of-action of drugs and associated pathways can yield pharmaceutical successes that meet regulatory requirements. It is this mind-set and systematic approach that led to the development of a predictive test for abacavir hyper-sensitivity reaction described above [16, 17] or the positioning of prasugrel in a highly competitive landscape against clopidogrel [18, 19]. It is also thanks to this approach and adoption of emerging scientific discoveries that enabled the refocusing of the development of crizotinib from a c-Met-inhibitor to an anaplastic lymphoma kinase (ALK)-inhibitor, and thus formed the target of a co-developed diagnostic for defining patient eligibility [20].

3.2 Sample Collection Strategy

PGx research depends on the collection of DNA samples to generate data. In order to respond to the regulatory authorities' guidelines associated with genetic analysis, most pharmaceutical companies are now devoting resources within their clinical trial programs to enable the collection and storage of DNA samples. These DNA samples provide the pharmaceutical industry with the opportunity to investigate drug response, thereby increasing the likelihood of developing better therapies for patients and enhancing our understanding of the of disease context (e.g. progression and subtype characteristics compounding PGx outcomes) [21]. The collection rates of optional DNA samples, however, remain below the ideal target rate of 90–100% which appropriately represents the PGx population out of the overall clinical trial ITT (intention to treat) dataset. This variable collection rate may be due to a variety of reasons as listed in Table 1.

Efforts should be made to mandate DNA sample acquisition across all programs where it is determined that DNA collection has a clear rationale and local laws/

Table 1 Common reasons for insufficient DNA sample collection rates in clinical trials and suggested mitigation plans [5, 22, 23]

	Issue	Mitigation plan
1	Insufficient understanding of the informed consent by clinical trial subjects	Ensure following best practices for informed consent writing [24], ensure site staff is knowledgeable and supportive (see 2 below)
2	Lack of support or interest by the site staff	Ensure communication to Principal Investigator clearly states the rationale and medical value of PGx testing in the study. Education program via Investigator Meeting, study newsletters as well as support and accessibility of knowledgeable PGx personnel
		Real-time. Incentivized DNA collection monitoring program Clear lab manual instructions that are easy to follow
		Mandatory DNA collection with clear underlying clinical justification is best practice.
		Incorporation of clear requirements in proficiency of DNA sample collection capabilities and attitudes should be incorporated a priori into site selection procedures
3	Reluctance of CROs to invest efforts in genetic study submission requirements	Select CROs experienced in DNA collection globally Include performance matrix of DNA collection as key elements of service contract Ensure communication to CRO clearly states the rationale and medical value of PGx testing in the study Mandate review by sponsor of country-specific submissions along with up-to-date regulatory guidelines in each recruiting country Establishing routine monitoring procedures for submission and sample collection
4	IRB/EC variation in interpretation of regulations	Clear protocol and ICF language on the purpose and rationale for DNA collection, adjusted to the specific requirements (in terms of detail and format) to each target country and recruiting center
5	Lack of logistical infrastructure	Select central labs with proven capabilities in collection and handling of samples intended for DNA collection (including tumor source) Consider providing refrigerators, centrifuges, dry ice, etc. as needed to ensure quality of samples maintained throughout the custody chain
6	Perception that DNA samples are associates with greater privacy violation risks than the collection of other types of samples during the clinical trial	Dialogue with Key Stakeholders regarding coding practices such that equal standards are applied to DNA and non DNA samples

regulations permit, and ideally from all mid and late phase programs as means for risk mitigation. DNA collection at baseline allows appropriate regulatory utility if and when needed [5]. The benefits of DNA sampling and storage are evident in drug labels, and contribute to internal decision making and regulatory filings [21]. Collecting DNA samples at >90 % rate is key to successful and effective translation of findings into improved performance, given that otherwise any such attempt would be significantly compromised by the requirement to conduct new confirmatory prospective studies [21]. The underlying working assumption of PGx, in drug development terms, is valuable *only* when it is delivered in time for project team decision-making. Ultimately, timeliness of results is what facilitates achieving the objectives of each drug development program [14].

A DNA sample collection strategy requires the following key elements:

DNA Sampling Strategy Senior management within the company must provide explicit support that will allow for a clearly defined process to collect DNA samples within clinical trials to address clinical, scientific and regulatory issues in drug development [23]. Ensuring open communications and responsiveness to IRBs, ECs and other Regulatory bodies in the collection process will help to fully utilize the value of PGx research [22]. Funding will also be required in order to create the appropriate infrastructure to not only collect the sample, but to track the collected DNA samples to allow for timely and complete reconciliation (i.e. matching signatures on consent forms with acquired samples at the storage site). An integrated sample management process ensures efficient access to the samples to support the PGx analysis as well as ensures a method to keep the samples secure and private, allow for the tracking of the DNA sample from collection through to genotyping, storage, utility, destruction throughout the chain of custody to support the PGx analysis.

Training Education and training on the value of PGx and why there is the need to achieve optimal DNA collection rates must be provided to both key internal stakeholders (clinical project teams and their operationally focused colleagues) and external collaborators (such as contracted clinical research organizations (CROs) and clinical trial site staff) [14]. A patient's level of understanding of how these samples will be used can be influenced by the level of the investigator's enthusiasm for genetic research.

Informed Consent To be able to use a DNA sample collected in a clinical trial there needs to be a consent form that pre-defines the genomic objective prior to sample acquisition. These objectives can include pre-planned analysis around known factors that are likely to influence the safety, efficacy and/or dosing of the drug [5]. These types of analyses often require access to individual clinical information, particularly in cases of safety investigations. Broader investigations of an open-ended nature can also be considered as long as the sponsor clearly states that intended research will be limited to PGx purposes, i.e. understanding the response profile of the drug. Sponsors wishing to engage in further unspecified broad research which is beyond the scope of PGx would need to separate this research objective from the

PGx objective, placed under strictly voluntary basis, and often commit to anonymizing samples before analysis.

There are some special considerations to take into account when developing the consent form. Regulations around the informed consent vary both globally and locally. To allow for the main study to move forward without any delays, many pharmaceutical companies have created a separate genetic consent form from the main study consent, due to the additional approvals that may be required for collection of genetic samples and PGx research [25]. There are other special considerations that may need to be addressed in the consent form, such as possible ethical implications of the collected data, security and privacy terms of the acquired genetic information and under what circumstances research results might be returned back to the study participants [25].

Sample Collection PGx samples should be collected from all subjects randomized to treatment in all cohorts and in all phases of clinical trials to ensure samples are collected from subjects who have the potential to have a variation in response to the drug [5]. Collection of these samples at the time of enrollment will ensure minimal bias (avoiding lack of representation of DNA samples from subjects who withdrew from the study for any reason) and importantly ensure coverage of sampling from subjects subsequently experiencing ADR during the course of the study. The sample set also needs to be representative of the targeted population for the therapy to cover genetic variation among individuals from different geographic locations [22]. The voluntary and incomplete nature of many exploratory genetic studies conducted in prior years has often raised concerns about potential bias and statistical power, which could compromise the scientific rigor of such studies [5]. There are multiple sample types that can be used for DNA analysis additional to blood (e.g. buccal swabs, hair follicles, etc.) and are particularly relevant to pediatric or other special populations. Furthermore, in oncology studies tumor source DNA and/or circulating tumor cells (CRC) are also required to fully capture the PGx associated variation source that can affect the studied endpoints. When considering DNA samples from sources other than blood a robust quality assurance and quality control programs must be put in place to ensure sufficient yield and DNA quality [14]. This is particularly important when considering tumor source DNA sampling, including aspiration, formalin fixed paraffin-embedded tissue (archived versus fresh), fresh frozen biopsy, etc. and likely to differ from one cancer type to another. In these cases it is beneficial to collect tumor DNA at treatment failure so as to investigate mechanisms of resistance to therapy which are often underlined by the tumor's rescue mutations.

Sample Retention The retention of the DNA sample allows for the opportunity to perform investigations that may occur after the completion of the studies. Samples should be retained for a time period that will permit post marketed analysis should the need arise (e.g., at least 15 years) [5]. Long term sample storage will allow for the investigation of not only observations that emerge during the trial, but also any observations that may occur in subsequent trials and in the first several years after the drug has been on the market. These can be used to investigate external claims generated by other groups once the drug is marketed, and may facilitate study of additional indications as part of the life cycle management of the product.

3.3 *Fit-for-Purpose Genetic Interrogation*

Traditionally, PGx studies were performed using a candidate-gene approach, often with genetic variants of the molecular drug target itself, or key polymorphic genes up or downstream in the drug target biological pathway. While candidate-gene hypotheses are statistically powerful, testing discrete genetic drug response hypotheses with a small number of variants, hypothesis-free approaches offer the opportunity for discovering novel genetic markers of drug response and revealing novel biological pathways. These genome-wide methodologies can be performed with custom or commercially available SNP arrays (genome-wide association studies/GWAS), and more recently have incorporated genome-wide sequencing technologies (whole genome sequencing/WGS or whole-genome exome sequencing/WGES). The shift to genome-wide genetic investigations has evolved as a consequence of several factors including lower costs for genotyping or sequencing, better statistical analysis methods and improved design of PGx clinical studies.

Historically, genome-wide association analyses of disease susceptibility have identified common sequence variants that impart modest, 10–20% increases in disease risk. In contrast, the genetic risk attributed to variants associated with drug response (safety or efficacy) has been much larger (300–2000%) [26, 27]. One explanation for this large difference in disease vs. drug-response genetic risk ratio could theoretically be attributed to the shorter period of evolutionary time that humans have been exposed to drugs, resulting in decreased selection pressure [28]. Leveraging this interaction of a patient's genome with drug response provides the potential to prescribe the right drug to the right patient (and at the right time for the right cost!). It should be noted that even though PGx science may lead to improvements in drug development, registration and patient health, its implementation has been hampered by the opinion that it might not be cost-effective [29]. However, this argument is becoming less relevant as costs of genotyping technologies drop and as central labs and medical centers increase their investment in genetic testing. Coupled with this is a robust improvement in the technology and breadth of gene tests available in a point-of-care instrumentation format that can provide the clinician with immediately actionable genetic information for personalized prescribing.

3.3.1 Technology of Choice, Genotyping and Sequencing

Candidate gene studies, utilizing either small number of often functionally significant SNPs in a key gene or a few genes (e.g., drug target or critical gene in drug target biological pathway) provide concise answers to specific gene association questions. They are usually employed if there is *a priori* genetic evidence that implicates a particular gene in drug disposition (ADME genes) or drug-response for efficacy/

safety purposes. Targeted gene variant assay panels are widely available from commercial sources and validated for use in diagnostic applications [30, 31]. Candidate gene studies have the advantage of being technically robust and are generally used to confirm a genetic hypothesis derived from a preceding study or reported finding. The original study(ies) is thus referred to as "hypothesis-generating", and often relies on approaches like customized, therapeutically- or disease- focused arrays or GWAS. The results of later confirmatory candidate gene studies often form the basis for development of a genetic companion diagnostic(s) co-development program, temporally synchronized with registration studies for a specific therapeutic.

In contrast, larger customized-array approaches or genome scans are undertaken when little or no genetic information exists, linking the clinical phenotype of interest to specific gene(s). Until recently, whole genome genotyping was usually more expensive than a candidate gene/SNP approach and results were limited to fairly common genetic variants that were selected for coverage across the entire human genome. Recently however, high-density arrays with tagging SNPs capable of assaying genetic variation down to ~1 % minor allele frequency (MAF) have been combined with custom arrays allowing the examination of groups of genetic variants with particular functional significance (e.g. exome arrays, ADME arrays, HLA arrays) [32].

Whole-genome sequencing (WGS) [33, 34] and whole-genome exome-sequencing (WGES) [35–37] costs are also plummeting and these technologies will ultimately replace array-based genotyping approaches in the near future. Advantageously, cheaper WGS and WGES [38] will permit transition away from GWAS-common variants to inclusion of rare genetic variants with potentially greater clinical effects. While accounting for a lower number of patients per specific variant, phylogenetic and coalescence methods are enabling the clustering of evolutionary-related variants into powerful genomic associations [39]. These WGS off-the-shelf products now widely validated for accuracy and coverage, also possess the advantage of condensed order-to-result timelines, since customized array solutions typically require 12–16 weeks for array design and manufacturing. These timelines are often incompatible with clinical development deadlines and force pharmaceutical companies to revert to pre-designed solutions in many cases. In fact, the high cost and complex logistics of obtaining properly consented DNA samples from well-phenotyped clinical trial subjects coupled with the ever decreasing costs of genotyping or sequencing on a genome-wide scale mean that GWAS or WGS/WGES is often cheaper than a candidate gene approach. Thus in practice a large database can be created of genetic variation across the genomes of the entire clinical trial cohort and then sequentially queried *in silico*, starting with a concise candidate gene analysis (hypothesis testing) and ending with a genome-wide screen for genetic variants with large effect (hypothesis-generation) [40, 41].

3.3.2 Statistical Analysis Considerations

The major objective of PGx analysis is to identify genetic marker(s) that can differentiate distinct subgroups of patients in a clinical trial based upon their drug response. Additionally, the pharmaceutical industry is also interested in discovering genetic variants that are prognostic of a specific disease state or rate of progression of a pathological phenotype. Analytical models for predictive genetic markers include an interaction effect between genotype and treatment while prognostic markers are generally a main effect; where "response" is independent of drug therapy [42].

Early exploratory PGx studies generally analyze many potential genetic variants (candidates) or even scan entire genomes (GWAS, WGS, WES) to identify genetic markers, but small sample size/power, multiple testing, and a high false discovery rate can constrain the ability to discern valid, statistically significant results [43–47]. One key approach to screen out false positive results is to replicate results from the initial exploratory study in a separate clinical trial with similar patients and treatment. Lastly, a prospective, confirmatory study is necessary to test hypotheses related to specific genetic effects and evaluation of the clinical utility of the genetic markers (e.g. specificity, sensitivity, positive predictive value and negative predictive value), establishing the qualifying performance characteristics of the genetic diagnostic (Dx) as a basis for its regulatory approval. Therefore, three separate clinical trials (exploratory, replication, confirmatory) are necessary to go from discovery of a genetic marker to a companion diagnostic, reinforcing the need to start a PGx strategy early in the drug development pipeline.

Study design considerations are important at all steps of the PGx pipeline process. For confirmatory studies, consideration of targeted, enriched or stratified trial designs can be advantageous [48], but are usually only employed when there is an abundance of *a priori* information on a particular genetic marker. Adaptive studies or "gated" approaches permit the analysis of particular genetically defined subgroups when a study fails to meet its primary objective(s), and statistical concerns about multiple testing can be controlled by judicious "alpha-spend" [49]. For exploratory studies, weaker genetic effects can be revealed by using an extreme-phenotype approach that accentuates the differences between subgroups (e.g. super-responders vs. non-response) [50], and variations of this approach may be of particular importance for the study of genetic markers related to serious adverse events (SAEs). Lastly, improvements in the integration of genetic, genomic and clinical information, coupled with newer analytical techniques like Bayesian approaches, multivariate analysis of genetic "features" (SNPs, CNVs, SNVs, etc.) [51, 52] or phylogenetic analysis of sequence data [53], will create new ways to evaluate PGx study data and discover and develop more robust genetic markers of disease and drug-response.

The cost of functional validation can be high if a large number of gene associations emerge from GWAS or sequencing studies, and predefined lists of candidate genes in biological pathways of interest are often chosen for follow-up association studies. Approaches that combine GES with functional genomic bioinformatics filters (e.g. protein folding, gain/loss-of function predictions) or systems biology approaches (genetic, genomic, proteomic, metabolomics, etc.) [54] can also be used to prioritize results for wet-lab functional validation and may uncover novel pathways of biological relevance that are missed in pre-determined analyses.

In conclusion, drug trials of the future will be focused on genomically-targeted patients; identifying those most likely to respond to treatment and least likely to have an adverse event [55–57]. Synergistic effects of high-resolution genomic data (e.g. DNA/RNA sequence), better statistical analysis methods, rapid testing, as well as cheaper genomic analyses will translate into substantial savings in drug development cost and greater patient benefit.

3.4 Integrated Execution Methodology

Opportunities for PGx and the value to the portfolio exist throughout the development process from preclinical through Marketing/Pharmacovigilence, as long as PGx is in lock step with discovery and clinical development milestones. For this value to be realized, PGx objectives must be integrated into study protocols from early drafting to ensure that the proper support framework and budget are in place for sample collection, data management, and statistical analysis. In addition, experienced PGx personnel should be fully integrated into the clinical development teams from their inception point. The PGx team should be led by a scientist and consist of contract and vendor manager, genomic data manager, statistician, bioinformatician, and PGx project manager.

3.5 Communicating with Stakeholders

Managing the exchange of information and expectations across and outside the organization is challenging, though essential, for a successful drug development program. The internal and external stakeholders for PGx information are similar to other elements of the Clinical program, though some specific considerations are noteworthy for a PGx program.

Internal Stakeholders

Drug Discovery teams:

Disease genetics can be critically important in target and lead identification and validation, making PGx involvement at the earliest stages of discovery highly valuable.

Clinical trial design teams:

Integrating clinical objectives (primary, secondary, exploratory, gated) in clinical studies is the key to generating both retrospective, as well as prospective, actionable genomics results, tailored specifically for the enrolled population.

Clinical operations teams:

Once PGx is built into the clinical program, managing sample collection and clinical data availability is necessary.

Drug program/management teams:

Overarching program teams defining the overall strategy for the compound and evaluating novel indication or combination strategies for the compound, need to be informed of the PGx progress and results, especially if and when unfavorable safety and efficacy results emerge in a study. If an integrated, prospective approach is taken, PGx information can be used to save some programs in light of results that would initially seem to kill a program.

Senior management (technical and non-technical):
Decisions of funding and ultimately the fate of programs facing unfavorable results are generally in the hands of senior managers that may not have specific and technical PGx background. Keeping management informed of the PGx strategy and value proposition, as well as current results is essential.

External Stakeholders

Regulatory Agencies:
As with much of the work companies plan and execute to develop drugs, communicating PGx plans and results to regulatory agencies in a timely manner is critical, especially at key clinical milestones. Agencies endorse the use of PGx information to increase the understanding of patient safety and drug efficacy as part of the benefit-risk assessment [58]. Furthermore, several communication routes are possible to convey PGx related information and should be chosen as appropriate, including for instance in the FDA the voluntary exploratory data submission (VXDS) route (non-trial specific), the "conventional" submission route to CDER, and co-development route to CDER and CDRH simultaneously.

Academic collaborators:
Trial recruitment rate is often better when key opinion leaders in the relevant therapeutic areas are involved in research and development of drugs. Including participation of academic collaborators in the PGx aspects of projects can often provide added benefit. This is also key for smooth introduction into the clinic and correlates well with market adoption at commercialization.

Payers:
Optimizing the health outcomes of patients is the primary goal of payers. Understanding payer's willingness and overall market drivers for drugs with PGx opportunities and label information will aid in developing a realistic value proposition, especially with companion diagnostic opportunities.

Physicians:
Beyond the physician's involvement in clinical trials, increasing the physician-wide knowledge of PGx and drug safety and efficacy will ultimately lead to better adoption by patients.

Patients:
Patients demand personalized approaches to many projects and services, and have increased willingness to provide genetic information when participating in clinical trials. Reaching out to patients or advocacy groups with regards to the opportunities to improve health and wellness through PGx is essentials. Recently patient advocacy groups have shown to be instrumental in targeted therapeutic approaches to drug development, for instance in the case of Cystic Fibrosis and Vertex's Kalydeco.

4 Pipeline Pharmacogenetic: Practical Application

For PGx to be successful, the objectives, tasks, and supporting roles (internally and externally) must be managed with a systematic methodology. Employing the established framework of formal Program and Project Management will maximize the delivered value of PGx. Since new molecular entities and drug candidates are considered *Program-level* effort due to the long timeframe (> 10–15 years from candidate selection to end of patent protection, plus possible product line extension), the corresponding PGx effort integrated in the development of these assets should be managed as a Program. The key deliverable emerging from a properly managed PGx Program is a PGx Strategy that is fully integrated and aligned with the asset development program. In a similar manner, individual preclinical and clinical studies that support assets are considered projects since they are a "temporary endeavor undertaken to create a unique product, service or result" [59], so the corresponding PGx experiments and studies should be managed as projects that are arranged and executed to secure the goals set within the PGx Strategy.

4.1 PGx Program Stages

The ideal PGx program would start very early in the asset life cycle, possibly pre-candidate selection or even at or as part of biological target identification. However, even mature assets with established clinical programs in Phase 1–3, possibly even approved and marketed assets, can initiate a PGx program. There are 3 key stages in the life cycle of a PGx Program illustrated in Table 2 [14].

Confirm When PGx is first considered for a drug candidate, the PGx team should be gathering and evaluating information related to disease biology, existing genomic factors for the biological target and potential patient populations, competitive landscape, early safety signals, available information on ADME, and other information useful to start formulating a PGx strategy. This early exploratory program stage results in the confirmation that there is indeed a PGx opportunity for a particular asset.

Integrate The chief purpose of this program phase is to establish the initial integrated PGx strategy, and to convey the value that PGx will bring to the particular asset and overall portfolio. It is recommended that this guiding information be recorded in the PGx Strategy and Value Proposition document (SVP) at this phase. The SVP is an overarching, "living" document that would serve as a reference point for all tactical decisions related individual PGx projects (Sect. 3.5.2). Stakeholders and funding sources (e.g. clinical teams, senior management) should be in agreement with the PGx strategy at this point.

Implement and Refine Once the PGx Strategy is established, this final Program phase is essentially the PGx program at "steady state" and is the longest phase,

Table 2 Pipeline Pharmacogenetics (PGx) Program Methodology

I. Confirm	II. Integrate	III. Implement and refine
Purpose:	*Purpose:*	*Purpose:*
Understand molecule, gather information and confirm PGx opportunity	Integrate with development team, create initial PGx strategy, identify value proposition	Implement PGx strategy, execute PGx experimental projects and deliver refined PGx strategy
Activities:	*Activities:*	*Activities:*
Molecule investigation via review of: - Preclinical data - Intended therapy - Target and pathway - External literature - Portfolio priority - Existing Clinical data, if available	Engage clinical team via: - Detailed molecule investigation - Review safety signals - Understand label, differentiation goals, development plan	Operationalize PGx via: - Protocol development, regulatory planning, and trial execution - Experimental project execution - Results interpretation - Strategy refinement
Key deliverable:	*Key deliverable:*	*Key deliverables:*
PGx molecule assessment	PGx strategy and value proposition	Experimental data and interpretation PGx strategy refinement

where individual projects are executed in alignment with the strategy. New information is gathered from external sources and results from implemented PGx projects, and the PGx strategy and corresponding SVP document are updated and refined to adapt to the changing situation of the asset.

4.2 PGx Project Stages

Once the PGx strategy is developed and clinical integration points are established, PGx projects should be implemented within the program by using the following 5 stages illustrated in Table 3 [14].

Scope A considerable part of "scoping" a project is in the gathering of specific clinical trial information from which genomic samples will be used along with the available clinical data. If the clinical team incorporated PGx objectives in the protocol prospectively and clinical samples and data were collected in preparation for PGx analysis, then this stage will largely be focused determining specific genomic assay platforms, vendor selection, and cost estimates. Also during this stage, the PGx lead should have specific engagements with the clinical team and funding sponsors (e.g. senior management) to reacquaint internal stakeholders to the purpose of the project and secure funding support. The final objective of this stage is to clearly delineate and document the objectives and boundaries of the project. This is critical to prevent "project creep" without deliberate and controlled scope revision,

Table 3 Pipeline Pharmacogenetics (PGx) Project Methodology

I. Scope	II. Plan	III. Execute	IV. Interpret	V. Close
Purpose:	*Purpose:*	*Purpose:*	*Purpose:*	*Purpose:*
Determine if a project will contribute to the PGx strategy, is feasible, and will be timely	Develop the project plan, identify deliverables, acquire resources, and create work breakdown and schedule	Execute the defined work to meet the project deliverables	Interpret the results of execution and recommend necessary next steps	Actively close the project, archive records, and perform post-project assessment
Activities:	*Activities:*	*Activities:*	*Activities:*	*Activities:*
Investigate: - Samples - Phenotype - Genotype - Statistical power - Technical feasibility - Strategy alignment	- Select deliverables - Create Project plan - Select vendor and technology - Work breakdown - Create schedule	- Genotyping - Genotyping data delivery - Statistical and power analyses	Statistical analyses results interpretation	Perform document quality checks, collate project archive, and represent findings in updated PGx strategy
Key deliverables:	*Key deliverables:*	*Key deliverables:*	*Key deliverables:*	*Key deliverables:*
Scope summary Project charter	Deliverables list Genetic variant list Genotyping contract PGx statistical analyses plan Project schedule	Genetic data and QA results Statistical requirements, output and report	PGx results interpretation and recommendation	Project archive binder Refined PGx strategy and value proposition

which is inevitable when clinical results emerge, organization priorities shift, and new genomic techniques/approaches are considered.

Plan Once the scope of the PGx project is finalized and approved, the planning of the project is initiated. The detailed project schedule is established and the overall operations and expectations of the project, including expected activities, deliverables and special considerations are documented in a Project Charter.

Execute Most of the expected activities defined during planning occur during the execute stage of the PGx project, usually starting with the planned genomic assays, including sample shipment and vendor management (if applicable). Other activities may include genomic data QC, development of statistical analysis plan and defining the expected table/lists/figures, genomic data transmission and merging with clinical data, and performing statistical analysis.

Interpret After the statistical analyses evaluating genetic associations with clinical responses/outcomes have been completed, the interpretation of the data and development of a recommended next step occurs. This important stage is led by the PGx

Scientist in consultation with the PGx statistician and clinician/clinical team, where necessary. The output of this stage is usually the results interpretation and recommendation document or report that can be summarized and incorporated into regulatory submission documents, manuscripts, etc.

Close In the final stage of the project, all PGx related documentation is stored and archived to retain the necessary information for regulatory review and future projects as part of the same program or for other programs with similar strategy and implementation.

5 Specific Examples in Early-Development

5.1 *OPRM1 PGx and Alcohol Dependence*

Pharmacotherapy of alcohol dependence shows widely divergent responses both within and between patients, and part of this variability can be attributed to the underlying genotype. Recently, treatment response to the opioid receptor antagonist naltrexone was shown to be predicted by a genetic variant of the OPRM1 gene (rs1799971) [60]. In a recent study, the effect of two genetic variants in OPRM1 and a variable-nucleotide tandem repeat (VNTR) in the dopamine receptor gene (DRD4) were evaluated for association with the clinical efficacy of a novel opioid receptor antagonist for the treatment of alcohol dependence [61]. Asp-carriers of the OPRM1/rs1799971 genetic variant did not demonstrate an enhanced response to LY2196044 treatment when evaluated by changes in % heavy drinking days (HDD), % days abstinent, or drinks per day. Surprisingly, however, placebo-treated Asp-carriers demonstrated a blunted response to standard medical management versus Asp-non-carriers by all efficacy measures. This Asp-carrier dependent "placebo-effect" reached statistical significance for change in % days abstinent and drinks per day ($p=0.0202$ and $p=0.0093$, respectively) but not change in % HDD ($p=0.1261$). Val-carriers of the OPRM1/rs1799972 variant treated with LY2196044 consistently had greater reduction in % HDD, % days abstinent, and drinks per day, but none of these reached statistical significance ($p=0.0653$, 0.8895 and 0.1073. LY2196044-treated patients who were DRD4-VNTR L-carriers had greater reductions in % HDD ($p=0.0565$), increased % days abstinent ($p=0.0496$), and reduced drinks per day ($p=0.0069$) than placebo-treated L-carriers.

In this study, Asp-carriers did not show a greater response to LY2196044 treatment, but instead had a blunted response to medical management in the placebo group. The difference between this result and earlier reports may be due to the differences in pharmacological profiles between LY2196044 and naltrexone, trial designs, definition of clinical endpoints and/or response, or unknown phenotypic differences within this trial population. The DRD4 L-carriers comprised >39% of the trial participants and showed statistically significantly superior treatment response. DRD4 L-carriers have demonstrated better response to other treatments for alcohol consumption including olanzapine [62] and naltrexone [63]. Thus, DRD4-L may represent a common, robust genetic marker of opioid receptor antagonist response and form the basis for a potential tailored drug development program and companion diagnostic.

5.2 Oncology and Rare Diseases/Early Phase

One of the most vibrant and successful areas for implementation of PGx has been oncology. Recent years have seen development of novel therapeutics that is almost exclusively a "targeted therapeutic" approach, requiring a co-developed test to identify the target responder population. The greatest successes in this realm over the last couple of years encompass the ALK-inhibitor, crizotinib, and the B-Raf inhibitor, vemurafenib. Already at early phase I studies was a beneficial effect demonstrated in marker-positive carriers, which formed the basis for development decisions and study design for each of these molecules. Competitors are now developing second-generation BRAF and ALK inhibitors, benchmarked by the first-to-market compounds, both in terms of efficacy, as well as in terms of diagnostics and combination therapy.

Another immediate application for PGx early on relates to the growing clinical development field of rare diseases. Increasing in-depth characterization of the molecular biology of inherited disorders, fueled by financial incentives in the form of the Orphan Drug Act and expedited regulatory review processes, such as Fast Track and the Breakthrough Therapy designation, have led many biopharmaceutical companies to focus efforts on these ailments. Some of the successes in this field have revolutionized the care and life-expectancy of subjects with diseases such as Fabry disease (Fabrazyme), Pompe disease (Myozyme) and Cystic Fibrosis (Kalydeco) [64]. In these cases, the development is targeted for carriers of specific mutations and may employ comprehensive genetic and molecular screening already at early phases, followed by limited to no requirement for late stage registration studies prior to marketing approval.

6 Late Stage Drug Development and Pharmacogenetically-Enabled Clinical Trials: Rx/Dx co-Development

A drug development plan accompanied by pre-emptive Pipeline PGx approach from the get-go should culminate in late, Phase III clinical trials with a focused, well designed PGx component. It is not to claim that all drugs should be guided by a PGx designation, rather that by the time a drug is tested for registration purposes, the PGx characteristics of its efficacy and safety profile should be embedded into the program. The translation of this statement could mean a range of possibilities, depending on the specific drug and indication, starting with screening subjects for eligibility based on carrier status of a particular genetic variant (i.e. the genetic predictor will become a required biomarker for prescription purposes), through to exploratory study of potential findings as no large PGx effects are anticipated based on pre-clinical and early development studies. In the latter case, exploratory analyses (and integral sample collection) are pursued to account for unexpected adverse drug reaction and other unexpected findings, such as high PK variability. The recently published draft guidance from FDA on enrichment strategies in clinical trials is the agency's response to recent development programs that employed genetic and other biomarkers in order to demonstrate favorable and safe benefit-risk balance [48]. One of the fields that have seen most innovation and creativity in this aspect has been Alzheimer's disease (AD) clinical research. The first such late-phase trial employed genotype of the apolipoprotein E (APOE) epsilon 4 (E4) gene as stratification biomarker toward development of rosiglitazone for the indication of mild-to-moderate AD treatment. The design was based on a prior Phase II trial that showed efficacy in an exploratory PGx analysis in APOE E4 non-carriers. The main Phase III study failed to reach its co-primary endpoints. Unfortunately, the result does not necessarily reflect lack of efficacy in this target indication as a high proportion of the study participants were of Asian ancestry, unknown at the time to possess a genetic signature that is different than that possessed by Caucasians and other ancestries. To this end, the study was likely underpowered to detect the clinical effect. It did indicate potential efficacy in the low dose arm in APOE E4 non-carriers.

Another set of studies employed APOE E4 carrier status as a patient selection criterion into clinical trials testing the efficacy of bapinuzumab, a humanized monoclonal antibody targeted against extra-cellular amyloid plaques, for the treatment of mild-to-moderate AD. The biomarker was considered to be predictive of drug response based on exploratory analyses of Phase II data, which did not reach statistical significance for its primary endpoint. As a result, treatment response in patients with the APOE E4 genotype versus patients without the APOE E4 genotype, was assessed in two phase 3, multicenter, randomized, double-blind, placebo-controlled studies, which were completed in April and June 2012, each with > 1,100 participants. The initial plan included two active doses in each trial, with the higher dose discontinued in the two APOE E4 carrier studies due to increased risk of amyloid-related imaging abnormalities (ARIA). Neither one of these studies reached statistical significance for clinical endpoints.

It is, however, by now generally accepted by field experts that treating AD at the mild-to-moderate clinical stages is simply too late, as the overt cell death and overall brain damage accumulated exceeds the potential for recovery. Given that neurons do not regenerate, it is unlikely that disease could be reversed once it has passed a critical severity threshold. Instead, efforts are now invested in preserving neuronal capacity at early disease stages (terms Mild Cognitive Impairment, MCI) or, better yet, to delay the onset of first symptoms and possibly prevent AD altogether. However, the feasibility of conducting disease prevention studies in this highly prevalent, yet highly heterogeneous disease in terms of age of onset, progression and clinical course, is very low. It is therefore necessary to employ an enrichment strategy that can pinpoint individuals at high-risk of developing the first symptoms within a short time frame of several years. Furthermore, it is critical to demonstrate that the potential benefit (i.e. delay of onset) outweighs the risks (adverse events) in a cognitively normal elderly population. Thus, the clinical study design should randomize high-risk individuals into active versus placebo treatment, while the low-risk individuals (expected to live several years before potentially converting their risk status to the high level) should be administered placebo only, in a blinded fashion. This scheme allows for full evaluation of the treatment effects, parallel to qualification of the biomarker in a prospective, unbiased manner.

Other therapeutic areas are employing PGx at the registration phase for various purposes. One important goal is to ensure characterization of already-known biomarkers in the context of novel investigational drugs since, for the first time in the development process, large populations of patients are being exposed to these compounds. To this end, the FDA publishes a list of Pharmacogenomic Biomarkers in Drug Labels mentioned also above. Some, but not all, of these labels include specific actions to be taken based on genetic information, and the scope of biomarker type ranges between genetic sequence variation to expression changes and others. For those genes with known functional relevance to protein activity or/and to clinical outcomes, regulators require and/or encourage developers to evaluate them in the course of clinical development of investigational drugs.

7 Pipeline Pharmacogenetics: Summary

7.1 Barriers

The use of PGx is now fairly common within the pharmaceutical industry. Therefore it is not unreasonable to expect the delivery of tailored therapeutics across many disease areas. However, while PGx has had a dramatic effect on new personalized medicines for oncology, most of the other therapeutic areas seem to be lagging behind. One reason is the lack of organized, therapy-wide PGx strategies for assets at all stages of drug development carried out by skilled PGx scientists and project managers using a comprehensive Pipeline PGx methodology. As described in the sections and examples above, a valuable R&D PGx strategy starts with DNA collections from every subject in every clinical trial and integrates well-designed PGx scientific hypotheses into clinical study protocols. Delivery of time-driven PGx results permits R&D leaders to make key decisions and develop safe and effective tailored medicines.

Unfortunately, many barriers exist to successful implementation of the Pipeline PGx approach within the pharmaceutical industry. First, many argue that statistical significance of PGx effects are impossible to attain in phase 1 or 2 studies. Additionally, some contend that the size of drug-response genetic effects are too small and current studies will be unable to detect them. Both of these opinions are based on a confused understanding of the difference between disease genetics and PGx. There are many examples of very large genetic effects on both efficacy and AE's. In addition, specific genetic variants effects on disease are often quite distinct from those on drug response. Secondly, anecdotal organizational "opinions" can hinder the implementation of PGx. Many clinical project teams erroneously believe that collecting DNA samples during the course of a clinical trial will impede recruitment, but this has consistently been shown to not be true. Thirdly, many are concerned on how various global drug regulatory agencies will interpret PGx data, and the belief that PGx results may lead to label restrictions and a restricted commercial potential. In fact, regulators have published guidelines on how PGx approaches (therapeutic coupled with a companion diagnostic) can lead to faster regulatory approval, focused labels and safer, more efficacious treatments, personalized for specific patient subgroups.

7.2 Outlook and Recommendations

The application of PGx tools, technologies and strategies to understanding the genetic contribution to pathophysiology and therapeutic response has been successful, and key stakeholders (patients, physicians, regulators, payers) have recognized these achievements. Recent progress in understanding the science of the genome, technological developments and bioinformatic/analytical approaches demonstrate that we can identify genetic markers that contribute to the safe and efficacious use of therapeutics. The evolving regulatory and business climate is placing greater value on increasing specificity and certainty around therapeutic choice.

However, the high attrition rates and reduced productivity of the pharma industry R&D is unsustainable and new strategies for tailoring medicines are needed. Currently, pharmaceutical companies are rarely, and/or inefficiently, leveraging the value inherent in the science of PGx to assist with critical decision making during drug development. One of the reasons is the lack of a systematic approach to incorporate PGx into the standard drug development process. This chapter has described a coherent Pipeline PGx methodology, described the tactical elements of this method, and provided successful examples of its application to drug development. In addition, some of the organizational and conceptual barriers that exist within and outside the pharmaceutical industry have been described. Therefore we recommend routine implementation of the PGx methodology throughout the drug development continuum that will deliver safer, efficacious and valuable tailored therapies for the benefit of patients, healthcare providers, payers, and the pharmaceutical industry.

References

1. Spear BB, Heath-Chiozzi M, Huff J (2001) Clinical application of pharmacogenetics. Trends Mol Med 7(5):201–204
2. Personalized Medicine Coalition (2014) http://www.personalizedmedicinecoalition.org/. Accessed 7 Feb 2014

3. European Commission (2013) Use of '-omics' technologies in the development of personalised medicine. http://ec.europa.eu/health/files/latest_news/2013-10_personalised_medicine_en.pdf. Accessed 20 Feb 2014
4. Food and Drug Administration (2013) Table of pharmacogenomic biomarkers in drug labels. http://www.fda.gov/drugs/scienceresearch/researchareas/pharmacogenetics/ucm083378.htm. Accessed 3 Oct 2013
5. Food and Drug Administration (2013) Guidance for industry clinical pharmacogenomics: premarket evaluation in early-phase clinical studies and recommendations for labeling. http://www.fda.gov/downloads/Drugs/GuidanceComplianceRegulatoryInformation/Guidances/UCM337169.pdf. Accessed 20 Feb 2014
6. Haga SB, O'Daniel JM, Tindall GM, Lipkus IR, Agans R (2012) Survey of US public attitudes toward pharmacogenetic testing. Pharmacogenomics J 12:197–204
7. Battelle Technology Partnership Practice (2012) The economic and functional impact of genetic and genomic clinical laboratory testing in the United States. http://www.labresultsforlife.org/news/battelle_impact_report.pdf. Accessed 20 Feb 2014
8. Saukko P (2013) State of play in direct-to-consumer genetic testing for lifestyle-related diseases: market, marketing content, user experiences and regulation. Proc Nutr Soc 72(1):53–60
9. Wetterstrand KA (2014) DNA sequencing costs: data from the NHGRI Genome Sequencing Program (GSP). www.genome.gov/sequencingcosts. Accessed 6 Feb 2014
10. Kulkarni S, Ma P, Furstenthal L, Evers M (2013) McKinsey & Company: personalized medicine—The path forward. http://www.mckinsey.com/Search.aspx?q=personalized%20medicine%20the%20path%20forward. Accessed 6 Nov 2013
11. Wester K, Jonsson AK, Spigset O, Druid H, Hagg S (2007) Incidence of fatal adverse drug reactions: a population based study. Br J Clin Pharmacol 65(4):573–579
12. Lazarou J, Pomeranz BH, Corey PN (1998) Incidence of adverse drug reactions in hospitalized patients: a meta-analysis of prospective studies. JAMA 279(15):1200–1205
13. Akkari PA, Swanson TW, Crenshaw DG, Grossman I, Sundseth S, Burns DK, Roses AD (2009) Pipeline pharmacogenetics: a novel approach to integrating pharmaco-genetics into drug development. Curr Pharm Des 15(32):3754–3763
14. Swanson TW, Akkari PA, Arbuckle JB, Grossman I, Sundseth S, Roses AD (2013) Methodology to enable integration of genomic knowledge into drug development. In: Vizirianakis IS (ed) Handbook of personalized medicine: advances in nanotechnology, drug delivery and therapy. CRC Press, Boca Raton. doi:10.4032/9789814411202
15. Ma Q, Lu AY (2011) Pharmacogenetics, pharmacogenomics, and individualized medicine. Pharmacol Rev 63(2):437–459
16. Hughes AR, Brothers CH, Mosteller M, Spreen WR, Burns DK (2009) Genetic association studies to detect adverse drug reactions: abacavir hypersensitivity as an example. Pharmacogenomics 10(2):225–233. doi:10.2217/14622416.10.2.225
17. Hughes S, Hughes A, Brothers C, Spreen W, Horborn D, CNA106030 Study Team (2008) PREDICT-1 (CNA106030): the first powered, prospective trial of pharmacogenetic screening to reduce drug adverse events. Pharm Stat 7(2):121–129
18. Mega JL, Close SL, Wiviott SD, Shen L, Walker JR, Simon T, Antman EM, Braunwald E, Sabatine MS (2010) Genetic variants in ABCB1 and CYP2C19 and cardiovascular outcomes after treatment with clopidogrel and prasugrel in the TRITON-TIMI 38 trial: a pharmacogenetic analysis. Lancet 376(9749):1312–1319
19. Roberts JD, Wells GA, Le May MR, Labinaz M, Glover C, Froeschl M, Dick A, Marquis JF, O'Brien E, Goncalves S, Druce I, Stewart A, Gollob MH, So DY (2012) Point-of-care genetic testing for personalisation of antiplatelet treatment (RAPID GENE): a prospective, randomised, proof-of-concept trial. Lancet 379(9827):1705–1711
20. Food and Drug Administration (2011) Clinical pharmacology and biopharmaceutics review(s). http://www.accessdata.fda.gov/drugsatfda_docs/nda/2011/202570Orig1s000ClinPharmR.pdf. Accessed 7 Feb 2014

21. Franc MA, Warner AW, Cohen N, Shaw PM, Groenen P, Snapir A (2011) Current practices for DNA sample collection and storage in the pharmaceutical industry, and potential areas for hamonization: perspective of the I-PWG. Clin Pharm Ther 89:546–553
22. Warner AW, Bhathena A, Gilardi S, Mohr D, Leong D, Bienfait KL, Sarang J, Duprey S, Franc MA, Nelsen A, Snapir A (2011) Challenges in obtaining adequate genetic sample sets in clinical trials: the perspective of the industry pharmacogenomics working group. Clin Pharmacol Ther 89(4):529–536
23. Franc MA, Cohen N, Warner AW, Shaw PM, Groenen P, Snapir A (2011) Industry pharmacogenomics working group. Coding of DNA samples and data in the pharmaceutical industry: current practices and future directions-perspective of the I-PWG. Clin Pharmacol Ther 89(4):537–545
24. Pharmacogenetics Working Group (2002) Elements of informed consent for pharmacogenetic research; perspective of the pharmacogenetics working group. Pharmacogenomics J 2:284–292. doi:10.1038/sj.tpj.6500131
25. Ricci DS, Broderick ED, Tchelet A, Hong F, Mayevsky S, Mohr DM, Schaffer ME, Warner AW, Hakkulinen P, Snapir A (2011) Global requirements for DNA sample collections: results of a survey of 204 ethics committees in 40 countries. Clin Pharm Ther 89:554–561
26. Hakonarson H (2013) Ask the experts: pharmacogenomics and genome-wide association studies. Pharmacogenomics 14(4):365–368
27. Ni X, Zhang W, Huang RS (2013) Pharmacogenomics discovery and implementation in genome-wide association studies era. Wiley Interdiscip Rev Syst Biol Med 5(1):1–9
28. Barrett PM, Topol EJ (2013) Pharm-econogenomics: a new appraisal. Clin Chem 59(4):592–594
29. Arnaout R, Buck TP, Roulette P, Sukhatme VP (2013) Predicting the cost and pace of pharmacogenomic advances: an evidence-based study. Clin Chem 59(4):649–657
30. Voight BF, Kang HM, Ding J, Palmer CD, Sidore C, Chines PS, Burtt NP, Fuchsberger C, Li Y, Erdmann J, Frayling TM, Heid IM, Jackson AU, Johnson T, Kilpeläinen TO, Lindgren CM, Morris AP, Prokopenko I, Randall JC, Saxena R, Soranzo N, Speliotes EK, Teslovich TM, Wheeler E, Maguire J, Parkin M, Potter S, Rayner NW, Robertson N, Stirrups K, Winckler W, Sanna S, Mulas A, Nagaraja R, Cucca F, Barroso I, Deloukas P, Loos RJ, Kathiresan S, Munroe PB, Newton-Cheh C, Pfeufer A, Samani NJ, Schunkert H, Hirschhorn JN, Altshuler D, McCarthy MI, Abecasis GR, Boehnke M (2012) The metabochip, a custom genotyping array for genetic studies of metabolic, cardiovascular, and anthropometric traits. PLoS Genet 8(8):e1002793
31. Johnson JA, Burkley BM, Langaee TY, Clare-Salzler MJ, Klein TE, Altman RB (2012) Implementing personalized medicine: development of a cost-effective customized pharmacogenetics genotyping array. Clin Pharmacol Ther 92(4):437–439
32. Illumina Inc. Omni Array Family http://www.illumina.com/applications/genotyping/omni_family.ilmn. Accessed 7 Feb 2014
33. Urban TJ (2013) Whole-genome sequencing in pharmacogenetics. Pharmacogenomics 14(4):345–348
34. Goldstein DB, Allen A, Keebler J, Margulies EH, Petrou S, Petrovski S, Sunyaev S (2013) Sequencing studies in human genetics: design and interpretation. Nat Rev Genet 14(7):460–470
35. Kiezun A, Garimella K, Do R, Stitziel NO, Neale BM, McLaren PJ, Gupta N, Sklar P, Sullivan PF, Moran JL, Hultman CM, Lichtenstein P, Magnusson P, Lehner T, Shugart YY, Price AL, de Bakker PI, Purcell SM, Sunyaev SR (2012) Exome sequencing and the genetic basis of complex traits. Nat Genet 44(6):623–630
36. Fromer M, Moran JL, Chambert K, Banks E, Bergen SE, Ruderfer DM, Handsaker RE, McCarroll SA, O'Donovan MC, Owen MJ, Kirov G, Sullivan PF, Hultman CM, Sklar P, Purcell SM (2012) Discovery and statistical genotyping of copy-number variation from whole-exome sequencing depth. Am J Hum Genet 91(4):597–607
37. Wang C, Krishnakumar S, Wilhelmy J, Babrzadeh F, Stepanyan L, Su LF, Levinson D, Fernandez-Viña MA, Davis RW, Davis MM, Mindrinos M (2012) High-throughput, high-fidelity HLA genotyping with deep sequencing. Proc Natl Acad Sci U S A 109(22):8676–8681

38. Mohamed S, Syed BA (2013) Commercial prospects for genomic sequencing technologies. Nat Rev Drug Discov 12(5):341–342
39. Roses AD, Saunders AM, Lutz MW, Zhang N, Hariri AR, Asin KE, Crenshaw DG, Budur K, Burns DK, Brannan SK (2014) New applications of disease genetics and pharmacogenetics to drug development. Curr Opin Pharmacol 14:81–89
40. Altman RB (2013) Personal genomic measurements: the opportunity for information integration. Clin Pharmacol Ther 93(1):21–23
41. Berger B, Peng J, Singh M (2013) Computational solutions for omics data. Nat Rev Genet 14(5):333–346
42. Flynn AA (2011) Pharmacogenetics: practices and opportunities for study design and data analysis. Drug Discov Today 16(19–20):862–866
43. Flannick J, Korn JM, Fontanillas P, Grant GB, Banks E, Depristo MA, Altshuler D (2012) Efficiency and power as a function of sequence coverage, SNP array density, and imputation. PLoS Comput Biol 8(7):e1002604
44. Lee SH, Wray NR (2013) Novel genetic analysis for case-control genome-wide association studies: quantification of power and genomic prediction accuracy. PLoS ONE 8(8):e71494
45. Wray NR, Yang J, Hayes BJ, Price AL, Goddard ME, Visscher PM (2013) Pitfalls of predicting complex traits from SNPs. Nat Rev Genet 14(7):507–515
46. Solovieff N, Cotsapas C, Lee PH, Purcell SM, Smoller JW (2013) Pleiotropy in complex traits: challenges and strategies. Nat Rev Genet 14(7):483–495
47. Howie B, Fuchsberger C, Stephens M, Marchini J, Abecasis GR (2012) Fast and accurate genotype imputation in genome-wide association studies through pre-phasing. Nat Genet 44(8):955–959
48. Food and Drug Administration (2012) Draft guidance for industry: enrichment strategies for clinical trials to support approval of human drugs and biological products. http://www.fda.gov/downloads/Drugs/GuidanceComplianceRegulatoryInformation/Guidances/UCM332181.pdf. Accessed 7 Feb 2014
49. Simon R, Roychowdhury S (2013) Implementing personalized cancer genomics in clinical trials. Nat Rev Drug Discov 12(5):358–369
50. Gurwitz D, McLeod HL (2013) Genome-wide studies in pharmacogenomics: harnessing the power of extreme phenotypes. Pharmacogenomics 14(4):337–339
51. Wittkowski KM, Sonakya V, Song T, Seybold MP, Keddache M, Durner M (2013) From single-SNP to wide-locus: genome-wide association studies identifying functionally related genes and intragenic regions in small sample studies. Pharmacogenomics 14(4):391–401
52. Motsinger-Reif A (2013) Interview: a discussion on genome-wide associations in pharmacogenomics. Pharmacogenomics 14(4):361–363
53. Roses AD1, Lutz MW, Amrine-Madsen H, Saunders AM, Crenshaw DG, Sundseth SS, Huentelman MJ, Welsh-Bohmer KA, Reiman EM (2010) A TOMM40 variable-length polymorphism predicts the age of late-onset alzheimer's disease. Pharmacogenomics J 10(5):375–384
54. Perez MV, Ashley EA (2010) Taming rare variation with known biology in long QT syndrome. Circ Cardiovasc Genet 6(3):227–229
55. Liao W, Tsai F (2013) Personalized medicine: a paradigm shift in healthcare. Biomedicine 3(2):66–72
56. Manolio TA (2013) Bringing genome-wide association findings into clinical use. Nat Rev Genet 14(8):549–558
57. Manolio TA, Chisholm RL, Ozenberger B, Roden DM, Williams MS, Wilson R, Bick D, Bottinger EP, Brilliant MH, Eng C, Frazer KA, Korf B, Ledbetter DH, Lupski JR, Marsh C, Mrazek D, Murray MF, O'Donnell PH, Rader DJ, Relling MV, Shuldiner AR, Valle D, Weinshilboum R, Green ED, Ginsburg GS (2013) Implementing genomic medicine in the clinic: the future is here. Genet Med 15(4):258–267
58. Food and Drug Administration (2013) Paving the way for personalized medicine—FDA's role in a new era of medical product development. http://www.fda.gov/downloads/scienceresearch/specialtopics/personalizedmedicine/ucm372421.pdf. Accessed 20 Feb 2014

59. Project Management Institute (2013) What is a Project. In: A guide to project management body of knowledge (PMBOK® Guide)—Fifth Edition. Project Management Institute Inc. (Newton Square, PA, USA). p3. ISDN: 978-1-935589-67-9
60. Anton RF (2008) Genetic basis for predicting response to naltrexone in the treatment of alcohol dependence. Pharmacogenomics 9(6):655–658
61. Wong CJ, Witcher J, Mallinckrodt C, Dean RA, Anton RF, Chen Y, Fijal BA, Ouyang H, Dharia S, Sundseth SS, Schuh KJ, Kinon BJ (2013) A phase 2, placebo-controlled study of the opioid receptor antagonist LY2196044 for the treatment of alcohol dependence. Alcohol Clin Exp Res 38(2):522–520
62. Hutchison KE, Wooden A, Swift RM, Smolen A, McGeary J, Adler L, Paris L (2003) Olanzapine reduces craving for alcohol: a DRD4 VNTR polymorphism by pharmacotherapy interaction. Neuropsychopharmacology 28(10):1882–1888
63. Tidey JW, Monti PM, Rohsenow DJ, Gwaltney CJ, Miranda R Jr, McGeary JE, MacKillop J, Swift RM, Abrams DB, Shiffman S, Paty JA (2007) Moderators of naltrexone's effects on drinking, urge, and alcohol effects in non-treatment-seeking heavy drinkers in the natural environment. Alcohol Clin Exp Res 32(1):58–66
64. Food and Drug Administration (2011) Improving the prevention, diagnosis, and treatment of rare and neglected diseases. http://www.fda.gov/downloads/ScienceResearch/SpecialTopics/CriticalPathInitiative/UCM265525.pdf. Accessed 20 Feb 2014

Pharmacogenetics—Statistical Considerations

Aiden Flynn, Craig Ledgerwood and Caroline O'Hare

Abstract The growth of Pharmacogenetics (PGx), using biomarkers to diagnose, prognose and identify patient subgroups most responsive to clinical intervention, heralds the possibility of more effectively targeted therapies and personalised medicine. Whilst demonstrating clinical significance in a number of studies, greater use of PGx has been limited by the need for further technological/methodological advancement together with a more integrated approach in study design and data analysis at the outset of clinical studies. Consideration of the statistical factors to be examined over the course of biomarker studies at the planning stage, instead of the current trend for retrospective analysis, will ensure that studies will be suitably powered to address specific questions and that subsequent data analysis will account appropriately for sources of variability. This will improve confidence levels in the conclusions drawn and the overall utility of PGx research. Greater use of PGx in the development of personalised medicine will require more guidance by statisticians and quantitative biologists in the handling and extraction of information derived from the data produced from large studies within the multidisciplinary network of researchers involved. This chapter highlights the key limiting statistical factors to be considered when embarking upon investigations using PGx, affecting the quality of information obtained from clinical data generated in personalised medicine research.

Keywords Pharmacogenetics (PGx) · Biomarkers · Data analysis · Statistics · Study design optimisation · Simulation · Modelling · Personalised medicine · Bioinformatics

A. Flynn (✉) · C. Ledgerwood · C. O'Hare
Exploristics Ltd, 55-59 Adelaide Street, BT2 8FE Belfast, UK
e-mail: aiden.flynn@exploristics.com

C. Ledgerwood
e-mail: craig.ledgerwood@exploristics.com

C. O'Hare
e-mail: caroline.ohare@exploristics.com

G. Grech, I. Grossman (eds.), *Preventive and Predictive Genetics: Towards Personalised Medicine,* Advances in Predictive, Preventive and Personalised Medicine 9,
DOI 10.1007/978-3-319-15344-5_4

1 Introduction

The development of Pharmacogenetics (PGx) using biological markers (biomarkers) to identify patient groups responsive to treatment during clinical trials promises a new era in personalised medicine. Its application within recent clinical development programmes has grown considerably as both healthcare providers and drug developers have recognised its importance in directing treatments to those most likely to benefit. Where PGx has been implemented it can be used to guide decision making in clinical studies. It offers additional options over the course of drug development by helping explain unexpected variability in safety and efficacy outcomes in clinical interventions. Previously, such variability would have resulted in the termination of costly research programmes. However, as PGx can identify patient subgroups which are most responsive to treatment it can be used to focus further studies within these subsets.

Despite the potential value of PGx in improving the benefits and reducing the risks of some drugs in development by targeting treatments more effectively, so far its successes within research have been limited. This may be due in part to its predominant use, at present, as a tool to re-evaluate development plans when study outcomes are negative or ambiguous rather than being integrated at the outset in a personalised medicine approach. However, although PGx is still an evolving strategy requiring further technological and methodological development to optimise its use in data analysis and study design, its uptake in research programmes at the prospective planning stage, aiding study design and data analysis, is likely to improve its utility. To this end, this chapter identifies and quantifies the key limiting statistical factors commonly encountered when using PGx within a research study, which affect the quality of the information derived from personalised medicine research.

2 Types of Biomarkers

The aim of PGx analysis is to identify and characterise clinical responses occurring in patients subject to a given clinical intervention. These effects can be traced through data sets acquired from a variety of biomarkers. The use of biomarkers to track disease and its treatment offers the future possibility of individualised therapies providing personalised medicine for each patient. The biomarkers observed are biological characteristics that may be detected and measured objectively and used as an indicator of normal biological, pathogenic or pharmacologic processes in response to therapeutic intervention. Identification of individual biomarkers in the form of chemical, physical or biological parameters can be used either to measure progress of a disease or the efficacy of its treatment. As a result, biomarkers may be used to diagnose or predict treatment or disease outcome.

There are different types of biomarkers, with each type requiring the application of distinct statistical methods depending on their relationship to the observed

Table 1 Examples of predictive and prognostic biomarkers in current use

Biomarker	Type	Associated biological process/ function	Indication
EGFR (ErbB-1)	Predictive	Signal transduction, cell proliferation, regulation of DNA replication/repair, stress response, cell adhesion, cell migration	Advanced non-small cell lung cancer, anal cancer glioblastoma multiforme
HER2/neu (ErbB-2)	Predictive	Transcriptional regulation, signal transduction, cell proliferation	Breast cancer
BluePrint ®	Predictive	80 gene panel for assessing molecular subtype of breast cancer	Breast cancer
MammaPrint®	Predictive	70 gene panel to categorise lymph node negative breast cancer	Breast cancer
OncoTypDX®	Predictive/ prognostic	21 gene panel for assessing response to chemotherapy of estrogen receptor (ER) positive tumours	Breast/colon cancer
HLA-B*5701	Predictive	Immune regulation	Hypersensitivity reaction to Abacavir
K-RAS	Predictive/ prognostic	Ras protein signal transduction, cell proliferation, gene expression regulation	Colorectal cancer
AB1-42	Prognostic	Protein component isoform of amyloid deposits associated with Alzheimer's Disease (AD)	Alzheimer's disease

treatment response. Therefore, the objective of a biomarker's use and its characteristics should be clear at the outset of analysis to ensure that the correct statistical approach is applied to the data. Some examples of biomarker types and their uses are given in Table 1. In this chapter, two types of biomarkers are considered, prognostic and predictive markers. For statistical purposes there is an important difference between these two marker types. Prognostic biomarkers, such as AB1-42, are linked to the prognosis or likely disease outcome in a defined patient group independent to the treatment given. As a consequence, they are usually identified with models where the biomarker is fixed as the main effect. In contrast, predictive biomarkers, including HER2, are able to help identify patients likely to respond to a given treatment but not to a comparator where response may be measured as efficacy or safety. Their identification requires the application of a statistical model which allows interaction between biomarker and treatment. In some instances, however, biomarkers may be both prognostic and predictive. An example of this is mutant K-ras which expressed in non-small cell lung tumours and can be used to predict responsiveness to EGFR Tyrosine Kinase Inhibitors.

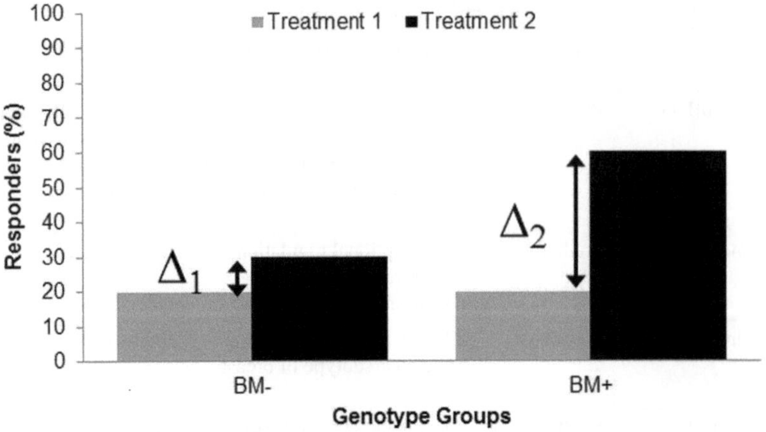

Fig. 1 An example of a predictive biomarker that is able to distinguish between groups of patients. In this case, the difference in the response rate in the BM+ group between treatment 2 and treatment 1 is greater than the equivalent difference in the BM− group. In other words, Δ_2 is greater than Δ_1

Identification of prognostic and predictive marker types is proving extremely useful in the development of personalised medicine. Prognostic markers can be used to segment populations by setting inclusion criteria at the start of a clinical study. This results in a reduction in the overall variability in the measure of response. In contrast, predictive markers are used to target treatments to patients more likely to derive benefit and are frequently further investigated as diagnostics for identifying responsive patient groups. An example of the use of a predictive biomarker to distinguish patient subgroups is shown in Fig. 1.

2.1 *Biomarker Platforms*

There are many different methods or platforms employed to measure biomarkers. These include a wide variety of technologies that can be used to produce biomarker data ranging from imaging modalities to the measurement of molecular biomarkers indicating gene expression, RNA expression, protein concentrations, single nucleotide polymorphisms (SNPs) or metabolites. Biomarker data can also take the form of continuous measurements, categories and ordinal scores. Different platforms may measure single markers or many thousands of markers simultaneously, generating data with specific attributes which will need to be accounted for in any statistical analysis. Data generated from some platforms may also require pre-processing steps such as scaling or normalisation [1] which must be considered prior to analysis. Consequently, it is important that the properties of the data obtained from each type of platform are factored into any statistical analysis.

2.2 *Variability and Data Quality*

Despite the accuracy of many of the biomarker platforms used, biomarker data can be prone to variability and bias. This can result in any subsequent analysis being subject to greater levels of statistical uncertainty leading to increased study failure rates. There are many factors which cause the observed variability and bias. These include the handling methods used for the tissue sample, when the sample was taken, patient factors such as drop-out rates, as well as inter-laboratory variation. If such factors are not addressed, inconsistent results are produced for the same biomarker across different biomarker studies [2, 3]. Therefore, it is critical that possible sources of variability should be evaluated during the development of a biomarker and suitable strategies for handling these sources and minimising their effect implemented.

2.3 *Sources of Missing Data*

Biomarker data often has a higher proportion of missing values than clinical data. These can arise as a result of numerous factors such as low consent rates for optional samples, patient drop-outs due to non-response or toxicity and measurements below the limits of detection of the biomarker assay. One key problem with these missing data is that the data are not missing at random. Indeed, the patients with missing values can be more likely to differ in their response to treatment compared to those with non-missing values. Therefore, it is important that missing biomarker data is not ignored as they are often informative. During any analysis involving missing biomarker data, it is important to compare key variables (e.g. those likely to impact response) between patients with and without biomarker data in order to understand differences between the missing group and the remainder of the study population. In addition, it is useful to understand reasons for missing data and take appropriate action. Where the pattern of missing data is understood, the implementation of models or imputation methods can help to recover the true underlying population statistics in the presence of missing data. On the latter point, information relating to the reason for missing data (e.g. below the limit of quantification) is often not recorded within the data set. This illustrates the need to improve on data standards and management practices relating to biomarker data.

2.4 *Dimensionality and False Positives*

Biomarker studies often involve the evaluation of numerous biomarkers in order to generate new hypotheses relating to the association between biomarker and response to treatment. This type of repetitive analysis results in a high number of false positive associations if the appropriate methods for controlling for the false positive rate are not used. Methods for adjusting for multiple testing have been reviewed elsewhere [4]. However, it should be noted the strategy for controlling false positives should be consistent with the aims of the experiment and the proposed use of the results. In hypothesis generating studies, it does not make sense to apply an overly conservative strategy that limits the likelihood of identifying plausible markers. Furthermore, exploratory studies do not end when a statistically significant p-value is generated. Indeed, there are often further steps in the evaluation process that will remove further spurious associations leaving those markers which are biologically plausible and have a clinically meaningful application.

3 Study Design Options

Good study design improves significantly the probability of meeting research objectives whilst minimising known sources of variability and bias. In PGx, the study design options depend on how and when PGx is being applied. At present, the early stages of PGx research is usually exploratory whereby many biomarkers are investigated, often using data collected as part of a study designed for another purpose. This is usually followed by confirmatory research where PGx becomes the primary objective in a prospectively designed study. To date, most methodological research into study designs for PGx has focused on the prospective, confirmatory applications.

3.1 Confirmatory Studies

Confirmatory studies are designed primarily to test a hypothesis based on observable pre-specified biological effects. Such studies are designed prospectively and measure markers of relatively known function which have previously been shown to explain variability in patient response. Several study designs that use prognostic and predictive markers to stratify the study population have been suggested and evaluated [5–8]. Three common designs used in confirmatory studies for predictive markers are an enriched design, a stratification design and an adaptive design, as shown in Fig. 2. The merits of each of these are discussed in the following sections.

In the targeted or enriched design (Fig. 2a) patients are selected for the study based on their biomarker status in a pre-screening step. This allows patients with the negative status to be excluded from the study. Positive status patients are then randomised to one of the treatment groups. The main advantage of this design is that a treatment effect can be observed within smaller studies. The disadvantage of this approach is that it does not provide information on the effect of treatment in the excluded population. As a result, it can only be used when there is already prior knowledge of the impact of a single biomarker.

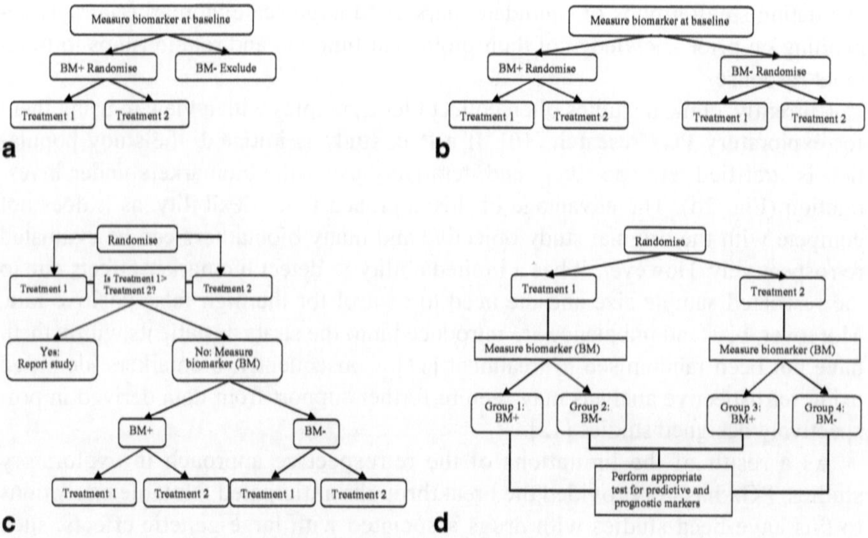

Fig. 2 Examples of study designs for Pharmacogenetics: Enriched design (**a**), Stratification design (**b**), Adaptive design (**c**), Retrospective design (**d**)

In contrast, the stratification design is less restrictive in its remit (Fig. 2b). It also has a pre-screening step whereby all the study subjects are stratified according to biomarker status and then randomised to treatment. The advantage of this design is that information can be collected on the treatment effect in the negative biomarker status group. Moreover, the performance characteristics of a diagnostic test, for example its sensitivity and specificity, can be estimated. However, as with enriched design, considerable prior knowledge about the biomarker is also required.

Both the enriched and the stratification designs are useful when studies are designed to test a single hypothesis relating to a given biomarker. However, more often studies have multiple objectives and involve evaluating a treatment effect in the entire study population as well as within sub-populations. In this instance an adaptive design is useful (Fig. 2c) [9]. With this design, patients are randomised to treatment groups and the treatments are compared. If there is no difference in the treatments, patients are stratified by biomarker status and a comparison of treatments is performed within these strata. This approach leads to a higher false-positive error rate, as multiple statistical tests are performed. Controlling for false-positives will result in larger studies. However, this design is more flexible than the targeted or stratified design as it allows the testing of multiple objectives and can be modified to include the evaluation of multiple biomarkers.

3.2 Exploratory Studies

PGx is used in exploratory studies for identifying useful biomarkers and to generate hypotheses for testing in further studies. Exploratory studies can range from evaluating small groups of candidate markers to large scale biomarker arrays, depending on prior knowledge of their biological function and relationships to treatment response.

Currently, clinical studies often collect blood samples with a view to using them for exploratory PGx research [10]. If a PGx study is initiated, the study population is stratified retrospectively and iteratively using the biomarkers under investigation (Fig. 2d). The advantage of this approach is its flexibility, as it does not compete with the original study objective and many biomarkers can be evaluated retrospectively. However, it has a limited ability to detect biomarker effects due to the restricted sample size and the need to control for the high false-positive rate. Moreover, bias and imbalance are introduced into the strata as patients within them have not been randomised to treatment [11]. Consequently, biomarkers identified using retrospective analysis may require further support from data derived in prospectively designed studies [12].

As a result of the limitations of the retrospective approach in exploratory studies, PGx has not provided the breakthroughs anticipated. Notable exceptions to this have been studies with drugs associated with large genetic effects, such as Abacavir [13] and Panitumumab [14]. This has highlighted the key challenge in PGx research. Where insufficient patient data has been available, it has been hard to detect more moderately sized effects and thus identify biomarkers with clinical utility. Nevertheless, retrospective PGx approaches will continue to play an important exploratory role. However, to improve the likelihood of successful exploratory PGx studies, a more integrated approach is required in research programmes at the outset of clinical study design and data analysis. Indeed, recent research using computer simulation to design studies addressing multiple objectives, including PGx investigations [15], showed that prospective planning is vital. This is particularly important when studies are designed for another purpose, so that useful PGx data can be generated without impacting the primary objectives of the study.

3.3 Data Analysis Methods

The main objective for PGx analysis is to identify and/or characterise genetic effects. Whilst there are too many methods to review adequately in this article, there are some general principles that are broadly followed in basic analyses, as shown in Fig. 3. Current approaches to biomarker or feature discovery involve a multi-step process whereby biomarkers are selected for further investigation based on the strength of association with an outcome; typically by setting an arbitrary limit on the likelihood of detecting false positives (e.g. p value <0.05). Evaluation of biomarkers involves the application of a statistical model comprising the factors that are thought to contribute to the observed variability in response. These models can include two types of effects: main effects where factors make a sole contribution to

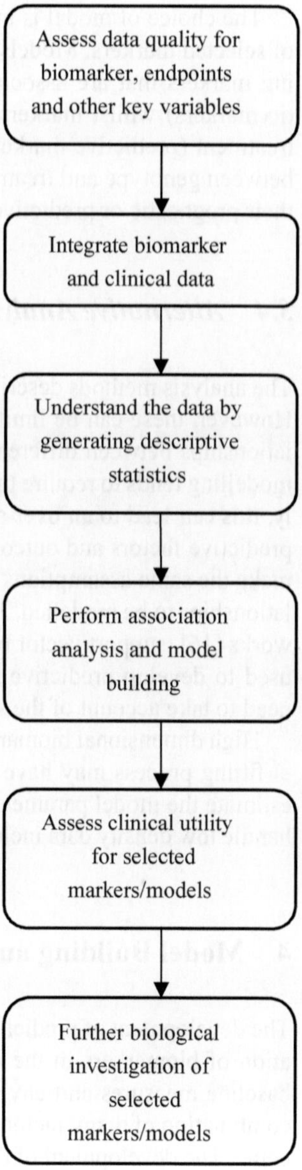

Fig. 3 An example of a statistical analysis strategy for a personalised medicine study. The analysis typically involves multiple steps that integrate different sources of information

the observed variability; and interaction effects where two or more factors make a combined contribution. This model can be written in the form of

$$R = B + T + B \times T$$

where R is the response to treatment, B is the biomarker measurement and T is the treatment group.

The choice of model is important as it will determine the utility and application of selected markers. Models with genotype as a main effect are useful for identifying markers that are associated with response, regardless of treatment (prognostic markers) whilst markers that are associated with response in the presence of treatment (predictive markers) can be identified using models with an interaction between genotype and treatment. Genetic markers are then selected on the basis of their prognostic or predictive utility.

3.4 Alternative Analysis Methods

The analysis methods described above relate in general to linear regression models. However, these can be limited in terms of their ability to incorporate complex relationships between different predictive and prognostic factors. In addition, linear modelling tends to require the pre-specification of the structure of the model. Clearly, this can lead to an over-simplification of the form of the relationships amongst predictive factors and outcomes. There are numerous other approaches that do not make the same assumptions and are more flexible in terms of enabling complex relationships to be modelled. There are too many to cover in any detail but neural networks [16], support vector machines [17] and random forests [18, 19] are regularly used to develop predictive models with some success. All modelling approaches need to take account of the study design and the biomarker utility and type.

High dimensional biomarker data sets are often sparse, in the sense that the model fitting process may have a limited number of observations that can be used to estimate the model parameters. There are a few useful methods that can be used to handle low density data including exact methods, lasso, elastic nets and others [20].

4 Model Building and Validation

The development of predictive and prognostic models generally involves the evaluation of biomarkers in the context of many other factors, such as demographics, baseline measures and environmental factors. As a result, these models include a combination of many factors that are additive in terms of their association with outcome. The development of these models is a multi-step process comprising variable selection and model evaluation followed by model validation [20].

Approaches to variable selection and model evaluation are generally well established. Typically, variable selection and model evaluation is an iterative process whereby variables are added or removed from a model following an evaluation of the contributions of those variables to the performance of the model. Following the model building process, the final model is the one that is considered to be the best performer.

One major problem with using high-dimensional data to build a predictive model is over-fitting of the data. In this instance, many variables are shown to have strong relationships with the outcome as a result of random selection. Consequently, any model that is based on these random relationships will not generalise to unseen data or an independent data set, highlighting the importance of model validation.

There are numerous ways to perform model validation [20]. A common approach is to train the model on data from one study and then use an independent dataset to validate the model by assessing its performance in the second dataset. One problem with this method is the lack of availability of a relevant independent dataset. An alternative approach is to split the data from one study into a training and validation set. The robustness of the model may be evaluated using an iterative procedure for selecting the test and validation set. This, however, relies on the availability of enough observations (patients) as splitting the data will reduce the power to identify useful markers. Where sample size is limited, another useful strategy is to use leave-one out cross-validation. In this case, the model is trained on all but one observation and the ability of the model to predict the outstanding observation is evaluated. This evaluation is performed repeatedly by randomly selecting the observation that is left out of the model building step.

Recent work has shown that the best approach to model building is to integrate the variable selection and the validation steps into one large iterative process [21]. The benefit of this method is that the performance of many models can be assessed at once whilst controlling for false positives. For all the cross-validation approaches described above, the model building and performance characterisation is performed in the same dataset using data that were collated under study-specific conditions. The most robust form of validation involves the use of completely independent data (external validation) to assess the performance of a model.

5 Diagnostic Development

The use of statistics and modelling is vital in demonstrating the utility of companion diagnostics, prior to regulatory approval. The sensitivity and specificity of a diagnostic in its target population, as well as its positive and negative predictive value need to be identified under the original conditions in which it has been evaluated and developed. There are also a range of criteria that need to be set [22], such as defining the optimal threshold for biomarkers on a continuous scale and evaluating the repeatability and reproducibility of the biomarker assay. In addition, the diagnostic development process can be validated by understanding and quantifying the factors that may impact its performance. When a diagnostic is being co-developed with a drug for regulatory approval, good coordination between these processes is critical. Diagnostics development may often fall behind that of its associated drug, due to identification of biomarkers over the course of a research programme. This can cause delays in drug approval unless both development programmes are well synchronised.

6 Visualisation and Presentation

A key component of any analysis in a personalised medicine study is a clear and simple visualisation of the results. The use of good graphical outputs able to display relevant information simply, help to place the results in a suitable context facilitating the interpretation of large data sets. Well-designed graphical displays can integrate information on the clinical utility of biomarkers along with biological information, such as the functional annotation of the gene region, by overlaying both sources of information on the same plot. Another important aspect of visualisation is the presentation of high dimensional data. In this case, the use of multi-panel plots, heat maps, contour and surface plots are extremely useful. In addition, it is common to reduce the dimensions of data using methods such as multi-dimensional scaling, principal components analysis and clustering. This enables the data to be displayed on standard plots in two or three dimensions and can also uncover hidden structures in the data.

The presentation of simple summary statistics can often mask effects and responses that are notably different to those of the broader population. Therefore, it is important to be able to distinguish those observations that differ in order to understand variability in the data and identify patients that derive benefit. Consequently, any analysis of personalised medicine research should include graphical displays that enable the visualisation of individual data points. It also presents an opportunity for the observations obtained from a biomarker study to be visualised alongside information derived from other sources, placing it within a wider biological context.

7 Bioinformatics and Biological Interpretation

Since the completion of the Human Genome Project (HGP) and the arrival of next generation sequencing (NGS), technological advances in genomic sequencing have increased the speed at which entire genomes can now be sequenced. In addition, the use of microarray gene chip technology to screen patient tissue samples for the presence of genetic biomarkers associated with some disease processes has become increasingly commonplace. These advances in the area of medical genetics have resulted in the generation of unprecedented volumes of raw biological data. The need to analyse this data in order to understand it and how it might be used for clinical applications has required the capabilities provided by the expanding field of bioinformatics. Bioinformatics combines the mathematics, computer sciences and statistics required for the collection, banking, deciphering, analysing and modelling that is necessary to analyse large amounts of biological information. Indeed, bioinformaticians continue to seek to address the pressing need for data analysis through the development of analytical tools that can be utilised on desktop systems to analyse and interpret the data collected.

The recent era of next generation sequencing and multiplex microarray platforms has allowed a vast expansion in the number of sequences able to be analysed in each experiment. Prior to the emergence of these technologies, the focus of molecular biology was on known sequences previously identified and attributed to a given protein and/or function. Complementary probes were used to identify the presence and abundance of those target sequences and determine differences between treated/non-treated or resistant/responsive groups. Performed initially in singleplex assays (PCR), this quickly progressed to multiplex microarrays which could simultaneously measure thousands of targets (genes, single nucleotide polymorphisms (SNP) or messenger RNA transcripts (mRNA)) thanks to the technologies developed by Affymetrix, Agilent and Illumina. However, whilst this has increased the number of sequences that can be analysed it has also raised problems in their analysis due to the high number of dimensions in the data produced and the relatively low number of observations in studies.

Bioinformaticians have played a key role in implementing these technologies and addressing the difficulty in dealing with high dimensional data. The pre-processing of data has become critical to the utility of high-throughput systems, with several normalisation techniques, such as Robust Multi-array Average (RMA) and the current Affymetrix algorithm MAS5, being developed and used routinely in both the proprietary software provided by the instrument manufacturers and in open source packages, such as that available on the Bioconductor software repository (http://www.bioconductor.org/). Following normalisation, the next problem is dealing with the high dimensional data and correcting for the false discovery rate in hypothesis testing. Both commercial and open source packages use standard statistical methods to make comparisons between groups. P-values are typically adjusted to correct for the number of tests being performed through methods such as Benjamini–Hochberg [23].

Whilst many of the current tools for analysing and interpreting microarray and next generation sequencing data are useful, the huge quantities of data they produce nevertheless continue to create new challenges in data analysis. The technical process of sequencing an entire genome may have become routine, however analysis of the data it generates remains problematic as it is very computationally intensive with over 3 billion base pairs and 50 million variations to consider. Although sequencing and microarray platforms have been around for some time, the ability to process the volume of data produced in a routine setting at an affordable cost has only become a relatively recent possibility due to the ability to store and process the terabytes of data produced.

Currently, a number of software tools exist which facilitate the process of interpreting sequencing and microarray data. Alignment tools, such as BLAST, have been used for years to identify proteins or genes from short amino acid or nucleotide sequences or to compare the similarity between two or more sequences. The concept of the algorithms used by FASTA/BLAST and software for NGS sequence alignment are similar, however, the alignment of hundreds of millions of short sequences (FASTQ) from the entire genome takes a lot longer to perform even with accelerated algorithms. The Bowtie sequence alignment tool is one of the most widely used aligners, due to its speed and the fact it is freely available to use. Even this can take several hours per sample to process and many researchers choose to run these on cloud-based platforms such as Amazon's Web Services. Other tools for analysing these alignments to identify variations, splice variants and differential expression of genes and isoforms are also freely available, such as the web-based application, Galaxy. However, these are still being actively developed and there are no clearly defined procedures as yet for the analysis and interpretation of alignments. Indeed, there are a plethora of tools both commercial and open-source for visualising, analysing and interpreting the large volume of information produced by each experiment.

Although the development of sequencing, storage and processing techniques has evolved concurrently, the ability to interpret the biological relevance of the information generated is still lagging. Some understanding of the underlying biological processes may be obtained through the mining of large gene sequence databases. These repositories of information encompass the knowledge gained to date regarding the biological relevance of genes and variations in sequences. Further biological context for biomarker studies might also be obtained through the use of data banked in public databases, such as ArrayExpress a functional genomics database containing data from both microarray and high-throughput sequencing studies. These databases now play a fundamental role in biological research and development, acting as a warehouse for storing, organising and providing large data sets relating to the occurrence and consequences of many biological processes, including gene variation, drug transport, drug targets, and other proteins of importance for drug response or toxicity [24–27]. Amongst the large number of databases generally available, some have become important bioinformatics tools within pharmacogenetics, such as the Human Genome Project (HGP) [28], Ensembl [29], the SNP databases dbSNP and JSNP [30], and HapMap [31]. These databases are rapidly expanding as they are continually updated with new submissions of genetic information, especially regarding the variation across the population.

The HGP demonstrated that the 20,000-plus genes expressed in humans only accounts for 1.5 % of the genome, with very little known about the function of the remaining 98.5 %. This is now being addressed through global collaborative projects such as ENCODE, which aims to completely annotate the non-protein encoding regions of the genome. Moreover, large sequencing projects like HapMap phase 3 and the 1000 Genomes Project, aim to give a clearer picture regarding the intrinsic genetic variation present within the human population. These projects in particular will provide a valuable resource for bioinformaticians, giving an important insight into the range of variation inherent in the human genome in general across multiple ethnicities. This will no doubt raise more considerations for analysis and interpretation of genomic data.

Possibly the most influential sequence database to date has been GenBank, an open access database storing known gene sequences from over 100,000 distinct species, along with their protein translations. It is run and maintained by the National Center for Biotechnology Information (NCBI) which plays an active and collaborative role in the development of computational biology. Their bioinformatics resources can be used to annotate and analyse an abundance of disparate data. Access to this important and expanding resource allows researchers to derive possible connections between the different aspects of biomarker data and thereby shape a more biologically meaningful view of it [32, 33].

The Ensembl database also provides genomic information with a rich source of gene variant data from humans and other species, including single nucleotide polymorphisms (SNPs). Alongside the HGP and Ensembl, dbSNP, Japan's JSNP and HapMap are another three of the more widely accessible and utilised bioinformatics resources. The Single Nucleotide Polymorphism Database, (dbSNP), created to supplement GenBank, is a public access archive for genetic variation within and between organisms developed by the NCBI. It comprises information on over 64 million distinct SNP variants in 55 species, including *Homo sapiens* [34]. Meanwhile, the HapMap project provides an alternative platform of information designed to enable researchers to carry out large scale studies to link genetic variants to the risk of specific diseases [31].

Other useful databases storing genetic data are the Gene Expression Omnibus (GEO), the Kyoto Encyclopaedia of Genes and Genomes (KEGG) [35] and PharmGKB [36]. The analysis of the data contained in these databases is now an integral element of the bioinformatics process. Many of the data storage facilities have analytical applications bolted on as add-ons, whilst others are standalone warehouses that store the information that is used in other analytical applications. As with genetic databases, there are many bioinformatics tools available for dissecting, analysing and visualising the data, a selection of these have been listed in Table 2. This is also an evolving field within bioinformatics with new software tools under development.

Another recent tool useful in understanding the biological complexity of differentially expressed genes and proteins and identifying statistically significant sets of genes, is Gene Set Enrichment Analysis (GSEA) [37]. This is a powerful analytical method for interpreting gene expression data, deriving its power by focusing on groups of genes or gene sets that share common biological function, chromosomal location or regulation. Whilst single-gene analysis is useful it can miss important effects on biological pathways which are distributed across large networks of genes and are hard to detect at individual gene level. In contrast, GSEA which examines sets of related genes can identify many common biological pathways as cellular processes often affect networks of interacting genes. The advantage of GSEA is that it facilitates interpretation of genome-wide expression data as it focuses on gene sets which give more reproducible and therefore interpretable data.

Undoubtedly, constructing biological meaning from lists of statistically significant markers remains a challenge to bioinformaticians. However the development of complex data warehouses and other resources, that detail the structures and processes and interactions where individual genes/proteins exist, have helped to overcome such difficulties through the grouping of long lists into smaller sets of related genes or proteins that share a similar physiological/pathological function, cellular localisation, position in the genome or can be defined by similar gene ontology terms. There is a plethora of gene set databases that use examples found in the

Table 2 A non-exhaustive list of bioinformatics tools available with a brief description of the tool and an external link

Tool	Description	Link
ArrayExpress	Functional genomics database containing data from microarray and high-throughput sequencing studies	http://www.ebi.ac.uk/arrayexpress/
BioMart	Search engine allowing generation of tables of terms linked to genes and SNPs	http://www.ensembl.org/biomart
BLAST	Database allowing searching and alignment of sequences to the RefSeq genome	http://www.blast.ncbi.nlm.nih.gov/Blast.cgi
GeneMania	Search engine to find related genes through linkage of association information	http://www.genemania.org/
GSEA	Tool to perform enrichment analysis on gene sets provided in MSigDB database	http://www.broadinstitute.org.gsea/index.jsp
Haploview	Range of tools for analysing linkage disequilibrium and haplotype patterns	http://www.broad.mit.edu/mpg/haploview/
KEGG mapper	Collection for mapping gene sets to the KEGG pathways	http://www.kegg.jp/kegg/kegg1b.html
KEGG pathway	R package for analysis and visualisation of expression data within KEGG pathway	http://www.bioconductor.org/packages/release/bioc/html/KEGGprofile.html
PathNet	Tool that performs pathway analysis using topological information from pathways	http://www.bioconductor.org/packages/release/bioc/html/PathNet.html
Pathway browser	Tool for visualising pathways	http://www.reactome.org/PathwayBrowser/
SNAP	SNP annotation/proxy search tool using linkage disequilibrium and physical distance	http://www.broadinstitute.org/mpg/snap/
Stitch	Tool for exploration of known/predicted molecular interactions	http://www.stitch.embl.de/
Sweep	Tool for large scale haplotype analysis	http://www.broadinstitute.org/mpg/sweep/index.html
topGO	Compares GO term representation in an expression set accounting for GO topology	http://www.bioconductor.org/packages/2.12/bioc/html/topGO.html
UCSC genome browser	Tool providing interactive graphical interface to visualise genome annotation and chromosomal position	http://www.genome.ucsc.edu/

literature and gene sets that have been computationally derived. Databases such as, MSigDB and ConsensusPathDB, have brought together a large collection of gene sets comprising gene regulation, protein interactions, genetic interactions, biochemical reactions, drug-target interactions, pathways, gene ontology, disease regulation and many more. As differences in database structures and terminology exist, moving towards large warehouses of this information and development of advanced tools for mining the information are crucial to their implementation. Tools have been developed as stand-alone applications or as add-ons in other packages such as R (PGSEA). The principles of these tools are largely similar, in that they comprise an annotation database, a process that can assign those annotations to a given gene list, a further process that performs a statistical test to identify annotations that are significantly represented in the gene list and a method to interpret this graphically.

Examples of statistical approaches used in gene enrichment analysis are over-representation analysis (ORA), functional class scoring and pathway topology. Using a statistical test, for example the hypergeometric or binomial, ORA evaluates whether a specified functionally defined group of genes/proteins is represented within a gene list, or if it occurs merely by chance [38]. One drawback, however, is that ORA treats each gene equally, losing any possible correlation and interaction between genes. Functional class scoring overcomes some of the limitations of ORA as it treats the genes differently depending on the strength of the individual raw microarray values [38]. The pathway topology approach has advantages over both of these methods, in that it does not only consider the number of genes in a pathway to identify significant pathways, but also utilises information about inter-pathway connectivity. GSEA utilises its own novel method for performing the analysis, this calculates an enrichment score using a weight Kolmogorov-Smirnov-like test. The enrichment score generally reflects the degree of over representation at the top or bottom of a ranked gene list [37].

Whilst it remains challenging, bioinformatics has made some progress in understanding and interpreting large data sets by exploiting alternative data resources. The effort to understand in more depth the implications of the data collected and analysed within an experiment has required input from other data sources enabling a fuller understanding of its meaning. It is becoming clear that results from primary biomarker analyses might need augmentation with additional information from other studies to provide a greater biological context for the role of the biomarker in question. Further biomarker context could be given by its characterisation as well as by pathway analysis. Additional data in this form would help authenticate the patient subgroups identified from biomarker analysis by providing further supporting evidence. Alternatively, longitudinal data would allow further characterisation of subgroups using variables that change over time. This would require the pre-selection of those variables, with the selection process used described.

8 Future Opportunities

There are a number of opportunities within statistics and modelling to improve the application and implementation of pharmacogenetics in future studies. Four key areas to be considered are improving study design, integration of analysis methods, use of disparate data sources to provide biological context and better multidisciplinary collaboration involving quantitative scientists. Addressing these factors will develop PGx by increasing the success rate of exploratory studies to identify new biomarkers. Additional use of computer simulation will enable the application of smarter clinical trials that optimise the likelihood of success of a study without prohibitively increasing its size.

At present integrative analysis methods, such as Bayesian methods, are based on the idea of obtaining a consensus by combining prior information and current opinion, thereby providing a statistical framework that enables the quantitative (probabilistic) integration of information across multiple analysis steps [39]. This approach can limit biomarker discovery as such studies may be underpowered to detect small to moderate (but biologically important) effects; it filters potentially useful information in variables that fail to reach significance and it ignores the additional control of false positives that naturally occurs in the subsequent analysis steps. To progress, there is a need to develop methods that do not filter out useful information and that enable the quantitative integration of information from additional analysis steps, such as clinical and biological pathway analysis and comparisons with literature. Furthermore, there is enormous scope for developing and applying statistical models that more closely reflect the underlying biology and patterns of response; using models that better describe the data will increase the power to detect genetic markers. The development of these capabilities will require extensive methodological research and development for the integration and application of disparate data sources.

It is clear that statisticians and quantitative biologists are of increasing importance in the multidisciplinary network of researchers involved in the development of personalised medicine. Bioinformatics has helped develop new algorithms and software to facilitate the analysis of complex data sets. The development of new computational data and analytical solutions are crucial to handling and extracting the information derived from large clinical studies, improving the understanding of disease progression and treatment. Nevertheless, a knowledge gap still exists between the exploratory world of bioinformatics and the rigour and regulation of clinical statistics. Closer collaboration between quantitative scientists will break down the barriers in communication that exist between the disciplines and will enable scientists to gain experience, knowledge and an appreciation of the skills and capabilities that exist in other fields. Deeper understanding of other capabilities and technologies will lead to new innovations that make use of the extensive information available and improve the application of Pharmacogenetics in the quest for personalised medicine.

References

1. Shi L, Campbell G, Jones WD, Campagne F, Wen Z, Walker SJ, Su Z, Chu TM, Goodsaid FM, Pusztai L, Shaughnessy JD Jr, Oberthuer A, Thomas RS, Paules RS, Fielden M, Barlogie B, Chen W, Du P, Fischer M, Furlanello C, Gallas BD, Ge X, Megherbi DB, Symmans WF, Wang MD, Zhang J, Bitter H, Brors B, Bushel PR, Bylesjo M, Chen M, Cheng J, Chou J, Da-

vison TS, Delorenzi M, Deng Y, Devanarayan V, Dix DJ, Dopazo J, Dorff KC, Elloumi F, Fan J, Fan S, Fan X, Fang H, Gonzaludo N, Hess KR, Hong H, Huan J, Irizarry RA, Judson R, Juraeva D, Lababidi S, Lambert CG, Li L, Li Y, Li Z, Lin SM, Liu G, Lobenhofer EK, Luo J, Luo W, McCall MN, Nikolsky Y, Pennello GA, Perkins RG, Philip R, Popovici V, Price ND, Qian F, Scherer A, Shi T, Shi W, Sung J, Thierry-Mieg D, Thierry-Mieg J, Thodima V, Trygg J, Vishnuvajjala L, Wang SJ, Wu J, Wu Y, Xie Q, Yousef WA, Zhang L, Zhang X, Zhong S, Zhou Y, Zhu S, Arasappan D, Bao W, Lucas AB, Berthold F, Brennan RJ, Buness A, Catalano JG, Chang C, Chen R, Cheng Y, Cui J, Czika W, Demichelis F, Deng X, Dosymbekov D, Eils R, Feng Y, Fostel J, Fulmer-Smentek S, Fuscoe JC, Gatto L, Ge W, Goldstein DR, Guo L, Halbert DN, Han J, Harris SC, Hatzis C, Herman D, Huang J, Jensen RV, Jiang R, Johnson CD, Jurman G, Kahlert Y, Khuder SA, Kohl M, Li J, Li M, Li QZ, Li S, Liu J, Liu Y, Liu Z, Meng L, Madera M, Martinez-Murillo F, Medina I, Meehan J, Miclaus K, Moffitt RA, Montaner D, Mukherjee P, Mulligan GJ, Neville P, Nikolskaya T, Ning B, Page GP, Parker J, Parry RM, Peng X, Peterson RL, Phan JH, Quanz B, Ren Y, Riccadonna S, Roter AH, Samuelson FW, Schumacher MM, Shambaugh JD, Shi Q, Shippy R, Si S, Smalter A, Sotiriou C, Soukup M, Staedtler F, Steiner G, Stokes TH, Sun Q, Tan PY, Tang R, Tezak Z, Thorn B, Tsyganova M, Turpaz Y, Vega SC, Visintainer R, von Frese J, Wang C, Wang E, Wang J, Wang W, Westermann F, Willey JC, Woods M, Wu S, Xiao N, Xu J, Xu L, Yang L, Zeng X, Zhang M, Zhao C, Puri RK, Scherf U, Tong W, Wolfinger RD, Consortium M (2010) The MicroArray quality control (MAQC)-II study of common practices for the development and validation of microarray-based predictive models. Nat Biotechnol 28(8):827–838. doi:10.1038/nbt.1665

2. Mattsson N, Blennow K, Zetterberg H (2010) Inter-laboratory variation in cerebrospinal fluid biomarkers for Alzheimer's disease: united we stand, divided we fall. Clin Chem Lab Med 48(5):603–607. doi:10.1515/CCLM.2010.131
3. Fenech M, Bonassi S, Turner J, Lando C, Ceppi M, Chang WP, Holland N, Kirsch-Volders M, Zeiger E, Bigatti MP, Bolognesi C, Cao J, De Luca G, Di Giorgio M, Ferguson LR, Fucic A, Lima OG, Hadjidekova VV, Hrelia P, Jaworska A, Joksic G, Krishnaja AP, Lee TK, Martelli A, McKay MJ, Migliore L, Mirkova E, Muller WU, Odagiri Y, Orsiere T, Scarfi MR, Silva MJ, Sofuni T, Surralles J, Trenta G, Vorobtsova I, Vral A, Zijno A, project HUM (2003) Intra- and inter-laboratory variation in the scoring of micronuclei and nucleoplasmic bridges in binucleated human lymphocytes. Results of an international slide-scoring exercise by the HUMN project. Mutat Res 534(1–2):45–64. doi:10.1016/S1383-5718(02)00248-6
4. Hsu JC (2010) Multiplicity adjustment big and small in clinical studies. Clin Pharmacol Ther 88(2):251–254. doi:10.1038/clpt.2010.122
5. Bromley CM, Close S, Cohen N, Favis R, Fijal B, Gheyas F, Liu W, Lopez-Correa C, Prokop A, Singer JB, Snapir A, Tchelet A, Wang D, Goldstaub D, Industry Pharmacogenomics Working G (2009) Designing pharmacogenetic projects in industry: practical design perspectives from the Industry Pharmacogenomics Working Group. Pharmacogenomics J 9(1):14–22. doi:10.1038/tpj.2008.11
6. Mandrekar SJ, Sargent DJ (2009) Clinical trial designs for predictive biomarker validation: one size does not fit all. J Biopharm Stat 19(3):530–542. doi:10.1080/10543400902802458
7. Simon R (2010) Clinical trial designs for evaluating the medical utility of prognostic and predictive biomarkers in oncology. Per Med 7(1):33–47. doi:10.2217/pme.09.49
8. Freidlin B, McShane LM, Korn EL (2010) Randomized clinical trials with biomarkers: design issues. J Natl Cancer Inst 102(3):152–160. doi:10.1093/jnci/djp477
9. Jiang W, Freidlin B, Simon R (2007) Biomarker-adaptive threshold design: a procedure for evaluating treatment with possible biomarker-defined subset effect. J Natl Cancer Inst 99(13):1036–1043. doi:10.1093/jnci/djm022
10. Goodsaid FM, Amur S, Aubrecht J, Burczynski ME, Carl K, Catalano J, Charlab R, Close S, Cornu-Artis C, Essioux L, Fornace AJ Jr, Hinman L, Hong H, Hunt I, Jacobson-Kram D, Jawaid A, Laurie D, Lesko L, Li HH, Lindpaintner K, Mayne J, Morrow P, Papaluca-Amati M, Robison TW, Roth J, Schuppe-Koistinen I, Shi L, Spleiss O, Tong W, Truter SL, Vonderscher J, Westelinck A, Zhang L, Zineh I (2010) Voluntary exploratory data submissions to the US FDA and the EMA: experience and impact. Nat Rev Drug Discov 9(6):435–445. doi:10.1038/nrd3116

11. Wang SJ, O'Neill RT, Hung HJ (2010) Statistical considerations in evaluating pharmacogenomics-based clinical effect for confirmatory trials. Clin Trials 7(5):525–536. doi:10.1177/1740774510375455
12. Burns DK, Hughes AR, Power A, Wang SJ, Patterson SD (2010) Designing pharmacogenomic studies to be fit for purpose. Pharmacogenomics 11(12):1657–1667. doi:10.2217/pgs.10.140
13. Hughes AR, Brothers CH, Mosteller M, Spreen WR, Burns DK (2009) Genetic association studies to detect adverse drug reactions: abacavir hypersensitivity as an example. Pharmacogenomics 10(2):225–233. doi:10.2217/14622416.10.2.225
14. Weber J, McCormack PL (2008) Panitumumab: in metastatic colorectal cancer with wild-type KRAS. BioDrugs 22(6):403–411. doi:10.2165/0063030-200822060-00006
15. Flynn AA (2011) Pharmacogenetics: practices and opportunities for study design and data analysis. Drug Discov Today 16(19–20):862–866. doi:10.1016/j.drudis.2011.08.008
16. Wei JS, Greer BT, Westermann F, Steinberg SM, Son CG, Chen QR, Whiteford CC, Bilke S, Krasnoselsky AL, Cenacchi N, Catchpoole D, Berthold F, Schwab M, Khan J (2004) Prediction of clinical outcome using gene expression profiling and artificial neural networks for patients with neuroblastoma. Cancer Res 64(19):6883–6891
17. Lee HS, Cho SB, Lee HE, Kim MA, Kim JH, Park do J, Yang HK, Lee BL, Kim WH (2007) Protein expression profiling and molecular classification of gastric cancer by the tissue array method. Clin Cancer Res 13(14):4154–4163
18. Patterson SD, Cohen N, Karnoub M, Truter SL, Emison E, Khambata-Ford S, Spear B, Ibia E, Sproule R, Barnes D, Bhathena A, Bristow MR, Russell C, Wang D, Warner A, Westelinck A, Brian W, Snapir A, Franc MA, Wong P, Shaw PM (2011) Prospective-retrospective biomarker analysis for regulatory consideration: white paper from the industry pharmacogenomics working group. Pharmacogenomics 12(7):939–951. doi:10.2217/pgs.11.52
19. Sanz-Pamplona R, Berenguer A, Cordero D, Riccadonna S, Sole X, Crous-Bou M, Guino E, Sanjuan X, Biondo S, Soriano A, Jurman G, Capella G, Furlanello C, Moreno V (2012) Clinical value of prognosis gene expression signatures in colorectal cancer: a systematic review. PLoS One 7(11):e48877. doi:10.1371/journal.pone.0048877
20. Taylor JM, Ankerst DP, Andridge RR (2008) Validation of biomarker-based risk prediction models. Clin Cancer Res 14(19):5977–5983. doi:10.1158/1078-0432.CCR-07-4534
21. Freidlin B, Jiang W, Simon R (2010) The cross-validated adaptive signature design. Clin Cancer Res 16(2):691–698. doi:10.1158/1078-0432.CCR-09-1357
22. FDA (2005) Drug diagnostics co-development concept paper. http://www.fda.gov/downloads/Drugs/ScienceResearch/ResearchAreas/Pharmacogenetics/UCM116689.pdf 2013. Accessed 29 May 2014
23. Benjamini Y, Hochberg Y (1995) Controlling the false discovery rate: a practical and powerful approach to multiple testing. J Roy Statist Soc Ser B 57(1):289–300
24. Chen YP, Chen F (2008) Identifying targets for drug discovery using bioinformatics. Expert Opin Ther Targets 12(4):383–389. doi:10.1517/14728222.12.4.383
25. Ginsburg GS, Willard HF (2009) Genomic and personalized medicine: foundations and applications. Transl Res 154(6):277–287. doi:10.1016/j.trsl.2009.09.005
26. Roos DS (2001) Computational biology. Bioinformatics–trying to swim in a sea of data. Science 291(5507):1260–1261
27. Sim SC, Altman RB, Ingelman-Sundberg M (2011) Databases in the area of pharmacogenetics. Hum Mutat 32(5):526–531. doi:10.1002/humu.21454
28. ORNL The Human Genome Management Information System (HGMIS) (2014) www.ornl.gov/sci/techresources/Human_Genome/project/about.shtml. Accessed 29 May 2014
29. Ensembl Ensembl Genome Browser (2014) http://hapmap.ncbi.nlm.nih.gov/. Accessed 29 May 2014

30. NCBI SNP—Short Genetic Variations (2014) http://www.ncbi.nlm.nih.gov/SNP. Accessed 29 May 2014
31. NCBI International HapMap Project (2014) http://hapmap.ncbi.nlm.nih.gov/. Accessed 29 May 2014
32. NCBI Human Genome Resources (2015) http://www.ncbi.nlm.nih.gov/projects/genome/guide/human/index.shtml. Accessed 23 Mar 2015
33. Roses AD (2000) Pharmacogenetics and the practice of medicine. Nature 405(6788):857–865. doi:10.1038/35015728
34. Johnson AD (2009) Single-nucleotide polymorphism bioinformatics: a comprehensive review of resources. Circ Cardiovasc Genet 2(5):530–536. doi:10.1161/CIRCGENETICS.109.872010
35. KEGG KEGG: Kyoto Encyclopedia of Genes and Genomes (2014) www.genome.jp/kegg/. Accessed 29 May 2014
36. PharmGKB PharmaGKB (2014) http://www.pharmagkb.org. Accessed 29 May 2014
37. Subramanian A, Tamayo P, Mootha VK, Mukherjee S, Ebert BL, Gillette MA, Paulovich A, Pomeroy SL, Golub TR, Lander ES, Mesirov JP (2005) Gene set enrichment analysis: a knowledge-based approach for interpreting genome-wide expression profiles. Proc Natl Acad Sci U S A 102(43):15545–15550. doi:10.1073/pnas.0506580102
38. Leong HS, Kipling D (2009) Text-based over-representation analysis of microarray gene lists with annotation bias. Nucleic Acids Res 37(11):e79. doi:10.1093/nar/gkp310
39. Heron EA, O'Dushlaine C, Segurado R, Gallagher L, Gill M (2011) Exploration of empirical Bayes hierarchical modeling for the analysis of genome-wide association study data. Biostatistics 12(3):445–461. doi:10.1093/biostatistics/kxq072

Pharmacokinetics and Pharmacogenetics: Bringing the Magic Bullet Closer to Reality

Janet Mifsud and Marc Maliepaard

Abstract Why do some patients respond positively to some drugs, while others may experience adverse effects? Can we predict which patients will react in which way? Does a magic bullet exist? The trend towards pharmacogenetics and personalised medicine in the last few years has somewhat sidelined the relevance of the traditional pharmaceutical sciences, such as pharmacokinetics and pharmacodynamics. Yet these actually are part and parcel of pharmacogenetics. Indeed understanding pharmacokinetics and pharmacodynamics in pharmacogenetics is essential in assessing the risk of new chemical entities (NCEs) in populations and individuals. Clinical pharmacokinetics, in fact, can be understood to have been a precursor to the implementation of pharmacogenetic understanding in the clinical setting. In this chapter, examples will be given of the strong interrelation between pharmacogenetics and the various pharmacokinetics processes i.e. absorption, distribution, metabolism and elimination. Reference will be made to studies which have shown how pharmacogenetics can be reinterpreted into pharmacokinetic principles, thus leading to the individualisation of drug therapy in the individual patient. The impact on recent regulatory guidelines published on the role of pharmacokinetics in pharmacogenetics and their impact on regulation of new medicinal drug development will also be discussed.

Keywords Pharmacokinetics · Pharmacodynamics · Pharmacogenetics · ADME · Drug transporters · Therapeutic drug monitoring · Individualised therapy · Drug development

J. Mifsud (✉)
Department of Clinical Pharmacology and Therapeutics,
University of Malta, Msida MSD 2040, Malta
e-mail: janet.mifsud@um.edu.mt

M. Maliepaard
Clinical Pharmacologist, Dutch Medicines Evaluation Board,
PO Box 8275, 3503 RG Utrecht, The Netherlands
e-mail: m.maliepaard@cbg-meb.nl

G. Grech, I. Grossman (eds.), *Preventive and Predictive Genetics: Towards Personalised Medicine,* Advances in Predictive, Preventive and Personalised Medicine 9,
DOI 10.1007/978-3-319-15344-5_5

1 Introduction: What is PK and What is its Role in PGx?

Drugs are, fundamentally, chemical compounds and drug therapy is a dynamic process. Thus the body handles the drug as it would any other chemical compound as it moves in the body. This fate of the drug proceeds over a certain time interval and the various, so called 'pharmacokinetic' processes (from Ancient Greek *pharmakon* "drug" and *kinetikos* "to do with motion") determine the time course of the drugs: drugs are *liberated* from the formulation; *absorbed* through the administration site; *distributed* through the body; *metabolised* mostly in the liver (but not only); *eliminated* mostly in the kidneys (but not only), often given the acronym of ADME [1].

Pharmacokinetics was first conceived as a term in 1950s by the German Professor Dost. Coincidently it was around the same time that Vogel Friedrich in 1959 published his key paper *Moderne probleme der Humangenetik*—the influence of genetic factors on the response to drug [2].

Pharmacokinetics was popularised by Holford in 1982 with his use of the aphorism 'pharmacokinetics is what the body does to the drug while pharmacodynamics is what the drug does to the body' [3]. Pharmacokinetic/pharmacodynamic (PK/PD) relationships are important in the drug therapy because they are predictive sciences (Fig. 1). They are essential to determine drug plasma/response relationships and thus the dose and dosage regimen needed in an individual patient, predict drug-drug and drug-food interactions and, in fact, form the basis of what has become more colloquially as *personalised medic*ine. It is mandatory for regulatory purposes that new chemical entities (NCEs) in various stages of drug development have their pharmacokinetic parameters well characterised with the use of software such as WINNONLIN® and NONMEM®, prior to marketing authorisation.

The *FDA Critical Path Initiative* and *NIH Roadmap* in 2004 changed the focus in innovation in drug development towards one based on translational science. It also led to a renewed understanding of the importance of PK/PD as quantitative pharmacology and led to terms such as "pharmacometrics", and "model-based drug development" [4]. This led to a shift from traditional paradigms, such as the ef-

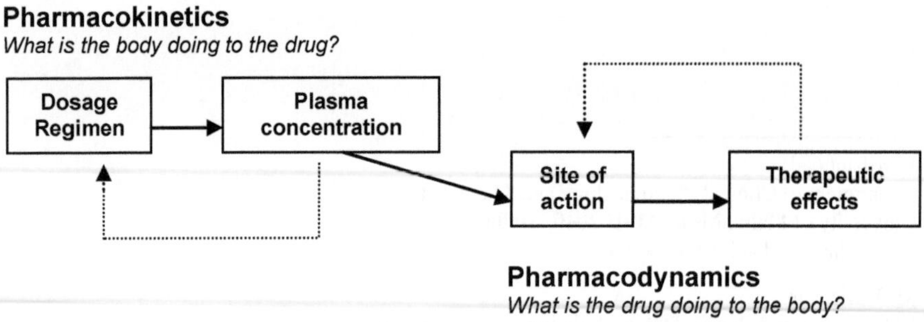

Fig. 1 The relationship between PK and PD in determining dosage regimens. The plasma–drug concentration data or effects produced are used via PK as a feedback (*dashed lines*) to modify dosage regimen to achieve optimal therapy

ficacy models in anticancer drug development area, to one where there is a truly scientific relation between preclinical and clinical pharmacology. The result thus expected would be one where there would be a reduction in the number of drugs being withdrawn from the market, due to unexpected and unacceptable adverse effects not picked up during drug development (cf. COX-II inhibitor class suits) [5].

At the same time, there was a parallel exponential shift in the understanding and role of genetics in pharmacology and the emergence of pharmacogenetics with the availability of cheaper and more rapid genetic analytical tests. This led to a new understanding that PK/PD differences may be due, in specific instances, to genetic variance [6]. Pharmacogenetics (PGx) was originally defined as the use of biological markers (DNA, RNA or protein) to predict the efficacy of a drug and the likelihood of the occurrence of an adverse event in individual patients. New definitions have emerged in recent years, yet the essential principles remain the same.

1.1 Succinylcholine: The First PK/PGx Study?

This parallel emergence of PK/PGx could not be better exemplified than in the emergence of the understanding of the differing mode of action of the muscle relaxant succinylcholine in some patients. In the 1950s it was clinically observed that patients differed in their response to this drug. It became known that the duration of this drug's action is determined by its enzymatic hydrolysis. It was only at a later stage that the genetic basis was discovered i.e. subjects homozygous for a gene encoding an atypical form of the hydrolysing enzyme, have a prolonged drug-induced muscle paralysis [7].

Other drugs followed suit. In the 1960s, it was discovered that a defect in N-acetylation metabolism of the drugs isoniazid (used for the treatment of tuberculosis), the antihypertensive drug hydralazine and the anti-arrythmic drug procainamide, could lead to prolonged half-life of these drugs in some patients, and this too was genetically determined [8].

The discovery in the mid-1970s, of the key role by the family of cytochromes, P450 system in drug metabolism, especially in hydroxylation, led, at the end of the 1980s, to an understanding of the molecular basis for slow metabolism in some patients, especially for those drugs which have a narrow therapeutic window [9].

Yet these developments remained only interesting academically. One of the key turning point in regulatory terms, however, was the discovery that irinotecan, which is used to treat bowel cancer, is metabolised by UDP-glucuronosyltransferase (UGT1A1). The *UGT1A1* promoter has a wild and variant copy (denoted *UGT1A1**28). This *UGT1A1**28 polymorphism is characterised by the presence of an additional TA repeat in the TATA sequence of the *UGT1A1* promoter, ((TA)7TAA, instead of (TA)6TAA). The latter causes a reduced UGT1A1 expression which leads to irinotecan toxicity. Today *UGT1A1* genotyping is mandatory in determining irinotecan dosage according to the FDA label for irinotecan, however in the EU the limitations of data available added to the reluctance to add a warning in several EU Member

States, i.e., it was not known what dose is the optimal one, in the *UGT1A1* WT as well as the *UGT1A1**28 [10].

More recently also for drug transporters the first genetic polymorphism was described which resulted in clinically relevant effects on the pharmacokinetics and adverse effects of some drugs, mainly statins, i.e., the *SLCO1B1* gene, encoding the OATP1B1 transporter protein [11].

1.2 Why Bother with PK/PGx?

Pharmacokinetic processes, like any aspect of human physiology, are inherently variable, and subject to both environmental and host-determined influences such as physiological and genetic factors. Genetic polymorphism (the occurrence in the same population of multiple allelic states) is responsible for a major proportion of the observed interindividual variability. These thus play a major role in PK/PD and therefore in dose response. Pharmacokinetic variation due to genetic factors can have the following consequences in any of the ADME processes such as:

- altered absorption/clearance
- difference in formation of active metabolites
- changes in drug interactions
- ethnic variation in drug response [12].

It is thus important to rationalise pharmacokinetic variation as exhibited by differing plasma concentration time curves in individual patients, on the basis of PGx effects and the interpretation of allelic changes. These need to be put into the context of PGx effect upon function e.g. one needs to be aware of how much an individual enzyme like CYP2D6 which is responsible for the metabolism of many drugs, contributes to the overall drug clearance. It is also important to consider what is the active moiety, the pathways affected and whether competing pathways could be present. In fact, PK/PGx studies would be key for those drugs which have a high inter-individual variation in efficacy and/or potency and have a narrow therapeutic range [12].

Thus, the ultimate goal of PK/PGx is to identify the contribution of genetic variability to differences in PK/PD, which in turn lead to individual differences in drug responses. This would enable prescribers to utilise a patient's PGx profile in order to select the drug which would exhibit the greatest efficacy and the least adverse effects in that specific patient and/or prescribe the drug at a dose which is appropriate for that patient i.e. personalised medicine [13].

In this review, a systematic discussion will be given on the PGx effect on the various PK processes i.e. ADME (absorption, metabolism, distribution and elimination). It will be shown how the development of a better understanding of PK/PGx guided principles, provide a crucial basis for the development of personalised medicines in individuals or specific subpopulations, optimising risk/benefit relationships, by maximising therapeutic efficacy with minimal adverse effects. It should be kept in mind, however, that the fulcrum of PK/PGx relationships would be the clinical availability of reliable PGx test, and a strong relationship between genotype and phenotype. To date not all studies provide confirmatory evidence in this regard.

2 PK/PGx Concepts in the Absorption of Drugs

Absorption is no longer considered to be a passive mechanism and is now known to be the summation of extremely complex processes. Several membrane bound drug transporters, such as P-glycoprotein (P-gp, MDR1) and multidrug resistance (MDR) transporters, encoded by the *ABC* genes, have been identified as being responsible for the transport of drugs across membranes, especially those in the gastrointestinal tract following oral administration [14, 15].

These mechanisms have an important bearing on a drug's systemic bioavailability (F) which is used as a measure of how much drug eventually reaches the circulation after oral and any other non IV administration. Since the bioavailability of an intravenous drug dose is assumed to be 100 %, F is best calculated as the ratio of drug concentrations after giving the drug by the route of interest (usually oral) compared with the same dose given intravenously [15].

Sequencing of the *ABCB1* gene (which encodes P-gp) has shown that there are more than 50 single nucleotide polymorphisms (SNPs) for this gene, which vary in frequency according to ethnicity [16]. Wild-type *ABCB1* alleles have been associated with increased tissue expression of P-gp, and it has been suggested that the haplotype of three specific SNPs (1236C>T in exon 12; 3435C>T in exon 26 and SNP 2677G>(T, A) in exon 21) are more predictive of phenotype (i.e., reduced transport activity) than the individual SNP genotype.

A 3435C>T mutation rs1045642 linked to one of the other mutations has been found to result in a changed protein folding, which can change substrate binding[17]. However, the robustness of *ABCB1* genotype/phenotype association, has not yet been established despite many studies e.g. no direct influence has been found of the effect of MDR1 C3435T polymorphism on digoxin pharmacokinetics [18]. This greatly limits, to date, the use of *ABCB1* in PK/PGx.

3 PK/PGx Concepts in Distribution

Following administration, a drug is distributed into all of the body compartments and tissues that it is able to enter taking into account physical-chemical properties. The drug is said to distribute into an imaginary volume, called its volume of distribution, or Vd. This volume is imaginary because it is based on sampling drug concentrations in some reference fluid (usually serum or plasma) immediately after dosing, with the assumption that the entire dose of drug is uniformly distributed throughout the body. For drugs which partition into lipids, e.g. general anaesthetics, plasma concentrations immediately after dosing will be quite low and the volume of distribution may appear to be many times larger than the volume of an average human being. Vd is essential for understanding where the drug goes and for estimating key parameters such as dose [12].

Vd is normally understood to be dependent on physiological parameters such as body mass index and fat deposits which may not have immediate PGx relations. However Vd may be PGx dependent in that distribution to certain body compartments, such as the brain across the blood brain barrier (BBB) and breast milk, may be dependent on transporters dependent on *ABC* genes, as outlined above. Overexpression of these *ABC* genes in certain patients may lead to drug efflux and what is clinically described as drug resistance [12].

Some studies suggest that the *ABCB1* variant 3435C > T rs1045642 affects plasma drug levels and drug resistance for drugs such as phenytoin, by-inhibiting transport [19]. In a study of British persons with epilepsy, the rs1045642 CC genotype was associated with drug resistance. In addition, a study of Egyptian persons with epilepsy showed increased likelihood of resistance to phenytoin in C allele carriers [20]. However, a meta-analysis failed to replicate the association with rs1045642, although many of these studies comprised patients on a variety of AEDs rather than phenytoin alone [21].

Also for the *SLCO1B1* gene, encoding OATP1B1, polymorphisms were demonstrated to result in clinically relevant effects on the pharmacokinetics, and more specifically the distribution, and adverse effects of some drugs, mainly statins. The *SLCO1B1**15 variant, 521T>C (Val174Ala) rs4149056 significantly affects the pharmacokinetics and adverse effects of many statins, and in a GWAS study indeed appeared to be associated with simvastatin-induced myopathy in patients treated for hypercholesteraemia. The prevalence of the *SLCO1B1**15 allele in the Caucasian population is 18 % [11, 22].

Another key pharmacokinetic parameter in PK distribution is related to protein binding. Most drugs bind to plasma proteins to some extent and it may play a significant role in pharmacokinetics if it exceeds 80 % (e.g. warfarin or phenytoin) as it is only the free (non-bound) drug which can exert a therapeutic effect. However to date while protein binding is an important pharmacokinetic parameter, at present no examples exists which point at PG affecting protein binding [15].

4 PK/PGx Concepts in Metabolism

There are over 170 genes known or expected to have a role in drug disposition, with more than half known to be polymorphic [23]. In pharmacokinetics, the highest level of polymorphism is found in genes involved in drug metabolism, especially cytochrome (CYP) P450 enzymes. These, in fact, account for over 80 % of current PGx drug labelling requirements.

Several CYPs have been shown to be polymorphic as a consequence of single nucleotide polymorphisms (SNPs), gene deletions and gene duplications. Perhaps the most studied is CYP2D6, which is involved in metabolism of approximately 100 drugs. More than 80 variants of CYP2D6 have been identified (http://www.cypalleles.ki.se), resulting in CYP enzymes with varying activities [24].

Examples abound in the literature of studies carried out on CYP2D6. For example it is responsible for the metabolism of codeine to the active metabolite, morphine. Thus the pharmacological activity of codeine is regulated by *CYP2D6* polymorphisms. In fact codeine has little therapeutic effect in patients who are CYP2D6 poor metabolisers, whereas due to excessive prodrug activation, CYP2D6 ultrarapid metabolisers suffer from adverse events due to increased levels of active metabolites. Thus *CYP2D6* genotype test results can be used to guide the dosing of codeine [25], and recently, information regarding the consequences of *CYP2D6* polymorphism have been included in the labelling of codeine.

However, even within the same drug class, genotype does not always predict drug metabolism. As an example, CYP2C19 is important in the metabolism of drugs such as protein pump inhibitors and an apparent gene–dose effect has been shown for the *CYP2C19**17 allele for pantoprazole, which predicts the plasma elimination rate constant, but this was not found for omeprazole. This difference could be to the difference in the contributions of CYP2C19 and CYP3A4 in the respective metabolism of the two drugs [26]. In fact, dependence on CYP2C19 metabolism is now seen by some as an undesirable property for NCEs in developments [6].

Polymorphisms in *CYP2D6* and *CYP2C19* have also been found to impact the metabolism of tricyclic antidepressents, such as amitriptyline and imipramine. These are demethylated by CYP2C19 to pharmacologically active metabolites, but then undergo further hydroxylation by CYP2D6 to less active metabolites. Thus polymorphisms in *CYP2D6* and *CYP2C19* may change the drug clearance or the ratio of parent drug to metabolites and dose adjustments can be estimated from the metaboliser status [27].

In oncology, the use of tamoxifen has for long been a mainstay in the adjuvant treatment of oestrogen receptor-positive breast cancer. Activity of tamoxifen is generally acknowledged to be mediated by the active metabolite endoxifen, which formation is catalysed by CYP2D6 [28–30]. Though the formation of endoxifen in CYP2D6 poor metaboliser patients is shown to be reduced, the consequences of the polymorphic status of *CYP2D6* for the success rate of tamoxifen treatment in relation to breast cancer recurrence or survival is not settled yet. In most cases these important clinical parameters have been investigated in fairly small studies, with only a small proportion of the known *CYP2D6* polymorphisms taken into account, whereas in some cases tumoral *CYP2D6* variations were assessed instead of germline variations, leading to a lack of Hardy–Weinberg equilibrium [31]. Though the totality of data are suggestive for a relationship between breast cancer recurrence and *CYP2D6* polymorphic status, more confirmative studies are needed, in particular with respect to the relationship between *CYP2D6* polymorphism and survival.

For clopidogrel, being a prodrug needing activation to an active metabolite mediated by CYP2C19, the efficacy may vary depending on the presence of specific functional allelic variants in patients. The conversion of the clopidogrel prodrug to active drug is strongly reduced in about 20% of Asian patients being CYP2C19 poor metaboliser. This reduced metabolism results in less anti-coagulation and less protection against cardiovascular events [32, 33].

On the other hand, *CYP2C9* has two common variant alleles (*2 and *3); which, unlike CYP2D6 and CYP2C19, retain enzymic activity albeit at a reduced rate [19]. Thus *CYP2C9* polymorphisms only have a minimal impact on pharmacokinetics and thus generally no significant effect on therapeutic outcome. However, in the case of a drug with a narrow therapeutic index, such as warfarin, *CYP2C9* genotype has been shown to correlate with the titrated dose in a population of 200 patients [34]. It was found that the highest titrated dose, was in patients homozygous for the wild-type *1 allele, which has the highest activity, whilst the lowest titrated dose was in patients with *3 homozygotes, which have the lowest enzyme activity.

There may also be key ethnic differences in these CYP variants. For phenytoin, *CYP2C9**3 (rs1057910 A>C) is associated with decreased metabolism of this drug. However, the *CYP2C9**2 variant was found to be associated with decreased metabolism in patients with epilepsy, but not with phenytoin dose in a study of white persons with epilepsy. There are different *CYP2C9* variants (*CYP2C9**5, *6, *8 and *11) in black populations, which are linked with a decreased phenytoin metabolism. In Asian Indians, increased free phenytoin was found in *CYP2C9**3 carriers which led to an increased risk for concentration-dependent toxicity compared with *1 homozygotes [19].

Furthermore, in white populations, the frequency of carriers of the wild-type *CYP3A5**1 allele (showing CYP3A5 activity) is only about 15 %, whereas it is up to 50 and 90 % in Asians and Blacks, respectively [35]. In a meta-analysis [36] a clear effect of CYP3A5 on rejection rates was indeed concluded after the first month of the treatment with the immunosuppressant tacrolimus.

Hundreds of studies were carried out in this area in recent years. This has led to pharmaceutical companies screening out compounds, in drug development, to assess whether they are substrates solely for a known polymorphic enzyme in order to avoid the wider intersubject variability in exposure [26]. However, by doing this, it may be argued that one may end up relatively often with drugs in development which may be substrates for less studied genetic polymorphisms. Therefore, the EMA advocates that the involvement of known polymorphic enzymes and transporters should not prohibit further development of the drug, but instead should be taken into account during this clinical development, in order to provide satisfactory efficacy and safety in genetic subpopulations that have variable systemic exposure of active [37].

5 PK/PGx Concepts in Elimination and Clearance

Immediately after a dose of drug is administered, the body begins to eliminate or clear it. Most drug elimination follows first-order kinetics. That is, a constant fraction of drug is eliminated from the body during each unit of time and it assumes the drug is uniformly distributed in a single body compartment with most of the drug eliminated from the body after four or five half-lives [12].

Clearance describes the rate at which the drug is eliminated from its volume of distribution, and its units are volume/time. Another important noncompartmental PK term is AUC, the area under the concentration-time curve. This term can be used to calculate overall clearance and half-life values for a drug. In addition, AUC is frequently used to compare drug exposures achieved with different drug doses, or to compare pharmacokinetics in the presence or absence of a drug with the potential to produce a PK drug interaction. It is to be pointed that drug systematic clearance is a summation of all the various organ clearances such as hepatic clearance renal clearance, salivary clearance, biliary clearance.

Renal clearance can be influenced by PGx differences especially for drugs which are eliminated mostly unchanged in urine. One such example is memantine, a frequently prescribed anti-dementia drug, which is mainly eliminated unchanged by the kidneys, partly via tubular secretion. Considerable inter-individual variability in plasma concentrations has been reported. A population pharmacokinetic study was performed in 108 patients who were genotyped for common polymorphisms in renal cation transporters (SLC22A1/2/5, SLC47A1, ABCB1). A SNP in NR1I2 (encoding the pregnane X receptor PXR) rs1523130 was identified as the unique significant genetic covariate for memantine clearance ($p = 0.006$), with carriers of the NR1I2 rs1523130 CT/TT genotypes presenting a 16 % slower memantine elimination than carriers of the CC genotype [38].

6 PK and PGx in dose prediction

As has been described in the previous sections, several genetic polymorphisms have been identified in drug targets, drug-metabolising enzymes and drug transporters. Thus individual patients could theoretically be screened for specific polymorphisms, effectively acting as biomarkers, facilitating more specific and individualised choice of drug and dose (see Fig. 2). This strategy may enable therapeutic concentrations to be attained more quickly. However, it should be kept in mind that for a PGx test to be useful, the genotype must have a major influence on the PK/PD of a drug with a narrow therapeutic index.

As discussed in previous sections, genetic polymorphisms can result in changes to functional activity and PK through changes in expression of enzymes such as CYP3A5, UGT1A3, UGT2B17 and CYP2D6. This may result in important changes in clinical outcome which need to be noted.

Such genetic tests may be particularly useful in certain patients, such as renal transplant patients where PK/PD are hard to predict. PGx can be used to reduce the wide interindividual variation in the dose of immunosuppressive drugs required to achieve target blood concentrations, since PGx can be used to predict metabolism of these drugs, improving graft outcome. Several clinically useful strategies have

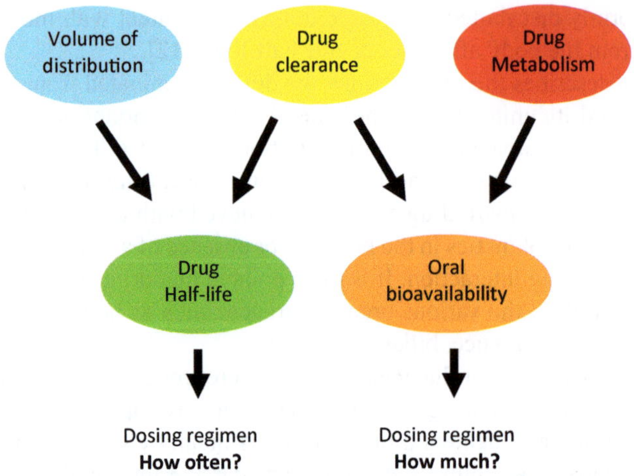

Fig. 2 The key pharmacokinetic parameters and their importance for designing dose regimen and dose size

emerged such as the use of the cytochrome P450 (CYP)3A5 (*CYP3A5*) genotype to predict the optimal initial dose for the immunosuppressant, tacrolimus [36, 39].

Warfarin has a narrow therapeutic index and a high dose variability. Thirty percent of this variance can be explained by SNPs in the warfarin drug target *VKORC1* and 12 % by two non-synonymous SNPs (*2, *3) in CYP2C9. Affected individuals require, on average, lower doses of warfarin to maintain a therapeutic INR and more time to achieve stable dosing. A PK/PD model for warfarin, with *CYP2C9* and *VKORC1* genotype, age and target international normalised ratio (INR) as dose predictors has been developed [40]. Such a dosing algorithm may yield a more rapid dosing at the appropriate level, which is expected to reduce mortality of warfarin treatment. The actual effect of genotype-based dosing of warfarin during the initiation of therapy in patients with atrial fibrillation or venous thromboembolism has recently been tested in prospective clinical trials. In these cases warfarin was either prescribed according to a *CYP2C9*, *VKORC1* based dosing algorithm or the standard dosing regimen. The percentage of time that patients were in the therapeutic range for the international normalised ratio (INR) during the first weeks after warfarin initiation was measured. Results of these prospective studies however were not consistent, with some studies showing that genotype guided dosing was associated with a higher percentage of time in the therapeutic INR range than was standard dosing, whereas other reported that genotype-guided dosing of warfarin did not improve anticoagulation control during the first weeks of therapy during [41, 42]. Currently, EU drug regulatory agencies do not require genotyping before initiation of warfarin therapy, however, the warfarin drug label in the USA (Coumadin, FDA) [43] presents dosing information on the combined *VKORC1* and *CYP2C9* status that should be considered if this genotype is known prior to treatment.

7 What is the Role of PK/PGx Relationship in the Development of New Drugs?

As a result of the large output of high through put screening (HTS) in the evaluation of a new chemical entities (NCE), PK/PD relationships are established early on the drug development. PK/PD data are also used in the evaluation of preclinical studies and in the prediction of these parameters in actual patients. This optimises drug screening and reduces the risk of late stage attrition due to poor pharmacokinetics (see Fig. 3).

In fact, *in vitro* screening of a broad panel of *in vitro* metabolic or transport pathway evaluations for NCEs generally is determined early in preclinical evaluations. Such PK/PGx evaluations may trigger subsequent clinical PG-related investigations, e.g. by the inclusion of various PG variant patients in the clinical studies, in order to obtain an appropriate dose advice for the different important phenotypes for a certain polymorphic enzyme shown to be important in the pharmacokinetics of the drug. The PK/PGx evaluations are also important in the evaluation of drug interactions and dosing paradigms for desirable agents [4].

Recently the characterisation and development of pharmacophore template model for many CYPs such as for the active site of CYP2D6 has taken large steps forward. For example, this model has in fact been used in drug development of a novel

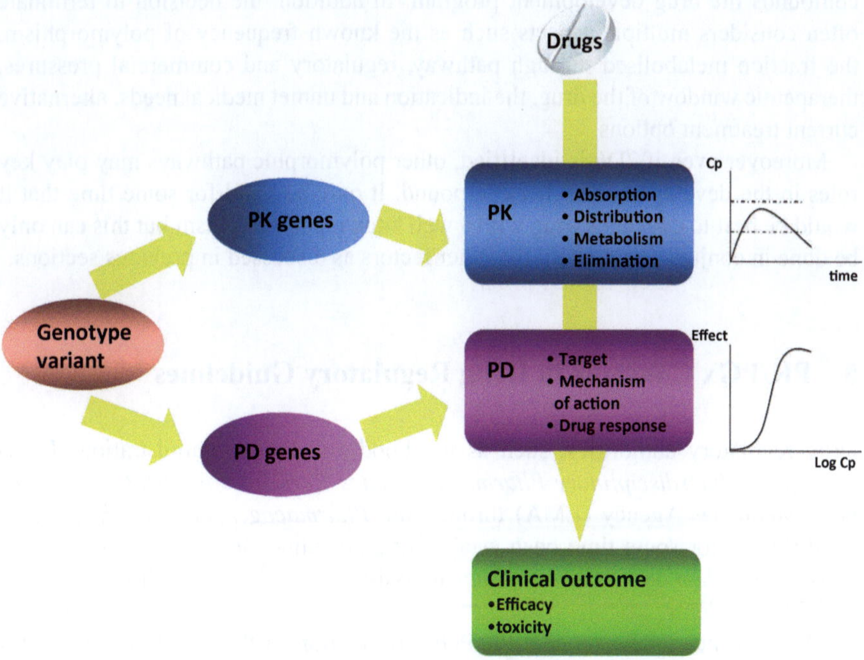

Fig. 3 Correlation of PK/PD and PGx in clinical outcome. By including PGx regarding PK and PD genes, PK and PD, and thus efficacy and safety of the drug may be optimised

calcium channel antagonist, together with in vitro data. These models indicated that the metabolism of this NCE was dependent on CYP2D6, and thus volunteers in the first in human study were genotyped for *CYP2D6*. Two volunteers were identified as poor metabolisers for CYP2D6 and were included in the study. The resulting PK data showed that the half-life of the drug was much higher in these individuals and it was decided to stop further development of the compound [6]. The current situation is that this is not considered a desirable approach anymore. Instead, a drug should be developed taking into account the various subpopulations. Indeed, stopping development because CYP2D6 is involved may seem wise on the short term, but on the longer term it may lead to the discovery that other, as yet unidentified polymorphic enzymes and transporters are important, which may appear only later in drug development, and was not anticipated. In that case, one is better off with a very well known polymorphic enzyme, to take into account just from the beginning of the development programme.

Go/no go decisions using such information from polymorphic enzymes, involves several approaches, many of them involving complex databases, drug–disease–trial models and simulation for the integration of information. Physiology-based pharmacokinetic (PBPK) models have also been developed which provide a very useful mechanistic approach to drug development. This plethora of information may result in a number of conflicting information which may determine issues such as when it may be best to stop the development especially if there are back ups without the potential issue? While data can be generated on different doses, this greatly confounds the drug development program. In addition, the decision to terminate often considers multiple aspects such as the known frequency of polymorphism, the fraction metabolised through pathway, regulatory and commercial pressures, therapeutic window of the drug, the indication and unmet medical needs, alternative current treatment options.

Moreover even if 2D6 is identified, other polymorphic pathways may play key roles in the development of that compound. It may be valid for some time that it would be best to develop a drug with a well known polymorphism but this can only be done in conjunction with all the other factors as discussed in previous sections.

8 PK/PGx Concepts in Drug Regulatory Guidelines

Drug regulatory authorities, such as the Food and Drug Administration (FDA) through the *Interdisciplinary Pharmacogenomics Review Group IPRG* and European Medicines Agency (EMA) through the *Pharmacogenomics Working Party PGWP* have for some time been establishing guidelines for submission of pharmacogenetic data on therapeutic drugs to assist in the tailoring of drug therapy to individual patients.

These agencies now request PK/PGx information in the labelling for several drugs. For example, in 2007, FDA (but to date not EMA) issued a labelling change advising physicians to consider the use of “genetic tests to improve their initial estimate" of the dose of anticoagulant drug warfarin, which is widely prescribed for reducing the risk of thrombosis and its complications [44]. This recommendation has set a precedent for the use of genetic technologies in clinical practice and now several bodies are pushing for such novel technology to enhance personalised medicine. Likewise, recently, information regarding the consequences of *CYP2D6* polymorphism have been included in the labelling of codeine. Presently, there are over 70 licensed drugs with PGx labels, where the polymorphic *CYP2C9, CYP2C19* and *CYP2D6* account for the majority of these labels. In fact IPRG and PGWP now liaise together with industry for combined VGDS—Voluntary Genomic Data Submissions/Pharmacogenomic briefing meetings in order to streamline regulatory requests in this fast developing area.

Further, the EMA PGWP published a guideline on the role of pharmacogenetics in PK in 2012 [37] soon followed by a guidance on this topic by the FDA (Guidance on Clinical Pharmacogenomics: Premarketing Evaluation in Early Phase Clinical Studies) [44]. These guidelines request the drug developers to identify those pharmacogenomic factors that may affect safety and/or efficacy of drugs that are currently being developed. For that purpose, the consequences of pharmacogenomic variation should be investigated if *in vitro* and/or clinical (*in vivo*) studies indicate that a known functionally polymorphic enzyme or drug transporter is likely to be important in the disposition of the drug, or if these represent an important factor in the formation, elimination or distribution of a pharmacologically active or toxic metabolite. Pharmacogenomic investigations are also required when clinical studies indicate that major interindividual differences in the pharmacokinetic properties (that cannot be explained by other intrinsic or extrinsic factors) are likely to influence the efficacy or safety of the drug in a genetically variable subpopulation. When looking at a global level, there appears broad agreement on the requirements with respect to pharmacogenomics related to pharmacokinetics in the EU, USA and Japan, though some divergence still exists on some areas, like the actual cut-off which would trigger the need for *in vivo* pharmacogenomic investigations, and the stringency by which banking of DNA samples from ongoing clinical studies is required [45]. Overall, however, it is clear that in the future, for new drugs, more pharmacogenomic data is expected to become available which will enable appropriate dosing in e.g. patients with a different metaboliser status, than has been in the past.

9 Outlook and Recommendations

The better understanding of pharmacogenetics on PK/PD inter-individual variability of drug disposition might be beneficial in the context of individual dose optimisation in personalised medicine. The greatest understanding has been in that of metabolic phenotyping especially of metabolising enzymes. The application of PGx to predict other PK processes and thus dosage regimens depends, however, on various other cofounding factors such as disease and co-administered drugs which limits the feasibility of clinical applications to date. This will determine how much a genetic variant contributes to a clinically significant pharmacokinetic variability overall.

There is now a growing recognition that the future of the pharmaceutical industry will depend a great deal on the integration of PGx and PK/PD data to guide drug decision-making. Novel PGx biomarkers are important to fill in gaps of uncertainty about therapeutic targets, variability in drug response; algorithm-based dose determination; response monitoring; early indicator/predictor of toxicity/adverse reactions. The debate still remains on the adoption of PGx assays in clinical examinations and the implications of reimbursement. Recent data on warfarin and clopidogrel have identified barriers to successful implementation [46]. The data available for such technology varies a great deal but used with agreed clinical guidelines, appears to be the strongest predictor of reimbursement. However bringing better clinical evidence is needed.

It important to note, however, that other factors can also influence PK/PD processes which may impact the predictability of outcomes in these patients, such as co morbid medical conditions, smoking, diet, drug interactions, race and frailty.

Integrating PK/PD with PGx can be our magic ball in the determination of drug doses, dosing intervals, titration regimens in order to decrease the risk of drug adverse events and toxicity and ensure successful outcomes in patients. Further knowledge is likely to add to our understanding of differences in sub-populations, but the potential limitations of these approaches should be recognised in order that they can be applied beneficially [6].

The availability of open access on-line resources such as PharmGKB, the pharmacogenomics database, and simulation models such as SIMCYP® have greatly facilitated the availability of resources which systematically assess the vast information now available on the impact of genetic variation on drug response for clinicians and researchers [47].

References

1. Rang H, Ritter J, Henderson G (2012) What is pharmacology? In: Rang H, Ritter J, Henderson G (eds) Rang and Dale's pharmacology, 7th edn. Churchill Livingston, London, pp 1–3
2. Vogel F (1959) The influence of genetic factors on the response to drugs: Moderne probleme der Humangenetik. Ergeb Inn Med Kinderheilkd 12:52–125
3. Holford NH, Sheiner LB (1982) Kinetics of pharmacologic response. Pharmacol Ther 16(2):143–166
4. Gallo JM (2010) Pharmacokinetic/pharmacodynamic-driven drug development. Mt Sinai J Med (New York) 77(4):381–388. doi:10.1002/msj.20193
5. Singh BK, Haque SE, Pillai KK (2014) Assessment of nonsteroidal anti-inflammatory drug-induced cardiotoxicity. Expert Opin Drug Metab Toxicol 10(2):143–156. doi:10.1517/17425255.2014.856881
6. Walker DK (2004) The use of pharmacokinetic and pharmacodynamic data in the assessment of drug safety in early drug development. Br J Clin Pharmacol 58(6):601–608. doi:10.1111/j.1365-2125.2004.02194.x

7. Gundert-Remy U, Dimovski A, Gajovic S (2012) Personalized medicine—where do we stand? Pouring some water into wine: a realistic perspective. Croat Med J 53(4):314–320
8. Evans WE, McLeod HL (2003) Pharmacogenomics–drug disposition, drug targets, and side effects. N Engl J Med 348 (6):538–549. doi:10.1056/NEJMra020526
9. Gonzalez FJ, Skoda RC, Kimura S, Umeno M, Zanger UM, Nebert DW, Gelboin HV, Hardwick JP, Meyer UA (1988) Characterization of the common genetic defect in humans deficient in debrisoquine metabolism. Nature 331(6155):442–446. doi:10.1038/331442a0
10. Iqbal S, Lenz HJ (2001) Determinants of prognosis and response to therapy in colorectal cancer. Curr Oncol Rep 3(2):102–108
11. Link E, Parish S, Armitage J, Bowman L, Heath S, Matsuda F, Gut I, Lathrop M, Collins R (2008) SLCO1B1 variants and statin-induced myopathy–a genomewide study. N Engl J Med 359(8):789–799. doi:10.1056/NEJMoa0801936
12. Rowland M, Tozer TN (2012). Clinical pharmacokinetics and pharmacodynamics: concepts and applications, 4th edn. Lippincott Williams and Wilkins, Baltimore, p 359, 367, 370, 300
13. Mayor S (2007) Fitting the drug to the patient. BMJ 334(7591):452–453. doi:10.1136/bmj.39133.452315.AD
14. Cropp CD, Yee SW, Giacomini KM (2008) Genetic variation in drug transporters in ethnic populations. Clin Pharmacol Ther 84(3):412–416. doi:10.1038/clpt.2008.98
15. Curry SH, Whelpton R (2011) Drug disposition and pharmacokinetics: from principles to applications. Wiley Blackwell, London, p 27, 29, 48
16. Marzolini C, Paus E, Buclin T, Kim RB (2004) Polymorphisms in human MDR1 (P-glycoprotein): recent advances and clinical relevance. Clin Pharmacol Ther 75(1):13–33. doi:10.1016/j.clpt.2003.09.012
17. Kimchi-Sarfaty C, Oh JM, Kim IW, Sauna ZE, Calcagno AM, Ambudkar SV, Gottesman MM (2007) A "silent" polymorphism in the MDR1 gene changes substrate specificity. Science 315(5811):525–528. doi:10.1126/science.1135308
18. Chowbay B, Li H, David M, Cheung YB, Lee EJ (2005) Meta-analysis of the influence of MDR1 C3435T polymorphism on digoxin pharmacokinetics and MDR1 gene expression. Br J Clin Pharmacol 60(2):159–171. doi:10.1111/j.1365-2125.2005.02392.x
19. Thorn CF, Whirl-Carrillo M, Leeder JS, Klein TE, Altman RB (2012) PharmGKB summary: phenytoin pathway. Pharmacogenet Genomics 22(6):466–470. doi:10.1097/FPC.0b013e32834aeedb
20. Siddiqui A, Kerb R, Weale ME, Brinkmann U, Smith A, Goldstein DB, Wood NW, Sisodiya SM (2003) Association of multidrug resistance in epilepsy with a polymorphism in the drug-transporter gene ABCB1. N Eng J Med 348(15):1442–1448. doi:10.1056/NEJMoa021986
21. Bournissen FG, Moretti ME, Juurlink DN, Koren G, Walker M, Finkelstein Y (2009) Polymorphism of the MDR1/ABCB1 C3435T drug-transporter and resistance to anticonvulsant drugs: a meta-analysis. Epilepsia 50(4):898–903. doi:10.1111/j.1528-1167.2008.01858.x
22. Pasanen MK, Neuvonen M, Neuvonen PJ, Niemi M (2006) SLCO1B1 polymorphism markedly affects the pharmacokinetics of simvastatin acid. Pharmacogenet Genomics 16(12):873–879. doi:10.1097/01.fpc.0000230416.82349.90
23. Katz DA, Murray B, Bhathena A, Sahelijo L (2008) Defining drug disposition determinants: a pharmacogenetic-pharmacokinetic strategy. Nat Rev Drug Discov 7(4):293–305. doi:10.1038/nrd2486
24. Crews KR, Gaedigk A, Dunnenberger HM, Klein TE, Shen DD, Callaghan JT, Kharasch ED, Skaar TC (2012) Clinical Pharmacogenetics Implementation Consortium (CPIC) guidelines for codeine therapy in the context of cytochrome P450 2D6 (CYP2D6) genotype. Clin Pharmacol Ther 91(2):321–326. doi:10.1038/clpt.2011.287
25. Preissner SC, Hoffmann MF, Preissner R, Dunkel M, Gewiess A, Preissner S (2013) Polymorphic cytochrome P450 enzymes (CYPs) and their role in personalized therapy. PloS one 8(12):e82562. doi:10.1371/journal.pone.0082562
26. Ward RM, Kearns GL (2013) Proton pump inhibitors in pediatrics: mechanism of action, pharmacokinetics, pharmacogenetics, and pharmacodynamics. Paediatr Drugs 15(2):119–131. doi:10.1007/s40272-013-0012-x

27. Hicks JK, Swen JJ, Thorn CF, Sangkuhl K, Kharasch ED, Ellingrod VL, Skaar TC, Muller DJ, Gaedigk A, Stingl JC (2013) Clinical Pharmacogenetics Implementation Consortium guideline for CYP2D6 and CYP2C19 genotypes and dosing of tricyclic antidepressants. Clin Pharmacol Ther 93(5):402–408. doi:10.1038/clpt.2013.2
28. Irvin WJ Jr, Walko CM, Weck KE, Ibrahim JG, Chiu WK, Dees EC, Moore SG, Olajide OA, Graham ML, Canale ST, Raab RE, Corso SW, Peppercorn JM, Anderson SM, Friedman KJ, Ogburn ET, Desta Z, Flockhart DA, McLeod HL, Evans JP, Carey LA (2011) Genotype-guided tamoxifen dosing increases active metabolite exposure in women with reduced CYP2D6metabolism: a multicenter study. J Clin Oncol 29(24):3232–3239. doi:10.1200/jco.2010.31.4427
29. Kiyotani K, Mushiroda T, Nakamura Y, Zembutsu H (2012) Pharmacogenomics of tamoxifen: roles of drug metabolizing enzymes and transporters. Drug Metab Pharmacokinet 27(1):122–131
30. Madlensky L, Natarajan L, Tchu S, Pu M, Mortimer J, Flatt SW, Nikoloff DM, Hillman G, Fontecha MR, Lawrence HJ, Parker BA, Wu AH, Pierce JP (2011) Tamoxifen metabolite concentrations, CYP2D6 genotype, and breast cancer outcomes. Clin Pharmacol Ther 89(5):718–725. doi:10.1038/clpt.2011.32
31. Sim SC, Kacevska M, Ingelman-Sundberg M (2013) Pharmacogenomics of drug-metabolizing enzymes: a recent update on clinical implications and endogenous effects. Pharmacogenomics J 13(1):1–11. doi:10.1038/tpj.2012.45
32. Shuldiner AR, O'Connell JR, Bliden KP, Gandhi A, Ryan K, Horenstein RB, Damcott CM, Pakyz R, Tantry US, Gibson Q, Pollin TI, Post W, Parsa A, Mitchell BD, Faraday N, Herzog W, Gurbel PA (2009) Association of cytochrome P450 2C19 genotype with the antiplatelet effect and clinical efficacy of clopidogrel therapy. JAMA 302(8):849–857. doi:10.1001/jama.2009.1232
33. Mega JL, Simon T, Collet JP, Anderson JL, Antman EM, Bliden K, Cannon CP, Danchin N, Giusti B, Gurbel P, Horne BD, Hulot JS, Kastrati A, Montalescot G, Neumann FJ, Shen L, Sibbing D, Steg PG, Trenk D, Wiviott SD, Sabatine MS (2010) Reduced-function CYP2C19 genotype and risk of adverse clinical outcomes among patients treated with clopidogrel predominantly for PCI: a meta-analysis. JAMA 304(16):1821–1830. doi:10.1001/jama.2010.1543
34. Higashi MK, Veenstra DL, Kondo LM, Wittkowsky AK, Srinouanprachanh SL, Farin FM, Rettie AE (2002) Association between CYP2C9 genetic variants and anticoagulation-related outcomes during warfarin therapy. JAMA 287(13):1690–1698
35. (NCBI) NCfBI (2014) dBSNP short genetic variations: Reference SNP (refSNP) Cluster Report: rs776746 NCBI. http://www.ncbi.nlm.nih.gov/SNP/snp_ref.cgi?rs=776746. Accessed July 2014
36. Tang HL, Xie HG, Yao Y, Hu YF (2011) Lower tacrolimus daily dose requirements and acute rejection rates in the CYP3A5 nonexpressers than expressers. Pharmacogenet Genomics 21(11):713–720. doi:10.1097/FPC.0b013e32834a48ca
37. EMA (2012) Guideline on the use of Pharmacogenetic Methodologies in the Pharmacokinetic Evaluation of Medicinal Products EMA/ CHMP/37646/2009. EU, London
38. Noetzli M, Guidi M, Ebbing K, Eyer S, Wilhelm L, Michon A, Thomazic V, Alnawaqil AM, Maurer S, Zumbach S, Giannakopoulos P, von Gunten A, Csajka C, Eap CB (2013) Population pharmacokinetic study of memantine: effects of clinical and genetic factors. Clin Pharmacokinet 52(3):211–223. doi:10.1007/s40262-013-0032-2
39. Ware N (2012) The role of genetics in drug dosing. Pediatr Nephrol (Berlin, Germany) 27(9):1489–1498. doi:10.1007/s00467-012-2105-0
40. Hamberg AK, Friberg LE, Hanseus K, Ekman-Joelsson BM, Sunnegardh J, Jonzon A, Lundell B, Jonsson EN, Wadelius M (2013) Warfarin dose prediction in children using pharmacometric bridging–comparison with published pharmacogenetic dosing algorithms. Eur J Clin Pharmacol 69(6):1275–1283. doi:10.1007/s00228-012-1466-4.
41. Kimmel SE, French B, Kasner SE, Johnson JA, Anderson JL, Gage BF, Rosenberg YD, Eby CS, Madigan RA, McBane RB, Abdel-Rahman SZ, Stevens SM, Yale S, Mohler ER 3rd,

Fang MC, Shah V, Horenstein RB, Limdi NA, Muldowney JA 3rd, Gujral J, Delafontaine P, Desnick RJ, Ortel TL, Billett HH, Pendleton RC, Geller NL, Halperin JL, Goldhaber SZ, Caldwell MD, Califf RM, Ellenberg JH (2013) A pharmacogenetic versus a clinical algorithm for warfarin dosing. N Eng J Med 369(24):2283–2293. doi:10.1056/NEJMoa1310669
42. Pirmohamed M, Burnside G, Eriksson N, Jorgensen AL, Toh CH, Nicholson T, Kesteven P, Christersson C, Wahlstrom B, Stafberg C, Zhang JE, Leathart JB, Kohnke H, Maitland-van der Zee AH, Williamson PR, Daly AK, Avery P, Kamali F, Wadelius M (2013) A randomized trial of genotype-guided dosing of warfarin. N Eng J Med 369 (24):2294–2303. doi:10.1056/NEJMoa1311386
43. FDA (2007) FDA Approves Updated Warfarin (Coumadin) Prescribing Information: New Genetic Information May Help Providers Improve Initial Dosing Estimates of he Anticoagulant for Individual Patients. U.S. Department of Health and Human Services
44. FDA (2013) Guidance on Clinical Pharmacogenomics: Premarketing Evaluation in Early Phase Clinical Studies
45. Maliepaard M, Nofziger C, Papaluca M, Zineh I, Uyama Y, Prasad K, Grimstein C, Pacanowski M, Ehmann F, Dossena S, Paulmichl M (2013) Pharmacogenetics in the evaluation of new drugs: a multiregional regulatory perspective. Nat Rev Drug Discov 12(2):103–115. doi:10.1038/nrd3931
46. Meckley LM, Neumann PJ (2010) Personalized medicine: factors influencing reimbursement. Health Policy 94(2):91–100. doi:10.1016/j.healthpol.2009.09.006
47. PharmGKB The Pharmacogenomics Database (2014) Stanford University. http://www.pharmgkb.org/

Pharmacogenetics of Adverse Drug Reactions

Richard Myles Turner and Munir Pirmohamed

Abstract A large variation in drug response exists between patients, with susceptible individuals being at risk of experiencing an adverse drug reaction (ADR). This susceptibility is attributable to environmental, clinical and genetic factors although the contribution of each varies with the drug, ADR and ethnicity. The variation in drug response makes personalisation of pharmacological therapy appealing to minimise ADRs whilst promoting efficacy. Pharmacogenetics seeks to contribute through genetic-guided drug and dose selection strategies. ADR pharmacogenetics was first highlighted in the 1950s, but it is only in the last decade that it has seen a rapid expansion, aided by significant advances in our knowledge of the human genome and improved genotyping technologies. ADRs can be classified according to whether the dominant mechanism is immune- or nonimmune-mediated. Several ADRs have been strongly associated with specific human leukocyte antigen (*HLA*) alleles. There is growing evidence for a central role of these alleles in the pathogenesis of immune-mediated delayed hypersensitivity ADRs through facilitation of 'off-target' interactions that lead to the presentation of 'altered self,' drugs and/or their metabolites to the T-cell receptor in an HLA-restricted fashion. Genetic variation can also predispose to nonimmune-mediated ADRs through perturbing drug pharmacokinetics or by altering nonimmune pharmacodynamic processes. In particular, genetic variants of phase I and phase II biotransformation enzymes and drug transporters alter the availability of a drug at the site(s) responsible for the ADR. Depending on the drug and ADR, these sites may be the therapeutic target site, the same molecular site in another tissue or distinct off-target sites. A prominent example of pharmacogenetics improving drug safety and enhancing the cost-effective use of limited healthcare resources is the reduction in the incidence of the abacavir hypersensitivity syndrome. It is apparent though that the success of ameliorating the abacavir hypersensitivity syndrome by genetic screening is proving difficult to emulate for other drug-ADR combinations. This highlights the considerable hurdles

R. M. Turner (✉) · M. Pirmohamed
The Wolfson Centre for Personalised Medicine, Institute of Translational Medicine, University of Liverpool, Block A: Waterhouse Building, 1-5 Brownlow Street, Liverpool L69 3GL, UK
e-mail: rt34@liverpool.ac.uk

M. Pirmohamed
e-mail: munirp@liverpool.ac.uk

G. Grech, I. Grossman (eds.), *Preventive and Predictive Genetics: Towards Personalised Medicine,* Advances in Predictive, Preventive and Personalised Medicine 9,
DOI 10.1007/978-3-319-15344-5_6

encountered in translating a pharmacogenetic association into a clinical test that benefits patient safety. The development of international consortia alongside the potential of next generation sequencing technologies and other innovations offer tantalising prospects for future advances in pharmacogenetics to reduce the burden of ADRs.

Keywords Adverse drug reaction · Pharmacogenetics · Predictive genotyping · Translation · Abacavir · Hypersensitivity · Malignant hyperthermia · Codeine · Warfarin · Statin

1 Introduction

For many drugs, substantial evidence exists at the population level to advocate their use. However there is considerable inter-individual variability in drug response, affecting both drug efficacy and safety [1]. Over 961.5 million prescription items were dispensed in England in 2011 [2]. This high drug usage and the individuality of drug response contribute to the high frequency of adverse drug reactions (ADRs). A prospective study in England estimated that 6.5 % of hospital admissions for patients > 16 years old were related to an ADR, with a median inpatient stay of 8 days [3]. This was contextualised through extrapolation to the entire hospital bed base of England to suggest that the equivalent of up to seven 800 bed hospitals in England could be occupied at any one time with patients admitted with ADRs [3]. Studies from other countries have reported similar ADR-related hospitalisation rates [4–8]. Clearly, ADRs pose a significant international challenge to the health and safety of individual patients and to the efficient use of limited resources by healthcare services.

An ADR, as defined by the World Health Organisation (WHO), is 'a response to a drug that is noxious and unintended and occurs at doses normally used in man for prophylaxis, diagnosis or therapy of disease or for the modification of physiologic function' [9]. Table 1 provides definitions and examples of the related pharmacovigilance terms [9–12]. In essence, an adverse event (AE) is an umbrella term for any harm occurring to a patient temporally associated with but not necessarily directly attributable to a therapeutic intervention [10, 13]. A subdivision of AE is an adverse drug event, which describes maleficence associated with the use of a drug and includes overdoses, medication error and ADRs [11, 14, 15].

There is considerable variability between ADRs in terms of presentation and level of current aetiological understanding. This poses a challenge to their accurate categorisation and so, different classifications have been developed. The most well-known system delineates ADRs into types A and B. Type A ('augmented') ADRs constitute over 80 % of ADRs; they are dose-dependent and predictable from the main pharmacological action of a drug [13, 16]. This is because Type A ADRs are 'on-target' and manifest through excessive drug action at the therapeutic target site. Type B ('bizarre') ADRs are dose-independent and are not predictable from a drug's conventional pharmacology [13, 16] as they represent idiosyncratic 'off-target' drug

Table 1 Pharmacovigilance terminology for adverse effects

Adverse effect term	Definition	Example(s)
Adverse event	Any untoward medical occurrence in a patient or clinical investigation subject administered a pharmaceutical product and which does not necessarily have to have a causal relationship with this treatment [10].	In a clinical trial for a topical emollient for psoriasis a road traffic accident could be a serious, unexpected, not study related adverse event
Adverse drug event	An injury resulting from the use of a drug [11].	i) Intentional overdose
		ii) Medication error
		iii) Adverse drug reaction
Medication error	A medication error is any preventable event that may cause or lead to inappropriate medication use or patient harm while the medication is in the control of the health care professional, patient, or consumer [12].	Decrease in consciousness following accidental insulin overdose due to a prescribing or administration error
Adverse drug reaction	A response to a drug that is noxious and unintended and occurs at doses normally used in man for prophylaxis, diagnosis, or therapy of disease or for the modification of physiologic function [9].	Hypersensitivity reaction to allopurinol through standard clinical use of the drug

effects. This classification was first defined in 1977 [17] and has been variously extended subsequently to include additional categories as shown in Table 2 [13, 18]. However, it can prove difficult to categorise ADRs using this system. For example, some type B ADRs, including statin-induced muscle toxicity, are clearly dose related and other type B ADRs, such as hypersensitivity to abacavir, are now predictable. A second system is the DoTS classification, which categorises ADRs according to dose relatedness, timing and patient susceptibility factors [19]. This descriptive system improves the accuracy of ADR classification, but its complexity makes it more difficult to use.

In this chapter, ADRs will be classified as immune- or nonimmune-mediated. Immune-mediated ADRs result principally from a deleterious immune reaction mounted following drug exposure. Nonimmune-mediated ADRs encompass all other ADRs and as a point of clarification, include infections that result from predictable immunosuppression by biologics and disease-modifying agents. This is a simple classification, but it reflects the clinical presentation and predominant pathogenic processes of many ADRs and is helpful when considering pharmacogenetics.

The reasons for the heterogeneity in inter-individual drug response are often not known but there are a trilogy of implicated factors: environmental (e.g. drug-drug and drug-food interactions), clinical (e.g. age, co-morbidities, body mass index (BMI), pregnancy) and genetic [20]. The contribution of each postulated factor

Table 2 Adverse drug reaction pharmacovigilance classification and characteristics [13, 18]

ADR classification	Characteristics	Example
Major types		
A	*Augmented*	Hypotension with iloprost therapy
	Due to main pharmacological action of a drug	
	Common	
	Dose-related	
	Predictable from conventional pharmacology	
	Severity variable, but usually mild	
B	*Bizarre*	Achilles tendonitis with quinolone therapy
	Associated with off-target drug effects	
	Uncommon	
	No clear dose relationship	
	Unpredictable from conventional pharmacology	
	Variable severity; proportionately more serious than type A	
Supplemental Types		
C	*Continuing*; time-related; ADR persistence for a long duration	Osteonecrosis of the jaw with bisphosphonate therapy
D	*Delayed*; time-related; ADR of slow onset	Tardive dyskinesia with antipsychotic therapy
E	*End-of-treatment*; associated with dose reduction or therapy discontinuation	Benzodiazepine withdrawal syndrome after abrupt drug cessation
F	*Failure of therapy*; inadequate therapeutic drug action so it does not achieve its intended purpose	Ischaemic stroke second to atrial fibrillation whilst on warfarin

likely varies with the drug, ADR and patient ethnicity [21]. A genetic basis for specific ADRs was first suggested in the 1950s when perturbed drug metabolism was associated with abnormal drug responses, such as butyrylcholinesterase deficiency and prolonged apnoea after succinylcholine administration. Over the last decade there has been a rapid growth in our understanding of ADR pharmacogenetics. This has been facilitated by an increased knowledge of the human genome and its variation through the Human Genome Project and HapMap Projects. Further, advances in genetic technologies and a reduction in processing costs have increased the volume of pharmacogenetic research conducted and its capacity to yield associations. At present, disproportionately more is known about associations of strong individual effect size with specific ADRs. However for some ADRs it may be that the genetic contribution is polygenic with distinct loci of individual low effect size collectively contributing. Despite the advances of the last decade, elucidation of potential complex interplays within and between different biological systems is proving challenging.

The rest of this chapter explores further the pharmacogenetics of immune- and nonimmune-mediated ADRs. A discussion about all known genetic associations is beyond the scope of this chapter and so a range of genetic associations with distinct ADRs have been selected, although the tables included provide additional examples. The selected associations facilitate expansion on the following key themes:

- the effect of specific genetic mutations on protein function,
- the variable extent of genetic contribution to ADRs,
- the pathogenesis of ADRs,
- the clinical application of specific genetic-ADR associations through predictive genotyping and
- the current variable evidence base supporting their use.

Lastly, the many challenges faced by pharmacogenetics in translating an observed genetic-ADR association from the 'bench' to the 'bedside' will be highlighted and contemporary strategies and future possibilities to overcome these obstacles and deepen our understanding of pharmacogenetics will be outlined.

2 Immune-Mediated Adverse Drug Reactions

Immune-mediated ADRs are off-target ADRs and more specifically, represent a form of hypersensitivity reaction. Hypersensitivity reactions can be classified according to the Gell and Coombs system into types I-IV representing IgE-mediated allergic reactions (type I), direct antibody-mediated (type II), immune complex-mediated (type III) and delayed-type hypersensitivity (DTH) reactions (type IV). At the time of writing, comparatively less is known about the pharmacogenetics of type I-III hypersensitivity reactions and therefore this section will concentrate on DTH reactions.

Over the last decade, the increasing use of genome-wide association studies (GWAS) in pharmacogenetic research has identified a growing number of ADRs that are strongly associated with specific human leukocyte antigen (*HLA*) haplotypes, genes and/or alleles. Table 3 provides an overview of *HLA*-ADR associations [22–59].

The *HLA* class I and II genes, located on chromosome 6, are the most polymorphic of the human genome and over 7000 classical alleles have been identified between them [60]. There is strong linkage disequilibrium between the alleles [61]. Classical HLA class I molecules (encoded on 3 loci: *HLA-A, -B, -C*) are expressed on the surface of most nucleated cells and present peptide antigen to the T-cell receptor (TCR) of CD8+ T-cells [62]. The peptides presented by HLA class I molecules are mostly derived from the degradation of intracellular proteins, although

Table 3 Examples of *HLA* associations to hypersensitivity adverse drug reactions

Reaction	Drug	HLA- association(s)	Reference(s)
Hypersensitivity syndrome/DRESS/ DIHS	Abacavir[a]	*B*57:01*	[22, 23]
	Allopurinol[a]	*B*58:01*	[24, 25]
	Carbamazepine	*A*31:01*	[26, 27]
	Nevirapine	*C*08:02-B*14:02* (*Italian*), *C*08* (*Japanese*), *B*35:05* (*Thai*)	[28–30]
Stevens-Johnson syndrome/Toxic epidermal necrolysis	Allopurinol[a]	*B*58:01*	[24, 31]
	Carbamazepine[a]	*B*15:02*[a], *A*31:01*	[32–34]
	Lamotrigine	*B*38*	[35]
	Methazolamide	*B*59:01*	[36]
	Nevirapine	*C*04:01 (Malawian)*	[37]
	Oxicam NSAIDs	*B*73:01*	[35]
	Phenytoin	*B*15:02*	[33, 38]
	Sulfamethoxazole	*B*38*	[35]
Delayed exanthem without systemic features	Allopurinol	*B*58:01* (*Han Chinese*)	[39]
	Aminopenicillins	*A2, DRw52*	[40]
	Carbamazepine	*A*31:01*	[27, 41]
	Nevirapine	*DRB1*01:01* (*French*) *B*35:05* (*Thai*) *C*04* (*Thai*)	[42] [30] [43]
Drug-induced liver injury	Antituberculosis drug therapy	*DQB1*02:01*	[44]
	Co-amoxiclav	*DRB1*15:01-DQB1*06:02*, *A*02:01*	[45, 46]
	Flucloxacillin	*B*57:01*	[47]
	Lapatinib	*DQA1*02:01*	[48]
	Lumiracoxib	*DQA1*01:02*	[49]
	Nevirapine	*DRB1*01*	[50]
	Ticlopidine	*A*33:03* *A*33:03* with *CYP2B6*1H* or **1J*[b]	[51, 52]
	Ximelagatran	*DRB1*07, DQA1*02*	[53]
Agranulocytosis	Clozapine	*DQB1* 6672G > C	[54]
	Levamisole	*B*27*	[55]
Asthma	Aspirin	*DPB1*03:01*	[56, 57]
Pneumonitis	Gold	*B*40, DRB1*01*	[58]
Proteinuria, Thrombocytopaenia	Gold	*DRB1*03*	[59]
Urticaria	Aspirin	*DRB1*13:02-DQB1*06:09*	[56]

DRESS drug reaction with eosinophilia and systemic symptoms, *DIHS* drug-induced hypersensitivity syndrome, *NSAID* non-steroidal anti-inflammatory drug

[a] odds ratio > 50 and reproduced in > 1 study. Adapted from Phillips et al. [78]

[b] *CYP2B6* is not an *HLA* gene

HLA class I molecules on specific dendritic cell subsets are additionally capable of presenting extracellular peptides through 'cross-presentation' [63]. Classical HLA class II molecule expression (encoded on 3 loci: *HLA-DP, -DQ, -DR*) is restricted to professional antigen-presenting cells (e.g. dendritic cells, macrophages, B-cells) and they present extracellular-derived peptides to the TCR of CD4+ T-cells [62]. *HLA* polymorphisms localise to the sequence motifs that encode residues of the peptide-binding groove [60, 64]. These polymorphisms alter the stereochemistry of pockets within the groove, creating individual HLA allotypes with distinct peptide-binding portfolios [62, 65]. The HLA system is integral to the development of T-cell tolerance to 'self' and to the development of adaptive immunity in response to 'non-self' peptide. *HLA* incompatibility is also known to be important in the pathogenesis of allogeneic transplant rejection and several HLA associations have been previously reported for autoimmune diseases including ankylosing spondylitis (with *HLA-B27*) and rheumatoid arthritis (e.g. with *HLA-DRB1* alleles [66]).

Most of the hypersensitivity ADRs with *HLA* associations, including the specific reactions to abacavir, carbamazepine, allopurinol and flucloxacillin discussed below, are considered DTH reactions. In keeping with DTH reactions, they normally present ≥72 h after drug exposure, may resolve with drug cessation and often re-present more rapidly and with a more severe phenotype following drug re-exposure. A T-cell mediated immunopathogenesis is thought to underlie this temporal pattern. Analogous to the development of pathogen-induced adaptive immune responses, it is thought that a T-cell clone(s) can be primed by presentation of culprit antigen on an HLA molecule during primary drug exposure and effector memory T-cells are rapidly activated with secondary exposure [62, 67, 68]. The isolation of drug-specific T-cells from patients that have suffered DTH ADRs supports T-cell involvement [69, 70].

Two hypotheses have conventionally been proposed to describe potential off-target pharmacodynamic processes that may lead to the neo-antigen formation necessary for DTH drug-specific T-cell development: the hapten (or pro-hapten) model and the pharmacologic interaction with immune-receptors (p-i) model [71]. The hapten model proposes that drugs and their metabolites are too small to be independently immunogenic and so covalently bind to self-protein and the resulting *de novo* hapten-self peptide adduct is antigenic [71, 72]. The p-i hypothesis proposes that drugs may interact directly with HLA molecules, without specific self-peptides, to elicit a T-cell response [73]. Regardless of the mechanism of neo-antigen formation, it is widely assumed that additional 'danger' signals are required to overcome the immune system's default tolerance and permit generation of an adaptive immune response. This concept is referred to as the 'danger hypothesis' [74]. Amongst the other key themes of this chapter, the following ADR examples illustrate how prior understanding of genetic susceptibility can facilitate elucidation of underlying mechanisms of antigen formation and presentation.

2.1 *HLA-B*57:01* and Abacavir Hypersensitivity Syndrome

Abacavir represents the epitome of translational pharmacogenetics as the loop from laboratory observation to improved patient care for the genetic association between *HLA-B*57:01* and the abacavir hypersensitivity syndrome (AHS) has been closed [75]. Abacavir is a nucleoside reverse transcriptase inhibitor indicated to treat HIV and is prescribed as a constituent of highly active antiretroviral treatment (HAART). AHS occurs in 2.3–9 % of patients [76] with a median time to onset of 8 days therapy [77]. The clinical diagnostic criteria require ≥2 of: fever, rash, nausea, vomiting, arthralgia, myalgia, headache, lethargy or gastrointestinal symptoms and importantly, onset must occur within 6 weeks of commencing therapy and remit within 72 h of abacavir cessation [76]. Unlike other drug hypersensitivity reactions, the mild to moderate rash is not a consistent feature [67] and eosinophilia is unusual [78]. Although the initial reaction is unpleasant, the significant morbidity and mortality occurs upon rechallenge [67, 78], consistent with a DTH reaction.

In 2002, two groups independently reported an association between AHS and *HLA-B*57:01* [22, 23] and subsequent further observational research confirmed the association [79, 80]. The Prospective Randomised Evaluation of DNA Screening in a Clinical Trial (PREDICT-1) study was a multicentre, double-blind randomised controlled trial (RCT) that demonstrated pre-therapy *HLA-B*57:01* screening significantly decreased the incidence of AHS [77]. Briefly, 1956 patients were enrolled and randomised on a 1:1 basis. The interventional group received pre-therapy *HLA-B*57:01* genotyping and either HAART with abacavir for *HLA-B*57:01* negative patients or HAART without abacavir for *HLA-B*57:01* positive patients. The control group received HAART with abacavir and retrospective *HLA-B*57:01* genotyping from blood samples taken pre-therapy. All participants with clinically diagnosed hypersensitivity reactions underwent skin patch testing for immunological corroboration to improve the specificity for the hypersensitivity phenotype. The study demonstrated that avoiding abacavir in *HLA-B*57:01* positive patients in the prospective screening interventional group eliminated immunologically confirmed hypersensitivity reactions (0 vs. 2.7 % in control group, $p<0.001$) with positive and negative predictive values of 47.9 % (PPV) and 100 % (NPV), respectively [77]. An estimate of the number needed to screen (NNS) to prevent one case of AHS, given an *HLA-B*57:01* carriage prevalence of 6 %, was ~25 [77]. However 84 % of participants were Caucasian, limiting generalisation. The Study of Hypersensitivity to Abacavir and Pharmacogenetic Evaluation (SHAPE) was a retrospective case-control study that addressed this and demonstrated that *HLA-B*57:01* has 100 % sensitivity for immunologically confirmed AHS in both US White and Black patients [81]. Pharmacoeconomic evaluations have demonstrated a cost effectiveness to pre-prescription *HLA-B*57:01* screening [80, 82, 83]. Observational data from open-screening studies has addressed practical matters of implementation [84–86] and shown genotyping to reduce the frequency of abacavir discontinuation due to clinically suspected as well as true immunological hypersensitivity reactions [85, 87]. This is most likely because the former clinician strategy of over-diagnosing AHS to ensure high sensitivity to avoid AHS maleficence at the expense of lower specificity [78] is no longer required given the exclusivity of the association between *HLA-B*57:01* and AHS. In accordance with the substantial evidence base, the drug label has been updated and clinical guidelines either mandate or strongly recommend prospective screening [76]. To summarise, abacavir represents a pioneering example of ADR translational pharmacogenetics and has charted a course that other genetic-ADR associations might follow from initial observations to a RCT to studies that address generalisation, pharmacoeconomics and applicability in widespread clinical practice.

Identifying the genetic basis for AHS has directly benefitted patient care but until recently, insight into the underlying immunopathogenesis has been limited. However 3 recent independent studies have begun to expose the pharmacodynamic off-target molecular mechanisms [88–90]. Native abacavir can bind non-covalently with exquisite specificity to *HLA-B*57:01* at the base of its peptide-binding groove, extending into the deep F pocket [88, 89]. The specificity for the interaction is accounted for by the F-pocket architecture and in particular residue 116 [88]. In the absence of abacavir, the C-terminus of peptides that bind to the F-pocket of *HLA-B*57:01* have large hydrophobic residues, such as tryptophan and less commonly phenylalanine [89]. In the presence of abacavir, the peptide repertoire of *HLA-B*57:01* shifts, so that around 20–25 % of recoverable peptides are novel [88] and have alternative residues including isoleucine or leucine at their C-terminus [88, 90]. In essence, abacavir alters the stereochemistry of *HLA-B*57:01* to create an HLA neo-allotype with a novel peptide portfolio. This model of antigen presentation is distinct from both the hapten and conventional p-i models. It is proposed that T-cells will not have been exposed to the novel range of *HLA-B*57:01*-restricted peptides during thymic maturation and so will lack tolerance. The formation of memory T-cells will lead to systemic AHS that is more deleterious upon abacavir re-exposure. In support of the large peptide shift and subsequent large array of 'altered immunological self', the observed CD8+ T-cell response is polyclonal [65]. Further, the effector T-memory cells from *HLA-B*57:01* positive patients with a clinical history of AHS respond preferentially in the presence of specific peptide and abacavir together rather than to peptide or abacavir alone [89]. This indicates that the memory T-cell response to self-peptide requires abacavir for efficient presentation.

Although the exact intracellular site(s) where abacavir associates with HLA-B*57:01-peptide complexes is currently unclear, there is evidence to suggest the endoplasmic reticulum [65]. However, a minority of T-cell clones *in vitro* appear to react to abacavir too quickly to be explained by *de novo* HLA-B*57:01-novel peptide assembly [91]. This suggests that abacavir may additionally bind to HLA-B*57:01-native peptide complexes already present on the cell surface, possibly distorting their stereochemistry [65]. This mechanism is in keeping with the conventional p-i hypothesis. An inadequately resolved question is why the PPV of *HLA-B*57:01* for AHS is < 50 % [77]. Postulated mechanisms to account for this include (a) the inter-individual polygenic influence on the novel peptide portfolio itself [89]; and (b) heterologous immunity as a result of pre-existing viral infections, in keeping with data which show that hypersensitivity reactions are often associated with reactivation of viruses such as Epstein-Barr virus [92].

2.2 *HLA-B*15:02, HLA-A*31:01 and Carbamazepine Hypersensitivity*

Carbamazepine is indicated in the treatment of epilepsy, trigeminal neuralgia and bipolar affective disorder but in up to 10% of patients, it can provoke a cutaneous ADR [93]. Drug-induced skin injury (DISI) encompasses a spectrum of manifestations and can be caused by a diverse range of drugs including anticonvulsants, allopurinol and β-lactam antibiotics [94]; Table 3 lists drugs with known associations between DISI and genetic variants. There exists both inter- and intra-drug heterogeneity in DISI presentation but fortunately most reactions are mild [94]. Standardising phenotypic definitions for serious DISI conditions has been challenging and required an international collaborative approach. Carbamazepine itself can cause DISI ranging from mild maculopapular exanthema (MPE) of increasing severity to the hypersensitivity syndrome (HSS), also referred to as drug reaction with eosinophilia and systemic symptoms (DRESS) or drug-induced hypersensitivity syndrome (DIHS) [94], to the distinct Stevens-Johnson syndrome (SJS) and toxic epidermal necrolysis (TEN) [95].

HSS/DRESS/DIHS (herein referred to as HSS) and SJS-TEN represent severe cutaneous adverse reactions (SCARs) [96]. HSS is a multisystem disorder that carries a mortality rate of 10% [96]. HSS can be diagnosed by the presence of at least 3 of: cutaneous involvement, internal organ involvement, fever, lymphadenopathy and eosinophilia (and/or atypical lymphocytes) with at least 1 of the first 2 criteria listed being present [94]. The skin manifestation is most commonly an exanthematous eruption and the internal organ involvement includes hepatic dysfunction and interstitial nephritis [94, 97]. SJS and TEN represent different severities along a spectrum of the same disease and are characterised by epidermal detachment involving the skin and mucous membranes with systemic manifestations including fever, intestinal and pulmonary involvement [94, 98]. SJS is diagnosed when epidermal detachment affects $\leq 10\%$ of the body surface area, TEN is diagnosed when $>30\%$ is affected and an overlap syndrome exists for 10–30% epidermal detachment [99]. SJS and TEN have estimated mortality rates of up to 5 and 50%, respectively [98].

Genetic association studies have shown *HLA-B*15:02* to be a susceptibility factor for carbamazepine-induced SJS-TEN in people of Han Chinese descent [32] and certain other Asian ethnicities including Thai, Malaysian and Indian [33, 100–102]; a meta-analysis of studies with Asian patients derived a pooled odds ratio (OR) of 113.4 (95% confidence interval (CI) 51.2–251.0) [34]. A recent open-label prospective study in Taiwan has demonstrated the beneficence of pre-therapy *HLA-B*15:02* screening in Han Chinese patients and reported no cases of SJS-TEN in both the carbamazepine-taking *HLA-B*15:02* negative cohort and *HLA-B*15:02* positive cohort administered alternative medication or advised to continue their pre-study medication [103]. Due to ethical and sample size considerations, an estimated historical annual incidence of carbamazepine SJS-TEN (0.23%) was used as the comparator rather than a prospective non-screened control group (with appropriate blinding), unlike the PREDICT-1 study for AHS [77]. Furthermore this genetic correlation is both phenotypically restricted to SJS-TEN [34, 41] and ethnically restricted: it has not been reproduced in Europeans [104, 105] or other specific Asian populations including South Koreans [26] and Japanese [106–108]. The reason(s) for the latter are not clear but allelic frequency should be considered since *HLA-B*15:02* is present in 8.6% of the Han-Chinese population but in < 1% of Europeans, Koreans and Japanese [26, 109], reducing the power of studies in these populations to detect statistically significant associations [109].

A more recently reported carbamazepine DISI association is with *HLA-A*31:01* and importantly, it is associated with MPE, HSS and SJS-TEN and is present in multiple diverse ethnicities, including Han Chinese [41], Koreans [26], Japanese [107] and Europeans [27]. Interestingly, the allele frequency of *HLA-A*31:01* for these ethnic groups is 1.8, 10.3, 9.1 and 2–5%, respectively [26, 109]. The magnitude of the association though is smaller than with *HLA-B*15:02*, with a meta-analysis pooled OR for *HLA-A*31:01* of 9.5 (95% CI 6.4–13.9) [34]. However, the estimated NNS to prevent one case of carbamazepine DISI with *HLA-A*31:01* is lower and ranges from 47–67 depending on ethnicity. This is in contrast to the estimated NNS of 461 Asian patients with *HLA-B*15:02* to specifically prevent one case of SJS-TEN [34]. This difference in NNS is largely attributable to the higher incidence of ADRs (up to 10% [93]) that are associated with *HLA-A*31:01* compared to *HLA*B15:02* (circa 0.23% [103]), due to the relationship of *HLA-A*31:01* with a broader range of phenotypes. These findings form a credible foundation for a future prospective study to assess the clinical benefit of pre-therapy *HLA-A*31:01* screening.

It has been shown that carbamazepine can non-covalently associate with HLA-B*15:02 molecules and in the presence of carbamazepine, there is a shift in the peptide repertoire of HLA-B*15:02 with a novel preference for smaller residues at the 4th and 6th peptide positions and an increase in hydrophobic residues at several positions [88]. The scale of this peptide shift is smaller (approximately 15%) than for abacavir [88]. This off-target pharmacodynamic effect does resonate with the novel model proposed for abacavir and *HLA-B*57:*01, but overall the mechanisms leading to carbamazepine hypersensitivity ADRs are likely to be more complex. This is because firstly, it has been shown that carbamazepine and its metabolite, carbamazepine-10,11-epoxide, can associate with other structurally related *HLA-B75* family members [110]. Secondly, the availability of a restricted range of T-cell clonotypes with specific TCR rearrangements has been demonstrated to be an important determinant in the pathogenesis of *HLA-B*15:02*-associated carbamazepine-induced SJS-TEN [111]. This indicates that both the TCR repertoire and *HLA* genotype modulate the risk of carbamazepine-induced SJS-TEN and likely explains why some *HLA-B*15:02* carriers tolerate carbamazepine [109]. Further research is still required to understand other complicating observations. These include how two seemingly disparate *HLA* alleles, *HLA-B*15:02* and *HLA-A*31:01*, are linked to carbamazepine ADRs and furthermore, how *HLA-A*31:01* can be associated with multiple phenotypes.

2.3 HLA-B*58:01 and Allopurinol Hypersensitivity

Allopurinol, an analogue of hypoxanthine, inhibits xanthine oxidase (XO) and is indicated in the management of gout and other hyperuricaemic conditions including tumour lysis syndrome. Although generally well tolerated, ~2–3 % of patients suffer mild hypersensitivity reactions including MPE [39, 112] and crucially, ~0.1–0.4 % of patients develop HSS or SJS-TEN [112]. The SCARs, HSS and SJS-TEN, normally present within weeks to months of commencing allopurinol but may take considerably longer [112]. Allopurinol is a major cause of SCARs [113, 114] and the combined mortality from allopurinol-induced SCARs approaches 25 % [112].

In 2005, a strong genetic association between *HLA-B*58:01* and allopurinol-induced SCARs (HSS/SJS-TEN) was described in the Taiwan Han-Chinese population (OR 580.3, 95 % CI 34.4–9780.9); all 51 cases (100 %) carried *HLA-B*58:01* in comparison to only 20 of 135 allopurinol-tolerant controls (15 %) [24]. This association has been replicated in Thai [31], Korean [25], European [35] and Japanese patients [115], although its magnitude was more modest for the latter 3 ethnic groups, possibly reflecting the lower prevalence of *HLA-B*58:01* in these populations. A meta-analysis has confirmed the association between allopurinol-induced SJS-TEN and *HLA-B*58:01* in both Asian and non-Asian patients compared to allopurinol-tolerant controls (combined OR 96.6, 95 % CI 24.5–381.0) [116]. Based on data from the Han Chinese and Thai populations, current estimates for the PPV and NPV for *HLA-B*58:01* are ~1.5 and 100 %, respectively [112], although these values will be lower for other ethnic groups. An exciting development is the ongoing prospective study in Taiwan to assess the clinical benefit of pre-therapy genotyping for *HLA-B*58:01* prior to commencing allopurinol [112]. Similarly to the prospective study discussed above for *HLA-B*15:02* screening prior to initiating carbamazepine [103], an estimated historical ADR incidence is being used for the control. At the time of writing, no results from this study have been published.

Interestingly, unlike carbamazepine and *HLA-A*31:01*, it is less clear at the current time whether *HLA-B*58:01* also predisposes to MPE in patients taking allopurinol. A study in Australia demonstrated no association between *HLA-B*58:01* and MPE [117], but a study of Han-Chinese patients in mainland China reported *HLA-B*58:01* as a risk factor for both allopurinol-induced MPE and SCARs [39]. More research into this area is required, but one hypothesis from the available literature is that the risk of MPE with *HLA-B*58:01* may be ethnically-restricted. If this association is confirmed, it will increase the PPV further for the affected ethnic groups and so augment the potential utility of pre-therapy *HLA-B*58:01* screening in these groups to reduce the burden of allopurinol-induced ADRs.

The exact underlying mechanism(s) by which allopurinol, or its long-circulating active metabolite oxypurinol, interact with *HLA-B*58:01* for the generation of drug-specific T-cells has yet to be elucidated. However, as the PPV of *HLA-B*58:01* for SCARs is low (~1.5%) [112], this alludes to other contributing factors in their pathogenesis. It has long been thought that viruses play a role in drug hypersensitivity and there is increasing recognition that the reactivation of herpes viridae is important in the aetiology of HSS [92, 118]. However, any interaction(s) between allopurinol and/or oxypurinol and viruses is poorly understood. Prior to the discovery of *HLA-B*58:01*, several non-genetic risk factors were espoused including renal dysfunction, higher allopurinol doses, diuretic use and concomitant antibiotic therapy [112]. Although verification of these variables is difficult as SCARs are fortunately rare events, patients on allopurinol with renal insufficiency have been shown to be almost 5 times more likely to develop SCARs [24]. In addition, patients on a daily dose of ≥200 mg allopurinol seem to be at an increased risk of SJS-TEN compared to lower doses [113]. By assimilation of these 2 observations, it can be hypothesised that increasing the plasma concentration of allopurinol and/or oxypurinol increases the risk of drug-specific T-cell development [109]. To mitigate the risk of ADRs the dose of allopurinol could be reduced, but it is well established that the most commonly used doses of allopurinol (≤300 mg daily) are frequently ineffective already for the long term treatment of hyperuricaemia in gout [119].

Nevertheless, *HLA-B*58:01* is the single largest predictor of allopurinol-induced SCARs and this makes genetic screening appealing to directly prevent *HLA-B*58:01*-associated SCARs. In addition, it is conceivable that a successful genetic screening programme may indirectly improve the overall benefit: harm ratio of allopurinol further. This is because genotyping may empower clinicians to titrate allopurinol doses up to optimise efficacy in *HLA-B*58:01* negative patients with normal renal function. However, any benefit derived from genetic screening in the ongoing Taiwan study will require follow up studies to determine the extent of generalisation. This is because for other ethnic groups and especially Europeans, allopurinol-induced SCARs also occur in *HLA-B*58:01* negative patients. Furthermore, the identification of other (non)-genetic risk factors may be required to improve the PPV of the test, as currently many *HLA-B*58:01* patients will be unnecessarily denied allopurinol in place of other urate-lowering therapies, with unmeasured effects as yet on cost-effectiveness and treatment efficacy.

2.4 *HLA-B*57:01* and Flucloxacillin-Induced Liver Injury

Flucloxacillin is a narrow-spectrum beta-lactam antibiotic indicated in Gram-positive bacterial infections and in particular, is used to treat non-methicillin resistant *Staphylococcus aureus* infections. In approximately 8.5 per 100,000 patients treated with flucloxacillin, cholestatic liver injury occurs [120]. Drug-induced liver injury (DILI) is associated with a structurally disparate range of drugs but notably these include non-steroidal anti-inflammatory drugs (NSAIDs) and certain antimicrobials including flucloxacillin [121]. Although rare, drug-induced liver injury (DILI) can be severe and accounts for up to 15% of all cases of acute hepatic failure [122–124].

Analogous to DISI, the type and severity of DILI vary between causative drugs and for a given drug, presentation is variable [121]. Consequently, standardising the DILI phenotype is not straightforward but the diagnosis can be made from clinical, biochemical and histopathological parameters [125].

The aetiology of DILI can be divided into immune- and nonimmune-mediated processes [121]. Pharmacogenetic associations with DILI have now been identified for several drugs and the associated genetic variants reflect both immune and non-immune aetiologies (see Tables 3 and 4, respectively for examples). However, DILI can be difficult to categorise by this means. This is because, although recognition of clinically suggestive features of hypersensitivity is relatively easy, the absence of such features, such as eosinophilia, does not preclude immune system involvement [126].

To date, the strongest DILI genetic association described is between flucloxacillin and *HLA-B*57:01* with an OR of 80.6 (95% CI 22.8–284.9) [47]. This is intriguing as the *HLA-B*57:01* allele is also strongly associated with AHS, yet AHS rarely involves hepatitis [78].

Unlike AHS, it is improbable that this association will lead to a screening test for clinical practice because, despite an adequate estimated sensitivity and specificity (84 and 94%, respectively) [47], the rarity of flucloxacillin-induced liver injury diminishes the PPV to 0.12% [127]. An estimate of the NNS to prevent one flucloxacillin-induced liver injury case is 13,513 and the screening approach would unnecessarily deny almost 7% of patients first line flucloxacillin therapy, with unmeasured adverse effects on infectious disease treatment efficacy and cost effectiveness [127]. However genetic testing may help establish the diagnosis of flucloxacillin-induced liver injury when the underlying cause of liver dysfunction is unclear [47].

The off-target pharmacodynamics that underpin flucloxacillin-induced liver injury are being unravelled. Flucloxacillin can adduct covalently to proteins to form neo-antigen drug-protein conjugates and specific flucloxacillin-modifiable lysine residues on albumin, the major circulating protein, have been identified [128]. It is predicted that several albumin-derived peptides containing flucloxacillin-modifiable lysine residues have high-affinity for HLA-B*57:01 [129] and could be presented on HLA-B*57:01 by professional antigen presenting cells through cross-presentation. Flucloxacillin-responsive CD4+ and CD8+ T-cells have been characterised *in vitro* from patients who have previously suffered flucloxacillin cholestatic liver injury and their activation is dependent on peptide processing. *In vitro*, the CD8+ T-cell activation is restricted to HLA-B*57:01 and the very similar allotype, HLA-B*58:01 [129]. The proposed model, which aligns with the hapten hypothesis, suggests that immunogenic peptide neo-antigens are derived from natural processing of flucloxacillin-protein conjugates and can be presented on HLA-B*57:01 to generate an adaptive immune response [129]. Interestingly, besides this proposed alternative mechanism of neo-antigen formation, this immunopathogenesis differs to that of abacavir in at least 2 ways. Firstly, CD4+ as well as CD8+ flucloxacillin-responsive T-cells have been cloned from patients and secondly, the T-cell clones

Table 4 Examples of associations between adverse drug reactions and nonimmune-related genetic variants

Reaction	Drug	Gene association(s)	Variant(s)	Reference(s)
i) Drug metabolizing enzyme and drug transporter variants				
Increased risk of bleeding	Clopidogrel	*CYP2C19*	*17	[130, 131]
	Warfarin	*CYP2C9*	*3	[132–135]
Increased risk of opioid toxicity	Codeine	*CYP2D6*	Ultrarapid metabolisers	[136, 137]
	Tramadol	*CYP2D6*	Ultrarapid metabolisers	[138, 139]
Drug-induced liver injury	Antituberculosis drug therapy	*NAT2*	Slow acetylator	[140–142]
		GSTM1	null/null	
		CYP2E1 (East Asians)	*1A/*1A	
	Diclofenac	*UGT2B7*	*2	[143]
		ABCC2	rs717620	
		CYP2C8	Different haplotypes	
	Tacrine	*GST T1* *GST M1*	Double null/null	[144]
	Troglitazone	*GST T1* *GST M1*	Double null/null	[145]
Diarrhoea, neutropaenia	Irinotecan	*UGT1A*	*Poor metabolisers*	[146]
Drug discontinuation	Risperidone	*CYP2D6*	Poor metabolisers	[147]
Muscle toxicity	Simvastatin	*SLCO1B1*	rs4149056	[148]
Myelosuppression	Azathioprine, 6-mercaptopurine, thioguanine	*TPMT*	Poor metabolisers	[149]
Peptic ulcer disease	NSAIDs	*CYP2C19*	*17	[150]
Prolonged apnoea	Succinylcholine, mivacurium	*BCHE*	rs1799807, other variants	[151]
Stent thrombosis	Clopidogrel	*CYP2C19*	*2	[152]
Therapy-induced toxicity[a]	5-fluorouracil/ capecitabine	*DPD*	rs3918290, rs55886062, rs67376798	[153]
ii) Other variants				
Drug-induced liver injury	Methotrexate	*MTHFR*	rs1801133	[154]
Malignant hyperthermia	Halogenated inhalation anaesthetics	*RYR1*	rs118192163 >30 other variants	[155, 156]
		CACNA1S	rs1800559 rs80338782	[157, 158]

Table 4 (continued)

Reaction	Drug	Gene association(s)	Variant(s)	Reference(s)
i) Drug metabolizing enzyme and drug transporter variants				
Metabolic syndrome	Clozapine, risperidone	*5HTR2C*	rs1414334	[159]
Nonimmune haemolytic anaemia	Primaquine, dapsone, methylene blue, others	*G6PD*	Mediterranean, A-(202A), >150 other variants	[160, 161]
Therapy-induced toxicity[a]	5-fluorouracil/ capecitabine	*TYMS*	rs45445694	[162]

NSAID non-steroidal anti-inflammatory drug

[a] Toxicity from 5-fluorouracil-based therapy includes diarrhoea, mucositis, nausea, neutropaenia

show cross-reactivity *in vitro* with other commonly prescribed beta-lactam antibiotics including amoxicillin and piperacillin [129].

In summary, this section illustrates that immune-mediated DTH reactions are an emerging prominent type of off-target ADR with the potential for significant morbidity and mortality. However, pharmacogenetics has been pivotal in reducing the healthcare burden associated with abacavir, may have important future roles in the prevention of carbamazepine and allopurinol DISI and is facilitating elucidation of underlying immune-mediated aetiologies.

3 Nonimmune-mediated Adverse Drug Reactions

Nonimmune-mediated ADRs are a heterogeneous group in aetiology and presentation. However, over the last decade it has been increasingly recognised that susceptibility to many nonimmune ADRs is associated with gene variants of drug metabolising enzymes (DMEs) and less frequently, with drug transporters. It is thought that perturbed pharmacokinetics increases the availability of drug/metabolite(s) at the target site(s), increasing the likelihood of developing an ADR. The sites that mediate nonimmune ADRs include both on-target and off-target sites. On-target ADRs manifest through excessive drug/metabolite(s) action either at the therapeutic target site or at the same molecular site located in other tissues. The latter occurs for instance with NSAID-induced upper gastrointestinal ADRs.

It is important to note that, although the majority of ADRs with a genetically-influenced pharmacokinetic-mediated susceptibility found to date are nonimmune ADRs, perturbed pharmacokinetics is also relevant in the genesis of a few immune-mediated ADRs. This was described earlier for the case of allopurinol-induced SCARs and non-genetic pharmacokinetic factors. Furthermore, genetic susceptibility to ticlopidine-induced hepatotoxicity has been demonstrated to be greatest in patients with *HLA-A*33:03* in combination with variants of a DME (Table 3).

In the following section, the effects of gene variants of phase I and phase II biotransformation enzymes on susceptibility to ADRs will be discussed in the context of codeine/warfarin and azathioprine, respectively. Then, the effects of gene variation for a drug transporter will be illustrated for statin-induced muscle toxicity. However as the first example of malignant hyperthermia shows, genetic susceptibility to nonimmune-mediated ADRs can occur through plausible pharmacodynamic mechanisms too. Table 4 lists examples of ADRs associated with nonimmune-related genetic variants.

3.1 *RYR1* and Anaesthesia-Induced Malignant Hyperthermia

In 1962, a paper was published about a pedigree that contained 10 relatives who had unfortunately and unexpectedly died during or shortly following general anaesthesia [163]. The deaths were associated with core body temperatures, when measured, in excess of 41 °C and followed an autosomal dominant inheritance pattern [163]. Other pedigrees have since been described [164, 165] and over 500 cases of malignant hyperthermia (MH) have now been reported in the medical literature [166].

MH is precipitated by volatile anaesthetics in genetically susceptible individuals. All halogenated inhalation anaesthetics have been implicated including halothane, isoflurane, sevoflurane and desflurane [167]. The depolarising neuromuscular blocker, succinylcholine, augments the adverse response to these potent inhalation anaesthetics but its role as an independent precipitant of fulminant MH is controversial [167, 168]. Rarely, non-pharmacological stressors including environmental heat [169, 170], infections [170] and severe exercise or emotional strain [171] have been implicated in MH-like episodes.

The incidence of anaesthetic-induced MH is approximately 1 per 50,000 adults and 1 per 15,000 paediatric patients [172] and it occurs in all ethnic groups [173]. The basis of MH is hypermetabolism which can present as tachypnoea, a rise in end-tidal carbon dioxide exhalation, tachycardia, cyanosis, cardiac arrhythmias, skeletal muscle rigidity, hyperthermia [174], convulsions and eventual death [163]. Associated electrolyte complications include acidosis, hyperkalaemia, elevated creatine kinase (CK) and acute kidney injury (AKI) [174]. Timely intervention improves prognosis [175]. However, an early diagnosis of MH can be challenging as the initial clinical signs are nonspecific and variable in their time course, making them easily mistaken for other pathologies (e.g. sepsis, thyrotoxic crisis) [176]. Nevertheless, the mortality from MH has dramatically fallen from 70 % in the 1970s [169] to <5 % today [173]. This reduction has been aided by the introduction of the muscle relaxant dantrolene for treatment of suspected MH [166] and testing for susceptible relatives (see later) [169]. A clinical grading scale has been introduced to help researchers retrospectively assess the likelihood of MH following an adverse anaesthetic event, which enables accurate phenotyping and determination of future susceptibility [177].

RYR1 is located on chromosome 19 and encodes ryanodine receptor 1 (RyR1). There are 3 RyRs isoforms (RyR1–3) and each forms a homotetrameric assembly within the endoplasmic (or sarcoplasmic) reticulum and functions as a Ca^{2+} channel [178]. They have evolved into the largest ion channels found to date (~2.2MDa) [179]. This is undoubtedly to facilitate tight channel regulation through interaction with numerous regulatory small molecules and proteins, which is important as Ca^{2+} is a potent intracellular mediator of several cell processes [179]. RyR1 is widely expressed in skeletal muscle and is pivotal to excitation-contraction coupling [180]. RyR1 opens in response to nerve impulses and releases Ca^{2+}, from the sarcoplasmic reticulum where it has been sequestered, into the cytoplasm to drive muscle contraction. It is thought that a direct physical connection exists between the voltage-gated Ca^{2+} channel $Ca_V1.1$ (the skeletal dihydropyridine receptor) in the transverse tubule and RyR1 [181, 182], which induces conformational changes that open RyR1 when the wave of depolarisation from the neuromuscular endplate is detected by $Ca_V1.1$ [183].

Approximately 70% of MH susceptible families carry *RYR1* variants [156]. A nonsynonymous mutation of *RYR1* was found to cause the porcine stress syndrome in inbred pigs, which is an animal model of MH [184]. The analogous C1843T mutation in humans was subsequently identified in an analysis of 1 of 35 MH susceptible pedigrees [185]. Currently, over 200 *RYR1* mutants have been described and most are single nucleotide polymorphisms (SNPs), but only 31 have been designated as causative of MH according to the specific criteria set out by the European Malignant Hyperthermia Group (EMHG) [155].

Impaired Ca^{2+} homeostasis underlies the pathogenesis of MH [173]. Gain-of-function *RYR1* mutations have been shown *in vitro* to lead to RyR1 hyperactivation [186, 187]. The increase in intracellular Ca^{2+} concentration results in sustained muscle contraction and heat generation [173]. Attempts to restore the Ca^{2+} balance and the contracting muscle filaments deplete the cell of adenosine triphosphate resulting in muscle rigidity, loss of integrity to the sarcolemma and leakage of intracellular contents (e.g. K^+, myoglobin) out into the extracellular fluid predisposing to systemic sequelae [172]. However, the exact mechanism(s) by which volatile anaesthetics precipitate this potentially fatal cascade has not been clearly elucidated [188].

Interestingly, ~20% of patients have undergone previous uneventful general anaesthesia with potent inhalation agents before experiencing MH [166]. The reasons for this incomplete penetrance are not fully understood but hypotheses include dose and/or duration dependency effects of the volatile anaesthetic agents [167], the ambient temperature and the simultaneous use of possible mitigating drugs [173].

The majority of MH occurs in asymptomatic individuals and they are considered to have a genetically-determined subclinical myopathy [176]. However, there are at least 3 rare clinical myopathies likely associated with MH susceptibility: central core disease (CCD), multiminicore disease (MmD) and King-Denborough syndrome [189, 190]. Within each syndrome there is clinical, genetic and histological variability and considerable overlap exists, in particular, between CCD and MmD [189]. Importantly, the majority of CCD cases are associated with *RYR1* variants and furthermore, *RYR1* mutations have also been found in cases of MmD [189] and King-Denborough syndrome [191], although these links are less certain [192]. Of the 200 *RYR1* variants, ≥150 are associated with MH alone (subclinical myopathy), ~100 with CCD and ≥20 with both MH and CCD [192]. *RYR1* alleles are also implicated in instances of exercise-induced rhabdomyolysis [193, 194]. Clearly, *RYR1* is involved in a spectrum of muscle disorders, but at the present time the degree of genotype to phenotype concordance is incompletely understood.

The gold standard for MH diagnosis in patients and unaffected relatives is the *in vitro* muscle biopsy contracture test (IVCT), which assesses muscle contraction in response to caffeine and halothane [195]. However, the IVCT is invasive, costly and confined to specialist centres. Therefore, genetic testing has been increasingly used since 2001 [156] to determine MH susceptibility in family members of MH patients that have been shown to carry a causative *RYR1* mutation, as classified by the EMHG [196]. A relative not carrying the familial *RYR1* mutation should still undergo an IVCT though as the absence of a *RYR1* mutation does not exclude MH susceptibility [176].

~75% of MH events occur in patients with no reported family history [166] and therefore universal pre-anaesthetic genetic screening is appealing. However, genetic screening for MH is currently untenable, due to the heterogeneous and incompletely understood genetics underpinning MH susceptibility. The complexity of the RyR1 molecule makes structural and functional predictions of *RYR1* variants challenging [197] and regardless, 30% of MH cases are not associated with *RYR1*. At least 5 other genetic loci have been implicated [172] but of these to date, only nonsynonymous SNPs in *CACNA1S*, the gene encoding the α1 subunit of $Ca_V1.1$, have been linked to MH and in only 1% of cases [157, 158, 172].

In summary MH is a potentially fatal disorder with a strong genetic predisposition, although the full spectrum of genetic risk variants and associated genotype-phenotype correlations are incompletely characterised. However, this strong genetic susceptibility lends itself to the future prospect of successful genetic screening to reduce the incidence of drug-induced MH.

3.2 CYP2D6 and Codeine Analgesia and Safety

Codeine is a weak opioid that is indicated for analgesia in mild to moderately severe pain and as an antitussive and anti-diarrhoeal agent. Although it has been used for many years, recent concerns are mounting over its variable efficacy and safety.

Figure 1 shows the principal pharmacokinetic pathways for codeine. Codeine is considered a prodrug whose function is derived from conversion into 2 active metabolites: morphine and morphine-6-glucuronide (M6G). Both are agonists for the widespread μ-opioid receptor, which is largely responsible for the therapeutic effects and opioidergic ADRs [198, 199]. The affinity of morphine for μ-opioid receptors is 200-fold stronger than compared to codeine [200]. The polymorphic cytochrome 2D6 enzyme (CYP2D6) catalyses the O-demethylation of codeine into

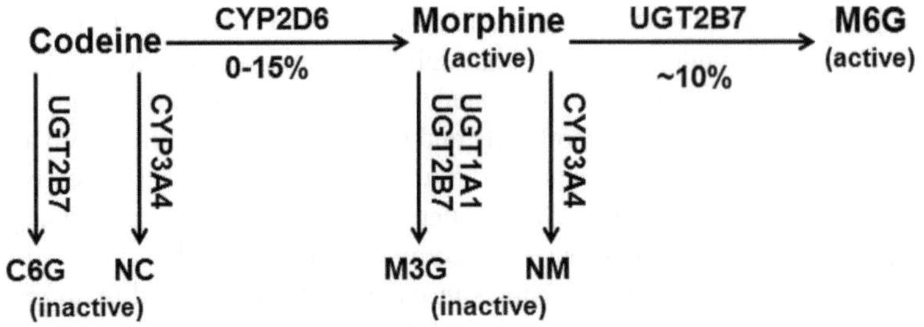

Fig. 1 Codeine metabolism. *C6G* codeine-6-glucuronide, *NC* norcodeine, *M6G* morphine-6-glucuronide, *M3G* morphine-3-glucuronide, *NM* normorphine, *CYP2D6* cytochrome P450 2D6, *CYP3A4* cytochrome P450 3A4, *UGT2B7* uridine diphosphate glucuronosyltransferase 2B7, *UGT1A1* uridine diphosphate glucuronosyltransferase 1A1

morphine. Uridine diphosphate glucuronosyltransferase 2B7 (UGT2B7) catalyses morphine into M6G and, in conjunction with UGT1A isoforms, into inactive morphine-3-glucuronide (M3G). Only a minority of codeine biotransformation ($\leq 15\%$) [201] is via CYP2D6; the majority of codeine is converted directly into the inactive metabolites codeine-6-glucuronide and norcodeine by UGT2B7 and CYP3A4, respectively. The major codeine metabolites, including morphine and M6G, are excreted renally [201].

CYP2D6 belongs to the superfamily of cytochrome P450 (*CYP*) genes. They encode haemoproteins that catalyse oxidative, phase I metabolism [202] and account for ~75% of all drug metabolism reactions [203]. Although 57 *CYP* genes have been identified, ~95% of these reactions are catalysed by just 5 isoenzymes including CYP2D6 [203]. Direct clinical measurement of CYP2D6 phenotypic activity is unfeasible as it is primarily expressed in the liver and indirect measurements of CYP2D6 metabolites in the plasma or urine are susceptible to other factors including renal dysfunction and drug interference. Consequently, *CYP2D6* genotyping as a phenotype surrogate is appealing for clinical practice.

CYP2D6 is located on chromosome 22 and over 80 alleles have been identified [204]. They are formed by a range of genetic alterations including SNPs, insertions and deletions and can be grouped functionally into increased, normal, reduced and non-functional alleles [137]. An individual's *CYP2D6* genotype can in turn be categorised into 1 of 4 predicted phenotype classes based on the combination of *CYP2D6* alleles they carry: an extensive, intermediate, poor or ultrarapid metaboliser (EM, IM, PM and UM, respectively) [205]. The EM is the wild-type CYP2D6 phenotype, IMs have reduced activity and PMs have no enzymatic activity as they carry no functional alleles. If multiple copies of functional alleles are detected this is denoted the UM phenotype as high enzymatic activity is expected [137]. There is considerable variability in the prevalence of *CYP2D6* alleles and in the prevalence of the extreme phenotypes in different ethnic groups (0–10 and 0–29% for PMs and UMs, respectively) [137].

It has been shown that following codeine administration, PMs have significantly lower plasma morphine concentrations, reduced urinary active metabolite excretion and decreased analgesia compared to EMs [206, 207]. Conversely, plasma morphine concentrations and urinary active metabolite excretion are significantly higher in UMs compared to EMs [208]. Furthermore although there is no definitive study, a growing series of case reports are documenting severe ADRs after standard codeine use associated with the UM phenotype [136, 209–213]. These case reports are from neonatal [209], paediatric [210–212] and adult populations [136, 213] and the documented on-target (opioidergic) ADRs include: severe epigastric pain, euphoria and dizziness [213], central nervous system/respiratory depression [136, 211] and death [209, 210, 212]. One especially poignant case was the death of a 13-day old neonate who was breastfed by a mother taking codeine (and paracetamol) for episiotomy pain [209]. The autopsy found an extremely high level of morphine in the neonate's blood and a sample of stored maternal breast milk from day 10 showed an elevated morphine concentration. The mother was found to have a *CYP2D6* gene duplication indicative of the UM phenotype [209]. Following this report, the US Food and Drug Administration (FDA) issued a warning on codeine use by nursing mothers [214].

Although there is increasing concern regarding the efficacy and safety of codeine, several barriers exist that hamper the translation of *CYP2D6* genotyping into widespread clinical practice. Firstly, the ADR profile of PMs is incompletely understood [137]. Secondly, when compared to the prevalence of the UM phenotype (0–10%), the documented case reports of severe ADRs are rare, suggesting that there are additional genetic and non-genetic susceptibility factors. The pharmacogenetic influence of *UGT2B7* is controversial at present [201]. Other risk factors may include renal dysfunction [136, 201, 215], drug inhibitors of CYP3A4 [136, 201], ontogeny [215, 216] and repeated episodes of hypoxia [215]. The paediatric case reports are from children receiving codeine after adeno(tonsillectomy) for recurrent tonsillitis and obstructive sleep apnoea (OSA) [210–212]. OSA leads to intermittent sleep hypoxia and it has been shown that opioid analgesia sensitivity increases in children after recurrent hypoxia [217]. Another factor is potential publication bias favouring selection of case reports documenting extreme but fortunately uncommon ADRs with codeine. 10 of 11 UM participants in a pharmacokinetics study felt sedation (91%) compared to 6 of 12 (50%) EMs ($p=0.03$) suggesting that ADRs in UMs may occur more frequently than is reported [208]. Other potential barriers include the absence of prospective studies that demonstrate clinical benefit of *CYP2D6* genotyping, scarce cost-effectiveness data, lack of clinician knowledge and no clear guidelines on what constitutes a suitable substitute for codeine in CYP2D6 PMs and UMs. This is important because CYP2D6 is involved in the metabolism of other opioid drugs including oxycodone, hydrocodone and tramadol. There is evidence at least for tramadol that CYP2D6 PMs experience reduced analgesia [218] and UMs a higher risk of nausea [138] when compared to EMs. There is also a case report of respiratory depression following tramadol in a UM patient with renal dysfunction [139]. Tramadol and codeine are step 2 'weak' opioid drugs on the WHO analgesia ladder [219] and are often used interchangeably in clinical practice for a patient that does not tolerate one. However if tramadol is also undesirable in CYP2D6 PMs and UMs, clinical guidance regarding suitable alternative analgesic agents is warranted.

3.3 CYP2C9 and the Risk of Haemorrhage with Warfarin

Warfarin is the most frequently prescribed oral anticoagulant worldwide [220] and is indicated in the prophylaxis and treatment of venous thromboembolism (VTE) and in the prophylaxis of systemic embolism in predisposing conditions such as atrial fibrillation and following mechanical heart valve insertion [221]. It is a coumarin-derived therapeutic that is administered as a racemic mixture; the S-warfarin enantiomer is more potent than R-warfarin [222]. They disrupt the vitamin K cycle by antagonising vitamin K epoxide reductase, resulting in a decrease in vitamin K-dependent post-translational γ-carboxylation of protein glutamate residues [223, 224]. This notably diminishes the activity of clotting cascade proteins including the procoagulant factors II, VII, IX and X and anticoagulant molecules protein C and protein S [225]. The overall anticoagulant effect is quantified by the prothrombin time-derived international normalised ratio (INR); the usual desired therapeutic INR is 2.5 [226]. However, certain high thrombotic risk conditions such as recurrent VTE(s) on warfarin and mechanical heart valves warrant higher anticoagulation levels (e.g. a desired INR range of 3.0–4.0) [226].

Epidemiological evidence has implicated warfarin as a major cause of ADRs; it is the therapeutic associated with the greatest number of preventable ADRs in Sweden [227] and the third most common cause of ADR-related hospitalisations in the UK [3]. Haemorrhage is an on-target ADR and is the predominant ADR associated with warfarin [221], especially during therapy initiation [228]. It is highly correlated to the intensity of anticoagulation [229, 230] and the risk of clinically significant bleeding increases when the desired INR range is higher [221]. The safe management of warfarin therapy is notoriously challenging because of the wide inter-individual range of optimal dose requirements (0.6–15.5 mg/day) and its narrow therapeutic index [231]. It is worth noting also that there is evidence to suggest a pharmacogenetic association between *CYP2C19*17* carriage and increased bleeding risk in patients taking clopidogrel (Table 4) [130, 131], although for now, the genetic susceptibility to haemorrhage on warfarin will be outlined.

CYP2C9, like CYP2D6, is 1 of the 5 main human CYP DMEs [203]. CYP2C9 is the principal enzyme involved in the metabolism of the potent S-warfarin stereoisomer, while R-warfarin is cleared via CYP1A1/CYP1A2/CYP3A4 [228]. Over 30 allelic variants of *CYP2C9* are known, but their relative prevalence varies with ethnicity [220]. The *CYP2C9* reference genotype **1/*1* produces the normal (EM) phenotype [220] and a resultant estimated warfarin half-life of 30–37 h [232]. The 2 most frequent reduction-of-function minor alleles amongst people with European ancestry are *CYP2C9*2* (rs1799853) and *CYP2C*3* (rs1057910) [202]. Both are characterised by one nonsynonymous SNP, prolong the half-life of warfarin (up to 92–203 h in **3/*3* homozygotes [233, 234]) and are associated with reduced maintenance warfarin dose requirements [235].

A recent meta-analysis has reported hazard ratios for the risk of bleeding in patients on warfarin with **1/*3* or **3/*3* genotypes, compared to **1/*1* patients, to be 2.05 (95% CI 1.36–3.10) and 4.87 (95% CI 1.38–17.14), respectively, suggestive of a gene-dose effect [135]. Although *CYP2C9*2* was also significantly associated

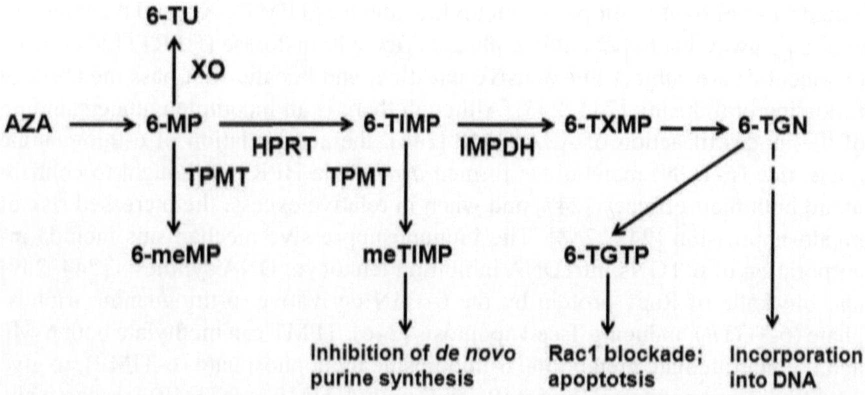

Fig. 2 Azathioprine (*AZA*) and 6-mercaptopurine (6-MP) metabolism (simplified). *me prefix* methyl, *XO* xanthine oxidase, *TU* thiouric acid, *TPMT* thiopurine methyltransferase, *HPRT* hypoxanthine phosphoribosyltransferase, *TIMP* thioinosine monophosphate, *IMPDH* inosine monophosphate dehydrogenase, *TXMP* thioxanthosine monophosphate, *TGN* thioguanine nucleotides, *TGTP* thioguanine triphosphate

with bleeding, albeit with a lower pooled effect size than *CYP2C9*3*, after stratification into **1/*2* and **2/*2* genotypes through synthesis of studies that reported these individual genotypes, neither genotype was significantly associated with bleeding. Overall, *CYP2C9*3* is the main risk factor for bleeding on warfarin, which is biologically plausible as the **3* allele has a more deleterious effect than **2* on CYP2C9 enzyme function [135]. For a more comprehensive account of overall warfarin pharmacogenetics, dosing strategies to incorporate multiple environmental, clinical and genetic factors and a discussion regarding the recently published prospective warfarin pharmacogenetic RCTs, the reader at this point is referred to Chap. 11.

3.4 TPMT, Azathioprine- and 6-Mercaptopurine-Induced Myelosuppression

The immunosuppressive agent azathioprine (AZA) is a pro-drug of 6-mercaptopurine (6-MP). AZA is indicated in both the prophylaxis of transplant rejection and in the treatment of many autoimmune conditions including inflammatory bowel disease (IBD), rheumatoid arthritis and severe eczema [236]. 6-MP is conventionally used with haematological malignancies and in particular acute lymphoblastic leukaemia [237], although it also has a role in IBD [238]. AZA/6-MP can induce several ADRs including myelosuppression (predisposing to neutropaenic sepsis), DILI, pancreatitis, nausea and vomiting [239]. Although ADRs occur in 10–28 % of patients [240], the rate of fatal ADRs among AZA users is estimated at 1 in 10,000 [241].

Approximately 90 % of AZA is converted to 6-MP by ubiquitous non-enzymatic processes [242, 243]. Figure 2 depicts the 3 main competing enzyme pathways for metabolism of 6-MP: thiopurine methyltransferase (TPMT), XO and the main anabolic pathway via hypoxanthine phosphoribosyltransferase (HPRT) [244]. Both therapeutics are subject to extensive intestinal and hepatic first pass metabolism following oral dosing [244, 245]. Although there is an incomplete understanding of the modes of action of AZA/6-MP [246], the accumulation of 6-thioguanine nucleotide (6-TGN) metabolites formed *in vivo* via HPRT is thought to contribute to both their efficacy [247] and when in relative excess, the increased risk of myelosuppression [242, 248]. The immunosuppressive mechanisms include incorporation of 6-TGNs into DNA inhibiting leukocyte DNA synthesis [244, 249] and blockade of Rac1 protein by the 6-TGN derivative, 6-thioguanine triphosphate (6-TGTP), inducing T-cell apoptosis [246]. TPMT can methylate both 6-MP and the intermediate metabolite, 6-thioinosine monophosphate (6-TIMP), to give 6-methylmercaptopurine (6-meMP) and methyl-TIMP (meTIMP), respectively. meTIMP may be efficacious through *de novo* purine synthesis inhibition [240, 250] whilst high levels of TPMT methylated thiopurine metabolites (and further phosphorylated metabolites) may be associated with DILI [251–255].

TPMT is a phase II biotransformation enzyme, encoded by *TPMT* on chromosome 6 [243], and is a major pharmacokinetic determinant for active 6-TGN metabolite levels [240], which are inversely related to TPMT activity [244, 256, 257]. It is variably expressed in several tissues; the highest levels of TPMT are present in the liver and the lowest in the brain and lung [240]. Erythrocyte TPMT activity correlates with hepatic TPMT activity [258] permitting direct TPMT phenotypic assessment of patients in clinical practice, which is unusual for a DME [259]. TPMT enzymatic activity follows a trimodal distribution; ~90 % of individuals have high activity, ~10 % intermediate and 0.3 % low/undetectable enzyme activity [260, 261].

Around 30 allelic variants of *TPMT* have been reported [20] and despite ethnic variability, 3 account for >90 % of the minor alleles: *TPMT*2*, *TPMT*3A* and *TPMT*3C* [254]. They are caused by one (*TPMT*2*, *TPMT*3C*) or two (*TPMT*3A*) nonsynonymous SNPs that reduce enzymatic activity through enhancing the rate that the TPMT variant is catabolised [262–264]. Analogous to *CYP2D6* and *CYP2C9*, *TPMT* genotype correlates with the variable TPMT enzymatic activity levels: heterozygotes have intermediate activity (IM) and individuals carrying no normally functioning alleles have low/absent activity (PM) [254]. Like *CYP2D6* and *CYP2C9*, homozygous deficient individuals include both those homozygous for 1 variant allele and compound heterozygotes with 2 distinct inactivating alleles [243]. *TPMT *1/*1* individuals have normal phenotypic activity (EM).

Clinically, ~27 % of AZA/6-MP-induced myelosuppression cases are explained by inactivating *TPMT* alleles [265], although little correlation exists with other specific ADRs including DILI [239, 266]. A meta-analysis of patients with chronic inflammatory diseases has reported a gene-dose effect for this on-target ADR: homozygous deficient individuals carry a higher risk of leukopaenia (OR 20.84, 95 % CI 3.42–126.89) than heterozygotes (OR 4.29, 95 % CI 2.67–6.89) when compared with **1/*1* individuals [149] and in general the myelosuppression onset is earlier [265, 267] and more severe [267]. A second systematic review, not limited to a specific class of disease, has reported that 86 % of *TPMT* homozygous deficient patients develop myelosuppression and the pooled OR for patients with intermediate TPMT activity or one *TPMT* variant allele, compared with wild-type, was 4.19 (95 % CI 3.20–5.48) [268]. For both studies, their results were primarily derived from synthesis of observational studies.

As clinical evidence has grown, consensus national clinical guidelines have been published that recommend and interpret pre-therapy TPMT testing, including the UK dermatology [269] and rheumatology guidelines [270]. In patients identified as TPMT deficient (by either genotyping of homozygous deficiency or TMPT phenotypic analysis of low/absent activity), guidance advises selection of alternative immunosuppressive therapy in non-malignant conditions and a reduction in starting dose to 10 % of normal when treating malignancy [254]. For heterozygous variant/intermediate activity patients commencing AZA/6-MP therapy, a dose reduction of 30–70 % is suggested [254]. TPMT analysis has been adopted into clinical practice and a national survey reported that 94 % of dermatologists, 60 % of gastroenterologists and 47 % of rheumatologists in England requested TPMT testing [271].

Despite the relatively high, albeit variable, clinical uptake of TPMT testing, outstanding issues remain. Firstly, there is a lack of robust prospective randomised evidence assessing the utility of pre-therapy TPMT analysis in reducing myelosuppression. An RCT (n = 333) was undertaken but the recruitment target (n = 1000) was not met due to guideline-driven pre-existing routine TPMT testing at some centres adversely impacting study recruitment [272]. The one patient in the non-genotyped arm found at study completion to be *TPMT* homozygous deficient developed severe, early onset neutropaenia. However overall, the study found no difference in the rates of AZA cessation due to ADRs between the *TPMT* genotyped arm (with recommended AZA dose reduction and avoidance in heterozygous and homozygous *TPMT* deficient patients, respectively) and the non-genotyped arm, and no increase in AZA cessation in *TPMT* heterozygous patients compared to wild-type patients [272].

Secondly, whilst the evidence and recommendations for *TPMT* homozygous deficient individuals are relatively clear, the optimal management strategy for heterozygous patients is less certain. Although overall they appear to be at a modest increased risk of myelosuppression [149, 268], complicating factors include the observation that only ~30–60 % of heterozygous patients do not tolerate full doses of AZA/6-MP [254, 257, 273] and the benefit: harm ratio attributable to different thiopurine starting doses for heterozygotes likely varies depending on the disease-specific necessity for rapid therapeutic action. A higher risk of myelosuppression with a higher starting dose in a heterozygote might be justifiable for treating malignancy, but not chronic, stable immunological disease.

Thirdly, *TPMT* can be analysed by phenotype or genotype and the screening test protocol remains incompletely standardised. Erythrocyte TPMT activity is predominantly offered to clinicians in the UK, but it can be affected by patient ethnicity, concurrent use of interacting drugs (e.g. mesalazine, sulfasalazine, allopurinol), allogeneic erythrocyte transfusions during the preceding 120 days, and in haematological malignancies, it can be affected by disease-related influences [274]. Whilst the overall genotype to phenotype test concordance is 98.4 % in healthy volunteers, it decreases to 86 % in the intermediate TPMT activity range, attributable to both non-genetic influences on TPMT activity, as described above, and to a lesser extent, novel mutations [275]. Therefore, neither test is 100 % sensitive to correctly identify TPMT deficiency, but research from a National Centre suggests that genotyping is more accurate and should be used as the primary test, in contrast to current UK practice [276].

Therefore, a pharmacogenetic association exists between *TPMT* and myelosuppression and there is strong evidence, affirmed by clinical guidelines, for avoiding thiopurine drugs or significantly reducing their dose in *TPMT* homozygous deficient patients, given their near universal experience of myelosuppression at conventional doses [254]. Further research is required to clarify optimal management for heterozygous patients. However, it is already cost-effective to routinely test TPMT status to identify homozygous deficient patients alone [274]. Pre-therapy TPMT testing is not a substitute for routine on-therapy blood test monitoring, given that several thiopurine ADRs are not associated with TPMT and the majority of myelosuppression cases are still not accounted for by *TPMT* variants [265]. Finally, in addition to TPMT testing, there is also a growing role for thiopurine metabolite level monitoring (e.g. 6-TGNs) to individualise thiopurine doses soon after starting treatment; prospective studies to evaluate this proactive approach are ongoing [277].

3.5 SLCO1B1 and Statin-Induced Muscle Toxicity

Statins are the most commonly prescribed class of medication worldwide [278] and are highly efficacious in the primary and secondary prevention of cardiovascular disease [1]. They reduce plasma low-density lipoprotein (LDL) cholesterol through competitive inhibition of 3-hydroxy-3-methylglutaryl-coenzyme A (HMG-CoA) reductase, the rate limiting enzyme in *de novo* cholesterol synthesis. This in turn leads to an upregulation of hepatic LDL receptors, increasing cholesterol influx into hepatocytes and reducing the plasma burden [279].

The currently licensed statins have a good safety profile, but carry a small risk of skeletal muscle toxicity [280]. The spectrum of muscle pathology varies from the most common manifestation of asymptomatic elevations in plasma CK level, to myopathies with pain and high plasma CK levels through to rhabdomyolysis with the potential sequelae of AKI and death. Alternatively, statin therapy can cause myalgias with no detectable plasma CK rise [21]. Depending on precise definitions, myopathy and rhabdomyolysis occur at frequencies of ~ 1/1000 and ~ 1/100,000, respectively [281], although this is modulated by other risk factors including higher statin dose, female gender, older age, low BMI, untreated hypothyroidism and other drug therapies, for example concomitant use of gemfibrozil [281].

The solute carrier organic anion transporter family member 1B1 (*SLCO1B1*) belongs to the superfamily of solute carrier (*SLC*) influx transporter genes and encodes the organic anion-transporting polypeptide 1B1 (OATP1B1) [282]. OATP1B1 is one of the most highly expressed influx transporters within the human liver [283]. It facilitates hepatic uptake of a variety of xenobiotic compounds and endogenous substances [284] and so affects the level of exposure of substrate drugs to intracellular hepatic DMEs [285].

Although the effects of statins on the off-target muscle tissue are incompletely defined at present [286], there exists a significant association between gene variants of *SLCO1B1* and the risk of statin-induced muscle ADRs. A seminal statin GWAS used data from the Study of the Effectiveness of Additional Reductions in Cholesterol and Homocysteine (SEARCH) RCT in the UK; 85 cases of definite or incipient myopathy were contrasted with 90 controls [148]. Both the cases and controls for the GWAS had been prescribed 80 mg simvastatin daily. Only an intronic SNP variant, rs4363657, was strongly correlated with myopathy and further regional genetic analysis showed it to be in near complete linkage disequilibrium with the nonsynonymous SNP, rs4149056, in exon 6 (*SLCO1B1*5*; 521T > C; V174A). Further, a gene-dose relationship was demonstrated for rs4149056: the OR for myopathy in heterozygotes and homozygotes for the minor C allele was 4.5 (95% CI 2.6–7.7) and 16.9 (95% CI 4.7–61.1), respectively, when compared to the ancestral TT genotype. Overall, greater than 60% of the myopathy cases in this study were attributable to the C variant [148]. The association with rs4149056 has been replicated [148, 287, 288] but the incidence of severe myopathy and the magnitude of correlation were lower in a second UK randomised trial population [289], attributable to the smaller 40 mg daily simvastatin dose used [148]. The rs4149056 variant has been subsequently associated with more mild, statin-induced muscle ADRs [290], reduced simvastatin adherence [290] and general intolerance to simvastatin defined as a composite endpoint of prescribing +/− mild biochemical changes [291]. The weight of evidence to date for rs4149056 is with simvastatin and the evidence with other statins is less compelling [287, 290, 292], suggesting that rs4149056 may represent a simvastatin-specific effect.

Mechanistically, rs4149056 may interfere with localisation of the transporter to the hepatic plasma membrane reducing its activity [284]. It is associated with higher statin, and especially simvastatin acid, plasma concentrations [293–295] that conceivably increase skeletal muscle drug exposure. However, the relationship between plasma simvastatin acid concentration and muscle toxicity is not straightforward. Clinically, current FDA guidance recommends against the 80 mg simvastatin dose unless a patient has tolerated the higher dose for over 12 months [296].

Overall, the rs4149056 variant is a plausible candidate for a predictive test to reduce simvastatin-induced skeletal muscle ADRs. Current guidance suggests that when initiating simvastatin therapy in CT or CC genotype patients, simvastatin 20 mg daily is selected rather than the normal 40 mg daily dose, possible routine CK surveillance is utilised and alternative statin therapy is commenced rather than increasing the dose of simvastatin if lipid goals are not reached. However, the effects of these recommendations on the incidence of simvastatin ADRs and adherence are currently unknown [281].

4 Outlook and Recommendations

The aspiration of pharmacogenetics is to individualise drug treatment to minimise harm and promote efficacy. Pre-therapy predictive genetic testing seeks to tailor therapy to reduce ADRs primarily through guiding drug or dose selection and has impacted positively upon clinical practice, notably with abacavir. Genetic screening may also find a role in identifying patients for whom regular biomarker surveillance may be indicated to minimise the incidence of severe ADRs. In addition to the direct patient benefit of reducing ADRs, there are at least 3 other potentially favourable spin-offs from understanding the pharmacogenetics of ADRs. Firstly, genetic-ADR associations provide novel insights that facilitate investigation into underlying pathological processes and the extrapolation of new knowledge regarding hypersensitivity reactions may have implications for cancer, autoimmune and infectious disease management. Secondly, the safety profile of new therapeutics may be improved through screening of drug candidates for affinity to high risk *HLA* alleles, for example *HLA-B*57:01* and *HLA-B*58:01* [71]. Thirdly, the beneficial side effects of some drugs have resulted in new therapeutic indications, for example with sidenafil (Viagra) and its fortuitous alleviation of erectile dysfunction. Pharmacogenetics has the potential to increase this 'drug repositioning' through identifying novel off target pharmacodynamic sites.

Abacavir has provided a blueprint for translational pharmacogenetics, but it has yet to be emulated. This is partly due to certain 'favourable' characteristics of AHS including: the high relative prevalence of AHS [76], the exclusivity of the association between *HLA-B*57:01* and immunologically-mediated AHS, the reduction of false-positive clinical diagnoses mediated by the screening programme [78], the vocal patient lobby, and a physician community who were relatively amenable to changing their prescribing and clinical behaviour. It is also because there are multiple obstacles encountered when attempting translation. It is important to first understand these hurdles, and then to have a systematic approach to both developing the ADR-genotype evidence base and to implementing it in clinical practice [297].

Many ADRs are rare and some, such as the HSS, consist of varying constellations of non-specific features. As a result, international consortia using standardised definitions for these ADRs are advisable so patient samples of sufficient size with well demarcated phenotypes that are generalisable across ethnic groups can be pooled together. The 'International Serious Advent Consortium' and their 'Phenotype Standardisation Project' are both steps in the right direction [298]. These coordinated efforts are a prerequisite to reducing the risk of type I and type II errors in genetic association studies of rare and variable ADRs.

Pharmacogenetics has traditionally harnessed the candidate gene approach, whereby genes predicted to be relevant, typically through knowledge of a drug's pharmacology, are selectively studied. However, this approach is limited to contemporary knowledge and so has largely been superseded by GWAS, which has no stipulation for *a priori* hypotheses [20] and can test at least 10^6 SNPs concurrently. However GWAS increases sample size requirements and data capture, increasing the complexity of study data management and statistical processes and potentiates the threat of selective publication reporting. Further, the lack of a preformed hypothesis augments the importance of confirming biological causality for GWAS putative associations.

Nevertheless, GWAS is a valuable asset: it can confirm in a 'blinded' fashion the results of previous candidate gene studies [20] and offer a novel foothold into the idiosyncratic processes of off-target ADRs. For polygenic ADRs, GWAS may detect new loci of individual small effect size and assess genotype-phenotype associations of larger haplotype signatures. The '1000 Genomes Project,' which has recently described the genomes of 1092 individuals, is in turn increasing the resolution of GWAS [299]. The 1000 Genomes Project should additionally provide a baseline reference for normal human genetic variation, enable fine mapping of existing GWAS associations and aid discovery of new genetic associations, partly through its detailed identification of indels and larger deletions as well as contemporary SNPs [299]. In the near future, next generation sequencing technologies that provide high throughput whole genome capability will offer the pinnacle of DNA resolution whilst advances in our understanding of epigenetic imprinting and microRNA regulation promise new directions for the study of ADR pharmacogenetics. As genetic variation does not usually account for all of the inter-individual variation in drug response, incorporation of data from transcriptomics, metabolomics and proteomics may further improve predictive values [127].

After identification and validation of a statistically significant genetic association(s) for an ADR, several hurdles still bar adoption into clinical practice. Large, well-conducted prospective studies represent the gold standard to confirm clinical outcome benefit, although given the rarity of some ADRs these are not always practical. For other ADRs, genetic sub-studies of clinical trials and registries will likely offer the highest attainable level of evidence [300]. Subsequent pharmacoeconomic studies should base their analyses on this high quality data rather than expert opinion and retrospective data [301].

Logistical and knowledge barriers to the implementation of ADR pharmacogenetics also exist. On-demand genotyping, where the treating physician requests a specific pharmacogenetic test for a patient when seeking to prescribe a drug with a clinically established ADR-genotype association, relies on both a physician's knowledge of pharmacogenetics and a system for following-up and acting on the pharmacogenetic test result. Robust and validated point-of-care genotyping tests may be necessary. An alternative proposed method is pre-emptive genotyping, where multiple relevant SNPs are routinely genotyped together and this genetic data is incorporated into a patient's electronic medical record, with subsequent access by automated clinical decision support (CDS) algorithms to provide a clinically relevant alert regarding a potential drug-genotype interaction specific to the individual patient, at the point in time when the physician is seeking to prescribe the drug of interest. This approach provides the pharmacogenetic information at the most pertinent time and secondly, the CDS approach is likely better suited to keep up with our rapidly expanding understanding of

ADR pharmacogenetics. However, the associated computational challenges are considerable [302].

Finally, a genetic test should be ethically acceptable to patients, clinicians and society. The emphasis of pharmacogenetics is for the beneficial personalisation of medicine, yet paradoxically the realisation of this goal requires not only very large international research collaborations but also active engagement with society as a whole. This is not least because genetic information harbours potential adverse implications, such as individual discrimination by insurance firms based on high risk genotype carriage and neglect of ethnic minorities by pharmaceuticals opting to segregate research initiatives to benefit the majority to maximise profit margins [303]. Open dialogue between patients, healthcare services, insurance providers, pharmaceuticals and the wider public is required to address these risks. If society chooses pharmacogenetics, it must safeguard against encroachment on the rights of individuals and minority groups. Ultimately, the widespread application of pharmacogenetics throughout clinical practice to ameliorate ADRs remains far off, but the examples in this chapter and the promises inherent in the new technologies foreshadow a future potential.

References

1. Roden DM, Johnson JA, Kimmel SE, Krauss RM, Medina MW, Shuldiner A, Wilke RA (2011) Cardiovascular pharmacogenomics. Circ Res 109(7):807–820. doi:10.1161/circresaha.110.230995
2. Trevelyan Square BL (2010) Search catalogue. http://www.ic.nhs.uk/searchcatalogue?productid=7930&q=general+pharmaceutical+services&topics=1%2fPrimary+care+services%2fCommunity+pharmacy+services&sort=Most+recent&size=10&page=1#top. Accessed 17 Dec 2012
3. Pirmohamed M, James S, Meakin S, Green C, Scott AK, Walley TJ, Farrar K, Park BK, Breckenridge AM (2004) Adverse drug reactions as cause of admission to hospital: prospective analysis of 18 820 patients. BMJ 329(7456):15–19. doi:10.1136/bmj.329.7456.15
4. Ahern F, Sahm LJ, Lynch D, McCarthy S (2014) Determining the frequency and preventability of adverse drug reaction-related admissions to an Irish University Hospital: a cross-sectional study. Emerg Med J 31(1):24–29. doi:10.1136/emermed-2012-201945
5. Conforti A, Costantini D, Zanetti F, Moretti U, Grezzana M, Leone R (2012) Adverse drug reactions in older patients: an Italian observational prospective hospital study. Drug Healthc Patient Saf 4:75–80. doi:10.2147/dhps.s29287
6. Hopf Y, Watson M, Williams D (2008) Adverse-drug-reaction related admissions to a hospital in Scotland. Pharm World Sci 30(6):854–862. doi:10.1007/s11096-008-9240-5
7. Miranda V, Fede A, Nobuo M, Ayres V, Giglio A, Miranda M, Riechelmann RP (2011) Adverse drug reactions and drug interactions as causes of hospital admission in oncology. J Pain Symptom Manage 42(3):342–353. doi:10.1016/j.jpainsymman.2010.11.014
8. Ruiter R, Visser LE, Rodenburg EM, Trifirò G, Ziere G, Stricker BH (2012) Adverse drug reaction-related hospitalizations in persons aged 55 years and over: a population-based study in the Netherlands. Drugs Aging 29(3):225–232. doi:10.2165/11599430-000000000-00000
9. World Health Organisation, WHO Collaborating Centre for International Drug Monitoring (2002) The importance of pharmacovigilance–safety monitoring of medicinal products. http://whqlibdoc.who.int/hq/2002/a75646.pdf. Accessed 13 March 2015

10. European Medicines Agency (1995) Clinical safety data management: definitions and standards for expedited reporting. London CPMP/ICH/377/95. http://www.ema.europa.eu/docs/en_GB/document_library/Scientific_guideline/2009/09/WC500002749.pdf. Accessed 13 March 2015
11. Gurwitz JH, Field TS, Avorn J, McCormick D, Jain S, Eckler M, Benser M, Edmondson AC, Bates DW (2000) Incidence and preventability of adverse drug events in nursing homes. Am J Med 109(2):87–94
12. National Coordinating Council for Medication Error Reporting and Prevention. What is a Medication Error? http://www.nccmerp.org/about-medication-errors. Accessed 13 March 2015
13. Edwards IR, Aronson JK (2000) Adverse drug reactions: definitions, diagnosis, and management. Lancet 356(9237):1255–1259. doi:10.1016/s0140-6736(00)02799-9
14. Bates DW, Cullen DJ, Laird N, Petersen LA, Small SD, Servi D, Laffel G, Sweitzer BJ, Shea BF, Hallisey R (1995) Incidence of adverse drug events and potential adverse drug events. Implications for prevention. ADE prevention study group. JAMA 274(1):29–34
15. Nebeker JR, Barach P, Samore MH (2004) Clarifying adverse drug events: a clinician's guide to terminology, documentation, and reporting. Ann Intern Med 140(10):795–801
16. Davies EC, Green CF, Mottram DR, Pirmohamed M (2007) Adverse drug reactions in hospitals: a narrative review. Curr Drug Saf 2(1):79–87
17. Rawlins MD, Thompson JW (1977) Pathogenesis of adverse drug reactions. In: Davies DM (ed) Textbook of adverse drug reactions. Oxford University Press, Oxford, p. 10
18. Medicines, Healthcare products Regulatory Agency wmgu Adverse drug reactions. http://www.mhra.gov.uk/Safetyinformation/Reportingsafetyproblems/Reportingsuspectedadversedrugreactions/Healthcareprofessionalreporting/Adversedrugreactions/index.htm#1. Accessed 18 Dec 2012
19. Aronson JK, Ferner RE (2003) Joining the DoTS: new approach to classifying adverse drug reactions. BMJ 327(7425):1222–1225. doi:10.1136/bmj.327.7425.1222
20. Evrard A, Mbatchi L (2012) Genetic polymorphisms of drug metabolizing enzymes and transporters: the long way from bench to bedside. Curr Top Med Chem 12(15):1720–1729
21. Pirmohamed M (2010) Pharmacogenetics of idiosyncratic adverse drug reactions. Handb Exp Pharmacol 196:477–491. doi:10.1007/978-3-642-00663-0_17
22. Mallal S, Nolan D, Witt C, Masel G, Martin AM, Moore C, Sayer D, Castley A, Mamotte C, Maxwell D, James I, Christiansen FT (2002) Association between presence of HLA-B*5701, HLA-DR7, and HLA-DQ3 and hypersensitivity to HIV-1 reverse-transcriptase inhibitor abacavir. Lancet 359(9308):727–732
23. Hetherington S, Hughes AR, Mosteller M, Shortino D, Baker KL, Spreen W, Lai E, Davies K, Handley A, Dow DJ, Fling ME, Stocum M, Bowman C, Thurmond LM, Roses AD (2002) Genetic variations in HLA-B region and hypersensitivity reactions to abacavir. Lancet 359(9312):1121–1122. doi:10.1016/s0140-6736(02)08158-8
24. Hung S-I, Chung W-H, Liou L-B, Chu C-C, Lin M, Huang H-P, Lin Y-L, Lan J-L, Yang L-C, Hong H-S, Chen M-J, Lai P-C, Wu M-S, Chu C-Y, Wang K-H, Chen C-H, Fann CSJ, Wu J-Y, Chen Y-T (2005) HLA-B*5801 allele as a genetic marker for severe cutaneous adverse reactions caused by allopurinol. Proc Natl Acad Sci U S A 102(11):4134–4139. doi:10.1073/pnas.0409500102
25. Kang H-R, Jee YK, Kim Y-S, Lee CH, Jung J-W, Kim SH, Park H-W, Chang Y-S, Jang I-J, Cho S-H, Min K-U, Kim S-H, Lee KW (2011) Positive and negative associations of HLA class I alleles with allopurinol-induced SCARs in Koreans. Pharmacogenet Genomics 21(5):303–307. doi:10.1097/FPC.0b013e32834282b8
26. Kim S-H, Lee KW, Song W-J, Kim S-H, Jee Y-K, Lee S-M, Kang H-R, Park H-W, Cho S-H, Park S-H, Min K-U, Chang Y-S (2011) Carbamazepine-induced severe cutaneous adverse reactions and HLA genotypes in Koreans. Epilepsy Res 97(1–2):190–197. doi:10.1016/j.eplepsyres.2011.08.010
27. McCormack M, Alfirevic A, Bourgeois S, Farrell JJ, Kasperavičiūtė D, Carrington M, Sills GJ, Marson T, Jia X, de Bakker PIW, Chinthapalli K, Molokhia M, Johnson MR, O'Connor GD, Chaila E, Alhusaini S, Shianna KV, Radtke RA, Heinzen EL, Walley N, Pandolfo M,

Pichler W, Park BK, Depondt C, Sisodiya SM, Goldstein DB, Deloukas P, Delanty N, Cavalleri GL, Pirmohamed M (2011) HLA-A*3101 and carbamazepine-induced hypersensitivity reactions in Europeans. N Engl J Med 364(12):1134–1143. doi:10.1056/NEJMoa1013297
28. Littera R, Carcassi C, Masala A, Piano P, Serra P, Ortu F, Corso N, Casula B, Nasa G L, Contu L, Manconi PE (2006) HLA-dependent hypersensitivity to nevirapine in Sardinian HIV patients. AIDS 20(12):1621–1626. doi:10.1097/01.aids.0000238408.82947.09
29. Gatanaga H, Yazaki H, Tanuma J, Honda M, Genka I, Teruya K, Tachikawa N, Kikuchi Y, Oka S (2007) HLA-Cw8 primarily associated with hypersensitivity to nevirapine. AIDS 21(2):264–265. doi:10.1097/QAD.0b013e32801199d9
30. Chantarangsu S, Mushiroda T, Mahasirimongkol S, Kiertiburanakul S, Sungkanuparph S, Manosuthi W, Tantisiriwat W, Charoenyingwattana A, Sura T, Chantratita W, Nakamura Y (2009) HLA-B*3505 allele is a strong predictor for nevirapine-induced skin adverse drug reactions in HIV-infected Thai patients. Pharmacogenet Genomics 19(2):139–146. doi:10.1097/FPC.0b013e32831d0faf
31. Tassaneeyakul W, Jantararoungtong T, Chen P, Lin P-Y, Tiamkao S, Khunarkornsiri U, Chucherd P, Konyoung P, Vannaprasaht S, Choonhakarn C, Pisuttimarn P, Sangviroon A, Tassaneeyakul W (2009) Strong association between HLA-B*5801 and allopurinol-induced Stevens-Johnson syndrome and toxic epidermal necrolysis in a Thai population. Pharmacogenet Genomics 19(9):704–709. doi:10.1097/FPC.0b013e328330a3b8
32. Chung W-H, Hung S-I, Hong H-S, Hsih M-S, Yang L-C, Ho H-C, Wu J-Y, Chen Y-T (2004) Medical genetics: a marker for Stevens-Johnson syndrome. Nature 428(6982):486. doi:10.1038/428486a
33. Locharernkul C, Loplumlert J, Limotai C, Korkij W, Desudchit T, Tongkobpetch S, Kangwanshiratada O, Hirankarn N, Suphapeetiporn K, Shotelersuk V (2008) Carbamazepine and phenytoin induced Stevens-Johnson syndrome is associated with HLA-B*1502 allele in Thai population. Epilepsia 49(12):2087–2091. doi:10.1111/j.1528-1167.2008.01719.x
34. Yip VL, Marson AG, Jorgensen AL, Pirmohamed M, Alfirevic A (2012) HLA genotype and carbamazepine-induced cutaneous adverse drug reactions: a systematic review. Clin Pharmacol Ther 92(6):757–765. doi:10.1038/clpt.2012.189
35. Lonjou C, Borot N, Sekula P, Ledger N, Thomas L, Halevy S, Naldi L, Bouwes-Bavinck J-N, Sidoroff A, de Toma C, Schumacher M, Roujeau J-C, Hovnanian A, Mockenhaupt M (2008) A European study of HLA-B in Stevens-Johnson syndrome and toxic epidermal necrolysis related to five high-risk drugs. Pharmacogenet Genomics 18(2):99–107. doi:10.1097/FPC.0b013e3282f3ef9c
36. Kim S-H, Kim M, Lee KW, Kim S-H, Kang H-R, Park H-W, Jee Y-K (2010) HLA-B*5901 is strongly associated with methazolamide-induced Stevens-Johnson syndrome/toxic epidermal necrolysis. Pharmacogenomics 11(6):879–884. doi:10.2217/pgs.10.54
37. Carr DF, Chaponda M, Jorgensen AL, Castro EC, van Oosterhout JJ, Khoo SH, Lalloo DG, Heyderman RS, Alfirevic A, Pirmohamed M (2013) Association of human leukocyte antigen alleles and nevirapine hypersensitivity in a Malawian HIV-infected population. Clin Infect Dis 56(9):1330–1339. doi:10.1093/cid/cit021
38. Hung S-I, Chung W-H, Liu Z-S, Chen C-H, Hsih M-S, Hui RC-y, Chu C-Y, Chen Y-T (2010) Common risk allele in aromatic antiepileptic-drug induced Stevens-Johnson syndrome and toxic epidermal necrolysis in Han Chinese. Pharmacogenomics 11(3):349–356. doi:10.2217/pgs.09.162
39. Cao Z-h, Wei Z-y, Zhu Q-y, Zhang J-y, Yang L, Qin S-y, Shao L-y, Zhang Y-t, Xuan J-k, Q-l L, Xu J-H, Xu F, Ma L, Huang H-y, Xing Q-h, Luo X-q (2012) HLA-B*58:01 allele is associated with augmented risk for both mild and severe cutaneous adverse reactions induced by allopurinol in Han Chinese. Pharmacogenomics 13(10):1193–1201. doi:10.2217/pgs.12.89
40. Romano A, De Santis A, Romito A, Fonso M D, Venuti A, Gasbarrini GB, Manna R (1998) Delayed hypersensitivity to aminopenicillins is related to major histocompatibility complex genes. Ann Allergy Asthma Immunol 80(5):433–437
41. Hung S-I, Chung W-H, Jee S-H, Chen W-C, Chang Y-T, Lee W-R, Hu S-L, Wu M-T, Chen G-S, Wong T-W, Hsiao P-F, Chen W-H, Shih H-Y, Fang W-H, Wei C-Y, Lou Y-H, Huang Y-L, Lin J-J, Chen Y-T (2006) Genetic susceptibility to carbamazepine-induced cu-

taneous adverse drug reactions. Pharmacogenet Genomics 16(4):297–306. doi:10.1097/01.fpc.0000199500.46842.4a
42. Vitezica ZG, Milpied B, Lonjou C, Borot N, Ledger TN, Lefebvre A, Hovnanian A (2008) HLA-DRB1*01 associated with cutaneous hypersensitivity induced by nevirapine and efavirenz. AIDS 22(4):540–541. doi:10.1097/QAD.0b013e3282f37812
43. Likanonsakul S, Rattanatham T, Feangvad S, Uttayamakul S, Prasithsirikul W, Tunthanathip P, Nakayama EE, Shioda T (2009) HLA-Cw*04 allele associated with nevirapine-induced rash in HIV-infected Thai patients. AIDS Res Ther 6:22. doi:10.1186/1742-6405-6-22
44. Sharma SK, Balamurugan A, Saha PK, Pandey RM, Mehra NK (2002) Evaluation of clinical and immunogenetic risk factors for the development of hepatotoxicity during antituberculosis treatment. Am J Respir Crit Care Med 166(7):916–919
45. Lucena MI, Molokhia M, Shen Y, Urban TJ, Aithal GP, Andrade RJ, Day CP, Ruiz-Cabello F, Donaldson PT, Stephens C, Pirmohamed M, Romero-Gomez M, Navarro JM, Fontana RJ, Miller M, Groome M, Bondon-Guitton E, Conforti A, Stricker BHC, Carvajal A, Ibanez L, Yue Q-Y, Eichelbaum M, Floratos A, Pe'er I, Daly MJ, Goldstein DB, Dillon JF, Nelson MR, Watkins PB, Daly AK (2011) Susceptibility to amoxicillin-clavulanate-induced liver injury is influenced by multiple HLA class I and II alleles. Gastroenterology 141(1):338–347. doi:10.1053/j.gastro.2011.04.001
46. O'Donohue J, Oien KA, Donaldson P, Underhill J, Clare M, MacSween RN, Mills PR (2000) Co-amoxiclav jaundice: clinical and histological features and HLA class II association. Gut 47(5):717–720
47. Daly AK, Donaldson PT, Bhatnagar P, Shen Y, Pe'er I, Floratos A, Daly MJ, Goldstein DB, John S, Nelson MR, Graham J, Park BK, Dillon JF, Bernal W, Cordell HJ, Pirmohamed M, Aithal GP, Day CP (2009) HLA-B*5701 genotype is a major determinant of drug-induced liver injury due to flucloxacillin. Nat Genet 41(7):816–819. doi:10.1038/ng.379
48. Spraggs CF, Budde LR, Briley LP, Bing N, Cox CJ, King KS, Whittaker JC, Mooser VE, Preston AJ, Stein SH, Cardon LR (2011) HLA-DQA1*02:01 is a major risk factor for lapatinib-induced hepatotoxicity in women with advanced breast cancer. J Clin Oncol 29(6):667–673. doi:10.1200/jco.2010.31.3197
49. Singer JB, Lewitzky S, Leroy E, Yang F, Zhao X, Klickstein L, Wright TM, Meyer J, Paulding CA (2010) A genome-wide study identifies HLA alleles associated with lumiracoxib-related liver injury. Nat Genet 42(8):711–714. doi:10.1038/ng.632
50. Yuan J, Guo S, Hall D, Cammett AM, Jayadev S, Distel M, Storfer S, Huang Z, Mootsikapun P, Ruxrungtham K, Podzamczer D, Haas DW (2011) Toxicogenomics of nevirapine-associated cutaneous and hepatic adverse events among populations of African, Asian, and European descent. AIDS 25(10):1271–1280. doi:10.1097/QAD.0b013e32834779df
51. Hirata K, Takagi H, Yamamoto M, Matsumoto T, Nishiya T, Mori K, Shimizu S, Masumoto H, Okutani Y (2008) Ticlopidine-induced hepatotoxicity is associated with specific human leukocyte antigen genomic subtypes in Japanese patients: a preliminary case-control study. Pharmacogenomics J 8(1):29–33. doi:10.1038/sj.tpj.6500442
52. Ariyoshi N, Iga Y, Hirata K, Sato Y, Miura G, Ishii I, Nagamori S, Kitada M (2010) Enhanced susceptibility of HLA-mediated ticlopidine-induced idiosyncratic hepatotoxicity by CYP2B6 polymorphism in Japanese. Drug Metab Pharmacokinet 25(3):298–306
53. Kindmark A, Jawaid A, Harbron CG, Barratt BJ, Bengtsson OF, Andersson TB, Carlsson S, Cederbrant KE, Gibson NJ, Armstrong M, Lagerström-Fermér ME, Dellsén A, Brown EM, Thornton M, Dukes C, Jenkins SC, Firth MA, Harrod GO, Pinel TH, Billing-Clason SME, Cardon LR, March RE (2008) Genome-wide pharmacogenetic investigation of a hepatic adverse event without clinical signs of immunopathology suggests an underlying immune pathogenesis. Pharmacogenomics J 8(3):186–195. doi:10.1038/sj.tpj.6500458
54. Athanasiou MC, Dettling M, Cascorbi I, Mosyagin I, Salisbury BA, Pierz KA, Zou W, Whalen H, Malhotra AK, Lencz T, Gerson SL, Kane JM, Reed CR (2011) Candidate gene analysis identifies a polymorphism in HLA-DQB1 associated with clozapine-induced agranulocytosis. J Clin Psychiatry 72(4):458–463. doi:10.4088/JCP.09m05527yel
55. Diez RA (1990) HLA-B27 and agranulocytosis by levamisole. Immunol Today 11(8):270

56. Lee HY, Lee JW, Lee KW, Park MH, Park HS (2009) The HLA allele marker for differentiating ASA hypersensitivity phenotypes. Allergy 64(9):1385–1387. doi:10.1111/j.1398-9995.2009.02048.x
57. Kim S-H, Hur G-Y, Choi J-H, Park H-S (2008) Pharmacogenetics of aspirin-intolerant asthma. Pharmacogenomics 9(1):85–91. doi:10.2217/14622416.9.1.85
58. Hakala M, van Assendelft AH, Ilonen J, Jalava S, Tiilikainen A (1986) Association of different HLA antigens with various toxic effects of gold salts in rheumatoid arthritis. Ann Rheum Dis 45(3):177–182
59. Shah P, Griffith SM, Shadforth MF, Fisher J, Dawes PT, Poulton KV, Thomson W, Ollier WER, Mattey DL (2004) Can gold therapy be used more safely in rheumatoid arthritis? Adverse drug reactions are more likely in patients with nodular disease, independent of HLA-DR3 status. J Rheumatol 31(10):1903–1905
60. Erlich H (2012) HLA DNA typing: past, present, and future. Tissue Antigens 80(1):1–11. doi:10.1111/j.1399-0039.2012.01881.x
61. Horton R, Wilming L, Rand V, Lovering RC, Bruford EA, Khodiyar VK, Lush MJ, Povey S, Talbot CC Jr, Wright MW, Wain HM, Trowsdale J, Ziegler A, Beck S (2004) Gene map of the extended human MHC. Nat Rev Genet 5(12):889–899. doi:10.1038/nrg1489
62. Bharadwaj M, Illing P, Theodossis A, Purcell AW, Rossjohn J, McCluskey J (2012) Drug hypersensitivity and human leukocyte antigens of the major histocompatibility complex. Annu Rev Pharmacol Toxicol 52:401–431. doi:10.1146/annurev-pharmtox-010611-134701
63. Joffre OP, Segura E, Savina A, Amigorena S (2012) Cross-presentation by dendritic cells. Nat Rev Immunol 12(8):557–569. doi:10.1038/nri3254
64. Reche PA, Reinherz EL (2003) Sequence variability analysis of human class I and class II MHC molecules: functional and structural correlates of amino acid polymorphisms. J Mol Biol 331(3):623–641
65. Illing PT, Vivian JP, Purcell AW, Rossjohn J, McCluskey J (2013) Human leukocyte antigen-associated drug hypersensitivity. Curr Opin Immunol 25(1):81–89. doi:10.1016/j.coi.2012.10.002
66. Kurkó J, Besenyei T, Laki J, Glant TT, Mikecz K, Szekanecz Z (2013) Genetics of rheumatoid arthritis—a comprehensive review. Clin Rev Allergy Immunol 45(2):170-179. doi:10.1007/s12016-012-8346-7
67. Hetherington S, McGuirk S, Powell G, Cutrell A, Naderer O, Spreen B, Lafon S, Pearce G, Steel H (2001) Hypersensitivity reactions during therapy with the nucleoside reverse transcriptase inhibitor abacavir. Clin Ther 23(10):1603–1614
68. Pichler W, Yawalkar N, Schmid S, Helbling A (2002) Pathogenesis of drug-induced exanthems. Allergy 57(10):884–893
69. Nassif A, Bensussan A, Boumsell L, Deniaud A, Moslehi H, Wolkenstein P, Bagot M, Roujeau J-C (2004) Toxic epidermal necrolysis: effector cells are drug-specific cytotoxic T cells. J Allergy Clin Immunol 114(5):1209–1215. doi:10.1016/j.jaci.2004.07.047
70. Phillips EJ, Sullivan JR, Knowles SR, Shear NH (2002) Utility of patch testing in patients with hypersensitivity syndromes associated with abacavir. AIDS 16(16):2223–2225
71. Alfirevic A, Park BK, Pirmohamed M, Naisbitt DJ (2012) Research highlights: explanation for HLA-B*57:01-linked immune-mediated abacavir-induced hypersensitivity. Pharmacogenomics 13(14):1567–1569. doi:10.2217/pgs.12.146
72. Park BK, Naisbitt DJ, Gordon SF, Kitteringham NR, Pirmohamed M (2001) Metabolic activation in drug allergies. Toxicology 158(1–2):11–23
73. Pichler WJ, Beeler A, Keller M, Lerch M, Posadas S, Schmid D, Spanou Z, Zawodniak A, Gerber B (2006) Pharmacological interaction of drugs with immune receptors: the p-i concept. Allergol Int 55(1):17–25. doi:10.2332/allergolint.55.17
74. Pirmohamed M, Naisbitt DJ, Gordon F, Park BK (2002) The danger hypothesis-potential role in idiosyncratic drug reactions. Toxicology 181-182:55–63
75. Alfirevic A, Pirmohamed M (2010) Drug-induced hypersensitivity reactions and pharmacogenomics: past, present and future. Pharmacogenomics 11(4):497–499. doi:10.2217/pgs.10.12

76. Chaponda M, Pirmohamed M (2011) Hypersensitivity reactions to HIV therapy. Br J Clin Pharmacol 71(5):659–671. doi:10.1111/j.1365-2125.2010.03784.x
77. Mallal S, Phillips E, Carosi G, Molina J-M, Workman C, Tomazic J, Jägel-Guedes E, Rugina S, Kozyrev O, Cid JF, Hay P, Nolan D, Hughes S, Hughes A, Ryan S, Fitch N, Thorborn D, Benbow A (2008) HLA-B*5701 screening for hypersensitivity to abacavir. N Engl J Med 358(6):568–579. doi:10.1056/NEJMoa0706135
78. Phillips EJ, Mallal SA (2010) Pharmacogenetics of drug hypersensitivity. Pharmacogenomics 11(7):973–987. doi:10.2217/pgs.10.77
79. Hughes AR, Mosteller M, Bansal AT, Davies K, Haneline SA, Lai EH, Nangle K, Scott T, Spreen WR, Warren LL, Roses AD (2004) Association of genetic variations in HLA-B region with hypersensitivity to abacavir in some, but not all, populations. Pharmacogenomics 5(2):203–211. doi:10.1517/phgs.5.2.203.27481
80. Hughes DA, Vilar FJ, Ward CC, Alfirevic A, Park BK, Pirmohamed M (2004) Cost-effectiveness analysis of HLA B*5701 genotyping in preventing abacavir hypersensitivity. Pharmacogenetics 14(6):335–342
81. Saag M, Balu R, Phillips E, Brachman P, Martorell C, Burman W, Stancil B, Mosteller M, Brothers C, Wannamaker P, Hughes A, Sutherland-Phillips D, Mallal S, Shaefer M (2008) High sensitivity of human leukocyte antigen-b*5701 as a marker for immunologically confirmed abacavir hypersensitivity in white and black patients. Clin Infect Dis 46(7):1111–1118. doi:10.1086/529382
82. Dervieux T, Bala MV (2006) Overview of the pharmacoeconomics of pharmacogenetics. Pharmacogenomics 7(8):1175–1184. doi:10.2217/14622416.7.8.1175
83. Schackman BR, Scott CA, Walensky RP, Losina E, Freedberg KA, Sax PE (2008) The cost-effectiveness of HLA-B*5701 genetic screening to guide initial antiretroviral therapy for HIV. AIDS 22(15):2025–2033. doi:10.1097/QAD.0b013e3283103ce6
84. Lalonde RG, Thomas R, Rachlis A, Gill MJ, Roger M, Angel JB, Smith G, Higgins N, Trottier B (2010) Successful implementation of a national HLA-B*5701 genetic testing service in Canada. Tissue Antigens 75(1):12–18. doi:10.1111/j.1399-0039.2009.01383.x
85. Rauch A, Nolan D, Martin A, McKinnon E, Almeida C, Mallal S (2006) Prospective genetic screening decreases the incidence of abacavir hypersensitivity reactions in the Western Australian HIV cohort study. Clin Infect Dis 43(1):99–102. doi:10.1086/504874
86. Waters LJ, Mandalia S, Gazzard B, Nelson M (2007) Prospective HLA-B*5701 screening and abacavir hypersensitivity: a single centre experience. AIDS 21(18):2533–2534. doi:10.1097/QAD.0b013e328273bc07
87. Zucman D, Truchis Pd, Majerholc C, Stegman S, Caillat-Zucman S (2007) Prospective screening for human leukocyte antigen-B*5701 avoids abacavir hypersensitivity reaction in the ethnically mixed French HIV population. J Acquir Immune Defic Syndr 45(1):1–3. doi:10.1097/QAI.0b013e318046ea31
88. Illing PT, Vivian JP, Dudek NL, Kostenko L, Chen Z, Bharadwaj M, Miles JJ, Kjer-Nielsen L, Gras S, Williamson NA, Burrows SR, Purcell AW, Rossjohn J, McCluskey J (2012) Immune self-reactivity triggered by drug-modified HLA-peptide repertoire. Nature 486(7404):554–558. doi:10.1038/nature11147
89. Ostrov DA, Grant BJ, Pompeu YA, Sidney J, Harndahl M, Southwood S, Oseroff C, Lu S, Jakoncic J, de Oliveira CAF, Yang L, Mei H, Shi L, Shabanowitz J, English AM, Wriston A, Lucas A, Phillips E, Mallal S, Grey HM, Sette A, Hunt DF, Buus S, Peters B (2012) Drug hypersensitivity caused by alteration of the MHC-presented self-peptide repertoire. Proc Natl Acad Sci U S A 109(25):9959–9964. doi:10.1073/pnas.1207934109
90. Norcross MA, Luo S, Lu L, Boyne MT, Gomarteli M, Rennels AD, Woodcock J, Margulies DH, McMurtrey C, Vernon S, Hildebrand WH, Buchli R (2012) Abacavir induces loading of novel self-peptides into HLA-B*57: 01: an autoimmune model for HLA-associated drug hypersensitivity. AIDS 26(11):F21–29. doi:10.1097/QAD.0b013e328355fe8f
91. Adam J, Eriksson KK, Schnyder B, Fontana S, Pichler WJ, Yerly D (2012) Avidity determines T-cell reactivity in abacavir hypersensitivity. Eur J Immunol 42(7):1706–1716. doi:10.1002/eji.201142159

92. Picard D, Janela B, Descamps V, D'Incan M, Courville P, Jacquot S, Rogez S, Mardivirin L, Moins-Teisserenc H, Toubert A, Benichou J, Joly P, Musette P (2010) Drug reaction with eosinophilia and systemic symptoms (DRESS): a multiorgan antiviral T cell response. Sci Transl Med 2(46):46ra62. doi:10.1126/scitranslmed.3001116
93. Marson AG, Al-Kharusi AM, Alwaidh M, Appleton R, Baker GA, Chadwick DW, Cramp C, Cockerell OC, Cooper PN, Doughty J, Eaton B, Gamble C, Goulding PJ, Howell SJL, Hughes A, Jackson M, Jacoby A, Kellett M, Lawson GR, Leach JP, Nicolaides P, Roberts R, Shackley P, Shen J, Smith DF, Smith PEM, Smith CT, Vanoli A, Williamson PR (2007) The SANAD study of effectiveness of carbamazepine, gabapentin, lamotrigine, oxcarbazepine, or topiramate for treatment of partial epilepsy: an unblinded randomised controlled trial. Lancet 369(9566):1000–1015. doi:10.1016/s0140-6736(07)60460-7
94. Pirmohamed M, Friedmann PS, Molokhia M, Loke YK, Smith C, Phillips E, Grenade L L, Carleton B, Papaluca-Amati M, Demoly P, Shear NH (2011) Phenotype standardization for immune-mediated drug-induced skin injury. Clin Pharmacol Ther 89(6):896–901. doi:10.1038/clpt.2011.79
95. Roujeau JC, Stern RS (1994) Severe adverse cutaneous reactions to drugs. N Engl J Med 331(19):1272–1285. doi:10.1056/nejm199411103311906
96. Profaizer T, Eckels D (2012) HLA alleles and drug hypersensitivity reactions. Int J Immunogenet 39(2):99–105. doi:10.1111/j.1744-313X.2011.01061.x
97. Criado PR, Avancini J, Santi CG, Medrado ATA, Rodrigues CE, de Carvalho JF (2012) Drug reaction with eosinophilia and systemic symptoms (DRESS): a complex interaction of drugs, viruses and the immune system. Isr Med Assoc J 14(9):577–582
98. Borchers AT, Lee JL, Naguwa SM, Cheema GS, Gershwin ME (2008) Stevens-Johnson syndrome and toxic epidermal necrolysis. Autoimmun Rev 7(8):598–605. doi:10.1016/j.autrev.2008.06.004
99. Bastuji-Garin S, Rzany B, Stern RS, Shear NH, Naldi L, Roujeau JC (1993) Clinical classification of cases of toxic epidermal necrolysis, Stevens-Johnson syndrome, and erythema multiforme. Arch Dermatol 129(1):92–96
100. Ding WY, Lee CK, Choon SE (2010) Cutaneous adverse drug reactions seen in a tertiary hospital in Johor, Malaysia. Int J Dermatol 49(7):834–841. doi:10.1111/j.1365-4632.2010.04481.x
101. Kulkantrakorn K, Tassaneeyakul W, Tiamkao S, Jantararoungtong T, Prabmechai N, Vannaprasaht S, Chumworathayi P, Chen P, Sritipsukho P (2012) HLA-B*1502 strongly predicts carbamazepine-induced Stevens-Johnson syndrome and toxic epidermal necrolysis in Thai patients with neuropathic pain. Pain Pract 12(3):202–208. doi:10.1111/j.1533-2500.2011.00479.x
102. Mehta TY, Prajapati LM, Mittal B, Joshi CG, Sheth JJ, Patel DB, Dave DM, Goyal RK (2009) Association of HLA-B*1502 allele and carbamazepine-induced Stevens-Johnson syndrome among Indians. Indian J Dermatol Venereol Leprol 75(6):579–582. doi:10.4103/0378-6323.57718
103. Chen P, Lin J-J, Lu C-S, Ong C-T, Hsieh PF, Yang C-C, Tai C-T, Wu S-L, Lu C-H, Hsu Y-C, Yu H-Y, Ro L-S, Lu C-T, Chu C-C, Tsai J-J, Su Y-H, Lan S-H, Sung S-F, Lin S-Y, Chuang H-P, Huang L-C, Chen Y-J, Tsai P-J, Liao H-T, Lin Y-H, Chen C-H, Chung W-H, Hung S-I, Wu J-Y, Chang C-F, Chen L, Chen Y-T, Shen C-Y (2011) Carbamazepine-induced toxic effects and HLA-B*1502 screening in Taiwan. N Engl J Med 364(12):1126–1133. doi:10.1056/NEJMoa1009717
104. Alfirevic A, Jorgensen AL, Williamson PR, Chadwick DW, Park BK, Pirmohamed M (2006) HLA-B locus in Caucasian patients with carbamazepine hypersensitivity. Pharmacogenomics 7(6):813–818. doi:10.2217/14622416.7.6.813
105. Lonjou C, Thomas L, Borot N, Ledger N, de Toma C, LeLouet H, Graf E, Schumacher M, Hovnanian A, Mockenhaupt M, Roujeau JC (2006) A marker for Stevens-Johnson syndrome…: ethnicity matters. Pharmacogenomics J 6(4):265–268. doi:10.1038/sj.tpj.6500356
106. Ikeda H, Takahashi Y, Yamazaki E, Fujiwara T, Kaniwa N, Saito Y, Aihara M, Kashiwagi M, Muramatsu M (2010) HLA class I markers in Japanese patients with carbamaze-

pine-induced cutaneous adverse reactions. Epilepsia 51(2):297–300. doi:10.1111/j.1528-1167.2009.02269.x
107. Ozeki T, Mushiroda T, Yowang A, Takahashi A, Kubo M, Shirakata Y, Ikezawa Z, Iijima M, Shiohara T, Hashimoto K, Kamatani N, Nakamura Y (2011) Genome-wide association study identifies HLA-A*3101 allele as a genetic risk factor for carbamazepine-induced cutaneous adverse drug reactions in Japanese population. Hum Mol Genet 20(5):1034–1041. doi:10.1093/hmg/ddq537
108. Kaniwa N, Saito Y, Aihara M, Matsunaga K, Tohkin M, Kurose K, Furuya H, Takahashi Y, Muramatsu M, Kinoshita S, Abe M, Ikeda H, Kashiwagi M, Song Y, Ueta M, Sotozono C, Ikezawa Z, Hasegawa R (2010) HLA-B*1511 is a risk factor for carbamazepine-induced Stevens-Johnson syndrome and toxic epidermal necrolysis in Japanese patients. Epilepsia 51(12):2461–2465. doi:10.1111/j.1528-1167.2010.02766.x
109. Yun J, Adam J, Yerly D, Pichler WJ (2012) Human leukocyte antigens (HLA) associated drug hypersensitivity: consequences of drug binding to HLA. Allergy 67(11):1338–1346. doi:10.1111/all.12008
110. Wei C-Y, Chung W-H, Huang H-W, Chen Y-T, Hung S-I (2012) Direct interaction between HLA-B and carbamazepine activates T cells in patients with Stevens-Johnson syndrome. J Allergy Clin Immunol 129(6):1562–1569.e1565. doi:10.1016/j.jaci.2011.12.990
111. Ko T-M, Chung W-H, Wei C-Y, Shih H-Y, Chen J-K, Lin C-H, Chen Y-T, Hung S-I (2011) Shared and restricted T-cell receptor use is crucial for carbamazepine-induced Stevens-Johnson syndrome. J Allergy Clin Immunol 128(6):1266–1276.e11. doi:10.1016/j.jaci.2011.08.013
112. Hershfield MS, Callaghan JT, Tassaneeyakul W, Mushiroda T, Thorn CF, Klein TE, Lee MTM (2013) Clinical Pharmacogenetics Implementation Consortium guidelines for human leukocyte antigen-B genotype and allopurinol dosing. Clin Pharmacol Ther 93(2):153–158. doi:10.1038/clpt.2012.209
113. Halevy S, Ghislain P-D, Mockenhaupt M, Fagot J-P, Bouwes Bavinck JN, Sidoroff A, Naldi L, Dunant A, Viboud C, Roujeau J-C (2008) Allopurinol is the most common cause of Stevens-Johnson syndrome and toxic epidermal necrolysis in Europe and Israel. J Am Acad Dermatol 58(1):25–32. doi:10.1016/j.jaad.2007.08.036
114. Cacoub P, Musette P, Descamps V, Meyer O, Speirs C, Finzi L, Roujeau JC (2011) The DRESS syndrome: a literature review. Am J Med 124(7):588–597. doi:10.1016/j.amjmed.2011.01.017
115. Kaniwa N, Saito Y, Aihara M, Matsunaga K, Tohkin M, Kurose K, Sawada J-i, Furuya H, Takahashi Y, Muramatsu M, Kinoshita S, Abe M, Ikeda H, Kashiwagi M, Song Y, Ueta M, Sotozono C, Ikezawa Z, Hasegawa R (2008) HLA-B locus in Japanese patients with antiepileptics and allopurinol-related Stevens-Johnson syndrome and toxic epidermal necrolysis. Pharmacogenomics 9(11):1617–1622. doi:10.2217/14622416.9.11.1617
116. Somkrua R, Eickman EE, Saokaew S, Lohitnavy M, Chaiyakunapruk N (2011) Association of HLA-B*5801 allele and allopurinol-induced Stevens Johnson syndrome and toxic epidermal necrolysis: a systematic review and meta-analysis. BMC Med Genet 12:118. doi:10.1186/1471-2350-12-118
117. Lee MH, Stocker SL, Anderson J, Phillips EJ, Nolan D, Williams KM, Graham GG, Sullivan JR, Day RO (2012) Initiating allopurinol therapy: do we need to know the patient's human leucocyte antigen status? Intern Med J 42(4):411–416. doi:10.1111/j.1445-5994.2011.02567.x
118. Shiohara T, Inaoka M, Kano Y (2006) Drug-induced hypersensitivity syndrome (DIHS): a reaction induced by a complex interplay among herpesviruses and antiviral and antidrug immune responses. Allergol Int 55(1):1–8. doi:10.2332/allergolint.55.1
119. Chao J, Terkeltaub R (2009) A critical reappraisal of allopurinol dosing, safety, and efficacy for hyperuricemia in gout. Curr Rheumatol Rep 11(2):135–140
120. Russmann S, Kaye JA, Jick SS, Jick H (2005) Risk of cholestatic liver disease associated with flucloxacillin and flucloxacillin prescribing habits in the UK: cohort study using

data from the UK general practice research database. Br J Clin Pharmacol 60(1):76–82. doi:10.1111/j.1365-2125.2005.02370.x
121. Daly AK, Day CP (2012) Genetic association studies in drug-induced liver injury. Drug Metab Rev 44(1):116–126. doi:10.3109/03602532.2011.605790
122. Ostapowicz G, Fontana RJ, Schiødt FV, Larson A, Davern TJ, Han SHB, McCashland TM, Shakil AO, Hay JE, Hynan L, Crippin JS, Blei AT, Samuel G, Reisch J, Lee WM (2002) Results of a prospective study of acute liver failure at 17 tertiary care centers in the United States. Ann Intern Med 137(12):947–954
123. Russo MW, Galanko JA, Shrestha R, Fried MW, Watkins P (2004) Liver transplantation for acute liver failure from drug induced liver injury in the United States. Liver Transpl 10(8):1018–1023. doi:10.1002/lt.20204
124. Sgro C, Clinard F, Ouazir K, Chanay H, Allard C, Guilleminet C, Lenoir C, Lemoine A, Hillon P (2002) Incidence of drug-induced hepatic injuries: a French population-based study. Hepatology 36(2):451–455. doi:10.1053/jhep.2002.34857
125. Aithal GP, Watkins PB, Andrade RJ, Larrey D, Molokhia M, Takikawa H, Hunt CM, Wilke RA, Avigan M, Kaplowitz N, Bjornsson E, Daly AK (2011) Case definition and phenotype standardization in drug-induced liver injury. Clin Pharmacol Ther 89(6):806–815. doi:10.1038/clpt.2011.58
126. Uetrecht J (2009) Immunoallergic drug-induced liver injury in humans. Semin Liver Dis 29(4):383–392. doi:10.1055/s-0029-1240007
127. Alfirevic A, Pirmohamed M (2012) Predictive genetic testing for drug-induced liver injury: considerations of clinical utility. Clin Pharmacol Ther 92(3):376–380. doi:10.1038/clpt.2012.107
128. Jenkins RE, Meng X, Elliott VL, Kitteringham NR, Pirmohamed M, Park BK (2009) Characterisation of flucloxacillin and 5-hydroxymethyl flucloxacillin haptenated HSA in vitro and in vivo. Proteomics Clin Appl 3(6):720–729. doi:10.1002/prca.200800222
129. Monshi M, Faulkner L, Gibson A, Jenkins RE, Farrell J, Earnshaw CJ, Alfirevic A, Cederbrant K, Daly AK, French N, Pirmohamed M, Park BK, Naisbitt DJ (2013) HLA-B*57:01-restricted activation of drug-specific T-cells provides the immunological basis for flucloxacillin-induced liver injury. Hepatology 57(2):727–739. doi:10.1002/hep.26077
130. Li Y, Tang HL, Hu YF, Xie HG (2012) The gain-of-function variant allele CYP2C19*17: a double-edged sword between thrombosis and bleeding in clopidogrel-treated patients. J Thromb Haemost 10(2):199–206. doi:10.1111/j.1538-7836.2011.04570.x
131. Zabalza M, Subirana I, Sala J, Lluis-Ganella C, Lucas G, Tomás M, Masiá R, Marrugat J, Brugada R, Elosua R (2012) Meta-analyses of the association between cytochrome CYP2C19 loss- and gain-of-function polymorphisms and cardiovascular outcomes in patients with coronary artery disease treated with clopidogrel. Heart 98(2):100–108. doi:10.1136/hrt.2011.227652
132. Higashi MK, Veenstra DL, Kondo LM, Wittkowsky AK, Srinouanprachanh SL, Farin FM, Rettie AE (2002) Association between CYP2C9 genetic variants and anticoagulation-related outcomes during warfarin therapy. JAMA 287(13):1690–1698
133. Aithal GP, Day CP, Kesteven PJ, Daly AK (1999) Association of polymorphisms in the cytochrome P450 CYP2C9 with warfarin dose requirement and risk of bleeding complications. Lancet 353(9154):717–719. doi:10.1016/s0140-6736(98)04474-2
134. Limdi NA, McGwin G, Goldstein JA, Beasley TM, Arnett DK, Adler BK, Baird MF, Acton RT (2008) Influence of CYP2C9 and VKORC1 1173C/T genotype on the risk of hemorrhagic complications in African-American and European-American patients on warfarin. Clin Pharmacol Ther 83(2):312–321. doi:10.1038/sj.clpt.6100290
135. Yang J, Chen Y, Li X, Wei X, Chen X, Zhang L, Zhang Y, Xu Q, Wang H, Li Y, Lu C, Chen W, Zeng C, Yin T (2013) Influence of CYP2C9 and VKORC1 genotypes on the risk of hemorrhagic complications in warfarin-treated patients: a systematic review and meta-analysis. Int J Cardiol 168(4):4234–4243. doi:10.1016/j.ijcard.2013.07.151

136. Gasche Y, Daali Y, Fathi M, Chiappe A, Cottini S, Dayer P, Desmeules J (2004) Codeine intoxication associated with ultrarapid CYP2D6 metabolism. N Engl J Med 351(27):2827–2831. doi:10.1056/NEJMoa041888
137. Crews KR, Gaedigk A, Dunnenberger HM, Klein TE, Shen DD, Callaghan JT, Kharasch ED, Skaar TC (2012) Clinical Pharmacogenetics Implementation Consortium (CPIC) guidelines for codeine therapy in the context of cytochrome P450 2D6 (CYP2D6) genotype. Clin Pharmacol Ther 91(2):321–326. doi:10.1038/clpt.2011.287
138. Kirchheiner J, Keulen J-THA, Bauer S, Roots I, Brockmöller J (2008) Effects of the CYP2D6 gene duplication on the pharmacokinetics and pharmacodynamics of tramadol. J Clin Psychopharmacol 28(1):78–83. doi:10.1097/JCP.0b013e318160f827
139. Stamer UM, Stüber F, Muders T, Musshoff F (2008) Respiratory depression with tramadol in a patient with renal impairment and CYP2D6 gene duplication. Anesth Analg 107(3):926–929. doi:10.1213/ane.0b013e31817b796e
140. Cai Y, Yi J, Zhou C, Shen X (2012) Pharmacogenetic study of drug-metabolising enzyme polymorphisms on the risk of anti-tuberculosis drug-induced liver injury: a meta-analysis. PLoS ONE 7(10):e47769. doi:10.1371/journal.pone.0047769
141. Huang Y-S, Chern H-D, Su W-J, Wu J-C, Chang S-C, Chiang C-H, Chang F-Y, Lee S-D (2003) Cytochrome P450 2E1 genotype and the susceptibility to antituberculosis drug-induced hepatitis. Hepatology 37(4):924–930. doi:10.1053/jhep.2003.50144
142. Roy B, Chowdhury A, Kundu S, Santra A, Dey B, Chakraborty M, Majumder PP (2001) Increased risk of antituberculosis drug-induced hepatotoxicity in individuals with glutathione S-transferase M1 'null' mutation. J Gastroenterol Hepatol 16(9):1033–1037
143. Daly AK, Aithal GP, Leathart JBS, Swainsbury RA, Dang TS, Day CP (2007) Genetic susceptibility to diclofenac-induced hepatotoxicity: contribution of UGT2B7, CYP2C8, and ABCC2 genotypes. Gastroenterology 132(1):272–281. doi:10.1053/j.gastro.2006.11.023
144. Simon T, Becquemont L, Mary-Krause M, de Waziers I, Beaune P, Funck-Brentano C, Jaillon P (2000) Combined glutathione-S-transferase M1 and T1 genetic polymorphism and tacrine hepatotoxicity. Clin Pharmacol Ther 67(4):432–437. doi:10.1067/mcp.2000.104944
145. Watanabe I, Tomita A, Shimizu M, Sugawara M, Yasumo H, Koishi R, Takahashi T, Miyoshi K, Nakamura K, Izumi T, Matsushita Y, Furukawa H, Haruyama H, Koga T (2003) A study to survey susceptible genetic factors responsible for troglitazone-associated hepatotoxicity in Japanese patients with type 2 diabetes mellitus. Clin Pharmacol Ther 73(5):435–455
146. Iyer L, King CD, Whitington PF, Green MD, Roy SK, Tephly TR, Coffman BL, Ratain MJ (1998) Genetic predisposition to the metabolism of irinotecan (CPT-11). Role of uridine diphosphate glucuronosyltransferase isoform 1A1 in the glucuronidation of its active metabolite (SN-38) in human liver microsomes. J Clin Invest 101(4):847–854. doi:10.1172/jci915
147. de Leon J, Susce MT, Pan R-M, Fairchild M, Koch WH, Wedlund PJ (2005) The CYP2D6 poor metabolizer phenotype may be associated with risperidone adverse drug reactions and discontinuation. J Clin Psychiatry 66(1):15–27
148. Link E, Parish S, Armitage J, Bowman L, Heath S, Matsuda F, Gut I, Lathrop M, Collins R (2008) SLCO1B1 variants and statin-induced myopathy-a genomewide study. N Engl J Med 359(8):789–799. doi:10.1056/NEJMoa0801936
149. Booth RA, Ansari MT, Loit E, Tricco AC, Weeks L, Doucette S, Skidmore B, Sears M, Sy R, Karsh J (2011) Assessment of thiopurine S-methyltransferase activity in patients prescribed thiopurines: a systematic review. Ann Intern Med 154(12):814–823. doi:10.1059/0003-4819-154-12-201106210-00009W-295-298
150. Musumba CO, Jorgensen A, Sutton L, Van Eker D, Zhang E, O'Hara N, Carr DF, Pritchard DM, Pirmohamed M (2013) CYP2C19*17 gain-of-function polymorphism is associated with peptic ulcer disease. Clin Pharmacol Ther 93(2):195–203. doi:10.1038/clpt.2012.215
151. Yen T, Nightingale BN, Burns JC, Sullivan DR, Stewart PM (2003) Butyrylcholinesterase (BCHE) genotyping for post-succinylcholine apnea in an Australian population. Clin Chem 49(8):1297–1308
152. Mega JL, Simon T, Collet J-P, Anderson JL, Antman EM, Bliden K, Cannon CP, Danchin N, Giusti B, Gurbel P, Horne BD, Hulot J-S, Kastrati A, Montalescot G, Neumann F-J,

Shen L, Sibbing D, Steg PG, Trenk D, Wiviott SD, Sabatine MS (2010) Reduced-function CYP2C19 genotype and risk of adverse clinical outcomes among patients treated with clopidogrel predominantly for PCI: a meta-analysis. JAMA 304(16):1821–1830. doi:10.1001/jama.2010.1543

153. Amstutz U, Froehlich TK, Largiadèr CR (2011) Dihydropyrimidine dehydrogenase gene as a major predictor of severe 5-fluorouracil toxicity. Pharmacogenomics 12(9):1321–1336. doi:10.2217/pgs.11.72
154. Dávila-Fajardo CL, Swen JJ, Cabeza Barrera J, Guchelaar H-J (2013) Genetic risk factors for drug-induced liver injury in rheumatoid arthritis patients using low-dose methotrexate. Pharmacogenomics 14(1):63–73. doi:10.2217/pgs.12.183
155. European Malignant Hyperthermia Group: Home. http://www.emhg.org/. Accessed 8 Feb 2013
156. Robinson R, Carpenter D, Shaw M-A, Halsall J, Hopkins P (2006) Mutations in RYR1 in malignant hyperthermia and central core disease. Hum Mutat 27(10):977–989. doi:10.1002/humu.20356
157. Monnier N, Procaccio V, Stieglitz P, Lunardi J (1997) Malignant-hyperthermia susceptibility is associated with a mutation of the alpha 1-subunit of the human dihydropyridine-sensitive L-type voltage-dependent calcium-channel receptor in skeletal muscle. Am J Hum Genet 60(6):1316–1325
158. Toppin PJ, Chandy TT, Ghanekar A, Kraeva N, Beattie WS, Riazi S (2010) A report of fulminant malignant hyperthermia in a patient with a novel mutation of the CACNA1S gene. Can J Anaesth 57(7):689–693. doi:10.1007/s12630-010-9314-4
159. Mulder H, Cohen D, Scheffer H, Gispen-de Wied C, Arends J, Wilmink FW, Franke B, Egberts ACG (2009) HTR2C gene polymorphisms and the metabolic syndrome in patients with schizophrenia: a replication study. J Clin Psychopharmacol 29(1):16–20. doi:10.1097/JCP.0b013e3181934462
160. Minucci A, Moradkhani K, Hwang MJ, Zuppi C, Giardina B, Capoluongo E (2012) Glucose-6-phosphate dehydrogenase (G6PD) mutations database: review of the "old" and update of the new mutations. Blood Cells Mol Dis 48(3):154–165. doi:10.1016/j.bcmd.2012.01.001
161. Youngster I, Arcavi L, Schechmaster R, Akayzen Y, Popliski H, Shimonov J, Beig S, Berkovitch M (2010) Medications and glucose-6-phosphate dehydrogenase deficiency: an evidence-based review. Drug Saf 33(9):713–726. doi:10.2165/11536520-000000000-00000
162. Jennings BA, Kwok CS, Willis G, Matthews V, Wawruch P, Loke YK (2012) Functional polymorphisms of folate metabolism and response to chemotherapy for colorectal cancer, a systematic review and meta-analysis. Pharmacogenet Genomics 22(4):290–304. doi:10.1097/FPC.0b013e328351875d
163. Denborough MA, Forster JF, Lovell RR, Maplestone PA, Villiers JD (1962) Anaesthetic deaths in a family. Br J Anaesth 34:395–396
164. Larard DG, Rice CP, Robinson R, Spencer RW, Westhead RA (1972) Malignant hyperthermia: a study of an affected family. Br J Anaesth 44(1):93–96
165. Matos AR, Sambuughin N, Rumjanek FD, Amoedo ND, Cunha LBP, Zapata-Sudo G, Sudo RT (2009) Multigenerational Brazilian family with malignant hyperthermia and a novel mutation in the RYR1 gene. Braz J Med Biol Res 42(12):1218–1224
166. Strazis KP, Fox AW (1993) Malignant hyperthermia: a review of published cases. Anesth Analg 77(2):297–304
167. Hopkins PM (2011) Malignant hyperthermia: pharmacology of triggering. Br J Anaesth 107(1):48–56. doi:10.1093/bja/aer132
168. Dexter F, Epstein RH, Wachtel RE, Rosenberg H (2013) Estimate of the relative risk of succinylcholine for triggering malignant hyperthermia. Anesth Analg 116(1):118–122. doi:10.1213/ANE.0b013e31826f5e3b
169. Denborough M (1998) Malignant hyperthermia. Lancet 352(9134):1131–1136. doi:10.1016/s0140-6736(98)03078-5
170. Groom L, Muldoon SM, Tang ZZ, Brandom BW, Bayarsaikhan M, Bina S, Lee H-S, Qiu X, Sambuughin N, Dirksen RT (2011) Identical de novo mutation in the type 1 ryanodine

receptor gene associated with fatal, stress-induced malignant hyperthermia in two unrelated families. Anesthesiology 115(5):938–945. doi:10.1097/ALN.0b013e3182320068
171. Gronert GA, Thompson RL, Onofrio BM (1980) Human malignant hyperthermia: awake episodes and correction by dantrolene. Anesth Analg 59(5):377–378
172. Correia ACdC, Silva PCB, da Silva BA (2012) Malignant hyperthermia: clinical and molecular aspects. Rev Bras Anestesiol 62(6):820–837. doi:10.1016/s0034-7094(12)70182-4
173. Rosenberg H, Davis M, James D, Pollock N, Stowell K (2007) Malignant hyperthermia. Orphanet J Rare Dis 2:21. doi:10.1186/1750-1172-2-21
174. Sinkovich DD, Mitch-Resignalo AE (1991) Malignant hyperthermia. Orthop Nurs 10(1):39–43
175. Larach MG, Gronert GA, Allen GC, Brandom BW, Lehman EB (2010) Clinical presentation, treatment, and complications of malignant hyperthermia in North America from 1987-2006. Anesth Analg 110(2):498–507. doi:10.1213/ANE.0b013e3181c6b9b2
176. Bandschapp O, Girard T (2012) Malignant hyperthermia. Swiss Med Wkly 142:w13652. doi:10.4414/smw.2012.13652
177. Larach MG, Localio AR, Allen GC, Denborough MA, Ellis FR, Gronert GA, Kaplan RF, Muldoon SM, Nelson TE, Ording H (1994) A clinical grading scale to predict malignant hyperthermia susceptibility. Anesthesiology 80(4):771–779
178. Lanner JT, Georgiou DK, Joshi AD, Hamilton SL (2010) Ryanodine receptors: structure, expression, molecular details, and function in calcium release. Cold Spring Harb Perspect Biol 2(11):a003996. doi:10.1101/cshperspect.a003996
179. Van Petegem F (2012) Ryanodine receptors: structure and function. J Biol Chem 287(38):31624–31632. doi:10.1074/jbc.R112.349068
180. Takeshima H, Nishimura S, Matsumoto T, Ishida H, Kangawa K, Minamino N, Matsuo H, Ueda M, Hanaoka M, Hirose T (1989) Primary structure and expression from complementary DNA of skeletal muscle ryanodine receptor. Nature 339(6224):439–445. doi:10.1038/339439a0
181. Block BA, Imagawa T, Campbell KP, Franzini-Armstrong C (1988) Structural evidence for direct interaction between the molecular components of the transverse tubule/sarcoplasmic reticulum junction in skeletal muscle. J Cell Biol 107(6 Pt 2):2587–2600
182. Inui M, Saito A, Fleischer S (1987) Purification of the ryanodine receptor and identity with feet structures of junctional terminal cisternae of sarcoplasmic reticulum from fast skeletal muscle. J Biol Chem 262(4):1740–1747
183. Meissner G, Lu X (1995) Dihydropyridine receptor-ryanodine receptor interactions in skeletal muscle excitation-contraction coupling. Biosci Rep 15(5):399–408
184. Fujii J, Otsu K, Zorzato F, de Leon S, Khanna VK, Weiler JE, O'Brien PJ, MacLennan DH (1991) Identification of a mutation in porcine ryanodine receptor associated with malignant hyperthermia. Science 253(5018):448–451
185. Gillard EF, Otsu K, Fujii J, Khanna VK, de Leon S, Derdemezi J, Britt BA, Duff CL, Worton RG, MacLennan DH (1991) A substitution of cysteine for arginine 614 in the ryanodine receptor is potentially causative of human malignant hyperthermia. Genomics 11(3):751–755
186. Fill M, Stefani E, Nelson TE (1991) Abnormal human sarcoplasmic reticulum Ca2+release channels in malignant hyperthermic skeletal muscle. Biophys J 59(5):1085–1090. doi:10.1016/s0006-3495(91)82323-2
187. Ohta T, Endo M, Nakano T, Morohoshi Y, Wanikawa K, Ohga A (1989) Ca-induced Ca release in malignant hyperthermia-susceptible pig skeletal muscle. Am J Physiol Cell Physiol 256(2):C358–C367
188. Kim D-C (2012) Malignant hyperthermia. Korean J Anesthesiol 63(5):391–401. doi:10.4097/kjae.2012.63.5.391
189. Klingler W, Rueffert H, Lehmann-Horn F, Girard T, Hopkins PM (2009) Core myopathies and risk of malignant hyperthermia. Anesth Analg 109(4):1167–1173. doi:10.1213/ANE.0b013e3181b5ae2d
190. Isaacs H, Badenhorst ME (1992) Dominantly inherited malignant hyperthermia (MH) in the King-Denborough syndrome. Muscle Nerve 15(6):740–742. doi:10.1002/mus.880150619

191. D'Arcy CE, Bjorksten A, Yiu EM, Bankier A, Gillies R, McLean CA, Shield LK, Ryan MM (2008) King-denborough syndrome caused by a novel mutation in the ryanodine receptor gene. Neurology 71(10):776–777. doi:10.1212/01.wnl.0000324929.33780.2f
192. Hirshey Dirksen SJ, Larach MG, Rosenberg H, Brandom BW, Parness J, Lang RS, Gangadharan M, Pezalski T (2011) Special article: future directions in malignant hyperthermia research and patient care. Anesth Analg 113(5):1108–1119. doi:10.1213/ANE.0b013e318222af2e
193. Capacchione JF, Sambuughin N, Bina S, Mulligan LP, Lawson TD, Muldoon SM (2010) Exertional rhabdomyolysis and malignant hyperthermia in a patient with ryanodine receptor type 1 gene, L-type calcium channel alpha-1 subunit gene, and calsequestrin-1 gene polymorphisms. Anesthesiology 112(1):239–244. doi:10.1097/ALN.0b013e3181c29504
194. Wappler F, Fiege M, Steinfath M, Agarwal K, Scholz J, Singh S, Matschke J, Schulte Am Esch J (2001) Evidence for susceptibility to malignant hyperthermia in patients with exercise-induced rhabdomyolysis. Anesthesiology 94(1):95–100
195. (1984) A protocol for the investigation of malignant hyperpyrexia (MH) susceptibility. The European Malignant Hyperpyrexia Group. Br J Anaesth 56 (11):1267–1269
196. Urwyler A, Deufel T, McCarthy T, West S (2001) Guidelines for molecular genetic detection of susceptibility to malignant hyperthermia. Br J Anaesth 86(2):283–287
197. Ramachandran S, Chakraborty A, Xu L, Mei Y, Samso M, Dokholyan NV, Meissner G (2013) Structural determinants of skeletal muscle ryanodine receptor gating. J Biol Chem 288(9):6154–6165. doi:10.1074/jbc.M112.433789
198. Dewire SM, Yamashita DS, Rominger DH, Liu G, Cowan CL, Graczyk TM, Chen X-T, Pitis PM, Gotchev D, Yuan C, Koblish M, Lark MW, Violin JD (2013) A G protein-biased ligand at the mu-opioid receptor is potently analgesic with reduced gastrointestinal and respiratory dysfunction compared to morphine. J Pharmacol Exp Ther 344(3):708–717. doi:10.1124/jpet.112.201616
199. Kieffer BL (1999) Opioids: first lessons from knockout mice. Trends Pharmacol Sci 20(1):19–26
200. Mignat C, Wille U, Ziegler A (1995) Affinity profiles of morphine, codeine, dihydrocodeine and their glucuronides at opioid receptor subtypes. Life Sci 56(10):793–799
201. Eissing T, Lippert J, Willmann S (2012) Pharmacogenomics of codeine, morphine, and morphine-6-glucuronide: model-based analysis of the influence of CYP2D6 activity, UGT2B7 activity, renal impairment, and CYP3A4 inhibition. Mol Diagn Ther 16(1):43–53. doi:10.2165/11597930-000000000-00000
202. Lee CR, Goldstein JA, Pieper JA (2002) Cytochrome P450 2C9 polymorphisms: a comprehensive review of the in-vitro and human data. Pharmacogenetics 12(3):251–263
203. Guengerich FP (2008) Cytochrome p450 and chemical toxicology. Chem Res Toxicol 21(1):70–83. doi:10.1021/tx700079z
204. Sim SC, Ingelman-Sundberg M (2010) The Human Cytochrome P450 (CYP) Allele Nomenclature website: a peer-reviewed database of CYP variants and their associated effects. Hum Genomics 4(4):278–281. doi:10.1186/1479-7364-4-4-278
205. Gaedigk A, Simon SD, Pearce RE, Bradford LD, Kennedy MJ, Leeder JS (2008) The CYP2D6 activity score: translating genotype information into a qualitative measure of phenotype. Clin Pharmacol Ther 83(2):234–242. doi:10.1038/sj.clpt.6100406
206. Eckhardt K, Li S, Ammon S, Schänzle G, Mikus G, Eichelbaum M (1998) Same incidence of adverse drug events after codeine administration irrespective of the genetically determined differences in morphine formation. Pain 76(1–2):27–33
207. Sindrup SH, Brøsen K, Bjerring P, Arendt-Nielsen L, Larsen U, Angelo HR, Gram LF (1990) Codeine increases pain thresholds to copper vapor laser stimuli in extensive but not poor metabolizers of sparteine. Clin Pharmacol Ther 48(6):686–693
208. Kirchheiner J, Schmidt H, Tzvetkov M, Keulen JTHA, Lötsch J, Roots I, Brockmöller J (2007) Pharmacokinetics of codeine and its metabolite morphine in ultra-rapid metabolizers due to CYP2D6 duplication. Pharmacogenomics J 7(4):257–265. doi:10.1038/sj.tpj.6500406

209. Koren G, Cairns J, Chitayat D, Gaedigk A, Leeder SJ (2006) Pharmacogenetics of morphine poisoning in a breastfed neonate of a codeine-prescribed mother. Lancet 368 (9536):704. doi:10.1016/s0140-6736(06)69255-6
210. Kelly LE, Rieder M, van den Anker J, Malkin B, Ross C, Neely MN, Carleton B, Hayden MR, Madadi P, Koren G (2012) More codeine fatalities after tonsillectomy in North American children. Pediatrics 129(5):e1343–e1347. doi:10.1542/peds.2011-2538
211. Voronov P, Przybylo HJ, Jagannathan N (2007) Apnea in a child after oral codeine: a genetic variant—an ultra-rapid metabolizer. Paediatr Anaesth 17(7):684–687. doi:10.1111/j.1460-9592.2006.02182.x
212. Ciszkowski C, Madadi P, Phillips MS, Lauwers AE, Koren G (2009) Codeine, ultrarapid-metabolism genotype, and postoperative death. N Engl J Med 361(8):827–828. doi:10.1056/NEJMc0904266
213. Dalén P, Frengell C, Dahl ML, Sjöqvist F (1997) Quick onset of severe abdominal pain after codeine in an ultrarapid metabolizer of debrisoquine. Ther Drug Monit 19(5):543–544
214. Food and Drug Administration (2007) FDA Warning on Codeine Use by Nursing Mothers. http://www.fda.gov/NewsEvents/Newsroom/PressAnnouncements/2007/ucm108968.htm. Accessed 26 Jan 2013
215. Niesters M, Overdyk F, Smith T, Aarts L, Dahan A (2013) Opioid-induced respiratory depression in paediatrics: a review of case reports. Br J Anaesth 110(2):175–182. doi:10.1093/bja/aes447
216. Tong TF, Ng KK (2001) Codeine ingestion and apparent life-threatening event in a neonate. Pediatr Int 43(5):517–518
217. Brown KA, Laferrière A, Lakheeram I, Moss IR (2006) Recurrent hypoxemia in children is associated with increased analgesic sensitivity to opiates. Anesthesiology 105(4):665–669
218. Stamer UM, Lehnen K, Höthker F, Bayerer B, Wolf S, Hoeft A, Stuber F (2003) Impact of CYP2D6 genotype on postoperative tramadol analgesia. Pain 105(1–2):231–238
219. Prommer E, Ficek B (2012) Management of pain in the elderly at the end of life. Drugs Aging 29(4):285–305. doi:10.2165/11599210-000000000-00000
220. Johnson JA, Gong L, Whirl-Carrillo M, Gage BF, Scott SA, Stein CM, Anderson JL, Kimmel SE, Lee MTM, Pirmohamed M, Wadelius M, Klein TE, Altman RB (2011) Clinical pharmacogenetics implementation consortium guidelines for CYP2C9 and VKORC1 genotypes and warfarin dosing. Clin Pharmacol Ther 90(4):625–629. doi:10.1038/clpt.2011.185
221. Hirsh J, Fuster V, Ansell J, Halperin JL (2003) American heart association/American college of cardiology foundation guide to warfarin therapy. Circulation 107(12):1692–1711. doi:10.1161/01.cir.0000063575.17904.4e
222. Choonara IA, Haynes BP, Cholerton S, Breckenridge AM, Park BK (1986) Enantiomers of warfarin and vitamin K1 metabolism. Br J Clin Pharmacol 22(6):729–732
223. Stenflo J, Fernlund P, Egan W, Roepstorff P (1974) Vitamin K dependent modifications of glutamic acid residues in prothrombin. Proc Natl Acad Sci U S A 71(7):2730–2733
224. Wallin R, Hutson SM (2004) Warfarin and the vitamin K-dependent gamma-carboxylation system. Trends Mol Med 10(7):299–302. doi:10.1016/j.molmed.2004.05.003
225. Ansell J, Hirsh J, Hylek E, Jacobson A, Crowther M, Palareti G (2008) Pharmacology and management of the vitamin K antagonists: American college of chest physicians evidence-based clinical practice guidelines (8th Ed). Chest 133(6 Suppl):160S–198S. doi:10.1378/chest.08-0670
226. Baglin TP, Keeling DM, Watson HG (2006) Guidelines on oral anticoagulation (warfarin): third edition-2005 update. Br J Haematol 132(3):277–285. doi:10.1111/j.1365-2141.2005.05856.x
227. Lövborg H, Eriksson LR, Jönsson AK, Bradley T, Hägg S (2012) A prospective analysis of the preventability of adverse drug reactions reported in Sweden. Eur J Clin Pharmacol 68(8):1183–1189. doi:10.1007/s00228-012-1237-2
228. Gage BF, Lesko LJ (2008) Pharmacogenetics of warfarin: regulatory, scientific, and clinical issues. J Thromb Thrombolysis 25(1):45–51. doi:10.1007/s11239-007-0104-y

229. Connolly SJ, Pogue J, Eikelboom J, Flaker G, Commerford P, Franzosi MG, Healey JS, Yusuf S (2008) Benefit of oral anticoagulant over antiplatelet therapy in atrial fibrillation depends on the quality of international normalized ratio control achieved by centers and countries as measured by time in therapeutic range. Circulation 118(20):2029–2037. doi:10.1161/circulationaha.107.750000
230. Reynolds MW, Fahrbach K, Hauch O, Wygant G, Estok R, Cella C, Nalysnyk L (2004) Warfarin anticoagulation and outcomes in patients with atrial fibrillation: a systematic review and metaanalysis. Chest 126(6):1938–1945. doi:10.1378/chest.126.6.1938
231. Owen RP, Gong L, Sagreiya H, Klein TE, Altman RB (2010) VKORC1 pharmacogenomics summary. Pharmacogenet Genomics 20(10):642–644. doi:10.1097/FPC.0b013e32833433b6
232. Eriksson N, Wadelius M (2012) Prediction of warfarin dose: why, when and how? Pharmacogenomics 13(4):429–440. doi:10.2217/pgs.11.184
233. Avery PJ, Jorgensen A, Hamberg AK, Wadelius M, Pirmohamed M, Kamali F (2011) A proposal for an individualized pharmacogenetics-based warfarin initiation dose regimen for patients commencing anticoagulation therapy. Clin Pharmacol Ther 90(5):701–706. doi:10.1038/clpt.2011.186
234. Linder MW, Bon Homme M, Reynolds KK, Gage BF, Eby C, Silvestrov N, Valdes R Jr (2009) Interactive modeling for ongoing utility of pharmacogenetic diagnostic testing: application for warfarin therapy. Clin Chem 55(10):1861–1868. doi:10.1373/clinchem.2009.125898
235. Jorgensen AL, FitzGerald RJ, Oyee J, Pirmohamed M, Williamson PR (2012) Influence of CYP2C9 and VKORC1 on patient response to warfarin: a systematic review and meta-analysis. PLoS ONE 7(8):e44064. doi:10.1371/journal.pone.0044064
236. Patel AA, Swerlick RA, McCall CO (2006) Azathioprine in dermatology: the past, the present, and the future. J Am Acad Dermatol 55(3):369–389. doi:10.1016/j.jaad.2005.07.059
237. Gervasini G, Vagace JM (2012) Impact of genetic polymorphisms on chemotherapy toxicity in childhood acute lymphoblastic leukemia. Front Genet 3:249. doi:10.3389/fgene.2012.00249
238. Manz M, Vavricka SR, Wanner R, Lakatos PL, Rogler G, Frei P, Safroneeva E, Schoepfer AM (2012) Therapy of steroid-resistant inflammatory bowel disease. Digestion 86(Suppl 1):11–15. doi:10.1159/000341952
239. Roberts RL, Barclay ML (2012) Current relevance of pharmacogenetics in immunomodulation treatment for Crohn's disease. J Gastroenterol Hepatol 27(10):1546–1554. doi:10.1111/j.1440-1746.2012.07220.x
240. Ford LT, Berg JD (2010) Thiopurine S-methyltransferase (TPMT) assessment prior to starting thiopurine drug treatment; a pharmacogenomic test whose time has come. J Clin Pathol 63(4):288–295. doi:10.1136/jcp.2009.069252
241. Gurwitz D, Rodríguez-Antona C, Payne K, Newman W, Gisbert JP, de Mesa EG, Ibarreta D (2009) Improving pharmacovigilance in Europe: TPMT genotyping and phenotyping in the UK and Spain. Eur J Hum Genet 17(8):991–998. doi:10.1038/ejhg.2009.10
242. Bradford K, Shih DQ (2011) Optimizing 6-mercaptopurine and azathioprine therapy in the management of inflammatory bowel disease. World J Gastroenterol 17(37):4166–4173. doi:10.3748/wjg.v17.i37.4166
243. Sahasranaman S, Howard D, Roy S (2008) Clinical pharmacology and pharmacogenetics of thiopurines. Eur J Clin Pharmacol 64(8):753–767. doi:10.1007/s00228-008-0478-6
244. Lennard L (1992) The clinical pharmacology of 6-mercaptopurine. Eur J Clin Pharmacol 43(4):329–339
245. Kurowski V, Iven H (1991) Plasma concentrations and organ distribution of thiopurines after oral application of azathioprine in mice. Cancer Chemother Pharmacol 28(1):7–14
246. Tiede I, Fritz G, Strand S, Poppe D, Dvorsky R, Strand D, Lehr HA, Wirtz S, Becker C, Atreya R, Mudter J, Hildner K, Bartsch B, Holtmann M, Blumberg R, Walczak H, Iven H, Galle PR, Ahmadian MR, Neurath MF (2003) CD28-dependent Rac1 activation is the molecular target of azathioprine in primary human CD4+ T lymphocytes. J Clin Invest 111(8):1133–1145. doi:10.1172/jci16432

247. Osterman MT, Kundu R, Lichtenstein GR, Lewis JD (2006) Association of 6-thioguanine nucleotide levels and inflammatory bowel disease activity: a meta-analysis. Gastroenterology 130(4):1047–1053. doi:10.1053/j.gastro.2006.01.046
248. Dubinsky MC (2004) Azathioprine, 6-mercaptopurine in inflammatory bowel disease: pharmacology, efficacy, and safety. Clin Gastroenterol Hepatol 2(9):731–743
249. Cuffari C, Hunt S, Bayless T (2001) Utilisation of erythrocyte 6-thioguanine metabolite levels to optimise azathioprine therapy in patients with inflammatory bowel disease. Gut 48(5):642–646. doi:10.1136/gut.48.5.642
250. Bökkerink JP, Stet EH, De Abreu RA, Damen FJ, Hulscher TW, Bakker MA, van Baal JA (1993) 6-Mercaptopurine: cytotoxicity and biochemical pharmacology in human malignant T-lymphoblasts. Biochem Pharmacol 45(7):1455–1463
251. Adam de Beaumais T, Fakhoury M, Medard Y, Azougagh S, Zhang D, Yakouben K, Jacqz-Aigrain E (2011) Determinants of mercaptopurine toxicity in paediatric acute lymphoblastic leukemia maintenance therapy. Br J Clin Pharmacol 71(4):575–584. doi:10.1111/j.1365-2125.2010.03867.x
252. Dubinsky MC, Lamothe S, Yang HY, Targan SR, Sinnett D, Théorêt Y, Seidman EG (2000) Pharmacogenomics and metabolite measurement for 6-mercaptopurine therapy in inflammatory bowel disease. Gastroenterology 118(4):705–713
253. Nygaard U, Toft N, Schmiegelow K (2004) Methylated metabolites of 6-mercaptopurine are associated with hepatotoxicity. Clin Pharmacol Ther 75(4):274–281. doi:10.1016/j.clpt.2003.12.001
254. Relling MV, Gardner EE, Sandborn WJ, Schmiegelow K, Pui CH, Yee SW, Stein CM, Carrillo M, Evans WE, Klein TE (2011) Clinical Pharmacogenetics Implementation Consortium guidelines for thiopurine methyltransferase genotype and thiopurine dosing. Clin Pharmacol Ther 89(3):387–391. doi:10.1038/clpt.2010.320
255. van Asseldonk DP, Seinen ML, de Boer NKH, van Bodegraven AA, Mulder CJ (2012) Hepatotoxicity associated with 6-methyl mercaptopurine formation during azathioprine and 6-mercaptopurine therapy does not occur on the short-term during 6-thioguanine therapy in IBD treatment. J Crohns Colitis 6(1):95–101. doi:10.1016/j.crohns.2011.07.009
256. Lennard L, Van Loon JA, Lilleyman JS, Weinshilboum RM (1987) Thiopurine pharmacogenetics in leukemia: correlation of erythrocyte thiopurine methyltransferase activity and 6-thioguanine nucleotide concentrations. Clin Pharmacol Ther 41(1):18–25
257. Relling MV, Hancock ML, Rivera GK, Sandlund JT, Ribeiro RC, Krynetski EY, Pui CH, Evans WE (1999) Mercaptopurine therapy intolerance and heterozygosity at the thiopurine S-methyltransferase gene locus. J Natl Cancer Inst 91(23):2001–2008
258. Szumlanski CL, Honchel R, Scott MC, Weinshilboum RM (1992) Human liver thiopurine methyltransferase pharmacogenetics: biochemical properties, liver-erythrocyte correlation and presence of isozymes. Pharmacogenetics 2(4):148–159
259. Wu AH (2011) Drug metabolizing enzyme activities versus genetic variances for drug of clinical pharmacogenomic relevance. Clin Proteomics 8(1):12. doi:10.1186/1559-0275-8-12
260. McLeod HL, Lin JS, Scott EP, Pui CH, Evans WE (1994) Thiopurine methyltransferase activity in American white subjects and black subjects. Clin Pharmacol Ther 55(1):15–20
261. Weinshilboum RM, Sladek SL (1980) Mercaptopurine pharmacogenetics: monogenic inheritance of erythrocyte thiopurine methyltransferase activity. Am J Hum Genet 32(5):651–662
262. Krynetski EY, Schuetz JD, Galpin AJ, Pui CH, Relling MV, Evans WE (1995) A single point mutation leading to loss of catalytic activity in human thiopurine S-methyltransferase. Proc Natl Acad Sci U S A 92(4):949–953
263. Tai HL, Fessing MY, Bonten EJ, Yanishevsky Y, d'Azzo A, Krynetski EY, Evans WE (1999) Enhanced proteasomal degradation of mutant human thiopurine S-methyltransferase (TPMT) in mammalian cells: mechanism for TPMT protein deficiency inherited by TPMT*2, TPMT*3A, TPMT*3B or TPMT*3C. Pharmacogenetics 9(5):641–650
264. Tai HL, Krynetski EY, Schuetz EG, Yanishevski Y, Evans WE (1997) Enhanced proteolysis of thiopurine S-methyltransferase (TPMT) encoded by mutant alleles in humans

(TPMT*3A, TPMT*2): mechanisms for the genetic polymorphism of TPMT activity. Proc Natl Acad Sci U S A 94(12):6444–6449

265. Colombel JF, Ferrari N, Debuysere H, Marteau P, Gendre JP, Bonaz B, Soulé JC, Modigliani R, Touze Y, Catala P, Libersa C, Broly F (2000) Genotypic analysis of thiopurine S-methyltransferase in patients with Crohn's disease and severe myelosuppression during azathioprine therapy. Gastroenterology 118(6):1025–1030
266. Dong X-W, Zheng Q, Zhu M-M, Tong J-L, Ran Z-H (2010) Thiopurine S-methyltransferase polymorphisms and thiopurine toxicity in treatment of inflammatory bowel disease. World J Gastroenterol 16(25):3187–3195. doi:10.3748/wjg.v16.i25.3187
267. Gisbert JP, Gomollón F (2008) Thiopurine-induced myelotoxicity in patients with inflammatory bowel disease: a review. Am J Gastroenterol 103(7):1783–1800. doi:10.1111/j.1572-0241.2008.01848.x
268. Higgs JE, Payne K, Roberts C, Newman WG (2010) Are patients with intermediate TPMT activity at increased risk of myelosuppression when taking thiopurine medications? Pharmacogenomics 11(2):177–188. doi:10.2217/pgs.09.155
269. Meggitt SJ, Anstey AV, Mohd Mustapa MF, Reynolds NJ, Wakelin S (2011) British Association of Dermatologists' guidelines for the safe and effective prescribing of azathioprine 2011. Br J Dermatol 165(4):711–734. doi:10.1111/j.1365-2133.2011.10575.x
270. Chakravarty K, McDonald H, Pullar T, Taggart A, Chalmers R, Oliver S, Mooney J, Somerville M, Bosworth A, Kennedy T, British Society for Rheumatology BHPiRSG, Audit Working G, British Association of D (2008) BSR/BHPR guideline for disease-modifying anti-rheumatic drug (DMARD) therapy in consultation with the British Association of Dermatologists. Rheumatology (Oxford) 47(6):924–925. doi:10.1093/rheumatology/kel216a
271. Fargher EA, Tricker K, Newman W, Elliott R, Roberts SA, Shaffer JL, Bruce I, Payne K (2007) Current use of pharmacogenetic testing: a national survey of thiopurine methyltransferase testing prior to azathioprine prescription. J Clin Pharm Ther 32(2):187–195. doi:10.1111/j.1365-2710.2007.00805.x
272. Newman WG, Payne K, Tricker K, Roberts SA, Fargher E, Pushpakom S, Alder JE, Sidgwick GP, Payne D, Elliott RA, Heise M, Elles R, Ramsden SC, Andrews J, Houston JB, Qasim F, Shaffer J, Griffiths CEM, Ray DW, Bruce I, Ollier WER, team Tsr (2011) A pragmatic randomized controlled trial of thiopurine methyltransferase genotyping prior to azathioprine treatment: the TARGET study. Pharmacogenomics 12(6):815–826. doi:10.2217/pgs.11.32
273. Stocco G, Cheok MH, Crews KR, Dervieux T, French D, Pei D, Yang W, Cheng C, Pui CH, Relling MV, Evans WE (2009) Genetic polymorphism of inosine triphosphate pyrophosphatase is a determinant of mercaptopurine metabolism and toxicity during treatment for acute lymphoblastic leukemia. Clin Pharmacol Ther 85(2):164–172. doi:10.1038/clpt.2008.154
274. Lennard L (2014) Implementation of TPMT testing. Br J Clin Pharmacol 77(4):704–714. doi:10.1111/bcp.12226
275. Schaeffeler E, Fischer C, Brockmeier D, Wernet D, Moerike K, Eichelbaum M, Zanger UM, Schwab M (2004) Comprehensive analysis of thiopurine S-methyltransferase phenotype-genotype correlation in a large population of German-Caucasians and identification of novel TPMT variants. Pharmacogenetics 14(7):407–417
276. Hindorf U, Appell ML (2012) Genotyping should be considered the primary choice for pre-treatment evaluation of thiopurine methyltransferase function. J Crohns Colitis 6(6):655–659. doi:10.1016/j.crohns.2011.11.014
277. Beswick L, Friedman AB, Sparrow MP (2014) The role of thiopurine metabolite monitoring in inflammatory bowel disease. Expert Rev Gastroenterol Hepatol 8(4):383–392. doi:10.1586/17474124.2014.894878
278. Postmus I, Verschuren JJW, de Craen AJM, Slagboom PE, Westendorp RGJ, Jukema JW, Trompet S (2012) Pharmacogenetics of statins: achievements, whole-genome analyses and future perspectives. Pharmacogenomics 13(7):831–840. doi:10.2217/pgs.12.25
279. Goldstein JL, Brown MS (2009) The LDL receptor. Arterioscler Thromb Vasc Biol 29(4):431–438. doi:10.1161/atvbaha.108.179564
280. Baigent C, Keech A, Kearney PM, Blackwell L, Buck G, Pollicino C, Kirby A, Sourjina T, Peto R, Collins R, Simes R (2005) Efficacy and safety of cholesterol-lowering treatment:

prospective meta-analysis of data from 90,056 participants in 14 randomised trials of statins. The Lancet 366(9493):1267–1278. doi:10.1016/s0140-6736(05)67394-1

281. Wilke RA, Ramsey LB, Johnson SG, Maxwell WD, McLeod HL, Voora D, Krauss RM, Roden DM, Feng Q, Cooper-Dehoff RM, Gong L, Klein TE, Wadelius M, Niemi M (2012) The clinical pharmacogenomics implementation consortium: CPIC guideline for SLCO1B1 and simvastatin-induced myopathy. Clin Pharmacol Ther 92(1):112–117. doi:10.1038/clpt.2012.57
282. Wei C-Y, Lee M-TM, Chen Y-T (2012) Pharmacogenomics of adverse drug reactions: implementing personalized medicine. Hum Mol Genet 21(R1):R58–65. doi:10.1093/hmg/dds341
283. Hilgendorf C, Ahlin G, Seithel A, Artursson P, Ungell A-L, Karlsson J (2007) Expression of thirty-six drug transporter genes in human intestine, liver, kidney, and organotypic cell lines. Drug Metab Dispos 35(8):1333–1340. doi:10.1124/dmd.107.014902
284. Tirona RG, Leake BF, Merino G, Kim RB (2001) Polymorphisms in OATP-C: identification of multiple allelic variants associated with altered transport activity among European- and African-Americans. J Biol Chem 276(38):35669–35675. doi:10.1074/jbc.M103792200
285. Karlgren M, Vildhede A, Norinder U, Wisniewski JR, Kimoto E, Lai Y, Haglund U, Artursson P (2012) Classification of inhibitors of hepatic organic anion transporting polypeptides (OATPs): influence of protein expression on drug-drug interactions. J Med Chem 55(10):4740–4763. doi:10.1021/jm300212s
286. Laaksonen R (2006) On the mechanisms of statin-induced myopathy. Clin Pharmacol Ther 79(6):529–531. doi:10.1016/j.clpt.2006.02.013
287. Brunham LR, Lansberg PJ, Zhang L, Miao F, Carter C, Hovingh GK, Visscher H, Jukema JW, Stalenhoef AF, Ross CJD, Carleton BC, Kastelein JJP, Hayden MR (2012) Differential effect of the rs4149056 variant in SLCO1B1 on myopathy associated with simvastatin and atorvastatin. Pharmacogenomics J 12(3):233–237. doi:10.1038/tpj.2010.92
288. Carr DF, O'Meara H, Jorgensen AL, Campbell J, Hobbs M, McCann G, van Staa T, Pirmohamed M (2013) SLCO1B1 genetic variant associated with statin-induced myopathy: a proof-of-concept study using the clinical practice research datalink. Clin Pharmacol Ther 94(6):695–701. doi:10.1038/clpt.2013.161
289. MRC/BHF Heart Protection Study of cholesterol lowering with simvastatin in 20,536 high-risk individuals: a randomised placebo-controlled trial (2002). Lancet 360(9326):7–22. doi:10.1016/s0140-6736(02)09327-3
290. Voora D, Shah SH, Spasojevic I, Ali S, Reed CR, Salisbury BA, Ginsburg GS (2009) The SLCO1B1*5 genetic variant is associated with statin-induced side effects. J Am Coll Cardiol 54(17):1609–1616. doi:10.1016/j.jacc.2009.04.053
291. Donnelly LA, Doney ASF, Tavendale R, Lang CC, Pearson ER, Colhoun HM, McCarthy MI, Hattersley AT, Morris AD, Palmer CNA (2011) Common non-synonymous substitutions in SLCO1B1 predispose to statin intolerance in routinely treated individuals with type 2 diabetes: a Go-DARTS study. Clin Pharmacol Ther 89(2):210–216. doi:10.1038/clpt.2010.255
292. Puccetti L, Ciani F, Auteri A (2010) Genetic involvement in statins induced myopathy. Preliminary data from an observational case-control study. Atherosclerosis 211(1):28–29. doi:10.1016/j.atherosclerosis.2010.02.026
293. Niemi M (2010) Transporter pharmacogenetics and statin toxicity. Clin Pharmacol Ther 87(1):130–133. doi:10.1038/clpt.2009.197
294. Pasanen MK, Fredrikson H, Neuvonen PJ, Niemi M (2007) Different effects of SLCO1B1 polymorphism on the pharmacokinetics of atorvastatin and rosuvastatin. Clin Pharmacol Ther 82(6):726–733. doi:10.1038/sj.clpt.6100220
295. Pasanen MK, Neuvonen M, Neuvonen PJ, Niemi M (2006) SLCO1B1 polymorphism markedly affects the pharmacokinetics of simvastatin acid. Pharmacogenet Genomics 16(12):873–879. doi:10.1097/01.fpc.0000230416.82349.90
296. Research C for DE and Drug Safety and Availability—FDA Drug Safety Communication: new restrictions, contraindications, and dose limitations for Zocor (simvastatin) to reduce the risk of muscle injury. http://www.fda.gov/Drugs/DrugSafety/ucm256581.htm. Accessed 30 Dec 2012

297. Pirmohamed M (2010) Acceptance of biomarker-based tests for application in clinical practice: criteria and obstacles. Clin Pharmacol Ther 88(6):862–866. doi:10.1038/clpt.2010.245
298. Pirmohamed M, Aithal GP, Behr E, Daly A, Roden D (2011) The phenotype standardization project: improving pharmacogenetic studies of serious adverse drug reactions. Clin Pharmacol Ther 89(6):784–785. doi:10.1038/clpt.2011.30
299. Abecasis GR, Auton A, Brooks LD, DePristo MA, Durbin RM, Handsaker RE, Kang HM, Marth GT, McVean GA (2012) An integrated map of genetic variation from 1,092 human genomes. Nature 491(7422):56–65. doi:10.1038/nature11632
300. Voora D, Ginsburg GS (2012) Clinical application of cardiovascular pharmacogenetics. J Am Coll Cardiol 60(1):9–20. doi:10.1016/j.jacc.2012.01.067
301. Payne K, Newman W, Fargher E, Tricker K, Bruce IN, Ollier WER (2007) TPMT testing in rheumatology: any better than routine monitoring? Rheumatology (Oxford) 46(5):727–729. doi:10.1093/rheumatology/kel427
302. Manolio TA, Chisholm RL, Ozenberger B, Roden DM, Williams MS, Wilson R, Bick D, Bottinger EP, Brilliant MH, Eng C, Frazer KA, Korf B, Ledbetter DH, Lupski JR, Marsh C, Mrazek D, Murray MF, O'Donnell PH, Rader DJ, Relling MV, Shuldiner AR, Valle D, Weinshilboum R, Green ED, Ginsburg GS (2013) Implementing genomic medicine in the clinic: the future is here. Genet Med 15(4):258–267. doi:10.1038/gim.2012.157
303. Hudson J (2012) Pharmacogenomics: where does Britain stand? Pharmacogenomics 13(1):1–3. doi:10.2217/pgs.11.159

Pharmacogenomics for Haemoglobinopathies Therapeutics

Aikaterini Gravia, Vasiliki Chondrou, Theodora Katsila and George P. Patrinos

Abstract Individual genetic composition has a fundamental role in the variations observed in drug response and tolerance. Pharmacogenomics aims to delineate the individual genetic profiles and drug response/toxicity. Nowadays, there are several medical disciplines where pharmacogenomics is readily applicable, while in others its usefulness is yet to be shown. Recent experimental evidence suggest that single nucleotide polymorphisms (SNPs) in modifier genes residing outside the human β-globin cluster are significantly associated with response to hydroxyurea (HU) treatment in β-type haemoglobinopathies patients, deducted from the increase in foetal haemoglobin levels. This chapter aims to provide an update and to discuss future challenges on the application of pharmacogenomics for β-type haemoglobinopathies therapeutics in relation to the current pharmacological treatment modalities for those disorders and the complexity of their pathophysiology.

Keywords β-thalassaemia · Sickle cell disease · Pharmacogenomics · Biomarkers · Single nucleotide polymorphisms · Hydroxyurea treatment

1 Haemoglobinopathies

Being easily isolated from blood, the study of haemoglobin has shed light on the understanding of the fundamental principles of gene regulation, gene function and the molecular basis of human genetic disorders. Haemoglobin is the key tetramer oxygen transport protein of red blood cells, consisting of two α-like and two β-like globin polypeptide chains [1–4]. Notably, various types of haemoglobins are found at different developmental stages; just before birth, foetal haemoglobin (HbF) represents the bulk of haemoglobin production, while ten months after birth it gradually declines to reach almost 1 % of the total haemoglobin production, being restricted to a distinct erythrocyte population, also known as F-cells [4]. In adults, both HbF and the number of F-cells share a genetic determination, and the latter varies among populations. The

G. P. Patrinos (✉) · A. Gravia · V. Chondrou · T. Katsila
School of Health Sciences, Department of Pharmacy, University of Patras,
University Campus, Rion, 26504 Patras, GR, Greece
e-mail: gpatrinos@upatras.gr

G. Grech, I. Grossman (eds.), *Preventive and Predictive Genetics: Towards Personalised Medicine,* Advances in Predictive, Preventive and Personalised Medicine 9,
DOI 10.1007/978-3-319-15344-5_7

primary adult and children haemoglobin is the HbA ($\alpha_2\beta_2$), while HbA$_2$ ($\alpha_2\delta_2$) and foetal haemoglobin HbF ($\alpha_2\gamma_2$) are found in amounts of less than 3 %, respectively.

Haemoglobinopathies are divided into two main categories: thalassaemia syndromes and structural haemoglobin variants, both resulting from genomic variations that can be found in cis and/or in trans of the globin gene cluster. Thalassaemia syndromes (namely, α-thalassaemia and β-thalassaemia) are inherited autosomal recessive disorders with defects in globin synthesis and not in haemoglobin structure. The structural haemoglobin variants (or abnormal haemoglobins) are disorders characterised by defects in haemoglobin structure. The majority of haemoglobin variants are extremely rare, with the exception of HbS, HbC and HbE, that are found in certain populations, presumably due to positive natural selection. Abnormal haemoglobins can cause: (a) sickle cell disease (SCD) that results from red cell membrane deformation, (b) haemolytic anaemia, as a result of unstable haemoglobins, (c) methaemoglobinemia owing to rapid oxidation of haemoglobin, and (d) erythrocytosis due to unnatural oxygen affinity [3–7]. Particularly, SCD and β-thalassaemia are one of the commonest single gene disorders and at the same time one of the most serious health problems worldwide. These disorders are most prevalent in Asian and African populations, as well as those residing in the Mediterranean basin. Nowadays, however, due to the international migration, haemoglobinopathies extend worldwide. These genetic disorders are caused by genetic quantitative and qualitative defects in haemoglobin production [4].

Nowadays, more than 1000 Hb variants have been discovered and characterised [8], being a milestone in the history of haemoglobin research, where T.H.J. Huisman could not be omitted. A registry of these Hb variants and related information has been available online, at HbVar database (http://globin.bx.psu.edu/hbvar). During the last century, major developments in Hb research have been made using physical, chemical, physiological and genetic methods, impacting our understanding and management of the thalassaemias and sickle cell disease.

2 Therapeutic Approaches

The β-thalassaemias are considered as one of a few clinical conditions in which a mutant gene that is normally expressed later in development can be functionally replaced by a gene, which is transiently expressed during foetal life [9]. Foetal γ globin expression can be re-activated, being an appealing therapeutic approach as the foetal globin genes are universally present and noteworthy, appropriately contextually integrated in the β-globin locus in human haematopoietic stem cells. However, anaemia in β-thalassaemia syndromes can be also due to the rapid cellular apoptosis (α-globin chain precipitation) and/or the relatively low levels of endogenous erythropoietin (EPO) [10]. Thus, the ultimate goal of transfusion independence for thalassaemia patients should be approached via the stimulation of both foetal globin gene expression and erythropoiesis. In this context, chemotherapeutic agents, erythropoietin (EPO) preparations and short chain fatty acid derivatives (SCFADs) have demonstrated proof-of-principle in animal models and clinical trials.

Focusing on the pharmacological reactivation of HbF to compensate for the loss of HbA, the only pharmacological agent recognised from FDA used to increase the HbF in adults is hydroxycarbamine (or hydroxyurea, HU). Even though the mechanism underlying the HU action still remains elusive, it has been shown to inhibit cellular ribonucleotide reductase, whose role is critical for the DNA synthesis in the dividing late progenitor cells. HU is broadly used for ameliorating SCD symptoms and to a lesser extend in β-thalassaemia patients, while it also serves as a chemotherapeutic agent for many myeloproliferative conditions [11]. Nevertheless, HU is cytotoxic and can lead to cytopenia, hyperpigmentation, weight gain, hypomagnesemia or it may have a teratogenic effect [5]. The patient response to HU as well as the HU toxicity incidents vary [4, 12–14]. Thus, the discrimination between responders and non-responders is of fundamental importance towards patient stratification.

Collins and colleagues have observed that haematologic responses to the foetal globin inducer sodium phenylbutyrate occurred only in those subjects who had high endogenous EPO levels, being unrelated to any particular pattern of globin gene mutation [15]. Hence, a red cell survival advantage of increased endogenous EPO in β-thalassaemia has been suggested that may facilitate an effective γ-globin induction. In agreement to the above, a subset of subjects with inappropriately low levels of endogenous EPO has responded to combined therapy with butyrate plus EPO, whereas each agent alone had a lesser or minimal effect in the same time frame [10]. It seems, therefore, that the exogenously administered EPO acts both as a survival factor and an erythropoietic stimulant. Prolonging erythroid precursor cell survival could be beneficial, allowing a foetal globin inducer to act towards the correction of the pro-apoptotic chain imbalance and hence, improving the anaemia found in β^+- thalassaemia patients. In thalassaemia intermedia and major, EPO has been combined with HU or arginine butyrate. Combinations of EPO with stem cell factor (SCF) to stimulate proliferation and transforming growth factor-β (TGF-β) to induce premature differentiation (like HU) could be also considered for more severe phenotypes, although pricing could be an obstacle for long-term use.

In other studies, SCFADs and hydroxamic acids have been shown to induce foetal globin gene expression, stimulate erythroid proliferation and prolong erythroblast survival [10, 16, 17]. In particular, they activate the $^{A}\gamma$ globin gene promoter in cells cultured from β-thalassaemia patients, without inhibiting the erythroid cell growth. Sodium 2,2-dimethylbutyrate and α-methyl-hydrocinnamate stimulate erythroid colony formation more than the optimal haematopoietic growth factors alone and at the same time, signal through STAT-5 phosphorylation that is common to EPO and IL-3 signalling pathways. Most importantly, some of these agents hold the promise of oral administration, being more tolerable for long-term treatments, providing two therapeutic effects via one tolerable agent [10, 17].

Although the development of drugs to increase foetal haemoglobin has been the major therapeutic strategy in the treatment of both disorders, SCD and thalassaemias, and new foetal haemoglobin–modulating agents have been studied, only HU has shown long-term benefit. To this end, the gradual elucidation of the pathophysiology of the disease(s) has led to alternative strategies, treating the associated complications (decreasing the iron overload and reducing the oxidative stress). More recently, novel agents have been developed targeting the multiple pathways causing vascular injury in haemoglobinopathies; the increased adhesion of cells to the vascular endothelium, the NO dysregulation, inflammation, oxidative injury and the altered iron metabolism. Such agents (propranolol, statins, niacin, curcuminoids, hepcidin agonists/antagonists) have reached phase 1 and 2 clinical trials [18].

3 Pathophysiological Features and Obstacles

The reduction of the globin chain imbalance has been well accepted as the way to improve red cell survival and blood counts in β-thalassaemia. Nevertheless, a number of factors seem to collectively contribute to anaemia, such as ineffective erythropoiesis and erythroid precursor apoptosis. Thus, combination therapy is required with more than one agent acting at various molecular levels to achieve a tolerable and long-term therapeutic response. Moreover, different magnitudes of therapeutic effect for different thalassaemia patients are required to achieve functional clinical endpoints that could result in abolishing or decreasing needs for regular red blood cell transfusion. In thalassaemia intermedia patients with basal total haemoglobin levels of 6–8 g/dL, a 1–2 g/dL increase would be quite adequate to prevent the need for a regular transfusion program, being highly beneficial, whereas thalassaemia major patients having baseline haemoglobin levels below 5 g/dL would require higher levels of foetal globin induction.

The paediatric pathophysiology of both SCD and thalassaemia should also be considered when haemoglobinopathies' therapeutics are in question. In SCD, acute chest syndrome, resulting from pulmonary microvascular occlusion and being a common cause of death, occurs in all age groups, but is most common in childhood. In children, acute sequestration of sickled cells in the spleen may also occur, exacerbating anaemia. Chronic spleen damage increases susceptibility to pneumococcal and *Salmonella* infections (including *Salmonella* osteomyelitis) that are especially common in early childhood and can be rapidly fatal [19]. Children with thalassaemia intermedia (mild to severe anaemia) or thalassaemia major need blood transfusions coupled to chelation therapy. Children with β-thalassaemia have elevated plasma levels of conjugated dienes and thiobarbituric acid-reactive substances (markers of lipid oxidation), while the RBC glutathione levels are much reduced [20]. There is evident oxidant injury to RBC haemoglobin and lipids.

Considering the overall complexity in the pathophysiology of haemoglobinopathies as well as the still unresolved pathophysiological issues in thalassaemias [20], patient stratification is critical towards effective cure or mitigation of the disease. In this context, pharmacogenomics are expected to have a fundamental role.

4 Pharmacogenomics for Haemoglobinopathies

Pharmacogenomics aims to determine how the genetic background of a patient influences his response to a drug or the probability to develop adverse drug reactions, via the correlation of gene expression or Single Nucleotide Polymorphisms

(SNPs) with drug efficacy and toxicity [21]. The application of pharmacogenomics in haemoglobinopathies is particularly appealing due to the limitation of the therapeutic approaches and the complexity of disease pathophysiology. So far, the role of HU treatment towards the cure or mitigation of the disease has been vital. HU increases the HbF levels mainly in SCD, but also in compound heterozygous SCD/β-thalassaemia patients, ameliorating their clinical manifestations. Additionally, β-thalassaemia intermedia patients have been also shown to respond to HU treatment. Herein, we summarise the current knowledge regarding the genetic factors that have been reported to influence HbF expression levels in relation to HU treatment, including a large number of genomic variations residing inside or outside the human β-globin gene cluster (Tables 1, 2).

Table 1 Summary of the various studies attempting to correlate genomic variations in genes located within the human globin gene cluster with HU treatment efficacy

Patient sample	No of patients	Origin	Gene involved	Association with HU treatment response	References
SCD	150	NA	*HBG2* (XmnI polymorphism)	YES	[23]
β-thalassaemia major	34	NA	*HBG2* (XmnI polymorphism)	YES	[25]
β-thalassaemia major	133	Iranian	*HBG2* (XmnI polymorphism); *HBB* framework 2	YES	[26]
β-thalassaemia major	143	NA	*HBG2* (XmnI polymorphism)	YES	[27]
β-thalassaemia major	54	Algerian	*HBG2* (XmnI polymorphism)	YES	[28]
β-thalassaemia major	18	NA	*HBG2* (XmnI polymorphism)	YES	[29]
β-thalassaemia intermedia	37	NA	*HBG2* (XmnI polymorphism)	YES	[30]
β-thalassaemia intermedia	16	NA	*HBG2* (XmnI polymorphism)	NO	[31]
β-thalassaemia major/ intermedia	38/41	Western Indian	*HBG2* (XmnI polymorphism); α-thalassaemia	YES	[13]
β-thalassaemia major	81	Iranian	*HBG2* (XmnI polymorphism)	YES	[38]
HbE/β-thalassaemia	13	Indian	*HBG2* (XmnI polymorphism); *HBB* genotype	NO	[32]
β-thalassaemia	232	Iranian	Hetero- and homozygous*HBB* null (β^0) mutation	YES	[33]
			HBG2 (XmnI polymorphism)	NO	
β-thalassaemia intermedia	24	NA	Hb Lepore; δβ-thalassaemia	YES	[34]

Table 2 Summary of the currently available studies aiming to correlate genomic variations located in genes outside the human globin gene cluster with HU treatment efficacy

Sample	No of patients	Origin	Chromosome	Gene and markers involved	Association with HU treatment response	References
SCD	137	NA	13	*FLT1* (rs2182008, rs9319428, rs8002446)	YES	[35]
			6	*MAP3K5*(rs9376230,rs9483947)		
			6	*PDE7B*(rs2327669, rs11154849, rs9376173, rs1480642, rs487278)		
			9	*ASS* (rs7860909, rs10793902, rs10901080)		
			8	*TOX* (rs826729, rs765587, rs9693712, rs172652, rs380620)		
			6	*ARG1* (rs17599586, rs2781667)		
			14	*ARG2*(rs10483801, rs10483802)		
			17	*NOS2A*(rs1137933, rs944725)		
			12	*NOS1* (rs816361, rs7977109)		
β-thalassaemia and SCD/ β-thalassaemia compound heterozygotes	38	Western Greek	6	*MAP3K5*(rs9483947, rs9376230)	YES	[37]
			6	*PDE7B*(rs2327669)	NO	
β-thalassaemia major	81	Iranian	2	*BCL11A*(rs766432,rs4671393)	YES	[38]
			6	*HBS1 L-MYB* (rs9399137, rs4895441)	NO	
SCD children	93	NA	2	*BCL11A*(rs1186868, rs1427407)	NO	[39]
β-thalassaemia and SCD/ β-thalassaemia compound heterozygotes	38	Western Greek	8	*KLF10* (rs3191333)	YES	[40]
SCD	386	NA	10	*SAR1A* (rs2310991, g.-809 C > T, g.-502G > T, g.-385 C > A, rs4282891)	YES	[41]

4.1 *Genomic Biomarkers Linked to the Human β-Globin Gene Cluster*

Patients' response to HU varies in terms of amplitude and velocity, leading to the in depth investigation of the determinants of this differential response. This investigation began years ago and continues until today, revealing several genetic factors including SNPs in various genes that are linked (or not) to the β-globin gene cluster. These genes are believed to act by modulating HbF levels.

Steinberg et al. [23] studied 150 HbS homozygous patients treated with HU. A group of them showed almost a 40 % HU-induced HbF levels, compared to the remaining of the study group. In order to define the genetic factors that may have influenced these responses, they focused on genetic factors that are believed to influence the foetal globin (*HBG1* and *HBG2*) gene expression, including the haplotypes of α- and β-globin gene clusters and the X-linked F-cell expression locus (FCP) [22]. As concluded, the FCP and the HbF levels before the treatment were not correlated with the HU-induced HbF response. Notably, the absence of a particular β-globin gene haplotype (namely, the Central African Republic-CAR) was related with higher HbF response [23].

Although HU is established to ameliorate the clinical manifestations of SCD, many patients die of this disease, mostly by acute chest syndrome (as it has been described earlier in the text, it occurs in all age groups, but it is most common in childhood), even upon HU treatment. Bakanay et al. [24] compared the β-globin gene cluster haplotype distributions (BAN, BEN, CAM, SEN) between the deceased and surviving patients treated with HU, concluding that homozygosity in the BAN haplotype or heterozygosity in the CAM haplotype were more likely to be observed in the deceased patient group [24].

In β-thalassaemia major, the increase of HbF expression by HU administration is not as effective as in the case of SCD. This may be due to the heterogeneity of the disease and also due to the complexity of the genetic elements involved in the HbF expression patterns of patients. Alebouyeh et al. [25] attempted to correlate the HU response in two different groups of β-thalassaemia major patients, from whom, 25 were blood transfusion-dependent and nine were non-dependent, in order to explore candidate genetic markers for the pharmacological HbF reactivation by HU. They have demonstrated that the XmnI polymorphism (*HBG2*: g.-158C > T) and the IVSII-1G > A mutation, both at the homozygous state, were found in the responders of both groups. On the contrary, these markers were either not present or in a heterozygous state in the non-responders groups tested. Nevertheless, the small number of patients included in the study as well as the fact that two siblings (responders group) were found to bear the common allele for both the XmnI polymorphism and the IVSII-1G > A mutation suggest that further research must be carried out in order to identify and elucidate the genetic modifiers for the HbF upregulation upon HU treatment [25].

Yavarian et al. [26] studied 133 Iranian transfusion-dependent β-thalassaemia patients, treated with HU in order to determine their response to the drug and the associated genetic background; 61 % of the patients became blood transfusion-independent, after 4 months of HU administration (good responders), 23 % of the patients, remained blood transfusion-dependent, albeit at a less frequent rate (moderate responders) than before HU treatment, whereas in the remaining 16 % of the patients HU, even after a year of administration, had no effect in their clinical manifestation and the frequency of blood transfusion. The authors examined the genetic aetiology of β-thalassaemia in these patients, the human β-gene cluster haplotype and their molecular background in the promoter region of the globin genes as well as at the HS2 hypersensitive site of the β-globin Locus Control Region (LCR). They concluded that the XmnI polymorphism was the most important genetic factor correlated with HU response and that its linkage with the human β-globin gene cluster haplotype I and with the *HBB* gene framework 2 is the "favourable genetic background" for good response to HU [26]. Also, Ansari et al. [27], studied 143 β-thalassaemia patients, treated with HU, confirming that XmnI polymorphism in homozygous or heterozygous state is a genomic marker to predict HU response. This finding was also demonstrated in two other studies including 54 Algerian β-thalassaemia patients [28] and 18 homozygous β-thalassaemia patients treated with HU for a period of 4 years [29].

From a number of studies, it is evident that HU is more promising treating β-thalassaemia intermedia due to the lesser imbalance of α/β-globin chain. Dixit et al. [30] studied the response of 37 β-thalassaemia intermedia patients to HU, from whom almost 70 % were categorised as responders. The response to HU was not associated with the β-thalassaemia mutation. On the other hand, a statistically insignificant correlation of HU response and the XmnI polymorphism was observed, suggesting that the combination of other genetic elements can possibly influence the final response to HU treatment [30]. In a similar survey of 18 homozygous β-thalassaemia patients, 11 of who were transfusion-dependent were treated with HU for 50 months in order to correlate their response to HU and their genetic background. The results showed that 82 % of transfusion-dependent patients who were treated with HU turned into transfusion-independent, while 78 % of them were found homozygous or heterozygous for the XmnI polymorphism. Interestingly, this genomic variation was not present in the HU non-responders. Nevertheless, there were two responding patients, who were negative for the XmnI polymorphism. In total, these data suggest that there may be other genetic elements, which could determine the HU response. In the same study, no correlation between response to HU and the nature of β-thalassaemia mutation or α-thalassaemia deletion was observed [29]. However, in a similar study of 16 transfusion-independent Iranian patients, treated with HU for 6 months, Ehsani et al. [31] could not establish any correlation between the XmnI polymorphism and the response to HU treatment. This may be due to the fact that in this study, as with the previous one, the number of patients was too small to reach any significant conclusion.

Similarly, Italia et al. [13] attempted to correlate the response to HU of 79 β-thalassaemia patients of western Indian origin, from which 38 were β-thalassaemia intermedia and 41 were β-thalassaemia major, treated with HU for almost a year. The correlation (if any) of the HU response to the genetic factors residing within the human β-globin gene cluster was investigated. As it was shown, in the presence of α-thalassaemia, β-thalassaemia patients showed a better HU response, and the presence of the XmnI polymorphism in homozygosity also resulted in a better clinical response to HU [13]. In a subsequent study of 13 Indian HbE/β-thalassaemia patients with severe clinical manifestations, from whom 36.3 % of the patients were good responders, 36.3 % were partial responders and 27.2 % showed no response, Italia et al. [32] failed to correlate the HU response with specific genetic factors, focusing mainly to the *HBB* genotype and the human β-globin cluster haplotype and XmnI polymorphism, most likely due to the small number of patients [32].

Karimi et al. [33] attempted to correlate the response to HU with *HBB* gene mutations and the XmnI polymorphism in a much larger patient sample, consisting of 232 β-thalassaemia patients of Iranian origin, upon HU treatment for a 13-year period. These authors showed that β-thalassaemia patients with homozygous or heterozygous for a β^0 mutation were better HU responders compared to patients who were homozygous for a β^+ mutation. Interestingly, though, these authors could not establish any correlation between the XmnI polymorphic site and HU response [33].

Finally, Rigano et al. [34] studied the HU efficiency in a long and short term treatment of 24 β-thalassaemia intermedia patients and concluded that the presence of Hb Lepore and δβ-thalassaemia genotypes were indicators of a better HU response [34].

All these studies are summarised in Table 1.

4.2 Genomic Biomarkers Non-Linked to the Human β-Globin Gene Cluster

Apart from the numerous studies presented above with the aim to delineate the response to HU and the genomic markers present in the human β-globin gene cluster, a number of studies have been recently conducted, attempting to implicate the genomic loci residing on other chromosomes with HU response. Some of these genes, particularly *BCL11A*, have been shown to be directly related with increasing HbF levels and as such, these genes might constitute excellent candidates for pharmacogenomic biomarkers to predict HU response (Table 2).

In the most comprehensive study so far, Ma et al. [35] studied 137 SCD patients in an effort to correlate the HU response to several genomic biomarkers that are linked trait loci (QTLs), located on chromosomes 6 and 8 as well as the X-chromosome (these genes have been previously linked to HU metabolism and to erythroid progenitor proliferation). In particular, the authors investigated the association of 327 tagSNPs within these loci to the HU response of patients, using HapMap data. It was concluded that the rs2182008 variation in the *FLT1* gene, either in homozygosity or heterozygosity, was correlated with an almost 6-fold increase in HbF expression levels, following HU treatment. Moreover, two other *FLT1* gene variants, namely the rs9319428 and rs8002446, were found to be associated with the HU response. Overall, there were various genomic variations residing in the *MAP3K5*, *PDE7B*, *ASS*, *TOX*, *ARG1*, *ARG2*, *NOS2A* and *NOS1* genes, found to be correlated with the HU influence to the HbF expression. Notably, the majority of the above mentioned SNPs were located in intronic or in untranslated regions of the candidate genes [35].

Similarly, Tafrali et al. [36] attempted to elucidate a probable association between the genetic variations in the *MAP3K5* and *PDE7B* gene with the β-thalassaemia disease severity and response to HU in two groups of 38 β-thalassaemia homozygous and SCD/β-thalassaemia compound heterozygous patients of western Greek origin. The authors showed that there is a significant correlation between two single nucleotide polymorphisms residing in the region of *MAP3K5* intron 1 (rs9483947, rs9376230) and improved HU response. Also, by comparative whole-transcriptome analysis in erythroid progenitor cell cultures from normal Maltese adults and Maltese HPFH haploinsufficient cases, bearing the *KLF1*:p.K288X nonsense mutation [37], before and after HU treatment, *MAP3K5* gene expression was increased upon HU treatment [36].

A retrospective association study was conducted with the purpose of detecting genetic determinants of the HU response in 81 transfusion-dependent β-thalassaemia patients from Iran. Genomic variants, located in three QTLs that have been previously shown to have an effect on HbF and F-cell levels, namely the XmnI polymorphism (rs7482144), two SNPs in the intron 2 of the *BCL11A* gene (rs766432, rs4671393) and two SNPs in the intergenic region of *HBS1L-MYB* gene (rs9399137, rs4895441) were investigated. The authors failed to obtain a correlation between the *HBSL1L-MYB* SNPs and HU response. On the contrary, the presence of the XmnI polymorphism, as well as the minor alleles of the *BCL11A* SNPs, namely rs766432C and rs4671393A, were significantly associated with good response to HU treatment [38]. Similarly, Flanagan et al. [39] attempted to elucidate the effect of HU on the erythroid gene expression in 93 children suffering from SCD, in order to explore how HU can influence both the red cell development and the HbF reactivation. Although the existence of the rs1186868 or rs1427407 SNPs in the *BCL11A* gene, in either homo- or heterozygosity, led to the down-regulation of the *BCL11A* expression and higher HbF levels, the authors failed to observe any difference in HbF levels, subject to HU treatment, between the patients who were homozygous or heterozygous for the above SNPs and the wild type ones [39].

Borg et al. [40] conducted a pharmacogenomic study on the HU effect on HbF levels of Hellenic SCD/β-thalassaemia compound heterozygotes, the first to be carried out using a whole transcriptome analysis approach. The authors have comparatively analysed, using whole transcriptome analysis, human erythroid progenitor cells, treated with HU, derived from SCD/β-thalassaemia patients that responded or not to HU. They also studied the effect of the HU on erythroid progenitor cells of healthy and *KLF1*-haploinsufficient Maltese adult patients ex vivo, expressing low and high HbF levels, respectively, aiming to reveal differential expression profiles in genes implicated in augmenting HbF levels. *KLF10* was shown to be the strongest candidate, among 43 identified genes, in both analyses [40]. Subsequently, the authors used an independent cohort of SCD/β-thalassaemia compound heterozygotes so as to corroborate their results. Their genotyping analysis demonstrated that the presence of the rs3191333 SNP in the 3′ UTR of the *KLF10* gene can be correlated with the severity of β-thalassaemia, as well as with efficacy to the HU therapy. In conclusion, *KLF10* has not only been shown to be a pharmacogenomic marker to predict β-thalassaemia patient response to HU, but has also been implicated for the first time in erythropoiesis [40].

Finally, Kumkhaek et al. [41] examined 386 SCD patients in an effort to correlate polymorphisms in the *SAR1A* gene promoter region with differential response to HU and differences in HbF levels among different patients. It was concluded that 5 SNPs in the *SAR1A* regulatory region were correlated with patients' response to HU and with different HbF levels, after a 2 year treatment with HU [41].

5 Concluding Remarks

In this chapter, we (i) provided an succinct overview of haemoglobinopathies, (ii) presented the challenges of their pathophysiology and the limitations of their therapeutics and (iii) summarised our current knowledge regarding the genetic factors that have been reported to influence HbF expression levels in relation to HU treatment, including a large number of genomic variations residing inside or outside the human β-globin gene cluster (Tables 1, 2).

Contrary to other medical specialties, such as oncology, and treatments, such as anticoagulation therapies, experimental data supporting the use of pharmacogenomics for haemoglobinopathies therapeutics using HU are currently very limited, and clearly, more pharmacogenomic studies are needed, not only in larger, but also in ethnically diverse β-thalassaemia and SCD patients groups. In this way, a better picture will be obtained as to whether it is possible to stratify those patients who are likely to benefit from HU therapy. In addition, similar studies may be also conducted for more pharmacological agents and different treatment modalities, such as decitabine and/or butyrate, although presently at experimental stage. However, although drug-induced augmentation therapies towards HbF levels have been demonstrated as a therapeutic modality for β-type haemoglobinopathies patients, it should be clarified that these cannot correct *per se* the numerous events that underlie the pathophysiology of this group of disorders. In addition, one should bear in mind that no straightforward correlation between HbF increase and clinical improvements in β-type haemoglobinopathies patients has been demonstrated. Therefore, all possible phenotype and clinical indicators should be determined to categorise the "responder" and the "non-responder" patient groups for pharmacogenomic studies, which do not necessarily have to be correlated with HbF increment alone, particularly in the case of SCD. Similar complexities also exist for other thalassaemia-related treatments, such as the use of iron chelators.

It should be also noted that not all genomic loci that have been shown to increase HbF levels can be also considered as pharmacogenomic markers for HU response. *KLF1*, one of the key players participating in *HBB* gene activation that is recently shown to be also indirectly involved in human foetal globin gene silencing [36] is not correlated with increased HbF levels upon HU treatment (Kaimakis and Patrinos, unpublished). The same is true for genomic variations in the *HBBP1* pseudogene and *PDE7B* gene that although recently shown to be related with β-thalassaemia disease severity [37, 42], genomic alterations in these genes cannot be correlated with response to HU treatment.

Whole genome association and whole transcriptome pharmacogenomic studies are only beginning and there are only few reports in the field [35, 37, 40]. Such studies may identify novel gene candidates that participate in different pathways related to HU treatment, such as stage-specific transcription factors, novel erythroid genes and/or genes involved in HbF-inducing HU metabolism. Also, the scarcity of β-thalassaemia intermedia patients and the need to stratify these patients not only according to their response status, but also, and most importantly, according to their *HBB* genotype, makes the formation of large multi-center consortia more than ever urging to better orientate pharmacogenomic marker identification in good and poor responders to HbF-inducing therapy. This will in turn facilitate the design of customised high throughput pharmacogenomic tests for β-type haemoglobinopathies.

Pharmacogenomic studies may be also extended to other therapeutic modalities for β-thalassaemia, such as iron chelation therapy. In particular, a fraction of β-thalassaemia patients present a number of adverse effects to iron chelators, which result in early death [43]. The correlation of genomic variations located in genes that influence, e.g. iron homeostasis with tolerance or response to iron chelation treatment would potentially better stratify patients for iron chelation therapies and enable the emergence of new and improved iron chelators. Similarly, as with β-thalassaemia and SCD patients, whole genome pharmacogenomic studies in these patient groups can also establish genes involved in iron chelators' metabolism pathways, hence allowing identifying putatively useful pharmacogenomic markers for iron chelation therapies, leading to the individual tailoring of chelation therapy to maximise iron excretion.

Pharmacogenomics in children bring on additional challenges. It is well established that there are differences in drug response among children and adults [44], especially in drug metabolism and gene expression, as the latter is a highly dynamic process functioning from the neonatal period over childhood and the adult life later on. Thus, the data quality and its analysis/ interpretation is challenging *per se*. Ethical and legal aspects also accompany this, since the child in question is incapable of giving informed consent himself [44]. Data interpretation difficulties and ethical considerations are clearly needed to be addressed.

In essence, although pharmacogenomics for β-type haemoglobinopathies is currently in its infancy, there is definitely a big potential to determine whether genomic biomarkers can be exploited in the clinic to stratify β-thalassaemia and SCD patients that are likely to benefit from therapy.

Acknowledgements We wish to thank Professors Frank Grosveld, Alex E. Felice, Sjaak Philipsen, Drs. Adamantia Papacatzopoulou, Joseph Borg, Sonja Pavlovic, Marios Phylactides, Marina Kleanthous, Alexandra Kourakli, Marianthi Georgitsi and Mrs. Christina Tafrali, Marina Bartsakou-lia, Arsinoi Paizi, Emily Giannopoulou, and Olga Giannakopoulou for their contribution at the various stages of our projects related to pharmacogenomics for haemoglobinopathies.

Our work is supported by a RDF (Cyprus, ΠΔΕ046_02) and European Commission (ITHANET Coordination action 026539) grants to GPP.

References

1. Pace BS, Zein S (2006) Understanding mechanisms of gamma-globin gene regulation to develop strategies for pharmacological foetal hemoglobin induction. Dev Dyn 235(7):1727–1737
2. Schechter AN (2008) Hemoglobin research and the origins of molecular medicine. Blood 112(10):3927–3938
3. Weatherall DJ (2001) Towards molecular medicine; reminiscences of the haemoglobin field, 1960–2000. Br J Haematol 115(4):729–738
4. Patrinos GP, Antonarakis SE (2010) Human hemoglobin, vol 11. Vogel and Motulsky's human genetics: problems and approaches. Springer, Berlin
5. Kohne E (2011) Hemoglobinopathies: clinical manifestations, diagnosis, and treatment. Dtsch Arztebl Int 108(31–32):532–540
6. Higgs DR (2013) The molecular basis of α-thalassaemia. Cold Spring Harb Perspect Med 3(1):a011718
7. Nienhuis AW, Nathan DG (2012) Pathophysiology and clinical manifestations of the β-thalassaemias. Cold Spring Harb Perspect Med 2(12):a011726
8. Thein SL (2011) Milestones in the history of hemoglobin research (in memory of professor Titus H.J. Huisman). Hemoglobin 35(5–6):450–462
9. Steinberg MH, Rodgers GP (2001) Pharmacologic modulation of foetal hemoglobin. Medicine (Baltimore) 80(5):328–344
10. Perrine SP (2005) Foetal globin induction-can it cure beta thalassaemia? Hematol Am Soc Hematol Educ Program 38–44
11. Karimi M, Darzi H, Yavarian M (2005) Haematologic and clinical responses of thalassaemia intermedia patients to hydroxyurea during 6 years of therapy in Iran. J Pediatr Hematol Oncol 27(7):380–385
12. Patrinos GP, Grosveld FG (2008) Pharmacogenomics and therapeutics of hemoglobinopathies. Hemoglobin 32(1–2):229–236
13. Italia KY, Jijina FJ, Merchant R, Panjwani S, Nadkarni AH, Sawant PM, Nair SB, Ghosh K, Colah RB (2009) Response to hydroxyurea in beta thalassaemia major and intermedia: experience in western India. Clin Chim Acta 407(1–2):10–15
14. Karimi M, Cohan N, Mousavizadeh K, Moosavizadeh K, Falahi MJ, Haghpanah S (2010) Adverse effects of hydroxyurea in beta-thalassaemia intermedia patients: 10 years' experience. Pediatr Hematol Oncol 27(3):205–211
15. Collins AF, Pearson HA, Giardina P, McDonagh KT, Brusilow SW, Dover GJ (1995) Oral sodium phenylbutyrate therapy in homozygous beta thalassaemia: a clinical trial. Blood 85(1):43–49
16. Boosalis MS, Bandyopadhyay R, Bresnick EH, Pace BS, Van DeMark K, Zhang B, Faller DV, Perrine SP (2001) Short-chain fatty acid derivatives stimulate cell proliferation and induce STAT-5 activation. Blood 97(10):3259–3267
17. Pace BS, White GL, Dover GJ, Boosalis MS, Faller DV, Perrine SP (2002) Short-chain fatty acid derivatives induce foetal globin expression and erythropoiesis in vivo. Blood 100(13):4640–4648
18. Vichinsky E (2012) Emerging 'A' therapies in hemoglobinopathies: agonists, antagonists, antioxidants, and arginine. Hematol Am Soc Hematol Educ Program 2012:271–275
19. Di Nuzzo DV, Fonseca SF (2004) [Sickle cell disease and infection]. J Pediatr (Rio J) 80(5):347–354
20. Schrier SL (2002) Pathophysiology of thalassaemia. Curr Opin Hematol 9(2):123–126

21. Roden DM, Altman RB, Benowitz NL, Flockhart DA, Giacomini KM, Johnson JA, Krauss RM, McLeod HL, Ratain MJ, Relling MV, Ring HZ, Shuldiner AR, Weinshilboum RM, Weiss ST, Network PR (2006) Pharmacogenomics: challenges and opportunities. Ann Intern Med 145(10):749–757
22. Chang YC, Smith KD, Moore RD, Serjeant GR, Dover GJ (1995) An analysis of foetal hemoglobin variation in sickle cell disease: the relative contributions of the X-linked factor, beta-globin haplotypes, alpha-globin gene number, gender, and age. Blood 85(4):1111–1117
23. Steinberg MH, Lu ZH, Barton FB, Terrin ML, Charache S, Dover GJ (1997) Foetal hemoglobin in sickle cell anaemia: determinants of response to hydroxyurea. Multicenter Study of Hydroxyurea. Blood 89(3):1078–1088
24. Bakanay SM, Dainer E, Clair B, Adekile A, Daitch L, Wells L, Holley L, Smith D, Kutlar A (2005) Mortality in sickle cell patients on hydroxyurea therapy. Blood 105(2):545–547
25. Alebouyeh M, Moussavi F, Haddad-Deylami H, Vossough P (2004) Hydroxyurea in the treatment of major beta-thalassaemia and importance of genetic screening. Ann Hematol 83(7):430–433
26. Yavarian M, Karimi M, Bakker E, Harteveld CL, Giordano PC (2004) Response to hydroxyurea treatment in Iranian transfusion-dependent beta-thalassaemia patients. Haematologica 89(10):1172–1178
27. Ansari SH, Shamsi TS, Munzir S, Khan MT, Erum S, Perveen K, Farzana T, Ashraf M, Mehboob T, Moinuddin M (2013) Gγ-Xmn I polymorphism: a significant determinant of β-thalassaemia treatment without blood transfusion. J Pediatr Hematol Oncol 35(4):e153–e156
28. Bradai M, Pissard S, Abad MT, Dechartres A, Ribeil JA, Landais P, de Montalembert M (2007) Decreased transfusion needs associated with hydroxyurea therapy in Algerian patients with thalassaemia major or intermedia. Transfusion 47(10):1830–1836
29. Koren A, Levin C, Dgany O, Kransnov T, Elhasid R, Zalman L, Palmor H, Tamary H (2008) Response to hydroxyurea therapy in beta-thalassaemia. Am J Hematol 83(5):366–370.
30. Dixit A, Chatterjee TC, Mishra P, Choudhry DR, Mahapatra M, Tyagi S, Kabra M, Saxena R, Choudhry VP (2005) Hydroxyurea in thalassaemia intermedia-a promising therapy. Ann Hematol 84(7):441–446
31. Ehsani MA, Hedayati-Asl AA, Bagheri A, Zeinali S, Rashidi A (2009) Hydroxyurea-induced haematological response in transfusion-independent beta-thalassaemia intermedia: case series and review of literature. Pediatr Hematol Oncol 26(8):560–565
32. Italia KY, Jijina FF, Merchant R, Panjwani S, Nadkarni AH, Sawant PM, Nair SB, Ghosh K, Colah RB (2010) Effect of hydroxyurea on the transfusion requirements in patients with severe HbE-beta-thalassaemia: a genotypic and phenotypic study. J Clin Pathol 63(2):147–150
33. Karimi M, Haghpanah S, Farhadi A, Yavarian M (2012) Genotype-phenotype relationship of patients with β-thalassaemia taking hydroxyurea: a 13-year experience in Iran. Int J Hematol 95(1):51–56
34. Rigano P, Pecoraro A, Calzolari R, Troia A, Acuto S, Renda D, Pantalone GR, Maggio A, Di Marzo R (2010) Desensitization to hydroxycarbamide following long-term treatment of thalassaemia intermedia as observed in vivo and in primary erythroid cultures from treated patients. Br J Haematol 151(5):509–515
35. Ma Q, Wyszynski DF, Farrell JJ, Kutlar A, Farrer LA, Baldwin CT, Steinberg MH (2007) Foetal hemoglobin in sickle cell anaemia: genetic determinants of response to hydroxyurea. Pharmacogenomics J 7(6):386–394
36. Borg J, Papadopoulos P, Georgitsi M, Gutiérrez L, Grech G, Fanis P, Phylactides M, Verkerk AJ, van der Spek PJ, Scerri CA, Cassar W, Galdies R, van Ijcken W, Ozgür Z, Gillemans N, Hou J, Bugeja M, Grosveld FG, von Lindern M, Felice AE, Patrinos GP, Philipsen S (2010) Haploinsufficiency for the erythroid transcription factor KLF1 causes hereditary persistence of foetal hemoglobin. Nat Genet 42(9):801–805
37. Tafrali C, Paizi A, Borg J, Radmilovic M, Bartsakoulia M, Giannopoulou E, Giannakopoulou O, Stojiljkovic-Petrovic M, Zukic B, Poulas K, Stavrou EF, Lambropoulou P, Kourakli A, Felice AE, Papachatzopoulou A, Philipsen S, Pavlovic S, Georgitsi M, Patrinos GP (2013) Genomic variation in the MAP3K5 gene is associated with β-thalassaemia disease severity and hydroxyurea treatment efficacy. Pharmacogenomics 14(5):469–483

38. Banan M, Bayat H, Azarkeivan A, Mohammadparast S, Kamali K, Farashi S, Bayat N, Khani MH, Neishabury M, Najmabadi H (2012) The XmnI and BCL11A single nucleotide polymorphisms may help predict hydroxyurea response in Iranian β-thalassaemia patients. Hemoglobin 36(4):371–380
39. Flanagan JM, Steward S, Howard TA, Mortier NA, Kimble AC, Aygun B, Hankins JS, Neale GA, Ware RE (2012) Hydroxycarbamide alters erythroid gene expression in children with sickle cell anaemia. Br J Haematol 157(2):240–248
40. Borg J, Phylactides M, Bartsakoulia M, Tafrali C, Lederer C, Felice AE, Papachatzopoulou A, Kourakli A, Stavrou EF, Christou S, Hou J, Karkabouna S, Lappa-Manakou C, Ozgur Z, van Ijcken W, von Lindern M, Grosveld FG, Georgitsi M, Kleanthous M, Philipsen S, Patrinos GP (2012) KLF10 gene expression is associated with high foetal hemoglobin levels and with response to hydroxyurea treatment in β-hemoglobinopathy patients. Pharmacogenomics 13(13):1487–1500
41. Kumkhaek C, Taylor JG, Zhu J, Hoppe C, Kato GJ, Rodgers GP (2008) Foetal haemoglobin response to hydroxycarbamide treatment and sar1a promoter polymorphisms in sickle cell anaemia. Br J Haematol 141(2):254–259
42. Giannopoulou E, Bartsakoulia M, Tafrali C, Kourakli A, Poulas K, Stavrou EF, Papachatzopoulou A, Georgitsi M, Patrinos GP (2012) A single nucleotide polymorphism in the HBBP1 gene in the human β-globin locus is associated with a mild β-thalassaemia disease phenotype. Hemoglobin 36(5):433–445
43. Hershko C, Link G, Konijn AM, Cabantchik ZI (2005) Objectives and mechanism of iron chelation therapy. Ann N Y Acad Sci 1054:124–135
44. Vanakker OM, De Paepe A (2013) Pharmacogenomics in children: advantages and challenges of next generation sequencing applications. Int J Pediatr 2013:136524

Pharmacogenetics of Neurodegenerative Disorders

Ramón Cacabelos, Clara Torrellas, Pablo Cacabelos, María J. Villanueva, Sergio Piñeiro and Cristina Solveira

Abstract Neurodegenerative disorders (NDDs) (Alzheimer's disease, Parkinson's disease, amyotrophic lateral sclerosis, multiple sclerosis, Huntington's disease) represent a major problem of health in developed countries, with an important repercussion in disability and health economics. NDDs pose several challenges to our society and the scientific community: they represent an epidemiological problem and a socio-economic, psychological and family burden; most of them have an obscure/complex pathogenesis; their diagnosis is not easy and lacks specific biomarkers; and their treatment is difficult and inefficient. Most NDDs share some common features: they are polygenic disorders in which genetic, epigenetic and environmental factors are involved; some of them follow a general rule in genomics related to disease onset, clinical course and prognosis; multifactorial dysfunctions in several metabolomic networks lead to functional damage to specific brain circuits; accumulation of toxic proteins (i.e. conformational changes) in the nervous tissue is involved in many cases of NDDs; all of them are costly for society, deteriorating the quality of life of sufferers and increasing disability; and although NDDs do not have a curative treatment, in practice available therapeutics is susceptible to pharmacogenomic intervention.

The genes involved in the pharmacogenomics of drugs to treat NDDs fall into five categories: (i) genes associated with disease pathogenesis (pathogenic genes); (ii) genes associated with the mechanism of action of drugs (mechanistic genes);

R. Cacabelos (✉) · C. Torrellas · P. Cacabelos
Genomic Medicine, Camilo José Cela University, Madrid, Spain
e-mail: rcacabelos@ucjc.edu

C. Torrellas
e-mail: neuropsicologia2@euroespes.com

P. Cacabelos
e-mail: asistentedireccion@euroespes.com

R. Cacabelos · C. Torrellas · P. Cacabelos · M. J. Villanueva · S. Piñeiro · C. Solveira
EuroEspes Biomedical Research Center, Institute of Medical Science and Genomic Medicine, Corunna, Spain

G. Grech, I. Grossman (eds.), *Preventive and Predictive Genetics: Towards Personalised Medicine,* Advances in Predictive, Preventive and Personalised Medicine 9,
DOI 10.1007/978-3-319-15344-5_8

(iii) genes associated with drug metabolism; (iv) genes associated with drug transporters; and (v) pleiotropic genes involved in multifaceted cascades and metabolic reactions. Pharmacogenomics accounts for 30–90 % variability in pharmacokinetics and pharmacodynamics. Only 20–30 % of the Caucasian population processes normally approximately 60 % of the current drugs which are metabolised via cytochromes CYP2D6, CYP2C9 and CYP2C19. Clinical pharmacogenomics may contribute to personalising pharmacological treatment, predicting patient/drug-dose selection, minimising drug interactions, increasing drug efficacy, and reducing unnecessary costs.

Keywords Neurodegenerative disorders · Alzheimer's disease · Parkinson's disease · Pharmacogenomics · CYP2D6 · CYP2C9 · CYP2C19 · CYP3A4/5 · APOE · Transporters

1 Introduction

Neurodegenerative disorders (NDDs) represent relatively frequent forms of central nervous system (CNS) disorders which can be classified according to their phenotypic expression affecting cognition, movement, strength, motor coordination or other activities/functions of the CNS (Table 1). Together with stroke and neuropsychiatric disorders, NDDs are the third greatest problem of health in developed countries, representing 10–15 % of deaths after cardiovascular disorders (25 %) and cancer (20 %) [11]. CNS disorders pose several challenges to our society and the scientific community: they represent an epidemiological problem and a socio-economic, psychological and family burden; most of them have an obscure/complex pathogenesis; their diagnosis is not easy and lacks specific biomarkers; and their treatment is difficult and inefficient. In terms of economic burden, approximately 10–20 % of direct costs are associated with their pharmacological treatment, with a gradual increase in parallel with the severity of the disease [11].

The pharmacological management of NDDs is an issue of special concern due to its complexity, poor cost-effectiveness, lack of specificity, and also the polymedication required to modulate the bulk of symptoms accompanying their clinical manifestations. A growing body of fresh knowledge on the pathogenesis of NDDs, together with data on neurogenomics and pharmacogenomics, is emerging in recent times. The incorporation of this new armamentarium of molecular pathology and genomic medicine to daily medical practice, together with educational programmes for the correct use of drugs, must help to: understand brain pathogenesis, establish an early diagnosis, and optimize therapeutics either as a preventive strategy or as a formal symptomatic treatment [12-16].

Table 1 Neurodegenerative disorders [1-10]

Disease	Subtypes	Disturbance					Prevalence
		Cognition	Movement	Strength	Coordination	Myelin	
Acute Disseminated Encephalomyelitis (ADEM)		+	+			+	N/A
Alzheimer disease (AD)	Alzheimer disease, early onset Alzheimer disease, late onset Dementia in Alzheimer disease, atypical or mixed type	+					5,000–7,000 per 100,000 individuals
Amyotrophic lateral sclerosis (ALS)	Sporadic ALS Familial ALS Western Pacific ALS Hiramaya disease		+	+			1–9 per 100,000 individuals
Ataxia with oculomotor apraxia (AOA)	Ataxia with oculomotor apraxia type 1 Ataxia with oculomotor apraxia type 2		+		+		0.4 per 100,000 individuals
Ataxia-telangiectasia (A-T) or Louis-Barr disease		+	+		+		0.1–0.9 per 100,000 individuals
Central pontine myelinolysis (CPM)		+	+	+	+	+	N/A
Cerebellar ataxia	Early-onset cerebellar ataxia Late-onset cerebellar ataxia		+		+		8.9 per 100,000 individuals
Cerebral amyloid angiopathy (CAA)	Icelandic type Dutch variant British type Familial Danish dementia Finnish type amyloidosis	+					N/A

Table 1 (continued)

Disease	Subtypes	Disturbance					Prevalence
		Cognition	Movement	Strength	Coordination	Myelin	
Charcot-Marie-Tooth (CMT)	Charcot-Marie-Tooth type 1 Charcot-Marie-Tooth type 2			+			10–50 per 100,000 individuals
Corticobasal degeneration (CBD)		+	+				1–9 per 100,000 individuals
Creutzfeldt-Jakob disease (CJD)		+			+		0.1–0.9 per 100,000 individuals
Dementia with Lewy bodies (DLB)		+	+				60–90 per 100,000 individuals
Dentatorubralpallido-luysian atrophy (DRPLA)		+	+		+		0.1–0.9 per 100,000 individuals
Diffuse cerebral sclerosis of Schilder		+	+			+	N/A
Fatal familial insomnia (IFF)		+					<0.1 per 100,000 individuals
Frontotemporal dementia (FTD)	Pick's disease Frontal dementia Primary progressive aphasia Semantic dementia	+					1–9 per 100,000 individuals
Gerstmann-Straüssler-Scheinker disorder (GSS)		+	+		+		0.001–0.01 per 100,000 individuals
Guillain-Barré syndrome (GBS)			+	+		+	1–9 per 100,000 individuals

Table 1 (continued)

Disease	Subtypes	Disturbance					Prevalence
		Cognition	Movement	Strength	Coordination	Myelin	
Hereditary spactic paraplegia (HSP)	Hereditary spactic paraplegia uncomplicated		+	+			4.1 per 100,000 individuals
	Hereditary spactic paraplegia complicated	+	+	+	+		
Huntington disease (HD)		+	+				1–9 per 100,000 individuals
Kuru			+		+		N/A
Leukodystrophies	Adrenoleukodystrophy Adrenomyeloneuropathy Krabbe disease Pelizaeus-Merzbacher disease Leukoencephalopathy with vanishing white matter or VWM disease Canavan disease Alexander disease	+	+			+	N/A
Multiple sclerosis (MS)	Disseminated MS Generalised MS MS no specified MS of brain stem MS of cord	+	+	+	+	+	60–90 per 100,000 individuals
Multiple system atrophy (MSA)		+	+	+	+		1–9 per 100,000 individuals
Neuromyelitis optica or Devic's disease (NMO)			+	+		+	1–9 per 100,000 individuals

Table 1 (continued)

Disease	Subtypes	Disturbance					Prevalence
		Cognition	Movement	Strength	Coordination	Myelin	
Parkinson's disease (PD)	PD no specified PD idiopathic PD primary	+	+	+	+		120 per 100,000 individuals
Progressive multifocal leukoencephalopathy (PML)		+		+	+		N/A
Progressive supranuclear palsy (PSP)	Richardson syndrome Progressive supranuclear palsy-parkinsonism pure-akinesia-with-gait-freezing (PAGF)	+	+		+		N/A
Spinal muscular atrophy (SMA)	SMA type I, or Werdnig-Hoffman disease), SMA type II, chronic infantile SMA type III chronic juvenile or Kugelberg-Welander disease SMA type IV Adult onset			+			11 per 100,000 individuals
Subacute sclerosing panencephalitis (SSPE) or Dawson Disease		+	+	+	+		<0.1 per 100,000 individuals
Transthyretin-related hereditary amyloidosis (TTR) (Familial amyloid polyneuropathy)			+	+			N/A

+ main pathologic features

Most NDDs share some common features: they are polygenic disorders in which genetic, epigenetic and environmental factors are involved; some of them follow a golden rule: the higher the number of genes affected, the earlier the onset of the disease, with a faster progression, and a poorer therapeutic response to conventional drugs; and the smaller the number of genes disturbed, the later the onset, with a slower progression and a more favourable therapeutic response to current treatments; multifactorial dysfunctions in several metabolomic networks lead to functional damage to specific brain circuits; accumulation of toxic proteins (i.e., conformational changes) in the nervous tissue is involved in many cases of NDDs; all of them are costly for society, deteriorating the quality of life of sufferers and increasing disability; and although NDDs do not have a curative treatment, in practice available therapeutics is susceptible to pharmacogenomic intervention [17-19].

The introduction of novel procedures into an integral genomic medicine protocol in CNS disorders is an imperative requirement for clinical practice and drug development in order to improve diagnostic accuracy (disease-specific biomarkers) and to optimise therapeutics (pharmacogenomics) [20-26]. Drug treatment has made remarkable strides, with the introduction of many new drugs; however, improvement in terms of clinical outcome has fallen short of expectations, with up to one third of the patients continuing to experience clinical relapse or unacceptable medication-related side-effects in spite of efforts to identify optimal treatment regimens. Potential reasons to explain this historical setback might be that: the molecular pathology of most NDDs is still poorly understood; drug targets are inappropriate, not fitting into the real aetiology of the disease; most treatments are symptomatic, but not anti-pathogenic; the genetic component of NDDs is poorly defined; and the understanding of genome-drug interactions is very limited [12, 15, 16, 24].

2 Pharmacogenomics

The genes involved in the pharmacogenomic response to drugs in CNS disorders may fall into five major categories: (i) genes associated with CNS pathogenesis (disease-specific genes) (Tables 2, 3, 4, 5, 6); (ii) genes associated with the mechanism of action of drugs; (iii) genes associated with drug metabolism; (iv) genes associated with drug transporters; and (v) pleiotropic genes involved in multifaceted cascades and metabolic reactions [11, 16, 18, 19] (Tables 7–12). The therapeutic outcome (efficacy and safety) is the result of the interplay of drugs with these different categories of gene products and epigenetic factors to reverse or modify the phenotypic expression of a given disease [57]. Pharmacogenomics account for 30–90 % variability in pharmacokinetics and pharmacodynamics.

2.1 Pathogenic Genes

Over 6,000 genes distributed across the human genome have been screened for associations with CNS disorders during the past 30 years. Studies of many candidate genes potentially associated with a particular NDD could not be replicated in different settings, cohorts, and geographical contexts due to methodological problems, sample selection and multi-ethnic genetic variation. Prototypical examples of NDDs are Alzheimer's disease (Table 2), Parkinson's disease (Table 3), amyotrophic lateral sclerosis (Table 4), multiple sclerosis (Table 5), and Huntington's disease (Table 6).

2.2 Genes Involved in the Mechanism of Action of CNS Drugs

Most genes associated with the mechanism of action of CNS drugs encode receptors, proteins, enzymes, and neurotransmitters on which these drugs act as ligands (agonists, antagonists), enzyme modulators (substrates, inhibitors, inducers) or neurotransmitter regulators (releasers, reuptake inhibitors) [17] (Tables 7, 8, 9, 10, 11, 12).

2.3 Genes Involved in Drug Metabolism

Drug metabolism includes phase I reactions (i.e. oxidation, reduction, hydrolysis) and phase II conjugation reactions (i.e. acetylation, glucuronidation, sulphation, methylation). The principal enzymes with polymorphic variants involved in phase I reactions are the following: Cytochrome P450 monooxygenases (CYP3A4/5/7, CYP2E1, CYP2D6, CYP2C19, CYP2C9, CYP2C8, CYP2B6, CYP2A6, CYP1B1, CYP1A1/2), EPHX1/EPHX2 (epoxide hydrolases), esterases, NQO1 (NADPH-quinone oxidoreductase), DPD (dihydropyrimidine dehydrogenase), ADH (alcohol dehydrogenase), and ALDH (aldehyde dehydrogenase); and major enzymes involved in phase II reactions include UGTs (uridine 5′-triphosphate glucuronosyl transferases), TPMT (thiopurine methyltransferase), COMT (catechol-O-methyltransferase), HMT (histamine methyl-transferase), STs (sulfotransferases), GST-A (glutathione S-transferase A), GST-P (glutathione S-transferase P), GST-T (glutathione S-transferase T), GST-M (glutathione S-transferase M), NAT1 (N-acetyltransferase 1), NAT2 (N-acetyltransferase 2), and others.

Patients with NDDs are chronic users of both specific pharmacotherapy and drugs for comorbid symptomatology (psychotropic agents) whose metabolism is mainly via the cytochrome P450 (CYP) gene superfamily (Tables 7, 8, 9, 10, 11, 12, 13). P450 enzymes convert xenobiotics into electrophilic intermediates which are then conjugated by phase II enzymes to hydrophilic derivatives that can be excreted, according to the database of the World Guide for Drug Use and Pharma-

Table 2 Alzheimer disease-related genes [15]

Subtype		Genes			
	OMIM	Name	Symbol	Locus	OMIM
Alzheimer disease 1, familial	104300	Amyloid beta (A4) precursor protein	APP	21q21.3	104760
		Granulin precursor	GRN	17q21.31	138945
		Microtubule-associated protein tau	MAPT	17q21.31	157140
Alzheimer disease 2	104310	Apolipoprotein E	APOE	19q13.32	107741
Alzheimer disease 3	607822	Presenilin 1	PSEN1	14q24.2	104311
Alzheimer disease 4	606889	Presenilin 2	PSEN2	1q42.13	600759
Alzheimer disease 5	104300	Alzheimer disease 5	AD5	12p11.23-q13.12	602096
Alzheimer disease 6	104300	Alzheimer disease 6	AD6	10q24	605526
Alzheimer disease 7	606187	Alzheimer disease 7	AD7	10p13	606187
Alzheimer disease 8	104300	Alzheimer disease 8	AD8	20p	607116
Alzheimer disease 9, late onset, susceptibility to	104300	Alzheimer disease 9	AD9	19p13.2	608907
Alzheimer disease 10	104300	Alzheimer disease 10	AD10	7q36	609636
Alzheimer disease 11	609790	Alzheimer disease 11	AD11	9p22.1	609790
Alzheimer disease 12	611073	Alzheimer disease 12	AD12	8p12-q22	611073
Alzheimer disease 13	611152	Alzheimer disease 13	AD13	1q21	611152
Alzheimer disease 14	611154	Alzheimer disease 14	AD14	1q25	611154
Alzheimer disease 15	611155	Alzheimer disease 15	AD15	3q22-q24	611155
Alzheimer disease 16	300756	Alzheimer disease 16	AD16	Xq21.3	300756
Alzheimer disease, late-onset	104300	Amyloid beta (A4) precursor protein-binding, family B, member 2	APBB2	4p14-p13	602710

Table 2 (continued)

Subtype		Genes			
	OMIM	Name	Symbol	Locus	OMIM
Alzheimer disease, late-onset, susceptibility to	104300	Nitric oxide synthase 3	NOS3	7q36.1	163729
		Plasminogen activator, urinary	PLAU	10q22.2	191840
Alzheimer disease, pathogenesis, association with	104300	Sortilin-related receptor	SORL1	11q24.1	602005
Alzheimer disease, susceptibility to	104300	Alpha-2-macroglobulin	A2M	12p13.31	103950
		ATP-binding cassette, subfamily A, member 7	ABCB7	19p13.3	605414
		Angiotensin I-converting enzyme	ACE	17q23.3	106180
		ADAM metallopeptidase domain 12	ADAM12	10q26.2	602714
		ADAM metallopeptidase domain 7	ADAM17	2p25	603639
		Advanced glycosylation end product-specific receptor	AGER	6p21.32	600214
		Angiotensinogen (serpin peptidase inhibitor, clade A, member 8)	AGT	1q42-q43	106150
		Arachidonate 5-lipoxygenase	ALOX5	10q11.21	152390
		Amyloid beta (A4) precursor protein-binding, family A, member 1	APBA1	9q21.11-q21.12	602414
		Amyloid beta (A4) precursor protein-binding, family A, member 2	APBA2	15q13.1	602712
		Amyloid beta (A4) precursor protein-binding, family A, member 3	APBA3	19p13.3	604262
		Amyloid beta (A4) precursor protein-binding, family B, member 1 (Fe65)	APBB1	11p15.4	602709
		Amyloid beta (A4) precursor protein-Binding ,family B, member 2	APBB2	15q22.2	602710

Table 2 (continued)

Subtype		Genes			
	OMIM	Name	Symbol	Locus	OMIM
Alzheimer disease, susceptibility to	104300	Amyloid beta (A4) precursor protein-binding, family B, member 3	APBB3	19q13.12	602711
		Amyloid P component, serum	APCS	1q21-q23	104770
		Anterior pharynx defective 1 homolog B (C. elegans)	APH1B	15q22.2	607630
		Amyloid beta (A4) precursor-like protein 1	APLP1	19q13.1	104775
		Amyloid beta (A4) precursor-like protein 2	APLP2	11q24	104776
		Apolipoprotein D	APOD	2p24-p23	107730
		Apolipoprotein C-I	APOC1	19q13.2	107710
		Amyloid beta precursor protein (cytoplasmic tail) binding protein 2	APPBP2	17q23.2	605324
		Ataxin 1	ATXN1	6p23	601556
		Beta-site APP-cleaving enzyme 1	BACE1	11q23.2-q23.3	604252
		Beta-site APP-cleaving enzyme 2	BACE2	21q22.3	605668
		Butyrylcholinesterase	BCHE	3q26.1-q26.2	177400
		Brain-derived neurotrophic factor	BDNF	11p13	113505
		Bridging integrator	BIN1	2q14	601248
		Bleomycin hydrolase	BLMH	17q11.2	602403
		Biogenesis of lysosomal organelles complex-1, subunit 3	BLOC1S3	19q13.32	609762
		Bromodomain PHD finger transcription factor	BPTF	17q24.3	601819
		Calcium homeostasis modulator 1	CALHM1	10q24.33	612234
		Calpastatin	CAST	5q15	114090
		Cholesteryl ester transfer protein, plasma	CETP	16q21	118470
		CD2-Associated protein	CD2AP	6p12	604241

Table 2 (continued)

Subtype		Genes			
	OMIM	Name	Symbol	Locus	OMIM
Alzheimer disease, susceptibility to	104300	CD33 antigen	CD33	19q13.3	159590
		Cholesterol 25-hydroxylase	CH25H	10q23	604551
		Cholinergic receptor, nicotinic, bet 2 (neuronal)	CHRNB2	1q21.3	118507
		Clusterin	CLU	8p21-p12	185430
		Catechol-O methyltransferase	COMT	22q11.21	116790
		COX10 homolog, cytochrome C oxidase assembly protein, heme A: farnesyltransferase (yeast)	COX10	17p12	602125
		COX15 homolog, cytochrome c oxidase assembly protein (yeast)	COX15	10q24	603646
		Complement component (3b/4b)receptor 1 (Knops blood group	CR1	1q32	120620
		cAMP responsive element binding protein 1	CREB1	2q34	123810
		Cystatin C	CST3	20p11.21	604312
		Cathepsin B	CTSB	8p22	116810
		Cathepsin D	CTSD	11p15.5	116840
		Cytochrome P450, family 19, Subfamily A, polypeptide 1	CYP19A1	15q21.1	107910
		Cytochrome P450, family 46, subfamily A, Polypeptide 1	CYP46A1	14q32.1	604087
		D-amino acid oxidase activator	DAOA	13q34	607408
		Death-associated protein kinase 1	DAPK1	9q21.33	600831
		Drebrin 1	DBN1	5q35.3	126660
		24-dehydrocholesterol reductase	DHCR24	1p32.3	606418
		Endothelin-converting enzyme 1	ECE1	1p36.1	600423
		ERI 1 exoribonuclease family member 3	ERI3	1p32	609917

Table 2 (continued)

Subtype		Genes			
	OMIM	Name	Symbol	Locus	OMIM
Alzheimer disease, susceptibility to	104300	Exocyst complex component 3-like 2	EXOC3L2	19q13.32	–
		FBJ murine osteosarcoma viral oncogene homolog	FOS	14q24.3	164810
		GRB2-associated binding protein 2	GAB2	11q14.1	606203
		Galanin-like peptide	GALP	19q13.43	611178
		Gliosis, familial progressive subcortical	GPSC	17q21-q22	–
		Glutamate receptor, ionotropic, N-methyl-D-aspartate 3A	GRIN3A	9q31.1	606650
		Glutamate receptor, ionotropic, N-methyl-D-aspartate 3B	GRIN3B	19p13.3	606651
		HECT domain containing 2	HECTD2	10q23.32	–
		HFE gene	HFE	6p21.3	613609
		Major histocompatibility complex, class I, A	HLA-A	6p21.3	142800
		Heat-shock 70-kd protein 5	HSPA5	9q33.3	138120
		Islet amyloid polypeptide	IAPP	12p12.1	147940
		Intercellular adhesion molecule 1	ICAM	19p13.3-p13.2	147840
		Insulin-degrading enzyme	IDE	10q23-q25	146680
		Interleukin-1-Alpha	IL1A	2q14	147760
		Interleukin-6 (interferon, beta 2)	IL6	7p21	147620
		Interleukin-10	IL10	1q31-q32	124092
		Interleukin 33	IL33	9p24.1	608678
		Kv channel interactin protein 3, calsenilin	KCNIP3	2q21.1	604662
		Lamin A/C	LMNA	1q22	150330
		Low-density lipoprotein-related protein 1 (alpha-2-macroglobulin receptor)	LRP1	12q13.3	107770

Table 2 (continued)

Subtype		Genes			
	OMIM	Name	Symbol	Locus	OMIM
Alzheimer disease, susceptibility to	104300	Low-density lipoprotein-related protein 1B (deleted in tumours)	LRP1B	2q21.2	608766
		Low-density lipoprotein-related protein 8, apolipoprotein E receptor	LRP8	1p34	602600
		MAP/microtubule affinity-regulating kinase 4	MARK4	19q13.3	606495
		MicroRNA 144	MIR144	17q11.2	612070
		Membrane metalloendopeptidase	MME	3q25.2	120520
		Myeloperoxidase	MPO	17q23.1	606989
		Membrane-spanning 4-domains, subfamily A, member 4E	MS4A4E	11q12.2	608401
		Membrane-spanning 4-domains, subfamily A, member 6A	MS4A6A	11q12.1	606548
		5,10-methylenetetrahydrofolate reductase (NADPH)	MTHFR	1p36.3	607093
		Myosin, heavy chain 13, skeletal muscle	MYH13	17p13	603487
		NEDD8 activating enzyme E1 subunit 1	NAE1	16q22	603385
		Nicastrin	NCSTN	1q22-q23	605254
		Notch homolog 3 (Drosophila)	NOTCH3	19p13.2-p13.1	600276
		Pax transcription activation domain-interacting protein 1	PAXIP1	7q36	608254
		Phosphoenolpyruvate carboxykinase 1 (soluble)	PCK1	20q13.31	614168
		Presenilin enhancer 2 homolog (C. elegans)	PSENEN	19q13.12	607632
		PiggyBac transposable element derived 1	PGBD1	6p22.1	–
		Phosphatidylinositol binding clathrin assembly protein	PICALM	11q14	603025
		Peptidylprolyl cis/trans isomerase, NIMA-interacting 1	PIN1	19p13	601052
		Pitrilysin metallopeptidase 1	PITRM1	10p15.2	–
		Paraoxonase 1	PON1	7q21.3	168820

Table 2 (continued)

Subtype		Genes			
	OMIM	Name	Symbol	Locus	OMIM
Alzheimer disease, susceptibility to	104300	Prion protein	PRNP	20p13	176640
		Prostaglandin-endoperoxide synthase 2 (prostaglandin G/H synthase and cyclooxygenase)	PTGS2	1q25.2-q25.3	600262
		Renal tumor antigen	RAGE	6p21.3	600214
		Reelin	RELN	7q22	600514
		Reticulon 4	RTN4	2p16.	604475
		Reticulon 4 receptor	RTN4R	22q11.21	605566
		S100-calcium-binding protein A1	S100A1	1q21	176940
		Serum amyloid A1	SAA1	11p15.1	104750
		Serpin peptidase inhibitor, clade A (alpha-1 antiproteinase, antitrypsin), member 3	SERPINA3	14q32.1	107280
		SH3 and PX domains 2A	SH3PXD2A	10q24.33	–
		Sterol O-acyltransferase 1	SOAT1	1q25	102642
		Sortilin-related VPS10 domain containing receptor 1	SORCS1	10q23-q25	606283
		Sortilin	SORT1	1p21.3-p13.1	602458
		Somastostatin	SST	3q28	182450
		Saitohin	STH	17q21.1	607067
		TARD DNA-binding protein	TARDBP	1p36.22	605078
		TM2 domain containing 1	TM2D1	1p31.3	610080
		Transcription factor A, mitochondrial	TFAM	10q21	600438
		Tumor necrosis factor (TNF superfamily, member 2)	TNF	6p21.3	191160
		Tyrosine kinase, nonreceptor, 1	TNK1	17p13.1	608076

Table 2 (continued)

Subtype		Genes			
	OMIM	Name	Symbol	Locus	OMIM
Alzheimer disease, susceptibility to	104300	Translocase of outer mitochondrial membrane 40 homolog (yeast)	TOMM40	19q13	608061
		Triggering receptor expressed on myeloid cells 2	TREM2	6p21.1	605086
		Transthyretin	TTR	18q12.1	176300
		Ubiquitin-conjugating enzyme E2I (UBC9 homolog, yeast)	UBE2I	16p13.3	601661
		Vascular endothelial growth factor	VEGFA	6p12	192240
		Zinc finger, FYVE domain containing 9	ZFYVE9	1p32.3	603755

Table 3 Parkinson disease-related genes

Subtype		Genes			
	OMIM	Name	Symbol	Locus	OMIM
Parkinson disease 1	168601	Synuclein, alpha (non A4 component of amyloid precursor)	SNCA	4q22.1	163890
Parkinson disease 2, juvenile	600116	Parkin	PARK2	6q26	602544
Parkinson disease 3	602404	Parkinson disease 3	PARK3	2p13	602404
Parkinson disease 4, autosomal dominant Lewy body	605543	Synuclein, alpha (non A4 component of amyloid precursor)	SNCA	4q22.1	163890
Parkinson disease 5, suscepti-bility to	613643	Ubiquitin carboxyl-terminal esterase L1	UCHL1	4p13	191342
Parkinson disease 6, early onset	605909	PTEN induced putative kinase 1	PINK1	1p36.12	608309
Parkinson disease 7, autosomal recessive early-onset	606324	Oncogene DJ1	DJ1	1p36.23	602533
Parkinson disease 8	607060	Leucine-rich repeat kinase 2 (dardarin)	LRRK2	12q12	609007
Parkinson disease 9	606693	ATPase type 13A2	ATP13A2	1p36.13	610513
Parkinson disease 10	606852	Parkinson disease 10	PARK10	1p32	606852
Parkinson disease 11	607688	GRB10-interacting GYF protein 2	GIGYF2	2q37.1	612003
Parkinson disease 12	300557	Parkinson disease-12	PARK12	Xq21-q25	300557
Parkinson disease 13	610297	HtrA serine peptidase 2	HTRA2	2p13.1	606441
Parkinson disease 14	612953	Phospholipase A2, group VI	PLA2G6	22q13.1	603604
Parkinson disease 15, autosomal recessive	260300	F-box only protein 7	FBOX7	22q12.3	605648
Parkinson disease 16	613164	Parkinson disease 16	PARK16	1q32	613164
Parkinson disease 17	614203	Vacuolar protein sorting 35, yeast, homolog of	VPS35	16q11.2	601501

Table 3 (continued)

Subtype		Genes			
	OMIM	Name	Symbol	Locus	OMIM
Parkinson disease 18	614251	Eukaryotic translation initiation factor 4-gamma, 1	EIF4G1	3q27.1	600495
Parkinson disease, late-onset, susceptibility to	168600	Glucosidase, beta, acid	GBA	1q22	606463
Parkinson disease, susceptibility to	168600	Alcohol dehydrogenase 1C (Class I), gamma polypeptide	ADH1C	4q23	103730
		Microtubule-associated protein tau	MAPT	17q21.31	157140
		TATA box binding protein	TBP	6q27	600075

Table 4 Amyotrophic lateral sclerosis disease-related genes

Subtype		Genes			
	OMIM	Name	Symbol	Locus	OMIM
Amyotrophic lateral sclerosis, susceptibility to	105400	Dynactin 1 (p150, glued, Drosophila, homolog of)	DCTN1	2p13.1	601143
		Neurofilament protein, heavy polypeptide	NEFH	22q12.2	162230
		Peripherin	PRPH	12q13.12	170710
Amyotrophic lateral sclerosis 1	105400	Superoxide dismutase 1	SOD1	21q22.11	147450
Amyotrophic lateral sclerosis 2, juvenile	205100	Alsin	ALS2	2q33.1	606352
Amyotrophic lateral sclerosis 3	606640	Amyotrophic lateral sclerosis 3	ALS3	18q21	606640
Amyotrophic lateral sclerosis 4, juvenile	602433	Senataxin	SETX	9q34.13	608465
Amyotrophic lateral sclerosis 5, juvenile recessive	602099	Amyotrophic lateral sclerosis 5, juvenile recessive	ALS5	15q15.1-q21.1	602099
Amyotrophic lateral sclerosis 6, autosomal recessive, with or without frontotemporal dementia	608030	Fused in sarcoma	FUS	16p11.2	137070
Amyotrophic lateral sclerosis 7	608031	Amyotrophic lateral sclerosis 7	ALS7	20p13	608031
Amyotrophic lateral sclerosis 8	608627	Vesicle-associated membrane protein-associated protein B	VAPB	20q13.32	605704
Amyotrophic lateral sclerosis 9	611895	Angiogenin	ANG	14q11.2	105850
Amyotrophic lateral sclerosis 10, with or without FTD	612069	TAR DNA-binding protein	TARDBP	1p36.22	605078
Amyotrophic lateral sclerosis 11	612577	FIG4, S. Cerevisiae, homolog of	FIG4	6q21	609390
Amyotrophic lateral sclerosis 12	613435	Optineurin	OPTN	10p13	602432
Amyotrophic lateral sclerosis 13, susceptibility to	183090	Ataxin-2	ATXN2	12q24.12	601517

Table 4 (continued)

Subtype		Genes			
	OMIM	Name	Symbol	Locus	OMIM
Amyotrophic lateral sclerosis 14, with or without frontotemporal dementia	613954	Valosin-containing protein	VCP	9p13.3	601023
Amyotrophic lateral sclerosis 15, with or without frontotemporal dementia	300857	Ubiquilin 2	UBQLN2	Xp11.21	300264
Amyotrophic lateral sclerosis 16, juvenile	614373	Sigma nonopioid intracellular receptor 1	SIGMAR1	9p13.3	601978
Amyotrophic lateral sclerosis 17	614696	CHMP family, member 2B	CHMP2B	3p11.2	609512
Amyotrophic lateral sclerosis 18	614808	Profilin 1	PFN1	17p13.2	176610
Amyotrophic lateral sclerosis-parkinsonism/dementia complex, susceptibility to	105500	Transient receptor potential cation channel, subfamily M, member 7	TRPM7	15q21.2	605692
Amyotrophic lateral sclerosis and/or frontotemporal dementia	105550	Chromosome 9 open reading frame 72	C9orf72	9p21.2	614260

Table 5 Multiple sclerosis disease-related genes [27]

Subtype		Genes			
	OMIM	Name	Symbol	Locus	OMIM
Multiple sclerosis, susceptibility to	126200	CD6 molecule	CD6	11q12.2	186720
		CD40 molecule, TNF receptor superfamily member 5	CD40	20q13.12	109535
		CD58 molecule	CD58	1p13.1	153420
		CD80 molecule	CD80	3q13.33	112203
		CD86 molecule	CD86	3q13.33	601020
		Cytochrome P450, family 27, subfamily B, polypeptide 1	CYP27B1	12q14.1	609506
		Major histocompatibility complex, class II, DP beta 1	HLA-DPB1	6p21.32	142858
		Major histocompatibility complex, class II, DQ beta-1	HLA-DQB1	6p21.32	604305
		Major histocompatibility complex, class II, DR beta-1	HLA-DRB1	6p21.32	142857
		Interleukin 2 receptor, alpha	IL2RA	10p15.1	147730
		Interleukin 7 receptor	IL7R	5p13.2	146661
		Interleukin 12A (natural killer cell stimulatory factor 1, cytotoxic lymphocyte maturation factor 1, p35)	IL12A	3q25.33	161560
		Interleukin 12B (natural killer cell stimulatory factor 2, cytotoxic lymphocyte maturation factor 2, p40)	IL12B	5q33.3	161561
		Interleukin 22 receptor, alpha 1	IL22RA1	1p36.11	605457
		SP140 nuclear body protein	SP140	2q37.1	608602
		Signal transducer and activator of transcription 3 (acute-phase response factor)	STAT3	17q21.2	102582

Table 5 (continued)

Subtype		Genes			
	OMIM	Name	Symbol	Locus	OMIM
Multiple sclerosis, susceptibility to	126200	Tyrosine kinase 2	TYK2	19p13.2	176941
		Multiple sclerosis, susceptibility to, 2	MS2	10p15.1	612594
		Multiple sclerosis, susceptibility to, 3	MS3	5p13.2	612595
		Multiple sclerosis, susceptibility to, 4	MS4	1p36	612596
		Tumor necrosis factor receptor superfamily, member 1A	TNFRSF1A	12p13.31	191190
Multiple sclerosis, disease progression, modifier of	126200	Programmed cell death 1	PDCD1	2q37.3	600244

Table 6 Huntington disease-related genes

Subtype		Genes			
	OMIM	Name	Symbol	Locus	OMIM
Huntington disease	143100	Huntingtin	HTT	4p16.3	613004
Huntington disease-like 1	603218	Prion protein	PRNP	20p13	176640
Huntington disease-like 2	606438	Junctophilin 3	JPH3	16q24.2	605268
Huntington disease-like 3	604802	Huntington-like neurodegenerative disorder 2	HDL3	4p15.3	604802
Huntington disease-like 4	607136	TATA box binding protein	TBP	6q27	600075

cogenomics [17]. Most CYP enzymes exhibit ontogenic-, age-, sex-, circadian-, and ethnic-related differences [58]. The practical consequence of this genetic variation is that the same drug can be differentially metabolised according to the genetic profile/expression during each subject's lifespan, and that knowing the pharmacogenomic profile of an individual, his/her pharmacodynamic response is potentially predictable to some extent.

As a general rule, the following phenotypes are differentiated: (a) extensive metabolisers (EMs) are individuals who respond normally to drugs at conventional doses; (b) intermediate metabolisers (IMs), carriers of a defective gene, which gives rise to an enzyme which is less efficient when metabolising drugs; (c) poor metabolisers (PMs), carriers of a defective gene whose mutation generates an enzyme with no drug-metabolising activity, or which simply has no enzyme; (d) ultra-rapid metabolisers (UMs), individuals whose gene is duplicated, triplicated or otherwise multiplied, bringing about an excessive enzymatic activity [17].

Among these enzymes, CYP2D6, CYP2C9, CYP2C19, and CYP3A4/5 are the most relevant in the pharmacogenetics of CNS drugs (Table 13).

2.3.1 *CYP2D6*

This drug metabolism gene is currently the most studied. Four RNA transcripts are expressed in the brain, liver, spleen and reproductive system, where 4 major proteins of 48–55 kDa (439–494 aa) are identified. This protein is a transport enzyme of the cytochrome P450 subfamily IID or multigenic cytochrome P450 superfamily of mixed-function monooxygenases which localises to the endoplasmic reticulum and is known to metabolise as many as 25 % of commonly-prescribed drugs and over 60 % of current psychotropics. This gene is highly polymorphic in the population. There are 141 *CYP2D6* allelic variants, of which the most important variants are listed in Table 13. 982 drugs are CYP2D6-related; 371 drugs are substrates, over 300 drugs are inhibitors, and 18 drugs are CYP2D6 inducers.

Table 7 Pharmacogenomics of Alzheimer's disease [17, 28-31].

Donepezil		
Category		Anti-Dementia agent; Cholinesterase inhibitor
Mechanism		Centrally active, reversible acetylcholinesterase inhibitor; increases the acetylcholine available for synaptic transmission in the CNS
Genes	Pathogenic	*APOE, CHAT*
	Mechanistic	*CHAT, ACHE, BCHE*
	Metabolism	
	Substrate	*CYP2D6* (major), *CYP3A4* (major), *UGTs, ACHE*
	Inhibitor	*ACHE, BCHE*
	Transporter	*ABCB1*
Galantamine		
Category		Anti-Dementia agent; Cholinesterase inhibitor
Mechanism		Reversible and competitive acetylcholinesterase inhibition leading to an increased concentration of acetylcholine at cholinergic synapses; modulates nicotinic acetylcholine receptor; may increase glutamate and serotonin levels
Genes	Pathogenic	*APOE, APP*
	Mechanistic	*ACHE, BCHE, CHRNA4, CHRNA7, CHRNB2*
	Metabolism	
	Substrate	*CYP2D6* (major), *CYP3A4* (major), *UGT1A1*
	Inhibitor	*ACHE, BCHE*
Memantine		
Category		Anti-Dementia drug; N-Methyl-D-Aspartate receptor antagonist
Mechanism		Binds preferentially to NMDA receptor-operated cation channels; may act by blocking actions of glutamate, mediated in part by NMDA receptors; Antagonist: GRIN2A, GRIN2B, GRIN3A, HTR3A, CHRFAM7A

Table 7 (continued)

Genes	Pathogenic	*APOE, PSEN1, MAPT*
	Mechanistic	*GRIN2A, GRIN2B, GRIN3A, HTR3A, CHRFAM7A*
	Metabolism nhibitor	
		CYP1A2 (weak), *CYP2A6* (weak), *CYP2B6* (strong), *CYP2C9* (weak), *CYP2C19* (weak), *CYP2D6* (strong), *CYP2E1* (weak), *CYP3A4* (weak)
	Pleiotropic	*APOE, MAPT, MT-TK, PSEN1*
Rivastigmine		
Category		Anti-Dementia agent; Cholinesterase inhibitor
Mechanism		Increases acetylcholine in CNS through reversible inhibition of its hydrolysis by cholinesterase
Genes	Pathogenic	*APOE, APP, CHAT*
	Mechanistic	*ACHE, BCHE,CHAT, CHRNA4, CHRNB2*
	Metabolism Inhibitor	
		ACHE, BCHE
	Pleiotropic	*APOE, MAPT*
Tacrine		
Category		Anti-Dementia agent; Cholinesterase inhibitor
Mechanism		Elevates acetylcholine in cerebral cortex by slowing degradation of acetylcholine
Genes	Pathogenic	*APOE*

Table 7 (continued)

	Mechanistic	*ACHE, BCHE,CHRNA4, CHRNB2*
	Metabolism	
	Substrate	*CYP1A2* (major), *CYP2D6* (minor), *CYP3A4* (major)
	Inhibitor	*ACHE, BCHE, CYP1A2* (weak)
	Transporter	*SCN1A*
	Pleiotropic	*APOE, MTHFR, CES1, LEPR, GSTM1, GSTT1*

ABCB1 ATP-binding cassette, sub-family B (MDR/TAP), member 1, *ACHE* Acetylcholinesterase (Yt blood group), *APOE* Apolipoprotein E, *APP* Amyloid beta (A4) precursor protein, *BCHE* Butyrylcholinesterase, *CES1* Carboxylesterase, *CHAT* Choline O-acetyltransferase, *CHRFAM7A* Cholinergic receptor, nicotinic, alpha 7, exons 5–10), *FAM7A* (family with sequence similarity 7A, exons A-E) fusion, *CHRNA4* Cholinergic receptor, nicotinic, alpha 4, *CHRNA7* Cholinergic receptor, nicotinic, alpha 7, *CHRNB2* Cholinergic receptor, nicotinic, beta 2 (neuronal), *CYP1A2* Cytochrome P450, family 1, subfamily A, polypeptide 2, *CYP2A6* Cytochrome P450, family 2, subfamily A, polypeptide 6, *CYP2B6* Cytochrome P450, family 2, subfamily B, polypeptide 6, *CYP2C19* Cytochrome P450, family 2, subfamily C, polypeptide 19, *CYP2C9* Cytochrome P450, family 2, subfamily C, polypeptide 9, *CYP2D6* Cytochrome P450, family 2, subfamily D, polypeptide 6, *CYP2E1* Cytochrome P450, family 2, subfamily E, polypeptide 1, *CYP3A4* Cytochrome P450, family 3, subfamily A, polypeptide 4, *GRIN2A* Glutamate receptor, ionotropic, N-methyl D-aspartate 2A, *GRIN2B* Glutamate receptor, ionotropic, N-methyl D-aspartate 2B, *GRIN3A* Glutamate receptor, ionotropic, N-methyl-D-aspartate 3A, *GSTM1* Glutathione S-transferase mu 1, *GSTT1* Glutathione S-transferase theta 1, *HTR3A* 5-Hydroxytryptamine (serotonin) receptor 3A, *LEPR* Leptin receptor, *MAPT* Microtubule-associated protein tau, *MT-TK* Mitochondrially encoded tRNA lysine, *MTHFR* 5,10 methylenetetrahydrofolate reductase (NADPH), *PSEN1* Presenilin 1, *SCN1A* sodium channel, voltage-gated, type I, alpha subunit, *UGT1A1* UDP glucuronosyltransferase 1 family, polypeptide A1, *UGTs* UDP glucuronosyltransferase families

Table 8 Pharmacogenomics of Parkinson disease [17, 30].

Amantadine		
Category		Antiparkinsonian agent, Adamantane. Dopamine agonist
Mechanism		Inhibition of dopamine reuptake into presynaptic neurons increase of dopamine release from presynaptic fibers. Antagonist: GRIN3A. Agonist: DRD2
Genes	Pathogenic	*DDC, DRDs, GRIN3A, PARK2*
	Mechanistic	*DDC, DRDs, GRIN3A*
	Metabolism	No appreciable metabolism, although negligible amounts of an acetyl metabolite have been identified
Apomorphine		
Category		Antiparkinsonian agent; Dopamine receptor agonists; Nonergot-derivative Dopamine receptor agonist
Mechanism		Stimulates postsynaptic D2-type receptors within the caudate putamen in the brain. Agonist: DRD2, DRD3
Genes	Pathogenic	*PARK2*
	Mechanistic	*DRD2, DRD3*
	Metabolism	
	Substrate	*CYP1A2* (minor), *CYP2C9* (minor), *CYP2C19* (minor), *CYP3A4* (minor) *SULT1A1, SULT1E1*
	Inhibitor	*CYP1A2* (weak), *CYP2C19* (weak), *CYP3A4* (weak)
Benztropine		
Category		Antiparkinsonian agent; Anticholinergic agent
Mechanism		Possesses both anticholinergic and antihistaminic effects. Inhibits the reuptake and storage of dopamine. Antagonist: CHRM1, HRH1
Genes	Mechanistic	*CHRM1, HRH1*
	Metabolism	
	Substrate	*ABCB1, CYP2C19* (minor), *CYP2D6* (major)
	Inhibitor	*SLC6A3*
	Transporter	*ABCB1*

Table 8 (continued)

Biperiden		
Category		Antiparkinsonian agent; Anticholinergic agent
Mechanism		This antimuscarinic agent inhibits acetylcholine or other cholinergic stimulus at autonomic effectors innervated by post-ganglionic nerves. Inhibition of striatal cholinergic receptors. Antagonist: CHRM1, CHRNA2
Genes	Pathogenic	*PARK2*
	Mechanistic	*ACHE, CHRM1, CHRNA2*
	Metabolism	
	Inhibitor	*CYP2D6* (weak)
Bromocriptine		
Category		Antiparkinsonian agent; Dopamine receptor agonist; Ergot-derivative Dopamine receptor agonist
Mechanism		Activates postsynaptic dopamine receptors in the tuberoinfundibular and nigrostriatal pathways. Agonist: DRD1, DRD2, DRD3
Genes	Pathogenic	*PARK2*
	Mechanistic	*DRD1, DRD2, DRD3*
	Metabolism	
	Substrate	*ABCG2, CYP3A4* (major)
	Inhibitor	*ABCB1, CYP1A2* (weak), *CYP3A4* (moderate)
	Inducer	*ABCB1*
	Transporter	*ABCB1, ABCG2*
	Pleiotropic	*DRD3, TGFB1, VEGFA*
Cabergoline		
Category		Antiparkinsonian agent; Dopamine receptor agonist; Ergot-derivative Dopamine receptor agonist
Mechanism		Long-acting dopamine receptor agonist with high binding affinity for dopamine D2 receptors and lesser affinity for D1, α1- and α2-adrenergic, and serotonin (5-HT1 and 5-HT2) receptors. Agonist: DRD1, DRD2, HTR1A, HTR1B, HTR2A, HTR2B, HTR2C. Antagonist: ADRA2A, ADRA2B, ADRA2C

Table 8 (continued)

Genes	Pathogenic	*PARK2*
	Mechanistic	*ADRA1s, ADRA2s, DRD1, DRD2, HTR1s, HTR2s*
	Metabolism	
	Substrate	*ABCB1, CYP3A4* (minor)
	Transporter	*ABCB1*
Carbidopa		
Category		Antiparkinsonian agent; Dopamine precursor
Mechanism		Inhibits the peripheral decarboxylation of levodopa to dopamine. Agonist: DRD2
Genes	Pathogenic	*PARK2*
	Mechanistic	*ACHE, COMT, DDC, DRD2*
	Pleiotropic	*ACE, BDNF, CHAT, OPRM1, PARK2*
Diphenhydramine		
Category		Antiparkinsonian agent; Anticholinergic agent
Mechanism		A potent antagonist to acetylcholine affecting muscarinic receptors; competes with free histamine for binding at HA-receptor sites. Antagonist: HRH1
Genes	Pathogenic	*PARK2*
	Mechanistic	*HRH1, CHRMs*
	Metabolism	
	Substrate	*CYP1A2* (minor), *CYP2C9* (minor), *CYP2C19* (minor), *CYP2D6* (major), *UGT1A3, UGT1A4*
	Inhibitor	*CYP2D6* (moderate)
Entacapone		
Category		Antiparkinsonian agent; Catechol-O-Methyltransferase (COMT) Inhibitor
Mechanism		Selective and reversible inhibitor of COMT. When entacapone is taken with levodopa, the pharmacokinetics is altered, resulting in more sustained levodopa serum levels compared to levodopa taken alone

Table 8 (continued)

Genes	Pathogenic	*PARK2*
	Mechanistic	*COMT*
	Metabolism	
	Substrate	*UGT1A1, UGT1A3, UGT1A4, UGT1A9, UGT1A10*
	Inhibitor	*COMT, CYP1A2* (weak), *CYP2A6* (weak), *CYP2C9* (weak), *CYP2C19* (weak), *CYP2D6* (weak), *CYP2E1* (weak), *CYP3A4* (weak)
Ethopropazine [32]		
Category		Antiparkinsonian agent; Anticholinergic agent
Mechanism		Partially blocks central cholinergic receptors, helping to balance cholinergic and dopaminergic activity in the basal ganglia. Ethopropazine 's local anesthetic effect is due to its antagonism of the NMDA glutamate receptor
Genes	Mechanistic	*CHRMs, NMDA*
	Metabolism	
	Inhibitor	*BCHE*
Levodopa		
Category		Antiparkinsonian agent; Dopamine precursor
Mechanism		Levodopa circulates in the plasma to the blood-brain-barrier, where it crosses, to be converted by striatal enzymes to dopamine. Agonist: DRD1, DRD2, DRD3, DRD4, DRD5
Genes	Mechanistic	*DRD1, DRD2, DRD3, DRD4, DRD5*
	Metabolism	
	Inhibitor	*COMT*
Lisuride		
Category		Antiparkinsonian agent; Ergot-derivative Dopamine receptor agonist
Mechanism		Displays dopaminergic and consequently prolactin-reducing properties. It has pronounced affinity for dopamine receptors in striatum and pituitary. Agonist: DRD2, HTR1A; Antagonist: DRD1

Table 8 (continued)

Genes	Mechanistic	*DRD1, DRD2, HTR1A*
	Metabolism	
	Substrate	*CYP2D6* (major), *CYP3A4* (major)
Pergolide		
Category		Antiparkinsonian agent; Ergot-derivative Dopamine receptor agonist
Mechanism		Stimulates postsynaptic dopamine receptors in corpus striatum, reduces serum prolactin concentrations, causes transient increases in serum somatotropin concentrations and decreases in serum luteinizing hormone concentrations. Agonist: *DRD1, DRD2*
Genes	Pathogenic	*PARK2*
	Mechanistic	*DRD1, DRD2, GH, LH, PRL*
	Metabolism	
	Substrate	*CYP3A4* (major)
	Inhibitor	*CYP2D6, CYP3A4*
Pramipexole		
Category		Antiparkinsonian agent; Nonergot-derivative Dopamine receptor agonist
Mechanism		Binds to D2 subfamily dopamine receptor and to D3 and D4 receptors; Stimulates dopamine activity on nerves of striatum and substantia nigra. Agonist: DRD2, DRD3, DRD4
Genes	Pathogenic	*PARK2*
	Mechanistic	*DRD2, DRD3, DRD4*
	Metabolism	
	Substrate	*CYP3A4*
Procyclidine		
Category		Antiparkinsonian agent; Anticholinergic agent

Table 8 (continued)

Mechanism		Might act blocking central cholinergic receptors, and thus balancing cholinergic and dopaminergic activity in the basal ganglia. Antagonist: CHRM1, CHRM2, CHRM4
Genes	Pathogenic	*PARK2*
	Mechanistic	*CHRM1, CHRM2, CHRM4*
Rasagiline		
Category		Antiparkinsonian agent; Monoamine Oxidase B inhibitor
Mechanism		MAO-B inhibitory activity; which causes an increase in extracellular levels of dopamine in the striatum
Genes	Pathogenic	*PARK2*
	Mechanistic	*MAO-B*
	Metabolism	
	Substrate	*CYP1A2* (major), *UGT1A1, UGT1A3, UGT1A4, UGT1A6, UGT1A7, UGT1A9, UGT1A10, UGT2B7, UGT2B15*
	Inhibitor	*MAOB*
Ropinirole		
Category		Antiparkinsonian agent; Nonergot-derivative Dopamine receptor agonist
Mechanism		Binds to dopamine receptors D2 (higher affinity), D3 and D4 in the striatum Agonist: *DRD2, DRD3, DRD4*
Genes	Pathogenic	*PARK2*
	Mechanistic	*DRD2, DRD3, DRD4*
	Metabolism	
	Substrate	*CYP1A2* (major), *CYP3A4* (minor)
	Inhibitor	*CYP1A2* (moderate), *CYP2D6* (moderate), *CYP3A4* (moderate)
Rotigotine		
Category		Antiparkinsonian agent; Nonergot-derivative Dopamine receptor agonist
Mechanism		Non-ergot dopamine agonist with specificity for D3, D2, and D1 dopamine receptors. Stimulation of postsynaptic dopamine D2-type auto-receptors within substantia nigra. Agonist: DRD1, DRD2, DRD3

Table 8 (continued)

Genes	Pathogenic	*PARK2*
	Mechanistic	*DRD1, DRD2, DRD3*
Selegiline [33]		
Category		Antiparkinsonian agent; Monoamine Oxidase B inhibitor
Mechanism		Binds to MAOB within the nigrostriatal pathways blocking microsomal metabolism of dopamine and enhancing the dopaminergic activity in the substantia nigra; at higher doses, it can also inhibit MAOA
Genes	Pathogenic	*PARK2*
	Mechanistic	*MAOA, MAOB*
	Metabolism	
	Substrate	*CYP1A2* (minor), *CYP2A6* (minor), *CYP2B6* (major), *CYP2C8* (minor), *CYP2C19* (major), *CYP2D6* (minor), *CYP2E1* (minor), *CYP3A4* (minor), *MAOA*
	Inhibitor	*CYP1A2* (weak), *CYP2A6* (weak), *CYP2C9* (weak), *CYP2C19* (weak), *CYP2D6* (weak), *CYP2E1* (weak), *CYP3A4* (weak), *MAOB, MAOA*
	Transporter	*SCNA*
Tolcapone [34]		
Category		Antiparkinsonian agent; Catechol-O-Methyltransferase (COMT) inhibitor
Mechanism		A selective and reversible inhibitor of COMT. In presence of a decarboxylase inhibitor, COMT is the major degradation pathway for levodopa. Inhibition of COMT leads to more sustained plasma levels of levodopa and enhanced central dopaminergic activity
Genes	Pathogenic	*PARK2*
	Mechanistic	*COMT*
	Metabolism	
	Substrate	*UGT1A*
	Inhibitor	*COMT*

Table 8 (continued)

Trihexyphenidyl		
Category		Antiparkinsonian agent; Anticholinergic agent
Mechanism		Selective M1 muscarinic acetylcholine receptor antagonist. It is able to discriminate between the M1 and the peripheral muscarinic subtypes. It also partially blocks cholinergic activity in the CNS and it is thought to increase the availability of dopamine. Antagonist: CHRM1
Genes	Pathogenic	*PARK2*
	Mechanistic	*CHRM1*

ABCB1 ATP-binding cassette, sub-family B (MDR/TAP), member 1, *ABCG2* ATP-binding cassette, sub-family G (WHITE), member 2, *ACE* Angiotensin I converting enzyme (peptidyl-dipeptidase A) 1, *ACHE* Acetylcholinesterase, *ADRA1s* Alpha 1 adrenergic receptors, *ADRA2s* Alpha 2 adrenergic receptors, *ADRA2A* Adrenoceptor alpha 2A, *ADRA2B* Adrenoceptor alpha 2B, *ADRA2C* Adrenoceptor alpha, *BCHE* Butyrylcholinesterase, *BDNF* Brain-derived neurotrophic factor, *CHAT* Choline O-acetyltransferase, *CHRM1* Cholinergic receptor, muscarinic 1, *CHRM2* Cholinergic receptor, muscarinic 2, *CHRM3* Cholinergic receptor, muscarinic 3, *CHRM4* Cholinergic receptor, muscarinic 4, *CHRM5* Cholinergic receptor, muscarinic 5, *CHRNA2* Cholinergic receptor, nicotinic, alpha 2 (neuronal), *COMT* Catechol-O-methyltransferase, *CYP1A2* Cytochrome P450, family 1, subfamily A, polypeptide 2, *CYP2A6* Cytochrome P450, family 2, subfamily A, polypeptide 6, *CYP2B6* Cytochrome P450, family 2, subfamily B, polypeptide 6, *CYP2C8* Cytochrome P450, family 2, subfamily C, polypeptide 8, *CYP2C19* Cytochrome P450, family 2, subfamily C, polypeptide 19, *CYP2C9* Cytochrome P450, family 2, subfamily C, polypeptide 9, *CYP2D6* Cytochrome P450, family 2, subfamily D, polypeptide 6, *CYP2E1* Cytochrome P450, family 2, subfamily E, polypeptide 1, *CYP3A4* Cytochrome P450, family 3, subfamily A, polypeptide 4, *DDC* Dopa decarboxylase (aromatic L-amino acid decarboxylase), *DRDs* Dopamine receptor, *DRD1* Dopamine receptor D1, *DRD2* Dopamine receptor D2, *DRD3* Dopamine receptor D3, *DRD4* Dopamine receptor D4, *DRD5* Dopamine receptor D5, *GRIN3A* Glutamate receptor, ionotropic, N-methyl-D-aspartate 3A, *HRH1* Histamine receptor H1, *HTR1s* 5-Hydroxytryptamine type 1 receptors, *HTR1A* 5-Hydroxytryptamine (serotonin) receptor 1A, G protein-coupled, *HTR1B* 5-Hydroxytryptamine (serotonin) receptor 1B, G protein-coupled, *HTR2s* 5-Hydroxytryptamine (serotonin) type 2 receptors, *HTR2A* 5-Hydroxytryptamine (serotonin) receptor 2A, G protein-coupled, *HTR2B* 5-Hydroxytryptamine (serotonin) receptor 2B, G protein-coupled, *HTR2C* 5-Hydroxytryptamine (serotonin) receptor, G protein-coupled, *MAOA* Monoamine oxidase A, *MAOB* Monoamine oxidase B, *OPRM1* Opioid receptor, mu 1, *PARK2* Parkinson protein 2, E3 ubiquitin protein ligase (parkin), *SCNA* sodium channel alpha-subunit gene family, *SLC6A3* Solute carrier family 6 (neurotransmitter transporter, dopamine), member 3, *SULT1A1* Sulfotransferase family, cytosolic, 1A, phenol-preferring, member 1, *SULT1E1* Sulfotransferase family 1E, estrogen-preferring, member 1, *TGFB1* Transforming growth factor, beta 1, *UGT1A* UDP glucuronosyltransferase 1 family, polypeptide A complex locus, *UGT1A1* UDP glucuronosyltransferase 1 family, polypeptide A1, *UGT1A3* UDP glucuronosyltransferase 1 family, polypeptide A3, *UGT1A4* UDP glucuronosyltransferase 1 family, polypeptide A4, *UGT1A6* UDP glucuronosyltransferase 1 family, polypeptide A6, *UGT1A7* UDP glucuronosyltransferase 1 family, polypeptide A7, *UGT1A9* UDP glucuronosyltransferase 1 family, polypeptide A9, *UGT1A10* UDP glucuronosyltransferase 1 family, polypeptide A10, *UGT2B7* UDP glucuronosyltransferase 2 family, polypeptide B7, *UGT2B15* UDP glucuronosyltransferase 2 family, polypeptide B15, *VEGFA* vascular endothelial growth factor A

Table 9 Pharmacogenomics of amyotrophic lateral sclerosis [17, 30].

Riluzole [35, 36]		
Category		Central nervous system agent, Glutamate inhibitor
Mechanism		Inhibition of glutamate release, inactivation of voltage-dependent sodium channels, and blocking the action of excitatory amino acid receptor agonists
Genes	Pathogenic	*GRIN3A*
	Mechanistic	*GRIN3A, SCN1A, SCN1B, SCN2A, SCN2B, SCN3A, SCN3B, SCN4A, SCN4B, SCN5A, SCN7A, SCN8A, SCN9A, SCN10A, SCN11A, SCNM1 SCNN1A, SCNN1B, SCNN1D, SCNN1G*
	Metabolism	
	Substrate	*ABCB1, CYP1A2* (major)
	Inhibitor	*CYP1A2* (major)
	Transporter	*ABCB1*

ABCB1 ATP-binding cassette, sub-family B (MDR/TAP), member 1, *CYP1A2* Cytochrome P450, family 1, subfamily A, polypeptide 2, *GRIN3A* Glutamate receptor, ionotropic, N-methyl-D-aspartate 3A, *SCN1A* sodium channel, voltage-gated, type I, alpha subunit, *SCN1B* sodium channel, voltage-gated, type I, beta subunit, *SCN2A* sodium channel, voltage-gated, type II, alpha subunit, *SCN2B* sodium channel, voltage-gated, type II, beta, *SCN3A* sodium channel, voltage-gated, type III, alpha subunit, *SCN3B* sodium channel, voltage-gated, type III, beta subunit, *SCN4A* sodium channel, voltage-gated, type IV, alpha subunit, *SCN4B* sodium channel, voltage-gated, type IV, beta subunit, *SCN5A* sodium channel, voltage-gated, type V, alpha subunit, *SCN7A* sodium channel, voltage-gated, type VII, alpha, *SCN8A* sodium channel, voltage-gated, type VIII, alpha, *SCN9A* sodium channel, voltage-gated, type IX, alpha subunit, *SCN10A* sodium channel, voltage-gated, type X, alpha subunit, *SCN11A* sodium channel, voltage-gated, type XI, alpha, *SCNM1* sodium channel modifier 1, *SCNN1A* sodium channel, non-voltage-gated 1 alpha subunit, *SCNN1B* sodium channel, non-voltage-gated 1, beta subunit, *SCNN1D* sodium channel, non-voltage-gated 1, delta subunit, *SCNN1G* sodium channel, non-voltage-gated 1, gamma subunit

Table 10 Pharmacogenomics of multiple sclerosis [17, 30, 37]

Relapsing MS treatment		
Prednisone		
Category	Hormone	
Mechanism	Decreases inflammation by suppression of migration of polymorphonuclear leukocytes and reversal of increased capillary permeability, suppresses immune system by reducing activity and volume of lymphatic system	
Genes	Metabolism	
	Substrate	*ABCB1, CYP3A4* (major)
	Inhibitor	*CYP3A4* (weak)
	Inducer	*ABCB1, CYP3A4*
	Transporter	*ABCB1*
	Pleiotropic	*AGT, AR, FKBP5, IL1B, IL4, IL6, IL8RB, IL10, MTHFR, NOS3, NPPA, NR1I2, NR3C1, SERPINA6*
Tecfidera (Dimethyl fumarate) [38-43]		
Category	Nervous system Drug; Methyl ester; Fumarate	
Mechanism	Primarily acts by triggering the activation of the Nuclear factor (erythroid-derived 2)-like 2 (Nrf2) transcriptional pathway	
Genes	Pathogenic	*CXCL11, TNF*
	Mechanistic	*CXCL2, CXCL11, HMOX1, MAPK1, MAPK3, NFKB1, NRF2, TNF*
	Pleiotropic	*TNF*
Drugs for slowing progression of MS		
Fingolimod [44]		
Category	Immunosuppressive agent; Sphingosine 1-phosphate receptor modulator	
Mechanism	The active metabolite (fingolimod phosphate) binds with high affinity to sphingosine 1-phosphate receptors 1, 3, 4, and 5. Blocks the capacity of lymphocytes to egress from lymph nodes, reducing the number of lymphocytes in peripheral blood. May involve reduction of lymphocyte migration into the central nervous system. Agonist: S1PR1, S1PR3, S1PR4, S1PR5. Antagonist: S1PR1	

Table 10 (continued)

Genes	Pathogenic	*CYP4F2*
	Mechanistic	*S1PR1, S1PR3, S1PR4, S1PR5, SPHK2*
	Metabolism	
	Substrate	*CYP2D6 (minor), CYP2E1* (minor), *CYP3A4* (minor), *CYP4F2* (major), *CYP4F12* (minor)
Glatiramer Acetate [45]		
Category	Biologic response modifier	
Mechanism	Binds to HLA-DRB1 variants and competes with various myelin antigens for their presentation to T cells; inhibition of dendritic cells and monocytes	
Genes	Pathogenic	*HLA-DRB1, MBP*
	Mechanistic	*HLA-DRB1, CD86, CTSS, IL1R1, IL12RB2, TRB@*
	Pleiotropic	*APP, FAS*
Interferon Beta-1a		
Category	Biologic response modifier	
Mechanism	Alters expression and response to surface antigens and can enhance immune cell activities. Agonist: IFNAR1	
Genes	Mechanistic	*IFNAR1, IFNAR2, PTGER3*
Interferon Beta-1b		
Category	Biologic response modifier	
Mechanism	Enhancement of suppressor T cell activity, reduction of proinflammatory cytokines, down-regulation of antigen presentation, and reduced trafficking of lymphocytes into CNS. Agonist: IFNAR1, IFNAR2	
Genes	Mechanistic	*IFNAR1, IFNAR2, PTGER3*
Mitoxantrone [46]		
Category	Antineoplastic agent; Anthracenedione	
Mechanism	Binds to nucleic acids and inhibits DNA and RNA synthesis by template disordering and steric obstruction	

Table 10 (continued)

Genes	Mechanistic	*ACSL3, CES2, CBR3*
	Metabolism	
	Substrate	*ABCC1, ABCG2 CYP2D6*
	Inhibitor	*CYP3A4* (weak), *TOP2A.*
	Transporter	*ABCB1, ABCC1, ABCG2, ABCA1, ABCC6, TAP1*
	Pleiotropic	*ALDH2, GSTA1, GSTM1, GSTP1, MSH2, RGS2*
Natalizumab		
Category	Biologic response modifier; Selective adhesion-molecule inhibitor; Monoclonal antibody	
Mechanism	Blockade of T-lymphocyte migration into CNS	
Genes	Pathogenic	*FCGR2C*
	Mechanistic	*ITGA4*
Teriflunomide (Aubagio) [47-55]		
Category	Immunomodulatory agent; Pyrimidine Synthesis Inhibitor	
Mechanism	Inhibits the dihydroorotate dehydrogenase which inhibits the proliferation of activated T cells. Decreases the amount of activated CNS lymphocytes, which results in anti-inflammatory and antiproliferative effects	
Genes	Mechanistic	*DHODH, MAPKs, RAF1, TCR*
	Metabolism	
	Substrate	*ABCG2*
	Transporter	*ABCG2*
	Pleiotropic	*RAF1*
Anti-symptomatic drugs		
Cyclophosphamide		
Category	Antineoplastic agent; Alkylating agents. Immunosuppressive agent	

Table 10 (continued)

Mechanism	Alkylating agent which prevents cell division by cross-linking DNA strands and decreasing DNA synthesis, also possesses potent immunosuppressive activity, has phosphorylating properties which enhance its cytotoxicity	
Genes	Metabolism	
	Substrate	*ALDH1A1, ALDH2, ALDH3A1, ABCC4, CYP2A6* (minor), *CYP2B6* (minor), *CYP2C9* (minor), *CYP2C19* (minor), *CYP3A4* (major), *CYP1A2, CYP1B1, CYP2D6, GSTA1, GSTM1, GSTP1*
	Inhibitor	*CYP3A4* (weak)
	Inducer	*ABCC4, CYP2B6, CYP2C8, CYP2C9, CYP3A4*
	Transporter	*ABCB1, ABCG2, ABCC1, ABCC4, SLC5A5*
	Pleiotropic	*CBR3, CRHR1, CRHR2, EGFR, ERBB2, ERBB4, ERCC1, ERCC2, ESR1, ESR2, FOS, HTR3B, HTR3C, ICAM1, IL1B, IL1RN, IL4, IL6, IL10, IL12B, MAOA, MGMT, MMP3, MTHFR, MSH2, NQO1, PTGS2, SOD2, SLC5A5, TGFB1, TNF, TP53, VCAM1*
Dalfampridine (4-Aminopyridine)		
Category	Potassium channel blocker	
Mechanism	Increases conduction of action potentials in demyelinated axons through inhibition of potassium channels	
Genes	Mechanistic	*KCN*
	Metabolism	
	Substrate	*CYP2E1* (major)
Dantrolene		
Category	Direct-acting skeletal muscle relaxant	
Mechanism	Acts directly on skeletal muscle by interfering with release of calcium ion from sarcoplasmic reticulum, prevents or reduces increase in myoplasmic calcium ion concentration, which activates the acute catabolic processes associated with malignant hyperthermia	

Table 10 (continued)

Genes	Mechanistic	*RYR1*
	Metabolism	
	Substrate	*CYP3A4* (major)

ABCA1 ATP-binding cassette, sub-family A (ABC1), member 1, *ABCB1* ATP-binding cassette, sub-family B (MDR/TAP), member 1, *ABCC1* ATP-binding cassette, sub-family C (CFTR/MRP), member 1, *ABCC6 ALDH2* aldehyde dehydrogenase 2 family (mitochondrial), ATP-binding cassette, sub-family C (CFTR/MRP), member 6, *ABCG2* ATP-binding cassette, sub-family G (WHITE), member 2, *ACSL3* Acyl-CoA synthetase long-chain family member 3, *ALDH2* Aldehyde dehydrogenase 2 family (mitochondrial), *APP* Amyloid beta (A4) precursor protein, *CBR3* Carbonyl reductase 3, *CD86* CD86 molecule, *CE2* Carboxylesterase 2, *CTSS* Cathepsin S, *CXCL2* Chemokine (C-X-C motif) ligand 2, *CXCL11*: Chemokine (C-X-C motif) ligand 11, *CYP2D6* Cytochrome P450, family 2, subfamily D, polypeptide 6, *CYP2E1* Cytochrome P450, family 2, subfamily E, polypeptide 1, *CYP3A4* Cytochrome P450, family 3, subfamily A, polypeptide 4, *CYP4F2* Cytochrome P450, family 4, subfamily F, polypeptide 2, *CYP4F12* Cytochrome P450, family 4, subfamily F, polypeptide 12, *DHODH* Dihydroorotate dehydrogenase (quinone), *FAS* Fas (TNF receptor superfamily, member 6), *FCGR2C* Fc fragment of IgG, low affinity IIc, receptor for (CD32) (gene/pseudogene), *GSTA1* Glutathione S-transferase alpha 1, *GSTM1* Glutathione S-transferase mu 1, *GSTP1* Glutathione S-transferase pi 1, *HLA-DRB1* Major histocompatibility complex, class II, DR beta 1, *HMOX1* Heme oxygenase (decycling) 1, *IFNAR1* Interferon (alpha, beta and omega) receptor 1, *IFNAR2* Interferon (alpha, beta and omega) receptor 2, *IL12RB2* Interleukin 12 receptor, beta 2, *IL1R1* Interleukin 1 receptor, type I, *ITGA4* Integrin, alpha 4 (antigen CD49D, alpha 4 subunit of VLA-4 receptor), *MAPK1* Mitogen-activated protein kinase 1, *MAPK3* Mitogen-activated protein kinase 3, *MAPKs* Mitogen-activated protein kinases, *MBP* Myelin basic protein, *MSH2* MutS homolog 2, colon cancer, nonpolyposis type 1 (E. coli), *NFKB1* Nuclear factor of kappa light polypeptide gene enhancer in B-cells 1, *NRF2* Nuclear factor (erythroid-derived 2)-like 2, *PTGER3* Prostaglandin E receptor 3 (subtype EP3), *RAF1* V-raf-1 murine leukemia viral oncogene homolog 1, *RGS2* Regulator of G-protein signaling 2, 24 kDa, *S1PR1* Sphingosine-1-phosphate receptor 1, *S1PR3* Sphingosine-1-phosphate receptor 3, *S1PR4* Sphingosine-1-phosphate receptor 4, *S1PR5* Sphingosine-1-phosphate receptor 5, *SPHK2* Sphingosine kinase 2, *TAP1* Transporter 1, ATP-binding cassette, sub-family B (MDR/TAP), *TCR* T cell receptor, *TNF* Tumor necrosis factor, *TOP2A* Topoisomerase (DNA) II alpha 170 kDa, *TRB@* T cell receptor beta locus

Table 11 Pharmacogenomics of Huntington disease [17, 30]

Tetrabenazine [56]		
Category	Central monoamine-depleting agent	
Mechanism	Within basal ganglia, interferes with and depletes monoamine neurotransmitters (including dopamine, serotonin, and norepinephrine) in presynaptic vesicles (probably through actions on vesicle monoamine transporter), inhibits presynaptic dopamine release and also blocks CNS dopamine receptors	
Genes	Mechanistic	*VMAT2, DRD1, DRD2, DRD3, DRD4*
	Metabolism	
	Substrate	*CYP2D6* (major)
	Inhibitor	*VMAT2*
	Transporter	*ABCB1, SLC18A1, SLC18A2, SLC6A3, SLC6A4*

ABCB1 ATP-binding cassette, sub-family B (MDR/TAP), member 1, *CYP2D6* Cytochrome P450, family 2, subfamily D, polypeptide 6, *DRD1* Dopamine receptor D1, *DRD2* Dopamine receptor D2, *DRD3* Dopamine receptor D3, *DRD4* Dopamine receptor D4, *SLC6A3* Solute carrier family 6 (neurotransmitter transporter, dopamine), member 3, *SLC6A4* Solute carrier family 6 (neurotransmitter transporter, serotonin), member 4, *SLC18A1* Solute carrier family 18 (vesicular monoamine), member 1, *SLC18A2* Solute carrier family 18 (vesicular monoamine), member 2, *VMAT2* Vesicle monoamine transporter type 2

Among the world healthy population, EMs account for 55.71 %, whereas IMs are 34.7 %, PMs 2.28 %, and UMs 7.31 %. Remarkable interethnic differences exist in the frequency of the PM and UM phenotypes among different societies all over the world [24, 59-61]. On average, approximately 6.28 % of the world population belongs to the PM category. Europeans (7.86 %), Polynesians (7.27 %), and Africans (6.73 %) exhibit the highest rate of PMs, whereas Orientals (0.94 %) show the lowest rate [61]. The frequency of PMs among Middle Eastern populations, Asians, and Americans is in the range of 2–3 %. *CYP2D6* gene duplications are relatively infrequent among Northern Europeans, but in East Africa the frequency of alleles with duplication of *CYP2D6* is as high as 29 % [62]. In Europe, there is a North-South gradient in the frequency of PMs (6–12 % of PMs in Southern European countries, and 2–3 % PMs in Northern latitudes) [17].

In Alzheimer's disease (AD), EMs, IMs, PMs, and UMs are 56.38, 27.66, 7.45, and 8.51 %, respectively, and in vascular dementia, 52.81, 34.83, 6.74, and 5.62 %, respectively. There is an accumulation of AD-related genes of risk in PMs and UMs. EMs and IMs are the best responders, and PMs and UMs are the worst responders to pharmacological treatment [63]. Patients with depression show significant differences in the genotypic and phenotypic profiles as compared to controls and also with respect to patients with psychosis, Parkinson's disease, or brain tumours. Patients with stroke show differences as compared to patients with brain tumours, and both patients with brain tumours or with cranial nerve neuropathies differ in their *CYP2D6* phenotype with regard to controls. These geno-phenotypic profiles might

Table 12 Pharmacogenomics of most frequently-used psychotropic drugs in NDDs [17, 30]

Antidepressants		
Amitriptyline		
Category	Antidepressant; Tricyclic and norepinephrine-reuptake Inhibitor	
Mechanism	Increases synaptic concentration of serotonin and/or norepinephrine in the central nervous system by inhibiting their reuptake in the presynaptic neuronal membrane. Agonist: NTRK1, NTRK2	
Genes	Pathogenic	*ABCB1, GNB3, HTRs, NTRK2, SLC6A4, TNF*
	Mechanistic	*ADRA1A, HTRs, NTRK1, NTRK2*
	Metabolism	
	Substrate	*ABCB1, CYP1A2* (minor), *CYP2B6* (minor), *CYP2C9* (minor), *CYP2C19* (minor), *CYP2D6* (major), *CYP3A4/5* (major), *GSTP1*
	Inhibitor	*ABCB1, ABCC2, ABCG2, CYP1A2* (moderate), *CYP2C9* (moderate), *CYP2C19* (moderate), *CYP2D6* (moderate), *CYP2E1* (weak)
	Transporter	*ABCB1, ABCC2, ABCG2, KCNE2, KCNH2, KCNQ1, SCN5A, LC6A4*
	Pleiotropic	*FABP1, GNAS, GNB3, NTRK1, TNF*
Citalopram		
Category	Antidepressant; Selective serotonin-reuptake inhibitor	
Mechanism	Selectively inhibits serotonin reuptake in the presynaptic neurons due to the S-enantiomer and enhances the actions of serotonin on 5HT1A autoreceptors. Displays little to no affinity for dopamine, adrenergic, histamine, GABA, or muscarinic receptor subtypes	
Genes	Pathogenic	*ABCB1, BDNF, CREB1, CRHR1, CRHR2, FKBP5, GRIA3, GRIK2, GRIK4, GSK3B, HTR1A, HTR1B, HTR2A, MAOA, SLC6A4, TPH1, TPH2*
	Mechanistic	*ADRs, CHRMs, DRDs, FKBP5, GABRs, GRIK4, HRHs, HTR1A, HTR1B, HTR1D, HTR2A, SLC6A4, TPH1*
	Metabolism	
	Substrate	*ABCC1, COMT, CYP2C19* (major), *CYP2D6* (minor), *CYP3A4* (major), *CYP3A5*
	Inhibitor	*ABCB1, CYP1A2* (weak), *CYP2B6* (weak), *CYP2C19* (weak), *CYP2D6* (weak), *MAOA, MAOB*
	Transporter	*ABCB1, SLC6A4*
	Pleiotropic	*BDNF*

Table 12 (continued)

Antidepressants		
Fluoxetine		
Category	Antidepressant; Selective Serotonin-reuptake Inhibitor	
Mechanism	Potentiates serotonergic activity in CNS inhibiting neuronal reuptake of serotonin (5-HT). Increases synaptic concentrations of serotonin in the CNS but has little or no effect on other neurotransmitters	
Genes	Pathogenic	*ABCB1, BDNF, CREB1, FKBP5, GSK3B, HTR1A, HTR2A, MAOA, NR3C1, NTRK2, SLC6A4, TBX21, TPH1, TPH2*
	Mechanistic	*BDNF, CHRMs, CREB1, DRD3, GSK3B, HTRs, MAOA, SLC6A4, TPH2*
	Metabolism	
	Substrate	*CYP1A2* (major), *CYP2B6* (major), *CYP2C8* (major), *CYP2C9* (major), *CYP2C19* (major), *CYP2D6* (major), *CYP2E1* (minor), *CYP3A4/5* (major), POR
	Inhibitor	*ABCB1*, *CYP1A2* (moderate), *CYP2B6* (weak), *CYP2C8* (moderate), *CYP2C9* (weak), *CYP2C19* (moderate), *CYP2D6* (strong), *CYP3A4* (moderate), *MAOA*, *SLC6A4*
	Transporter	*ABCB1, KCNH2*
	Pleiotropic	*DRD3, FABP1, HTR2A, IFNA1, NTRK2, TPH1*
Maprotiline		
Category	Antidepressant; Tricyclic and norepinephrine-reuptake Inhibitor	
Mechanism	Inhibits the norepinephrine reuptake. Exhibits anticholinergic activity. Strong inhibitor of the histamine H1 receptor, which explains its sedative actions. Antagonist: ADRA2s, ADRA1s	
Genes	Pathogenic	*ABCB1*
	Mechanistic	*ADRA2s, ADRA1s, CHRM4, CHRM5, HRH1*
	Metabolism	
	Substrate	*CYP1A2* (minor), *CYP2C19*, *CYP2D6* (major), *CYP3A4*,
	Inhibitor	*MAOB*
	Transporter	*ABCB1*, SLC6A2
Paroxetine		

Table 12 (continued)

Antidepressants		
Category	Antidepressant; Selective serotonin-reuptake inhibitor	
Mechanism	Acts by inhibiting serotonin reuptake from brain synapse stimulating its activity in the brain	
Genes	Pathogenic	*ABCB1, CREB1, HTR1B, HTR2A, HTR3B, MAOA, SLC6A3, SLC6A4, TNF, TPH1, TPH2*
	Mechanistic	*CREB1, HTR2A, HTR3A, STAT3, TNF*
	Metabolism	
	Substrate	*ABCB1, COMT, CYP1A2 (minor), CYP2C19* (minor), *CYP2D6* (major), *CYP3A4* (major), *MAOA, MAOB*
	Inhibitor	*ABCB1*, *CYP1A2* (weak), *CYP2B6* (moderate), *CYP2C9* (weak), *CYP2C19* (weak), *CYP2D6* (strong), *CYP3A4* (weak), *SLC6A3*, *SLC6A4*
	Transporter	*ABCB1, SLC6A3, SLC6A4*
	Pleiotropic	*HTR1D, HTR3C, HTR6, HTT, TPH1, TPH2*
Sertraline		
Category	Antidepressant; Selective Serotonin-reuptake Inhibitor	
Mechanism	Selective inhibitor of serotonin reuptake; very weak effects on norepinephrine and dopamine neuronal uptake	
Genes	Pathogenic	*ABCB1, CREB1, GNB3, HTR1B, MAOA, SIGMAR1, SLC6A4, TNF, TPH1, TPH2*
	Mechanistic	*HTR1B, HTR1D, SIGMAR1, SLC6A2, SLC6A3, SLC6A4, TNF*
	Metabolism	
	Substrate	*CYP2A6, CYP2B6* (minor), *CYP2C9* (minor), *CYP2C19* (major), *CYP2D6* (minor), *CYP3A4* (minor), *MAOA*, *MAOB*, *UGT1A1*, *UGT2B7*
	Inhibitor	*ABCB1*, *ACHE*, *CYP1A2* (weak), *CYP2B6* (moderate), *CYP2C8* (weak), *CYP2C9* (weak), *CYP2C19* (moderate), *CYP2D6* (moderate), *CYP3A4* (moderate), *SLC6A4*
	Transporter	*ABCB1, SCL6A2, SLC6A3, SLC6A4*
	Pleiotropic	*FABP1, FOS, GNB3, TPH1, TPH2*
Trazodone		
Category	Antidepressant; Serotonin Modulator	

Table 12 (continued)

Antidepressants		
Mechanism	Inhibits the reuptake of serotonin, causes adrenoreceptor subsensitivity, and induces significant changes in 5-HT presynaptic receptor adrenoreceptors. Blocks histamine (H1), α1 and α2 adrenergic receptors	
Genes	Pathogenic	*ABCB1, GNB3, HTR1A, HTR2A, SLC6A4*
	Mechanistic	*ADRA1A, ADRA2s, HRH1, HTR2A, HTR2C*
	Metabolism	
	Substrate	*CYP1A2* (minor), *CYP2D6* (minor), *CYP3A4* (major), *GSTs*, *SOD2*
	Inhibitor	*CYP2D6* (moderate), *CYP3A4* (weak), *SLC6A4*
	Inducer	*ABCB1*
	Transporter	*ABCB1, SLC6A4*
	Pleiotropic	*GNAS, GNB3, HTR2A*
Venlafaxine		
Category	Antidepressant; Selective serotonin and norepinephrine-reuptake inhibitor	
Mechanism	Venlafaxine and its active metabolite, O-desmethylvenlafaxine (ODV), are potent inhibitors of neuronal serotonin and norepinephrine reuptake and weak inhibitors of dopamine reuptake	
Genes	Pathogenic	*ABCB1, BDNF, CREB1, FKBP5, HTR1A, HTR2A* NR3C1, SLC6A3, SLC6A4, TPH2*
	Mechanistic	*BDNF, FKBP5*
	Metabolism	
	Substrate	*ABCB1, CYP2C9* (minor), *CYP2C19* (minor), *CYP2D6* (major), *CYP3A4* (major)
	Inhibitor	*ABCB1*, *CYP1A2* (weak), *CYP2B6* (weak), *CYP2D6* (weak), *CYP3A4* (weak), *SLC6A2*, *SLC6A3*, *SLC6A4*
	Transporter	*ABCB1, ABCG2, SLC6A2, SLC6A3*
	Pleiotropic	*DRD2, HTR2A, TPH2*

Table 12 (continued)

Benzodiazepines		
Alprazolam		
Category	Anxiolytic, Sedative, and hypnotic; Benzodiazepine	
Mechanism	Binds to the GABA benzodiazepine receptor complex, particularly in the limbic system and reticular formation. The inhibitory effect of GABA on neuronal excitability increases the neuronal membrane permeability to chloride ions resulting in hyperpolarization and stabilization	
Genes	Pathogenic	*GABAARs*
	Mechanistic	*CLCNs, GABRAs, GABRBs, GABRD, GABRE, GABRGs, GABRP, GABRQ, GABRRs*
	Metabolism	
	Substrate	*CYP1A1* (minor), *CYP1A2* (minor), *CYP2C19* (minor), *CYP2D6* (minor), *CYP3A4/5* (major)
	Transporter	*CLCNs*
Diazepam		
Category	Anticonvulsant; Anxiolytic, Sedative, and Hypnotic; Benzodiazepine	
Mechanism	Binds to GABAA receptors enhancing the effects of GABA by increasing GABA affinity for its receptor. This action increases neuronal membrane permeability to chloride ions thus resulting in hyperpolarization and stabilization	
Genes	Pathogenic	*BDNF, CNR1, GABAARs*
	Mechanistic	*ACHE, BCHE, BDNF, CHRMS*, CLCNs, GABRAs, GABRBs, GABRD, GABRE, GABRGs, GABRP, GABRQ, GABRRs, TSPO*
	Metabolism	
	Substrate	*CYP1A2* (minor), *CYP2B6* (minor), *CYP2C9* (minor), *CYP2C19* (minor), *CYP3A4/5* (major), *UGTs*
	Inhibitor	*CYP2C19* (weak), *CYP3A4* (weak), *UGT2B7*
	Transporter	*CLCNs*
	Pleiotropic	*ACHE, FOS, IL6, SPG7*
Lorazepam		
Category	Anticonvulsant; Benzodiazepine; Anxiolytic, Sedative, and Hypnotic; Benzodiazepine	
Mechanism	Binds to GABAA receptors enhancing the effects of GABA by increasing GABA affinity for its receptor. This action increases neuronal membrane permeability to chloride ions thus resulting in hyperpolarization and stabilization. Antagonist: TSPO	

Table 12 (continued)

Benzodiazepines		
Genes	Pathogenic	*GABAARs*
	Mechanistic	*CLCNs, GABRAs, GABRBs, GABRD, GABRE, GABRGs, GABRP, GABRQ, GABRRs, TSPO*
	Metabolism	
	Substrate	*CYP3A4* (minor), *UGT1A1, UGT1A3, UGT1A4, UGT1A6, UGT1A7, UGT1A9, UGT1A10, UGT2B7, UGT2B15*
	Transporter	*CLCNs*
Oxazepam		
Category	Anxiolytic, sedative, and hypnotic; benzodiazepine	
Mechanism	Binds to GABAA receptors enhancing the effects of GABA by increasing GABA affinity for its receptor. This action increases neuronal membrane permeability to chloride ions thus resulting in hyperpolarization and stabilization	
Genes	Pathogenic	*GABAARs*
	Mechanistic	*CLCNs, GABRAs, GABRBs, GABRD, GABRE, GABRGs, GABRP, GABRQ, GABRRs*
	Metabolism	
	Substrate	*CYP2D6, UGT2B7, UGT2B15*
	Inhibitor	*UGT2B7*
	Inducer	*CYP1A2*
	Transporter	*CLCNs*
Temazepam		
Category	Anxiolytic, Sedative, and Hypnotic; Benzodiazepine	
Mechanism	Binds to GABAA receptors enhancing the effects of GABA by increasing GABA affinity for its receptor. This action increases neuronal membrane permeability to chloride ions thus resulting in hyperpolarization and stabilization. Antagonist: *TSPO*	
Genes	Mechanistic	*CLCNs, GABRAs, GABRBs, GABRD, GABRE, GABRGs, GABRP, GABRQ, GABRRs, TSPO*
	Metabolism	
	Substrate	*CYP2B6* (major), *CYP2C9* (major), *CYP2C19* (major), *CYP3A4* (major), *UGT1A1, UGT1A3, UGT1A4, UGT1A6, UGT1A9, UGT1A10, UGT2B7, UGT2B15*
	Transporter	*CLCNs*

Table 12 (continued)

Neuroleptics		
Aripiprazole		
Category	Antimanic Agent; Atypical Antipsychotic	
Mechanism	Partial agonist of the D2 and 5-HT1A receptors and antagonist profile at the 5-HT2A receptor. Moderate affinity for histamine and alpha adrenergic receptors	
Genes	Pathogenic	*DRD2, DRD3, HTR2A*
	Mechanistic	*ADRA1A, DRD2, DRD3, HRHs, HTR1A, HTR2A, HTR2C*
	Metabolism	
	Substrate	*CYP2D6* (major), *CYP3A4* (major), *CYP3A5*
	Transporter	*ABCB1*
Haloperidol		
Category	Conventional Antipsychotic; First-Generation antipsychotic; Typical antipsychotic; Butyrophenone	
Mechanism	Blocks postsynaptic mesolimbic dopaminergic D1 and D2 receptors. Depresses release of hypothalamic and hypophyseal hormones. May depress reticular activating system. Antagonist: DRD2	
Genes	Pathogenic	*BDNF, DRD1, DRD2, DRD4, DTNBP1, GRIN2B, HTR2A*
	Mechanistic	*BDNF, DRD1, DRD2, DTNBP1*
	Metabolism	
	Substrate	*CBR, CYP1A1* (minor), *CYP1A2* (minor), *CYP2A6, CYP2C8* (minor), *CYP2C9* (minor), *CYP2C19* (minor), *CYP2D6* (major), *CYP3A4/5* (major), *UGTs*
	Inhibitor	*CYP2D6* (moderate), *CYP3A4* (moderate)
	Transporter	*ABCB1, ABCC1, KCNE1, KCNE2, KCNH2, KCNJ11, KCNQ1*
	Pleiotropic	*ADRA2A, CHRM2, FOS, GSK3B, GSTP1, HRH1, HTR2A, HTT, IL1RN*
Olanzapine		
Category	Atypical Antipsychotic. Antimanic agent	
Mechanism	Potent antagonism of serotonin 5-HT2A and 5-HT2C, dopamine D1–4, histamine H1 and α1-adrenergic receptors; moderate antagonism of 5-HT3 and muscarinic M1-5 receptors, and weak binding to GABA-A, BZD, and β-adrenergic receptors	

Table 12 (continued)

Neuroleptics		
Genes	Pathogenic	*COMT, DRD1, DRD2, DRD3, DRD4, GRM3, HTR2A, LPL*
	Mechanistic	*ADRA1A, ADRB3, CHRM1-5, COMT, DRD1, DRD2, DRD4, GABAARs, HRH1, HTR2A, HTR2C, HTR3A, LEP, RGS2, RGS7, STAT3*
	Metabolism	
	Substrate	*ABCB1, COMT, CYP1A2* (major), *CYP2D6* (major), *UGT1A4*
	Inhibitor	*ABCB1, CYP1A2* (weak), *CYP2C9* (weak), *CYP2C19* (weak), *CYP2D6* (weak), *CYP3A4* (weak), *FMO1, UGT1A4*
	Transporter	*KCNH2, SLC6A2*
	Pleiotropic	*APOA5, APOC3, GNB3, LEP, LEPR, LPL*
Quetiapine		
Category	Atypical Antipsychotic	
Mechanism	Antagonistic effect on dopaminergic (D1, D2), histaminergic (H1), sertoninergic (5-HT1A, 5HT2) and adrenergic (α1- and α2) receptors	
Genes	Pathogenic	*DRD1, DRD2, DRD4, HTR1A, HTR2A, RGS4*
	Mechanistic	*ADRA1s, ADRA2s, DRD1, DRD2, DRD4**, *HRH1, HTR1A, HTR2A, HTR2B*
	Metabolism	
	Substrate	*ABCB1, CYP2D6* (minor), *CYP3A4/5* (major), *CYP3A7, CYP2C19*
	Inhibitor	*ABCB1*
	Transporter	*ABCB1, KCNE1, KCNE2, KCNH2, KCNQ1, SCN5A*
Risperidone		
Category	Atypical Antipsychotic; Antimanic agent	
Mechanism	It has serotonin-dopamine antagonist activity. Antagonizes α1-, α2-adrenergic, and histaminergic receptors	

Table 12 (continued)

Neuroleptics		
Genes	Pathogenic	*BDNF, COMT, DRD1, DRD2, DRD3, DRD4, GRM3, HTR2A, PON1, RGS4*
	Mechanistic	*ADRA1A, ADRA1B, ADRA2s, DRD1, DRD2, DRD3, DRD4, FOS*, HTR2A, HTR2C, HTR3A, HTR3C, HTR6, STAT3*
	Metabolism	
	Substrate	*ABCB1, CYP2D6* (major), *CYP3A4/5* (minor), *COMT*
	Inhibitor	*ABCB1, CYP2D6* (weak), *CYP3A4* (weak)
	Transporter	*KCNH2, SLC6A4*
	Pleiotropic	*APOA5, BDNF, RGS2*

ABCB1 ATP-binding cassette, sub-family B (MDR/TAP), member 1, *ABCC1* ATP-binding cassette, sub-family C (CFTR/MRP), member 1, *ABCG2* ATP-binding cassette, sub-family G (WHITE), member 2, *ACHE* Acetylcholinesterase, *ADRA1A* Adrenoceptor alpha 1A, *ADRA1B* Adrenergic receptor, alpha 1b, *ADRA1s* Alpha 1 adrenoceptors, *ADRA2A* Adrenoceptor alpha 2A, *ADRA2s* Alpha 2 adrenoceptors, *ADRB3* Adrenoceptor beta 3, *ADRs* Adrenoceptors, *APOA5* Apolipoprotein A-V, *APOC3* Apolipoprotein C-III, *BCHE* Butyrylcholinesterase, *BDNF* Brain-derived neurotrophic factor, *CBR* NADH–cytochrome B5 reductase 1, *CHRM2* Cholinergic receptor, muscarinic 2, *CHRM4* Cholinergic receptor, muscarinic 4, *CHRM5* Cholinergic receptor, muscarinic 5, *CHRMs* Muscarinic cholinergic receptors, *CHRNs* Nicotinic cholinergic receptors, *CLCNs* Chloride channel, voltage-sensitive, *CNR1* Cannabinoid receptor 1 (brain), *COMT* Catechol-O-methyltransferase, *CREB1* cAMP responsive element binding protein 1, *CRHR1* Corticotropin releasing hormone receptor 1, *CRHR2* Corticotropin releasing hormone receptor 2, *CYP1A1* Cytochrome P450, family 1, subfamily A, polypeptide 1, *CYP1A2* Cytochrome P450, family 1, subfamily A, polypeptide 2, *CYP2A6* Cytochrome P450, family 2, subfamily A, polypeptide 6, *CYP2B6* Cytochrome P450, family 2, subfamily B, polypeptide 6, *CYP2C8* Cytochrome P450, family 2, subfamily C, polypeptide 8, *CYP2C9* Cytochrome P450, family 2, subfamily C, polypeptide 9, *CYP2C19* Cytochrome P450, family 2, subfamily C, polypeptide 19, *CYP2D6* Cytochrome P450, family 2, subfamily D, polypeptide 6, *CYP2E1* Cytochrome P450, family 2, subfamily E, polypeptide 1, *CYP3A4* Cytochrome P450, family 3, subfamily A, polypeptide 4, *CYP3A5* Cytochrome P450, family 3, subfamily A, polypeptide 5, *CYP3A7* Cytochrome P450, family 3, subfamily A, polypeptide 7, *DRD1* Dopamine receptor D1, *DRD2* Dopamine receptor D2, *DRD3* Dopamine receptor D3, *DRD4* Dopamine receptor D4, *DRDs* Dopamine receptors, *DTNBP1* Dystrobrevin binding protein 1, *FABP1* Fatty acid binding protein 1, *FKBP5* FK506 binding protein 5, *FMO1* Flavin containing monooxygenase 1, *FOS* FBJ murine osteosarcoma viral oncogene homolog, *GABAARs* Glutamate receptor, ionotropic, *AMPA 2, GABRAs* Gamma-aminobutyric acid (GABA) A receptors, *GABRBs* Gamma-aminobutyric acid (GABA) B receptors, *GABRs* Gamma-aminobutyric acid (GABA) receptors, *GABRD* Gamma-aminobutyric acid (GABA) A receptor, subunit delta, *GABRE* Gamma-aminobutyric acid (GABA) A receptor, epsilon, *GABRGs* Gamma-aminobutyric acid (GABA) A receptors, gamma, *GABRP* Gamma-aminobutyric acid (GABA) A receptor, pi, *GABRQ* Gamma-aminobutyric acid (GABA) A receptor, theta, *GABRRs* Gamma-aminobutyric acid (GABA) A receptors, rho subunits, *GNAS* GNAS complex locus, *GNB3* Guanine nucleotide binding protein (G protein), beta polypeptide 3, *GRIA3* Glutamate receptor, ionotropic, AMPA 3, *GRIK2* Glutamate receptor, ionotropic, kainate 2, *GRIK4* Glutamate receptor, ionotropic, kainate 4, *GRM3* Glutamate receptor, metabotropic 3, *GSK3B* Glycogen synthase kinase 3 beta, *GSTP1* Glutathione S-transferase

Table 12 (continued)

Neuroleptics
pi 1, *GSTs* Glutathione S-transferases, *HRH1* Histamine receptor H1, *HRHs* Histamine receptors, *HTR1A* 5-hydroxytryptamine (serotonin) receptor 1A, G protein-coupled, *HTR1B* 5-hydroxytryptamine (serotonin) receptor 1B, G protein-coupled, *HTR1D* 5-hydroxytryptamine (serotonin) receptor 1D, G protein-coupled, *HTR2A* 5-hydroxytryptamine (serotonin) receptor 2A, G protein-coupled, *HTR2B* 5-hydroxytryptamine (serotonin) receptor 2B, *HTR2C* 5-hydroxytryptamine (serotonin) receptor, G protein-coupled, *HTR3A* 5-hydroxytryptamine (serotonin) receptor 3A, ionotropic, *HTR3B* 5-hydroxytryptamine (serotonin) receptor 3B, ionotropic, *HTR3C* 5-hydroxytryptamine (serotonin) receptor, ionotropic, *HTR6* 5-hydroxytryptamine (serotonin) receptor protein-coupled, *HTRs* 5-hydroxytryptamine (serotonin) receptors, *HTT* Huntingtin, *IFNA1* Interferon, alpha 1, *IL1RN* Interleukin 1 receptor antagonist, *IL6* Interleukin 6 (interferon, beta 2), *KCNE1* Potassium voltage-gated channel, Isk-related family, member 1, *KCNE2* Potassium voltage-gated channel, Isk-related family, member 2, *KCNH2* Potassium voltage-gated channel, subfamily H (eag-related), member 2, *KCNJ11* Potassium inwardly-rectifying channel, subfamily J, member 11, *KCNQ1* Potassium voltage-gated channel, KQT-like subfamily, member 1, *LEP* Leptin, *LEPR* Leptin receptor,
LPL Lipoprotein lipase, *MAOA* Monoamine oxidase A, *MAOB* Monoamine oxidase B, *NR3C1* Nuclear receptor subfamily 3, group C, member 1 (glucocorticoid receptor), *NTRK2* Neurotrophic tyrosine kinase, receptor, type 2, *PON1* Paraoxonase 1, *POR* P450 (cytochrome) oxidoreductase, *RGS2* Regulator of G-protein signaling 2, 24 kDa, *RGS4* Regulator of G-protein signaling 4, *RGS7* Regulator of G-protein signaling 7, *RIN2B* Glutamate receptor, ionotropic, N-methyl D-aspartate 2B, *SIGMAR1* Sigma non-opioid intracellular receptor 1, *SCN5A* Sodium channel, voltage-gated, type V, alpha subunit, *SLC6A2* Solute carrier family 6 (neurotransmitter transporter, noradrenalin), member 2, *SLC6A3* Solute carrier family 6 (neurotransmitter transporter, dopamine), member 3, *SLC6A4* Solute carrier family 6 (neurotransmitter transporter, serotonin), member 4, *SOD2* Superoxide dismutase 2, mitochondrial, *SPG7* Spastic paraplegia 7 (pure and complicated autosomal recessive), *STAT3* Signal transducer and activator of transcription 3 (acute-phase response factor), *TBX21* T-box 21, *TNF* Tumor necrosis factor, *TPH1* Tryptophan hydroxylase 1, *TPH2* Tryptophan hydroxylase 2, *TSPO* Translocator protein (18 kDa), *UGT1A1* UDP glucuronosyltransferase 1 family, polypeptide A1, *UGT1A10* UDP glucuronosyltransferase 1 family, polypeptide A10, *UGT1A3* UDP glucuronosyltransferase 1 family, polypeptide A3, *UGT1A4* UDP glucuronosyltransferase 1 family, polypeptide A4, *UGT1A6* UDP glucuronosyltransferase 1 family, polypeptide A6, *UGT1A7* UDP glucuronosyltransferase 1 family, polypeptide A7, *UGT1A9* UDP glucuronosyltransferase 1 family, polypeptide A9, *UGT2B7* UDP glucuronosyltransferase 2 family, polypeptide B7, *UGT2B15* UDP glucuronosyltransferase 2 family, polypeptide B15, *UGTs* UDP glucuronosyltransferases

Table 13 Main characteristics of cytochrome P-450 genes [16, 17]

CYP2D6

OMIM	Size (kb)	Exons number	RNA tran-scripts (bp)	Locus	Representative allelic variants	Genotype	Phenotype	Frecuency (Caucasian population) (%)
124030	4.38	9	1684 1433 1190 2257	22q13.1	g.24825C > T g.25747C > T g.26385G > A g.12739delT g.26572G > A 2549delA 2613-2615delAGA g.27576T > C g.27714G > A g.27909G > A	*1/*1	EM	47,10
						*1/*3	IM	1,95
						*1/*4	IM	17,42
						*1/*5	IM	3,87
						*1/*6	IM	2,58
						*1/*7	IM	0,65
						*10/*10	IM	1,30
						*1xN/*1	UM	4,52
						*1xN/*4	UM	1,95
						*1/*10	EM	4,52
						*4/*4	PM	8,37
						*4/*10	IM	3,23
						*5/*5	PM	0,65
						*6/*10	IM	0,65
						*7/*10	IM	0,65

Table 13 (continued)

CYP2C9

Omin	Size (kb)	Exons number	RNA transcripts (bp)	Locus	Representative allelic variants	Genotype	Phenotype	Frecuency (Caucasian population) (%)
601130	50.71	9	1860	10q24	g.15489579A > C g.15450573C > T g.15456202T > G g.47506179T > C	*1/*1	EM	60.56
						*1/*2	IM	18.78
						*1/*3	IM	13.62
						*2/*2	PM	3.76
						*2/*3	PM	3.28
						*3/*3	PM	0

CYP2C19

Omin	Size (kb)	Exons number	RNA transcripts (bp)	Locus	Representative allelic variants	Genotype	Phenotype	Frecuency (Caucasian population) (%)
124020	90.21	9	1901 2395 1417	10q24.1-q24.3	g.47346080G > C g.15288936G > A g.15270989A > G	*1/*1	EM	68.54
						*1/*2	IM	30.05
						*2/*2	PM	1.41

CYP3A4/5

Omin	Size (kb)	Exons number	RNA transcripts (bp)	Locus	Representative allelic variants	Genotype	Phenotype	Frecuency (Caucasian population) (%)
124010	27.2	13	2153 651 564 2318 2519	7q21.1	g.24589084T > A g.24601703C > T 392A > G	*1/*1	UM	1.37
						*1/*3	IM	15.88
						*3/*3	EM	82.75

be important in the pathogenesis of some CNS disorders and in the therapeutic response to conventional psychotropic drugs as well [16].

2.3.2 CYP2C9

This gene is mainly expressed in hepatocytes, where a protein of 55.63 kDa (490 aa) can be identified. Over 600 drugs are CYP2C9-related, 311 acting as substrates (177 are major substrates, 134 are minor substrates), 375 as inhibitors (92 weak, 181 moderate, and 102 strong inhibitors), and 41 as inducers of the CYP2C9 enzyme [17]. There are 481 *CYP2C9* SNPs. The phenotypic distribution is presented in Table 13. No *CYP2C9-*3/*3* cases have been found in the control population; however, in patients with depression, psychosis, and mental retardation the frequency of this genotype is 0.91, 1.03, and 1.37%, respectively [16]. Significant variation has been found in *CYP2C9* genotypes among diverse brain diseases. The plethora of metabolising profiles in CNS disorders suggest a potential pathogenic role of *CYP2C9* in brain pathology and a very strong role of the CYP2C9 enzyme on drugs with deleterious effects on cerebrovascular function (e.g. nonsteroidal anti-inflammatory drugs (NSAIDs) and thromboembolic phenomena and/or bleeding (e.g. warfarin, coumarinics).

2.3.3 CYP2C19

This gene is expressed in liver cells where a protein of 55.93 kDa (490 aa) is identified. Nearly 500 drugs are CYP2C19-related, 281 acting as substrates (151 are major substrates, 130 are minor substrates), 263 as inhibitors (72 weak, 127 moderate, and 64 strong inhibitors), and 23 as inducers of the CYP2C19 enzyme [17]. About 541 SNPs have been detected in the *CYP2C19* gene [16]. Minor variation has been reported in different brain disorders.

2.3.4 CYP3A4/5

These genes are expressed in intestine, liver, prostate and other tissues where 4 protein variants of 57.34 kDa (503 aa), 17.29 kDa (153 aa), 40.39 kDa (353 aa), and 47.99 kDa (420 aa) are identified. The human *CYP3A* locus contains the three *CYP3A* genes (*CYP3A4, CYP3A5* and *CYP3A7*), three pseudogenes as well as a novel *CYP3A* gene termed *CYP3A43*. The gene encodes a putative protein with between 71.5 and 75.8% identity to the other CYP3A proteins. The predominant hepatic form is CYP3A4, but CYP3A5 contributes significantly to the total liver CYP3A activity. This enzyme metabolises over 1900 drugs, 1033 acting as substrates (897 are major substrates, 136 are minor substrates), 696 as inhibitors (118 weak, 437 moderate, and 141 strong inhibitors), and 241 as inducers of the CYP3A4 enzyme [17]. About 347 SNPs have been identified in the *CYP3A4* gene (*CYP3A4*1A*: Wild-type), 25 of which are of clinical relevance.

2.3.5 *CYP* Clustering

The construction of a genetic map integrating the most prevalent *CYP2D6* + *CYP2C19* + *CYP2C9* polymorphic variants in a trigenic cluster yields 82 different haplotype-like profiles. The most frequent trigenic genotypes are **1*1-*1*1-*1*1* (25.70 %), **1*1-*1*2-*1*2* (10.66 %), **1*1-*1*2-*1*1* (10.45 %), **1*4-*1*1-*1*1* (8.09 %), **1*4-*1*2-*1*1* (4.91 %), **1*4-*1*1-*1*2* (4.65 %), and **1*1-*1*3-*1*3* (4.33 %). These 82 trigenic genotypes represent 36 different pharmacogenetic phenotypes. According to these trigenic clusters, only 26.51 % of the population show a pure 3EM phenotype, 15.29 % are 2EM1IM, 2.04 % are pure 3IM, 0 % are pure 3PM, and 0 % are 1UM2PM (the worst possible phenotype). This implies that only one-quarter of the population processes normally the drugs which are metabolised via CYP2D6, CYP2C9 and CYP2C19 (approximately 60 % of the drugs of current use) [13, 16].

2.4 *Genes Encoding Drug Transporters*

ATP-binding cassette genes (ABC), especially *ABCB1, ABCC1, ABCG2* (White1), and other genes of this family encode proteins which are essential for drug metabolism and transport. The multidrug efflux transporters P-gp, multidrug-resistance associated protein 4 (MRP4) and breast cancer resistance protein (BCRP), located on endothelial cells lining brain vasculature, play important roles in limiting movement of substances into and enhancing their efflux from the brain. Transporters also cooperate with Phase I/Phase II metabolism enzymes by eliminating drug metabolites. Their major features are their capacity to recognise drugs belonging to unrelated pharmacological classes, and their redundancy, by which a single molecule can act as a substrate for different transporters. This ensures an efficient neuroprotection against xenobiotic invasions. The pharmacological induction of ABC gene expression is a mechanism of drug interaction, which may affect substrates of the up-regulated transporter, and overexpression of multidrug resistance (MDR) transporters confers resistance to anticancer agents and CNS drugs [64-66]. Also of importance for CNS pharmacogenomics are transporters encoded by genes of the solute carrier superfamily (SLC) and solute carrier organic (SLCO) transporter family, responsible for the transport of multiple endogenous and exogenous compounds, including folate (*SLC19A1*), urea (*SLC14A1, SLC14A2*), monoamines (*SLC29A4, SLC22A3*), aminoacids (*SLC1A5, SLC3A1, SLC7A3, SLC7A9, SLC38A1, SLC38A4, SLC38A5, SLC38A7, SLC43A2, SLC45A1*), nucleotides (*SLC29A2, SLC29A3*), fatty acids (*SLC27A1–6*), neurotransmitters (*SLC6A2* (noradrenaline transporter), *SLC6A3* (dopamine transporter), *SLC6A4* (serotonin transporter, SERT), *SLC6A5, SLC6A6, SLC6A9, SLC6A11, SLC6A12, SLC6A14, SLC6A15, SLC6A16, SLC6A17, SLC6A18, SLC6A19*), glutamate (*SLC1A6, SLC1A7*), and others [11, 18]. Some organic anion transporters (OAT), which belong to the solute carrier (SLC) 22A family, are also expressed at the blood-brain barrier (BBB), and regulate the excretion of endogenous and exogenous organic anions and cations [67]. The transport of amino acids and di- and tripeptides is mediated by a number of different transporter families, and the bulk of oligopeptide transport is attributable to the activity of members of the *SLC15A* superfamily (Peptide Transporters 1 and 2 [*SLC15A1* (PepT1) and *SLC15A2* (PepT2), and Peptide/Histidine Transporters 1 and 2 [*SLC15A4* (PHT1) and *SLC15A3* (PHT2)]. ABC and SLC transporters expressed at the BBB may cooperate to regulate the passage of different molecules into the brain [68]. Polymorphic variants in ABC and SLC genes may also be associated with pathogenic events in CNS disorders and drug-related safety and efficacy complications [17].

3 Alzheimer's Disease

Alzheimer's disease (AD) is the most prevalent form of dementia with a prevalence of about 1% at the age of 60–65 years and over 25% in patients older than 85 years of age. Identified by Alois Alzheimer in 1906, AD is characterised by memory disorders, behavioural changes and progressive functional decline. Its neuropathologic phenotype includes the presence of extracellular deposits of aberrant forms of amyloid-beta (Aβ) protein in senile plaques, intracellular deposits of hyperphosphorylated protein tau in neurofibrillary tangles (NFT), synaptic and neuronal loss in selected regions of the neocortex, neuroinflammatory reactions, neurotrophic dysfunction, and overproduction of oxidative stress reactions [20, 24]. In the Alzgene database [69] there are over 600 genes potentially associated with AD, of which the top ten are (in decreasing order of importance): *APOE, BIN1, CLU, ABCA7* (ATP-binding cassette, sub-family A (ABC1), member 7), *CR1, PICALM, MS4A6A, CD33, MS4A4E*, and *CD2AP*. Potentially defective genes associated with AD represent about 1.73% of the human genome, which is integrated by 20774 coding genes [70]. The highest number (> 5%) of AD genes concentrate on chromosomes 10 (15.69%), 1 (9.67%), 6 (7.61%), 19 (6.81%), 12 (6.50%), 11 (6.18%), and 17 (5.07%), with the highest proportion (related to the total number of genes mapped on a single chromosome) located on chromosome 10 and the lowest on chromosome X [11] (Table 2).

The genetic and epigenetic defects identified in AD can be classified into 4 major categories: Mendelian mutations, susceptibility SNPs, mtDNA mutations, and epigenetic changes. Mendelian mutations affect genes directly linked to AD, including 32 mutations in the amyloid beta (Aβ) (ABP) precursor protein (*APP*) gene (AD1); 165 mutations in the presenilin 1 (*PSEN1*) gene (AD3); and 12 mutations in the presenilin 2 (*PSEN2*) gene (AD4) [15, 16, 20, 71]. *PSEN1* and *PSEN2* are important determinants of γ-secretase activity responsible for proteolytic cleavage of APP and NOTCH receptor proteins. Mendelian mutations are very rare in AD (1:1000). Mutations in exons 16 and 17 of the *APP* gene appear with a frequency of 0.30 and 0.78%, respectively, in AD patients. Likewise, *PSEN1, PSEN2*, and microtubule-associated protein Tau (*MAPT*) mutations are present in less than 2% of the cases. Mutations in these genes confer specific phenotypic profiles to patients with dementia: amyloidogeneic pathology associated with *APP, PSEN1* and *PSEN2* mutations; and tauopathy associated with *MAPT* mutations, representing the two major pathogenic hypotheses for AD [20, 33, 72, 73].

Multiple polymorphic risk variants (Table 2) can increase neuronal vulnerability to premature death. Among these susceptibility genes, the apolipoprotein E (*APOE*) gene (19q13.2) (AD2) is the most prevalent as a risk factor for AD, especially in those subjects harboring the *APOE-4* allele, whereas carriers of the *APOE-2* allele might be protected against dementia [15, 16, 20, 71]. *APOE*-related pathogenic mechanisms are also associated with brain ageing, several CNS disorders and the neuropathological hallmarks of AD.

APOE is a pathogenic gene in dementia and the prototypical paradigm of a pleiotropic gene with multifaceted activities in physiological and pathological conditions, including cardiovascular disease, dyslipidemia, atherosclerosis, stroke, and AD [15, 16]. ApoE is consistently associated with the amyloid plaque marker for AD. APOE-4 may influence AD pathology interacting with APP metabolism and Aβ accumulation, enhancing hyperphosphorylation of tau protein and NFT formation, reducing choline acetyltransferase activity, increasing oxidative processes, modifying inflammation-related neuroimmunotrophic activity and glial activation, altering lipid metabolism, lipid transport and membrane biosynthesis in sprouting and synaptic remodelling, and inducing neuronal apoptosis [20].

The distribution of *APOE* genotypes in the Iberian peninsula is as follows: *APOE-2/2* 0.32 %, *APOE-2/3* 7.3 %, *APOE-2/4* 1.27 %, *APOE-3/3* 71.11 %, *APOE-3/4* 18.41 %, and *APOE-4/4* 1.59 % [15, 16]. These frequencies are very similar in Europe and in other Western societies. There is a clear accumulation of *APOE-4* carriers among patients with AD (*APOE-3/4* 30.30 %; *APOE-4/4* 6.06 %) and vascular dementia (*APOE-3/4* 35.85 %, *APOE-4/4* 6.57 %) as compared to controls. The distribution and frequencies of *APOE* genotypes in AD also differ from those of patients with anxiety, depression, psychosis, migraine, vascular encephalopathy, and post-traumatic brain injury syndrome [15, 16] (Fig. 3). Different *APOE* genotypes confer specific phenotypic profiles to AD patients. Some of these profiles may add risk or benefit when the patients are treated with conventional drugs, and in many instances the clinical phenotype demands the administration of additional drugs which increase the complexity of therapeutic protocols. From studies designed to define *APOE*-related AD phenotypes, several conclusions can be drawn: the age-at-onset is 5–10 years earlier in approximately 80 % of AD cases harboring the *APOE-4/4* genotype; the serum levels of ApoE are lowest in *APOE-4/4*, intermediate in *APOE-3/3* and *APOE-3/4*, and highest in *APOE-2/3* and *APOE-2/4*; serum cholesterol levels are higher in *APOE-4/4* than in the other genotypes; HDL-cholesterol levels tend to be lower in *APOE-3* homozygotes than in *APOE-4* allele carriers; LDL-cholesterol levels are systematically higher in *APOE-4/4* than in any other genotype; triglyceride levels are significantly lower in *APOE-4/4*; nitric oxide levels are slightly lower in *APOE-4/4*; serum and cerebrospinal fluid Aβ levels tend to differ between *APOE-4/4* and the other most frequent genotypes (*APOE-3/3, APOE-3/4*); blood histamine levels are dramatically reduced in *APOE-4/4* as compared with the other genotypes; brain atrophy is markedly increased in *APOE-4/4* > *APOE-3/4* > *APOE-3/3*; (xi) brain mapping activity shows a significant increase in slow wave activity in *APOE-4/4* from early stages of the disease; (xii) brain haemodynamics, as reflected by reduced brain blood flow velocity and increased pulsatility and resistance indices, is significantly worse in *APOE-4/4* (and in *APOE-4* carriers in general, as compared with *APOE-3* carriers); brain hypoperfusion and neocortical oxygenation is also more deficient in *APOE-4* carriers; lymphocyte apoptosis is markedly enhanced in *APOE-4* carriers; cognitive deterioration is faster in *APOE-4/4* patients than in carriers of any other *APOE* genotype; in approximately 3–8 % of the AD cases, the presence of some dementia-related metabolic dysfunctions accumulates more in *APOE-4* carriers than in *APOE-3* carriers; some behavioural disturbances, alterations in circadian rhythm patterns, and mood disorders are slightly more frequent in *APOE-4* carriers; aortic and systemic atherosclerosis is also more frequent in *APOE-4* carriers; liver metabolism and transaminase activity also differ in *APOE-4/4* with respect to other genotypes; hypertension and other cardiovascular risk factors also accumulate in *APOE-4*; and *APOE-4/4* carriers are the poorest responders to conventional drugs. These 20 major phenotypic features clearly illustrate the biological disadvantage of *APOE-4* homozygotes and the potential consequences that these patients may experience when they receive pharmacological treatment for AD and/or concomitant pathologies [12, 13, 15, 16, 20-24, 71].

When *APOE* and *CYP2D6* genotypes are integrated in bigenic clusters and the *APOE* + *CYP2D6*-related therapeutic response to a combination therapy is analysed in AD patients, it becomes clear that the presence of the *APOE-4/4* genotype is able to convert pure *CYP2D6*1/*1* extensive metabolisers into full poor responders to conventional treatments, indicating the existence of a powerful influence of the *APOE-4* homozygous genotype on the drug-metabolising capacity of pure *CYP2D6* extensive metabolisers. In addition, a clear accumulation of *APOE-4/4* genotypes is observed among *CYP2D6* poor and ultra-rapid metabolisers [21, 24].

Conventional treatments in AD include the cholinesterase inhibitors tacrine (suspended due to hepatotoxicity), donepezil, rivastigmine, and galantamine, and the N-methyl-D-aspartate (NMDA) receptor antagonist memantine. Donepezil (ACHE and BCHE inhibitor) is a major substrate of CYP2D6, CYP3A4 and UGTs, and is transported by ABCB1. Galantamine (ACHE and BCHE inhibitor) is a major substrate of CYP2D6, CYP3A4, and UGT1A1. Memantine is a strong inhibitor of CYP2B6 and CYP2D6, and a weak inhibitor of CYP1A2, CYP2A6, CYP2C9, CYP2C19, CYP2E1, and CYP3A4 [17] (Table 7).

4 Pharmacogenomics of Other NDDs

The pharmacogenomics of current drugs in use for prevalent NDDs are summarized in the following tables: Parkinson's disease (Table 8), amyotrophic lateral sclerosis (Table 9), multiple sclerosis (Table 10), and Huntington's disease (Table 11). At the present time, the pharmacogenomics of most drugs for the treatment of NDDs is poorly understood; however, recent data collected from the World Guide for Drug Use and Pharmacogenomics [17] allowed us to identify the most significant genes potentially involved in the pharmacogenomic output associated with these treatments.

4.1 *Parkinson's Disease*

Parkinson's disease (PD) is the most prevalent form of movement disorder (affects 1 % of people >60 years), characterised by tremor, bradykinesia, and rigidity due to neurodegeneration of dopaminergic neurons of the nigrostriatal pathway. PD is a paradigm of genome-environment interactions in which genetic defects may increase susceptibility to the deleterious effects of toxic agents, cerebrovascular microlesions, and other neurodegeneration-induced phenomena in selective neurons. Most mutations (*SNCA, PINK1, PARK2, PARK7), PLA2G6, FBXO7*, and *ATP13A2*) represent rare causes of PD, and one *LRRK2* mutation is relatively common in certain populations [74-77] (Table 3). Mutations in 7 genes are robustly associated with autosomal dominant (*SNCA, LRRK2, EIF4G1, VPS35*) or recessive (*parkin/PARK2, PINK1, DJ1/PARK7*) PD or parkinsonism. Different SNPs in other genes have been suggested as causes for parkinsonism or PD, including genes for hereditary ataxias (*ATXN2, ATXN3, FMR1*), frontotemporal dementia (*C9ORF72, GRN, MAPT, TARDBP),* Dopa-responsive dystonia *(DYT5, GCH1, TH, SPR*), and others (*ATP13A2, CSF1R, DNAJC6, FBXO, GIGYF2, HTRA2, PLA2G6, POLG, SPG11, UCHL1*) [78].

In PD, several categories of drugs are currently used: (a) dopamine precursors (carbidopa, levodopa), (b) dopamine receptor agonists (amantadine, apomorphine, bromocriptine, cabergoline, lisuride, pergolide, pramipexole, ropinirole, rotigotine), (c) anticholinergic agents (benztropine, biperiden, diphenhydramine, ethopropazine, procyclidine, trihexyphenidyl), (d) monoamine oxidase (MAO) inhibitors (rasagiline, selegiline), and (e) catechol-O-methyltransferase (COMT) inhibitors (entacapone, tolcapone) [17]. Levodopa still remains the gold standard for the treatment of motor symptoms of PD. Dopamine agonists, COMT inhibitors and MAO-B inhibitors have been developed to provide oral delivery of dopaminergic stimulation in order to improve motor outcomes and decrease the risk of levodopa-induced motor complications [79]. Levodopa crosses the blood-brain barrier to be converted into dopamine by striatal enzymes. The formation of dopamine from levodopa at functional nigrostiatal dopaminergic sites may correct akinesia via DRD1, DRD2, DRD3, DRD4, and DRD5, and may also inhibit COMT. Benserazide inhibits peripheral decarboxylation of levodopa without affecting its metabolism in brain. Benserazide exerts is effects via DRD2, COMT and MAOB. Carbidopa inhibits the plasma breakdown of levodopa by inhibiting its decarboxylation, and thereby increases available intracerebral levodopa. In addition to COMT and DRD2, some pleiotropic gene-related products, such as ACE, BDNF, ACHE, and OPRM1, are influenced by the co-administration of carbidopa and levodopa [17].

The amine oxidase inhibitors (MAOIs), of current use in neuropsychiatry, may prove useful for treating some NDDs. Some MAOIs are multifaceted compounds in which their neuroprotective properties seem to be independent of their MAO inhibition [46]. L-Deprenyl (l-N-propargyl, N-methylamphetamine, selegiline), a selective irreversible MAO-B inhibitor (and MAO-A inhibitor at higher doses) has been demonstrated to have neuroprotective or neurorescue properties in vivo and in vitro. Rasagiline, a structurally related drug, has an advantage over L-deprenyl of not being metabolised to L-amphetamine and L-methamphetamine. These two metabolites of L-deprenyl are potentially neurotoxic, whereas N-proparglyamphetamine, another metabolite, might be neuroprotective. In addition, L-amphetamine might interfere with the neuroprotective action of L-deprenyl, whereas the major metabolite of rasagiline, aminoindan, might be neuroprotective [80, 81]. L-Deprenyl and rasagiline downregulate proapoptotic proteins such as BC-associated death promoter (BAD) and BCL-associated protein X (BAX) and prevent the activation and nuclear localisation of glyceraldehyde-3-phosphate dehydrogenase (GAPDH) in response to neurotoxins and reactive oxygen species [81]. Increased MAO activity and expression of tranyclypromine, an irreversible, nonselective MAOI, may cause an increase in BDNF mRNA and AMP response element binding protein (CREB) in the hippocampus. Phenelzine (2-phenylethylhydrazine, PLZ) is an irreversible, nonselective MAO inhibitor (antidepressant) that also inhibits gamma-aminobutyric acid transaminase (GABA-T), increases brain GABA, and exerts a neuroprotective effect in cerebral ischaemia models [46]. Moclobemide, an MAO-A inhibitor, displays anti-Parkinsonian effects and neuroprotective effects which might be independent of MAO-A inhibition [81]. Clorgyline, an irreversible MAO-A inhibitor, is also neuroprotective under different experimental conditions.

4.2 Amyotrophic Lateral Sclerosis

Amyotrophic lateral sclerosis (ALS) (Lou Gehrig's disease) is a neurodegenerative disease, described by Charcot in 1874, characterised by progressive muscular atrophy and weakness resulting from loss of both upper and lower motor neurons. Its incidence ranges from 1.5 to 2.5 per 100,000 per year, with a lifetime risk of 1:400, a mean age of onset of 60 years, with a male predominance of 1.3:1 [51, 82]. Familial and sporadic forms of ALS are clinically indistinguishable. Mutations in different genes have been associated with ALS (Table 4). The most common ALS genes, *SOD1*, *TDP-43*, and *FUS* mutations and the C9orf72 hexanucleotide repeat, might account for 65 % of familial ALS cases in the United States, with great variability depending on the geographical region [83, 84]. Major pathogenic theories for ALS include glutamate toxicity, oxidative stress, autoinmune reactions, and protein aggregation. In terms of therapeutic intervention, mesenchymal stem cells (bone marrow-derived) and neural progenitor cells (spinal cord-derived) are the two cell types with the most evidence for use in ALS [85]; however, riluzole, an antiglutamate drug, is the most prevalent pharmacological treatment with poor results. Many mechanistic genes might be involved in riluzole-related effects (Table 9). Riluzole is a major substrate of CYP1A2 and ABCB1, a major inhibitor of CYP1A2, and is transported by ABCB1 [17]. Other antiglutamate drugs under investigation for the treatment of ALS include: talampanel (LY300164), a benzodiazepine, noncompetitive AMPA receptor antagonist with antiglutamate properties; ceftriaxone, a third-generation cephalosporin; and glutamate carboxypeptidase II (NAALADase) [85]. Recent studies search for the identification of novel pharmacological targets through genomic analysis of deregulated genomic pathways [86].

4.3 Multiple Sclerosis

Multiple sclerosis (MS) is the commonest demyelinating disorder (prevalence <5/100,000 in Africa, South America and Asia; >100/100 000 in Scotland, Scandinavia and Canada), potentially caused by the interaction of multiple genetic and environmental factors [87] (Table 5). The presence of oligoclonal bands (OCB) in cerebrospinal fluid is a typical finding in MS. SNPs from the HLA complex (rs3129871and rs3817963, correlating with the *HLA-DRB1*15* and the *HLA-DRB1*04* alleles, respectively) and 6 other loci were associated to OCB status in Scandinavian patients [88].

Different categories of drugs are currently used for the treatment of MS including: (a) hormones (prednisone), (b) immunosuppressive and inmunomodulatory agents (cyclophosphamide, fingolimod, teriflunomide), (c) potassium channel blockers (dalfampridine/4-aminopyridine), (d) sphingosine 1-phosphate receptor modulators (fingolimod), (e) biologic response modifiers (glatiramer acetate, interferon beta-1a, interferon beta-1b, natalizumab), (f) antineoplastic agents (cyclophosphamide, mitoxantrone), (g) skeletal muscle relaxants (dantrolene), (h) monoclonal antibodies with selective adhesion-molecule inhibitor activity (natalizumab) and (i) fumarates (tecfidera) (Table 10). Most of these compounds show complex pharmacogenomic profiles depending on their pharmacological category [17]. Interferon-beta (IFNβ) was the first immunomodulatory treatment for multiple sclerosis, with limited clinical efficacy. Whole-genome association studies have revealed that clusters of brain-specific genes may act as IFNβ response modifiers [89]. Furthermore, gene expression studies have shown that the expression levels of IFN response genes in the peripheral blood of MS patients prior to treatment could serve a role as a biomarker for the differential clinical response to IFNβ [90].

4.4 *Huntington's Disease*

Huntington's disease (HD) is a neurodegenerative autosomal dominant disease caused by 36 or more CAG triplet repeat expansions in exon 1 of the *HTT* gene, coding for huntingtin protein. These expansions lead to a substantial loss of spiny neurons in the striatum causing the abnormal involuntary movements, cognitive decline, and psychiatric disturbance which characterise this disease. The mean age of onset is 35–44 years and the median survival time is 15–18 years after onset [91].

The Guideline Development Subcommittee of the American Academy of Neurology [35] recommended to prescribe tetrabenazine (up to 100 mg/day) (Table 11), amantadine (300–400 mg/day) (Table 8), or riluzole (200 mg/day) (Table 9) for varying degrees of expected benefit in patients with Huntington's disease (Table 6). However, most patients with this type of coreiform movement disorder usually receive different types of neuroleptics and other psychotropic drugs [19, 58]. Tetrabenazine is a central monoamine-depleting agent with a complex mechanism of action, acting as a major CYP2D6 substrate, and a VMAT2 inhibitor, transported by ABCB1, SLC18A1, SLC18A2, SLC6A3, and SLC6A4 [17, 92].

5 Future Trends

Historically, the vast majority of pharmacogenetic studies of CNS disorders have been addressed to evaluate the impact of cytochrome P450 enzymes on drug metabolism, and conventional targets for psychotropic drugs were dopamine, serotonin, noradrenaline, GABA, ion channels, acetylcholine and their respective biosynthetic and catalysing enzymes, receptors and transporters; however, in the past few years many different genes have been associated with both pathogenesis and pharmacogenomics of neuropsychiatric disorders [12, 15, 16, 19, 58]. Some of these genes and their products constitute potential targets for future treatments. New developments in genomics, including whole genome genotyping approaches and comprehensive information on genomic variation across populations, coupled with large-scale clinical trials in which DNA collection is routine, now provide the impetus for a next generation of pharmacogenetic studies and identification of novel candidate drugs.

Priority areas for pharmacogenetic research are predicting serious adverse reactions and establishing variation in efficacy [93]. Both requirements are necessary in CNS disorders to cope with efficacy and safety issues associated with both current psychotropic drugs and new drugs. Since drug response is a complex trait, genome-wide approaches may provide new insights into drug metabolism and drug response. Of paramount importance is the identification of polymorphisms affecting gene regulation and mRNA processing in genes encoding cytochrome P450s and other drug-metabolising enzymes, drug transporters, and drug targets and receptors, with broad implication in pharmacogenetics since functional polymorphisms which alter gene expression and mRNA processing appear to play a critical role in shaping human phenotypic variability [94]. It is also most relevant, from a practical point of view, to understand the pharmacogenomics of drug transporters, especially *ABCB1* (P-glycoprotein/MDR1) variants, due to the pleiotropic activity of this gene on a large number of drugs [95]. It is necessary to have a better documentation related to the pharmacogenetic roles of the enormous number (>170) of human solute carrier transporters which transport a variety of substrates, including amino acids, lipids, inorganic ions, peptides, saccharides, metals, drugs, toxic xenobiotics, chemical compounds, and proteins [96]. RNAi pharmacogenomics will also bring new insights into the nature and therapeutic value of gene silencing in CNS disorders [1, 9, 97-99].

The optimisation of CNS therapeutics requires the establishment of new postulates regarding (i) the costs of medicines, (ii) the assessment of protocols for multifactorial treatment in chronic disorders, (iii) the implementation of novel therapeutics addressing causative factors, and (iv) the setting-up of pharmacogenomic strategies for drug development and drugs in the market [14-16].

By knowing the pharmacogenomic profiles of patients with complex disorders, it might be possible to obtain some of the following benefits related to efficacy and safety issues: to identify candidate patients with the ideal genomic profile to receive a particular drug; to adapt the dose in over 90% of the cases according to the condition of EM, IM, PM or UM (diminishing the occurrence of direct side-effects in 30–50% of the cases); to reduce drug interactions by 30–50% (avoiding the administration of inhibitors or inducers able to modify the normal enzymatic activity on a particular substrate); to enhance efficacy; and to eliminate unnecessary costs (>30% of pharmaceutical costs) derived from the consequences of inappropriate drug selection and the overmedication administered to mitigate adverse drug reactions [11, 18].

To make complex disorders a global health priority in the coming years, conceptual and procedural changes are needed on several grounds, such as political, administrative, economic, legal, ethical, industrial, regulatory and educational issues; the implantation of novel biomarkers (genomics, transcriptomics, proteomics, metabolomics) as diagnostic and therapeutic aids; the introduction of innovative therapeutics; the implementation of pharmacogenomics in the clinical practice in order to optimise therapeutics; and the promotion of selective preventive plans for the population at risk [11].

References

1. Boudreau RL, Spengler RM, Davidson BL (2011) Rational design of therapeutic siRNAs: minimizing off-targeting potential to improve the safety of RNAi therapy for Huntington's disease. Mol Ther 19(12):2169–2177
2. Coutinho P, Ruano L, Loureiro JL et al (2013) Hereditary ataxia and spastic paraplegia in portugal: a population-based prevalence study. JAMA Neurol 70(6):746–755
3. International Statistical Classification of Diseases and Related Health Problems (ICD-10) [homepage on the Internet]. Geneva, Switzerland: World Health Organization, 1948. http://apps.who.int/classifications/icd10/browse/2010/en. Accessed 19 July 2013

4. Liberski PP, Budka H (2004). Gerstmann-Sträussler-Scheinker disease. I. Human diseases. Folia Neuropathol 42(Suppl B):120–140
5. Medscape [homepage on the Internet]. New York, USA: WebMD, 1984. http://reference.medscape.com/. Accessed 19 July 2013
6. National Institute of Neurological Disorders and Stroke (NINDS) [homepage on the Internet]. Bethesda, Maryland, USA: National Institutes of Health (NIH), 1950. http://www.ninds.nih.gov/. Accessed 19 July 2013
7. Orphanet (version 4.10.2) [homepage on the Internet]. Paris, France: Orphanet. http://www.orpha.net/consor/cgi-bin/index.php. [updated 18 July 2013; Accessed 19 July 2013]
8. Prince M, Bryce R, Albanese E et al (2013) The global prevalence of dementia: a systematic review and metaanalysis. Alzheimers Dement 9(1):63–75.e2
9. Ying SY, Lin SL (2009) Intron-mediated RNA interference and microRNA biogenesis. Methods Mol Biol 487:387–413
10. Su YN, Hung CC, Lin SY et al (2011) Carrier screening for spinal muscular atrophy (SMA) in 107,611 pregnant women during the period 2005–2009: a prospective population-based cohort study. PLoS ONE 6(2):e17067
11. Cacabelos R (2012) Pharmacogenomics of central nervous system (CNS) drugs. Drug Dev Res 73(8):461–476.
12. Cacabelos R (2009) The path to personalized medicine in mental disorders. In: Ritsner MS (ed) The handbook of neuropsychiatric biomarkers, endophenotypes and genes, vol 4. Springer, Netherlands, pp 3–63
13. Cacabelos R (2009) Pharmacogenomics and therapeutic strategies for dementia. Expert Rev Mol Diag 9(6):567–611
14. Cacabelos R, Fernández-Novoa L, Martínez-Bouza R, McKay A, Carril JC, Lombardi V, Corzo L, Carrera I, Tellado I, Nebril L, Alcaraz M, Rodríguez S, Casas A, Couceiro V, Alvarez A (2010) Future trends in the pharmacogenomics of brain disorders and dementia: influence of APOE and CYP2D6 variants. Pharmaceuticals 3(10):3040–3100.
15. Cacabelos R, Martínez R, Fernández-Novoa L, Carril JC, Lombardi V, Carrera I, Corzo L, Tellado I, Leszek J, McKay A, Takeda M (2012). Genomics of dementia: APOE- and CYP2D6-related pharmacogenetics. Int J Alzheimer Dis 2012:518901. doi:10.1155/2012/518901
16. Cacabelos R, Martínez-Bouza R, Carril JC, Fernández-Novoa L, Lombardi V, Carrera I, Corzo L, McKay A (2012) Genomics and pharmacogenomics of brain disorders. Cur Pharm Biotech 13(5):674–725
17. Cacabelos R (ed) (2012) World guide for drug use and pharmacogenomics. EuroEspes Publishing, Corunna
18. Cacabelos R (2012) The metabolomics paradigm of pharmacogenomics in complex disorders. Metabolomics 2:e119. doi:10.4172/2153-0769.1000e119
19. Cacabelos R, Cacabelos P, Aliev G (2013) Genomics of schizophrenia and pharmacogenomics of antipsychotic drugs. Open J Psychiatry 3:46–139.
20. Cacabelos R, Fernández-Novoa L, Lombardi V, Kubota Y, Takeda M (2005) Molecular genetics of Alzheimer's disease and aging. Meth Find Exper Clin Pharmacol 27(Suppl A):1–573
21. Cacabelos R (2007) Pharmacogenetic basis for therapeutic optimization in Alzheimer's disease. Mol Diag Ther 11(6):385–405
22. Cacabelos R (2007) Molecular pathology and pharmacogenomics in Alzheimer's disease: polygenic-related effects of multifactorial treatments on cognition, anxiety, and depression. Meth Find Exper Clin Pharmacol 29(Suppl B):1–91
23. Cacabelos R (2008) Pharmacogenomics and therapeutic prospects in dementia. Eur Arch Psychiatry Clin Neurosci 258(Suppl 1):28–47
24. Cacabelos R (2008) Pharmacogenomics in Alzheimer's disease. Meth Mol Biol 448:213–357
25. Zhou SF, Di YM, Chan E, Du YM, Chow VD, Xue CC, Lai X, Wang JC, Li CG, Tian M, Duan W (2008) Clinical pharmacogenetics and potential application in personalized medicine. Curr Drug Metab 9(8):738–784
26. Malhotra AK, Lencz T, Correll CU, Kane JM (2007) Genomics and the future of pharmacotherapy in psychiatry. Int Rev Psychiatry 19(5):523–530

27. Vedeler CA, Myhr KM, Nyland H (2001) Fc receptors for immunoglobulin G–a role in the pathogenesis of Guillain-Barré syndrome and multiple sclerosis. J Neuroimmunol 118(2):187–193
28. DrugBank [homepage on the internet]. Alberta, Canada; University of Alberta. http://www.drugbank.ca/. Accessed 19 July 2013
29. EMEA [homepage on the internet]. London, UK: European Medicines Agency; 1995. http://www.ema.europa.eu/ema/. Accessed 19 July 2013
30. FDA [homepage on the internet]. Silver Spring, MD: U.S. Food and Drug Administration; 1906. http://www.fda.gov/. Accessed 19 July 2013
31. GeneReviews [homepage on the internet]. Seattle: University of Washington; 1993. http://www.ncbi.nlm.nih.gov/books/NBK1116/. Accessed 19 June 2013
32. Sanderink GJ, Bournique B, Stevens J, Petry M, Martinet M (1997) Involvement of human CYP1A isoenzymes in the metabolism and drug interactions of riluzole in vitro. J Pharmacol Exp Ther 282(3):1465–1472
33. Hardy J, Selkoe DJ (2002) The amyloid hypothesis of Alzheimer's disease: progress and problems on the road to therapeutics. Science 297(5580):353–356
34. Morrow CS, Peklak-Scott C, Bishwokarma B, Kute TE, Smitherman PK, Townsend AJ (2006) Multidrug resistance protein 1 (MRP1, ABCC1) mediates resistance to mitoxantrone via glutathione-dependent drug efflux. Mol Pharmacol 69(4):1499–1505
35. Armstrong MJ, Miyasaki JM, American Academy of Neurology (2012) Evidence-based guideline: pharmacologic treatment of chorea in Huntington disease: Report of the guideline development subcommittee of the American Academy of Neurology. Neurology 79(6):597–603.
36. Ries V, Selzer R, Eichhorn T, Oertel WH, Eggert K, German Tolcapone Study Group (2010) Replacing a dopamine agonist by the COMT-inhibitor tolcapone as an adjunct to L-dopa in the treatment of Parkinson's disease: a randomized, multicenter, open-label, parallel-group study. Clin Neuropharmacol 33(3):142–150
37. The Anatomical, Therapeutic, Chemical (ATC) classification system with Defined Daily Doses (DDDs) [homepage on the internet]. Geneva, Switzerland: World Health Organization 1948. http://www.whocc.no/atc_ddd_index. Accessed 19 July 2013
38. Messina S, Patti F (2013) The pharmacokinetics of glatiramer acetate for multiple sclerosis treatment. Expert Opin Drug Metab Toxicol 9(10):1349–1359
39. Scannevin RH, Chollate S, Jung MY et al (2012) Fumarates promote cytoprotection of central nervous system cells against oxidative stress via the nuclear factor (erythroid-derived 2)-like 2 pathway. J Pharmacol Exp Ther 341(1):274–284
40. Loewe R, Holnthoner W, Gröger M, Pillinger M, Gruber F, Mechtcheriakova D, Hofer E, Wolff K, Petzelbauer P (2002) Dimethylfumarate inhibits TNF-induced nuclear entry of NF-kappa B/p65 in human endothelial cells. J Immunol 168(9):4781–4787
41. Seidel P, Goulet S, Hostettler K, Tamm M, Roth M (2010) DMF inhibits PDGF-BB induced airway smooth muscle cell proliferation through induction of hemeoxygenase-1. Respir Res 11:145
42. Szczuciński A, Losy J (2011) CCL5, CXCL10 and CXCL11 chemokines in patients with active and stable relapsing-remitting multiple sclerosis. Neuroimmunomodulation 18(1):67–72
43. Valero T, Steele S, Neumüller K, Bracher A, Niederleithner H, Pehamberger H, Petzelbauer P, Loewe R (2010) Combination of dacarbazine and dimethylfumarate efficiently reduces melanoma lymph node metastasis. J Invest Dermatol 130(4):1087–1094
44. Sinko G, Kovarik Z, Reiner E, Simeon-Rudolf V, Stojan J (2011) Mechanism of stereoselective interaction between butyrylcholinesterase and ethopropazine enantiomers. Biochimie 93(10):1797–1807
45. Takuwa N, Du W, Kaneko E, Okamoto Y, Yoshioka K, Takuwa Y (2011) Tumor-suppressive sphingosine-1-phosphate receptor-2 counteracting tumor-promoting sphingosine-1-phosphate receptor-1 and sphingosine kinase 1—Jekyll Hidden behind Hyde. Am J Cancer Res 1(4):460–481
46. Baker GB, Sowa B, Todd KG (2007) Amine oxidases and their inhibitors: what can they tell us about neuroprotection and the development of drugs for neuropsychiatric disorders? J Psychiatry Neurosci 32(5):313–315

47. Caminero A, Comabella M, Montalban X (2011) Role of tumour necrosis factor (TNF)-α and TNFRSF1A R92Q mutation in the pathogenesis of TNF receptor-associated periodic syndrome and multiple sclerosis. Clin Exp Immunol 166(3):338–345
48. Somnay Y, Chen H, Kunnimalaiyaan M (2013). Synergistic effect of pasireotide and teriflunomide in carcinoids in vitro. Neuroendocrinology 97(2):183–192
49. Kis E, Nagy T, Jani M et al (2009) Leflunomide and its metabolite A771726 are high affinity substrates of BCRP: implications for drug resistance. Ann Rheum Dis 68(7):1201–1207
50. Latchoumycandane C, Seah QM, Tan RC, Sattabongkot J, Beerheide W, Boelsterli UA (2006) Leflunomide or A77 1726 protect from acetaminophen-induced cell injury through inhibition of JNK-mediated mitochondrial permeability transition in immortalized human hepatocytes. Toxicol Appl Pharmacol 217(1):125–133
51. McCombe PA, Henderson RD (2010) Effects of gender in amyotrophic lateral sclerosis. Gend Med 7(6):557–570
52. Márki-Zay J, Tauberné Jakab K, Szerémy P, Krajcsi P (2013) MDR-ABC transporters: biomarkers in rheumatoid arthritis. Clin Exp Rheumatol 31(5):779–787
53. Palmer AM (2010) Teriflunomide, an inhibitor of dihydroorotate dehydrogenase for the potential oral treatment of multiple sclerosis. Curr Opin Investig Drugs 11(11):1313–1323
54. Wiese MD, Rowland A, Polasek TM, Sorich MJ, O'Doherty C (2013) Pharmacokinetic evaluation of teriflunomide for the treatment of multiple sclerosis. Expert Opin Drug Metab Toxicol 9(8):1025–1035
55. Zeyda M, Poglitsch M, Geyeregger R, Smolen JS, Zlabinger GJ, Hörl WH, Waldhäusl W, Stulnig TM, Säemann MD (2005) Disruption of the interaction of T cells with antigen-presenting cells by the active leflunomide metabolite teriflunomide: involvement of impaired integrin activation and immunologic synapse formation. Arthritis Rheum 52(9):2730–2739
56. Goris A, Pauwels I, Dubois B (2012) Progress in multiple sclerosis genetics. Curr Genomics 13(8):646–663
57. Ingelman-Sundberg M, Sim SC, Gomez A, Rodríguez-Antona C (2007) Influence of cytochrome P450 polymorphisms on drug therapies: Pharmacogenetic, pharmacoepigenetic and clinical aspects. Pharmacol Ther 116(3):496–526
58. Cacabelos R, Martínez-Bouza R (2011) Genomics and pharmacogenomics of schizophrenia. CNS Neurosci Ther 17(5):541–565
59. Isaza CA, Henao J, López AM, Cacabelos R (2000) Isolation, sequence and genotyping of the drug metabolizer CYP2D6 gene in the Colombian population. Meth Find Exp Clin Pharmacol 22(9):695–705
60. Mizutani T (2003) PM frequencies of major CYPs in Asians and Caucasians. Drug Metab Rev 35(2–3):99–106
61. Xie HG, Kim RB, Wood AJ, Stein CM (2001) Molecular basis of ethnic differences in drug disposition and response. Ann Rev Pharm Toxicol 41:815–850
62. Ozawa S, Soyama A, Saeki M, Fukushima-Uesaka H, Itoda M, Koyano S, Sai K, Ohno Y, Saito Y, Sawada J (2004) Ethnic differences in genetic polymorphisms of CYP2D6, CYP2C19, CYP3As and MDR1/ABCB1. Drug Metab Pharmacokin 19(2):83–95
63. Cacabelos R, Llovo R, Fraile C, Fernández-Novoa L (2007) Pharmacogenetic aspects of therapy with cholinesterase inhibitors: the role of CYP2D6 in Alzheimer's disease pharmacogenetics. Curr Alzheimer Res 4(4):479–500
64. Haufroid V (2011) Genetic polymorphisms of ATP-binding cassette transporters ABCB1 and ABCC2 and their impact on drug disposition. Curr Drug Targets 12(5):631–646
65. Marquez B, Van Bambeke F (2011) ABC multidrug transporters: target for modulation of drug pharmacokinetics and drug-drug Interactions. Curr Drug Targets 12(5):600–620
66. Weinshilboum RM, Wang L (2006) Pharmacogenetics and pharmacogenomics: development, science, and translation. Ann Rev Genomics Hum Genet 7:223–245
67. Hosoya K, Tachikawa M (2011) Roles of organic anion/cation transporters at the blood-brain and blood-cerebrospinal fluid barriers involving uremic toxins. Clin Exp Nephrol 15(4):478–585
68. Ensembl [homepage on the internet]. Hinxton, UK: European Bioinformatics Institute and Wellcome Trust Sanger Institute. http://www.ensembl.org/index.html. Accessed 19 July 2013

69. Bertram L, McQueen MB, Mullin K, Blacker D, Tanzi RE (2007) Systematic meta-analyses of Alzheimer disease genetic association studies: the AlzGene database. Nat Genet 39(1):17–23. http://www.alzgene.org/. Accessed 2 April 2013
70. Carl SM, Lindley DJ, Couraud PO, Weksler BB, Romero I, Mowery SA, Knipp GT (2010) ANC and SLC transporter expression and Pot substrate characterization across the human CMEC/D3 blood-brain barrier cell line. Mol Pharm 7(4):1057–1068
71. Cacabelos R (2005) Pharmacogenomics and therapeutic prospects in Alzheimer's disease. Exp Opin Pharmacother 6:1967–1987
72. Selkoe DJ, Podlisny MB (2002) Deciphering the genetic basis of Alzheimer's disease. Annu Rev Genomics Hum Genet 3:67–99
73. Suh YH, Checler F (2002) Amyloid precursor protein, presenilins, and α synuclein: molecular pathogenesis and pharmacological applications in Alzheimer's disease. Pharmacol Rev 54(3):469–525
74. Lesage S, Le Ber I, Condroyer C, Broussolle E, Gabelle A, Thobois S, Pasquier F, Mondon K, Dion PA, Rochefort D, Rouleau GA, Dürr A, Brice A, French Parkinson's Disease Genetics (PDG) Study Group (2013) C9orf72 repeat expansions are a rare genetic cause of parkinsonism. Brain 136(Pt 2):385–391
75. Proukakis C, Dudzik CG, Brier T, Mackay DS, Cooper JM, Millhauser GL, Houlden H, Schapira AH (2013) A novel α-synuclein missense mutation in Parkinson disease. Neurology 80(11):1062–1064
76. Ross OA (2013) A prognostic view on the application of individualized genomics in Parkinson's disease. Curr Genet Med Rep 1(1):52–57.
77. Singleton AB, Farrer MJ, Bonifati V (2013) The genetics of Parkinson's disease: progress and therapeutic implications. Mov Disord 28(1):14–23
78. Puschmann A (2013) Monogenic Parkinson's disease and parkinsonism: clinical phenotypes and frequencies of known mutations. Parkinsonism Relat Disord 19(4):407–415
79. Sprenger F, Poewe W (2013) Management of motor and non-motor symptoms in Parkinson's disease. CNS Drugs 27(4):259–272
80. Am OB, Amit T, Youdin MBH (2004) Contrasting neuroprotective and neurotoxic actions of respective metabolites of anti-Parkinson drugs rasagiline and selegiline. Neurosci Lett 355(3):169–172
81. Youdim MBH, Edmondson D, Tipton KF (2006) The therapeutic potential of monoamine oxidase inhibitors. Nat Rev Neurosci 7(4):295–309
82. Hardiman O, van den Berg LH, Kiernan MC (2011) Clinical diagnosis and management of amyotrophic lateral sclerosis. Nat Rev Neurol 7(11):639–649
83. Andersen PM, Al-Chalabi A (2011) Clinical genetics of amyotrophic lateral sclerosis: what do we really know? Nat Rev Neurol 7(11):603–615
84. Majounie E, Renton AE, Mok K et al. Chromosome 9-ALS/FTD Consortium; French research network on FTLD/FTLD/ALS; ITALSGEN Consortium (2012) Frequency of the C9orf72 hexanucleotide repeat expansion in patients with amyotrophic lateral sclerosis and frontotemporal dementia: a cross-sectional study. Lancet Neurol 11(4):323–330
85. Gibson SB, Bromberg MB (2012) Amyotrophic lateral sclerosis: drug therapy from the bench to the bedside. Semin Neurol 32(3):173–178.
86. Paratore S, Pezzino S, Cavallaro S (2012) Identification of pharmacological targets in amyotrophic lateral sclerosis through genomic analysis of deregulated genes and pathways. Curr Genomics 13(4):321–333
87. Love S (2006) Demyelinating diseases. J Clin Pathol 59(11):1151–1159
88. Mero IL, Gustavsen MW, Sæther HS, Flåm ST, Berg-Hansen P, Søndergaard HB, Jensen PE, Berge T, Bjølgerud A, Muggerud A, Aarseth JH, International Multiple Sclerosis Genetics Consortium, Myhr KM, Celius EG, Sellebjerg F, Hillert J, Alfredsson L, Olsson T, Oturai AB, Kockum I, Lie BA, Andreassen BK, Harbo HF (2013) Oligoclonal band status in scandinavian multiple sclerosis patients is associated with specific genetic risk alleles. PLoS One 8(3):e58352. doi:10.1371/journal.pone.0058352

89. Vandenbroeck K, Urcelay E, Comabella M (2010) IFN-beta pharmacogenomics in multiple sclerosis. Pharmacogenomics 11(8):1137–1148
90. van Baarsen LGM, Vosslamber S, Tijssen M, Baggen JM, van der Voort LF, Killestein J, van der Pouw Kraan TC, Polman CH, Verweij CL (2008) Pharmacogenomics of interferon-ß therapy in multiple sclerosis: baseline IFN signature determines pharmacological differences between patients. PLoS One 3(4):e1927
91. Harris JM, Fraser J, Fahn S (2007) Genetics of movement disorders. In: Rosenberg RN, Dimauro S, Paulson HL, Ptácek L, Nestler EJ (eds) The molecular and genetic basis of neurologic and psychiatric disease, 4th ed. Lippincott Williams & Wilkins, Philadelphia, USA pp 346–349
92. Killoran A, Biglan KM (2012) Therapeutics in Huntington's disease. Curr Treat Options Neurol 14(2):137-149
93. Need AC, Motulsky AG, Goldstein DB (2005) Priorities and standards in pharmacogenetic research. Nat Genet 37(7):671–681
94. Johnson AD, Wang S, Sadee W (2005) Polymorphisms affecting gene regulation and mRNA processing: broad implications for pharmacogenetics. Pharmacol Ther 106(1):19–38
95. Ishikawa T, Onishi Y, Hirano H, Oosumi K, Nagakura M, Tarui S (2004) Pharmacogenomics of drug transporters: a new approach to functional analysis of the genetic polymorphisms of ABCB1 (P-glycoprotein/MDR1). Biol Pharm Bull 27(7):939–948
96. Nishimura M, Naito S (2008) Tissue-specific mRNA expression profiles of human solute carrier transporter superfamilies. Drug Metab Pharmacokin 23(1):22–44
97. Agrawal N, Dasaradhi PV, Mohmmed A, Malhotra P, Bhatnagar RK, Mukherjee SK (2003) RNA interference: biology, mechanism, and applications. Microbiol Mol Biol Rev 67(4):657–685
98. Leung RK, Whittaker PA (2005) RNA interference: from gene silencing to gene-specific therapeutics. Pharmacol Ther 107(2):222–239
99. Gonzalez-Alegre P (2007) Therapeutic RNA interference for neurodegenerative diseases: from promise to progress. Pharmacol Ther 114(1):34–55

Pharmacogenetics of Asthma

Anthony G. Fenech, Ian Sayers and Michael A. Portelli

Abstract The study of pharmacogenetics has expanded from what were initially casual drug response observations present in families, to a fully-fledged science with direct therapeutic applications, within a timespan of less than 60 years. Within the field of asthma therapeutics, heterogeneity in patient response to therapeutic agents has been reported from various studies, and several genotype associations with specific therapeutic-response phenotypes have been established. There is however much to be accomplished. The future of asthma pharmacogenetics lies in consolidating and validating clinically important pharmacogenes which are relevant to currently available therapy, and to integrate itself into day to day drug management and the drug development process, in order to streamline the discovery of the potential pharmacogenetic relevance of new drugs. This chapter reviews the pharmacogenetics of current asthma therapies, and discusses challenges which need to be addressed in order to enable the optimisation and personalisation of patient management in a genotype-dependent manner.

Keywords Pharmacogenetics · Asthma · Drug development · Genotype-guided prescribing · Predictive therapeutics

A. G. Fenech (✉)
Department of Clinical Pharmacology and Therapeutics, Faculty of Medicine and Surgery, University of Malta, Msida MSD2080, Malta
e-mail: anthony.fenech@um.edu.mt

I. Sayers · M. A. Portelli
Division of Respiratory Medicine, University of Nottingham, Queen's Medical Centre, Nottingham NG7 2UH, UK
e-mail: ian.sayers@nottingham.ac.uk

M. A. Portelli
e-mail: michael.portelli@nottingham.ac.uk

G. Grech, I. Grossman (eds.), *Preventive and Predictive Genetics: Towards Personalised Medicine,* Advances in Predictive, Preventive and Personalised Medicine 9,
DOI 10.1007/978-3-319-15344-5_9

1 Introduction

The internationally established Global Initiative for Asthma (GINA), defines asthma as a heterogeneous disease, characterised by chronic airway inflammation and variable airflow limitation, and described by a history of respiratory symptoms which may include wheeze, shortness of breath, chest tightness and cough that are variable over time and in intensity [1]. The joint European Respiratory Society/ American Thoracic Society guidelines on the definition of asthma add that "Asthma, and severe asthma in particular, are increasingly recognised as heterogeneous processes, not all of which may respond similarly to current therapies or have the same clinical course" [2].

Addressing inter-patient variability in response to treatment is the fulcrum of the gradual but expanding practice of personalised medicine. Research in pharmacogenetics is the driving force behind the evolution of the "one size fits all" paradigm to a scenario where specific patient parameters, including genetic factors, form part of a treatment strategy which is optimised on an individual basis.

Asthma, a condition which is multifactorial and polygenic in nature, is managed by a therapeutic repertoire of drugs which target multiple pathways, and are themselves subject to pharmacogenetic influence by multiple genes. The complex molecular interactions occurring within the disease pathology, its environmental and genetic determinants, the drugs administered as well as pharmacogenetic and epigenetic contributions, make the development of genotype guided prescribing protocols for asthma, a formidable activity. The identification of genotype-drug response associations and assessment of their clinical relevance, are critical steps towards the development of such protocols.

The genotype nomenclature, adopted for different alleles throughout this chapter, refers to that adopted by the respective cited authors.

2 Pharmacological Management of Asthma

Current asthma management guidelines, including those developed by the GINA as well as several other professional organisations such as the European Respiratory Society (www.ersnet.org), the American Thoracic Society (www.thoracic.org), the British Thoracic Society (www.brit-thoracic.org.uk), the Scottish Intercollegiate Guidelines Network (www.sign.ac.uk) and the National Institute for Health and Care Excellence (www.nice.org.uk) advocate for a stepwise management approach. This approach is based on clinical severity assessments, with regular re-evaluations of treatment resulting in a step up, when asthma is not adequately managed, or a step down if disease control can be maintained with less aggressive therapy. The drug classes currently in use for asthma comprise; β_2-adrenoceptor agonists, the less commonly used anti-muscarinic bronchodilators, glucocorticoids, anti-leukotriene drugs, theophylline, and the monoclonal anti IgE antibody omalizumab which is normally reserved for severe, atopic asthma patients with elevated serum IgE titres, who are refractory to other treatments. This chapter will focus on the current knowledge regarding the pharmacogenetics of medicines used clinically in the treatment of asthma and also review the emerging importance of both pharmacogenetics and stratified approaches to newer asthma drugs currently in Phase II development.

3 β_2-adrenoreceptor Agonists

β_2-adrenoceptor agonists act by binding to a 413-amino-acid G-protein coupled receptor (GPCR) encoded for by an intronless gene (*ADRB2*) located on chromosome 5q31.32. Binding of the ligand to the β_2-adrenoceptor results in stimulation of Gs proteins, activating protein kinase A, resulting in the activation of adenyl cyclase. This activation pathway enables the phosphorylation of several target proteins, followed by reduction in intracellular calcium which in turn causes the targeted β_2 agonist effect of airway smooth muscle relaxation.

ADRB2 is highly polymorphic containing 49 known and validated single nucleotide polymorphisms (SNPs) and two insertion/deletion variants [3]. There have been extensive studies of the pharmacogenetic effects of *ADRB2* on the efficacy of β_2-adrenoceptor agonist therapy. The vast majority of these studies have focussed on the role of four non-synonymous coding region polymorphisms, namely Arg16Gly, Gln27Glu, Val34Met and Thr164Ile [4]. The frequency of the position 16 and 27 polymorphism was found to be Arg16Glu 59 %, Gln27Glu 29 % in the Caucasian population [4]. The Val34Met and Thr164Ile polymorphisms are rare, with approximate frequencies of <0.001 % and 0.05 %, respectively. Although the Arg16 variant has been associated with an enhanced acute response to β_2-adrenoceptor agonists, a decline of asthma control following prolonged use and a subsensitivity of response for bronchoprotection, several studies have failed to reproduce these effects. Therefore a common consensus on the contribution of Arg16 has yet to be reached. At least twelve different haplotypes (combinations of polymorphisms) have been described in the *ADRB2* gene [5], suggesting a varied population-specific genotype for *ADRB2*. This may explain the lack of consensus when examining genotype effects in isolation [6].

Functional effects of these β_2-adrenoceptor coding region polymorphisms obtained from *in vivo* studies have shown the following: (a) the rare Thr164Ile variant receptor produces impairment in agonist binding and a reduction in adenyl cyclase activity, (b) an enhanced agonist-mediated receptor down-regulation occurs in the Gly16 receptor variants, and (c) a resistance to down-regulation is associated with the Glu27 variant [6–9]

Although cell studies have not identified a significant effect of the common Arg16Gly and Gln27Glu *ADRB2* coding region polymorphisms on the responses to a range of β_2-adrenoceptor agonists including the long-acting beta-adrenoceptor agonist indacaterol, they did however confirm that the Thr164Ile variant receptor produced an impaired agonist mediated cAMP production in recombinant cell lines [10].

A recent clinical study by Lee et al. reported associations between the Gln/Glu residue 27 status, and lung function in asthmatic children on nebulised terbutaline. More specifically, the peak expiratory flow rate (PEFR) improved by 23.4 ± 14.4 % in Glu/Glu homozygotes and by 8.2 ± 0.4 % in Gln/Gln individuals ($p < 0.05$), while Arg16Gly heterozygotes showed no bronchodilator effect [11]. The main limitation of this study was the small number of participating patients ($n = 27$). One of the largest studies to date ($n = 1182$), carried out by Basu et al. in asthmatic patients having an age range of 3–22 years, reported an increased risk of exacerbations per copy of Arg16 (odds ratio [OR], 1.30; CI 1.09–1.55 $p = 0.003$). Two patient groups were studied; one was receiving regular inhaled corticosteroid plus salbutamol *prn*, while the other was being managed with a regular inhaled corticosteroid/salmeterol combination, plus salbutamol *prn*. Interestingly, this pharmacogenetic effect was driven by those patients who used daily doses of at least one of the bronchodilators [12]. The study is however limited by the absence of control patients who were not on β_2-adrenoceptor agonist treatment. This missing control group may have provided a clearer interpretation for the detrimental effects of β_2-adrenoceptor agonists in Arg16 subjects. This is especially relevant in view of other studies where clinically relevant effects of these *ADRB2* variants could not be identified. For example, a large study conducted earlier by Bleecker et al., failed to identify any clinically relevant outcomes of the Gly16Arg genotype. The group studied 2250 asthma patients, who were assigned to three management groups for a 6 month period as follows: (a) budesonide/formoterol combination maintenance therapy plus reliever therapy *prn*, (b) fixed dose budesonide plus formoterol, and (c) fixed dose fluticasone plus salmeterol [13].

These contrasting outcomes may be rationaliszed by the different study designs employed. Furthermore, deeper data analyses (e.g. looking at data stratified by daily *prn* reliever use) was lacking. A more recent study conducted by the same group, explored the potential pharmaocogenetic effects of 11 *ADRB2* SNPs in patients administered a fluticasone/salmeterol combination ($n = 268$) or salmeterol alone ($n = 266$), Interestingly, this study found no Arg16Gly pharmacogenetic effects on clinical outcomes, including morning PEFR and number of symptom free days, while other *ADRB2* SNPs localised to regulatory regions of the gene appeared to influence clinical outcomes of bronchodilator treatment. It is however unlikely that these outcomes would survive correction for multiple testing [14]. A similar lack of clinical influence of the Arg16 genotype, on fluticasone/salmeterol clinical outcomes, has also been observed by other research workers. For example, a double blind, placebo controlled trial carried out by Wechsler et al., using Arg/Arg and Gly/Gly adult asthmatic subjects with matched morning PEF (peak expiratory flow), FEV_1 (forced expiratory volume in one second) and symptom scores (an approach undertaken in order to maximise the likelihood of identifying genotype-specific effects) failed to identify Arg or Gly related outcomes after an 18-week study period [15].

While the majority of studies have focussed on the Arg16Gly, it is important to emphasise that as in the majority of genes, *ADRB2* is highly polymorphic with multiple potential polymorphisms occurring in the gene's regulatory regions. Single SNP approaches on their own may therefore be ineffective at elucidating genotype-phenotype relationships. Haplotype analysis, or *in vitro* "whole gene transfection" may be a more appropriate approach, and one that approximates more the *in vivo* situation. Such an approach has been recently reported by Panebra et al. who studied eight common haplotypes based on 26 SNPs, and identified differential effects on receptor expression and down-regulation that are haplotype driven [16]. This identification that four common haplotypes are related to elevated receptor expression and two haplotypes are related to enhanced receptor down-regulation, suggests that haplotypes rather than single SNPs may be the driving force between these differential effects [6].

The lack of agreement between large prospective clinical studies which have identified no pharmacogenetic effects of the common Arg16Gly and Gln27Glu *ADRB2* polymorphisms, and *in vitro* molecular research which commonly identified receptor signalling effects, questions the validity of the direct translation of *in vitro* functional work to what happens in a clinical setting. Current evidence points to haplotype rather than individual SNP-driven effects. Future clinical research in this area should perhaps focus on carefully designed studies, which are suitably powered to enable identification of both haplotype-driven and SNP-driven pharmacogenetic effects where these actually exist.

4 Novel Regulators of β_2-adrenoceptor Agonist Responses

As the mechanism of action of β_2-adrenoceptor agonists is both complex and multifactorial, further genes are expected to influence β_2-adrenoceptor agonist efficacy. Indeed the estimated heritability for β_2-adrenoceptor agonist response is thought to be 28.5 %, a value too high to be driven by a single gene [17]. Using a novel algorithm implemented in a family-based association test (FBAT), Litonjua et al. screened the association of 844 genotyped SNPs in 111 candidate genes (42 involved in β_2-adrenoceptor signalling/regulation, 28 genes involved in glucocorticoid regulation, 41 genes from prior asthma association studies) in 209 children and their parents participating in the Childhood Asthma Management Program (CAMP) for their association with acute response to inhaled β_2-adrenoceptor agonist. This study identified the Arginase 1 (*ARG1*) SNP rs2781659 as being significantly associated with bronchodilator response (BDR) (p=0.047) [18]. Data from guinea pig models of allergic airway disease suggests that Arginase enzymes deplete stores of L-arginine in the airways, a nitric oxide synthase substrate leading to decreased nitric oxide which normally acts to relax smooth muscle resulting in airway hyper-responsiveness [19, 20]. The relationship between *ARG1* and β_2-adrenoceptor agonist response was also identified in a recent candidate gene study involving 221 asthma subjects [21]. The *ARG1* polymorphisms identified in both studies were in linkage disequilibrium (inherited together) suggesting a common causative mechanism involving potential transcriptional regulation as these polymorphisms were predominantly 5' to the gene. This alteration in transcription has now been confirmed in promoter-reporter studies with the key *ARG1* haplotype associated with improved

BDR driving the highest level of *ARG1* promoter activity [22]. Interestingly, in the study by Vonk et al., *ARG2* SNPs were also associated with patient responses to Salbutamol [21].

The S-nitrosoglutathione reductase (*GSNOR*) gene has recently been associated with β_2-adrenoceptor agonist (salbutamol) responses. Specifically, the *GSNOR* promoter SNP rs1154400, was associated with a decreased response to salbutamol in 107 African American children [23]. *GSNOR* is an alcohol dehydrogenase that breaks down GSNO, an endogenous bronchodilator [24]. In addition, GSNO regulates nitrosylation of proteins leading to alterations in function, including G protein–coupled receptor kinase 2 (GRK2) which phosphorylates and desensitises the β_2-adrenoceptor [25]. Within the same study, a post-hoc multi-locus analysis identified that a combination of rs1154400 with the *ADRB2* SNPs Arg16Gly and Gly27Glu and rs2230739, a SNP associated with the carbamoyl phosphate synthetase-1 (*CPS1*) gene, give a 70% predictive value for therapeutic unresponsiveness [23]. This implies that pharmacogenetic regulation of β_2-adrenoceptor agonist therapy may depend on several loci acting together via gene-gene interactions. In confirmation 4/5 SNPs tested within *GSNOR* were associated with asthma patient responses to salbutamol in 168 Puerto Rican asthma patients [26]. These SNPs were also associated with asthma susceptibility and the key risk haplotype was associated with increased transcriptional activity based on promoter-reporter studies [26]. The identification of *GSNOR* in two racially distinct populations underlines the importance of cross population studies to identify common regulators of β_2-adrenoceptor agonist activity, where different haplotypes of a gene may have a population dependant role on drug function, efficacy and side-effects.

A recent study by Himes et al. using RNA-seq, a cutting edge technology able to detect differential gene expression at the transcript level in a cell population, has identified the up-regulation of the cysteine-rich secretory protein LCCL domain containing 2 gene (*CRISPLD2*) on stimulation of a bronchial epithelial cell line with a long acting β_2-adrenoceptor agonist (formoterol) [27]. Although this association has not been replicated *in vivo*, we highlight this study as an example of potential novel genetic associations that remain to be discovered through developing laboratory based approaches. Although our understanding of the pharmacogenetics of β_2-adrenoceptor agonist response currently remains fragmented, a combination of laboratory based applications and larger better stratified population studies will help us to better understand this relationship.

5 Leukotriene Modifier Drugs

Leukotrienes were first recognised to be important modulators of airway constriction, by Kellaway and Trethewie, who in 1940 reported the release of a "slow-reacting smooth muscle-stimulating substance" from perfused guinea pig lung, following sensitisation with egg albumin [28]. This was later termed SRS-A (slow reacting substance of anaphylaxis) by Brocklehurst in 1960 and was eventually recognised to consist of a combination of the cysteinyl leukotrienes LTC_4, LTD_4 and LTE_4 [29]. Leukotrienes are generated via the 5-lipoxygenase arm of the arachidonic acid metabolism pathway, and they exert their main pro-inflammatory and bronchoconstrictive actions through their interaction with the G-protein coupled receptors $CysLTR_1$, and $CysLTR_2$. The dihydroxy-containing leukotriene LTB_4 acts via the $LTB4R_1$ and $LTB4R_2$ receptors, though there is evidence to suggest that these receptors may also exhibit some binding affinity for the cysteinyl-containing ligands. In the human lung, leukotriene receptors are primarily expressed on airway smooth muscle but *CYSLTR1* and *CYSLTR2* gene transcripts have also been identified in circulating immune cells including eosinophils, monocytes, T-cells and B-cells, where cytokines such as IL-4 and IFN-γ have been shown to upregulate their expression [30].

There are two classes of FDA approved therapeutic agents which target the leukotriene pathway; the leukotriene receptor antagonists (LTRA) zafirlukast, montelukast, pranlukast and cinalukast, which have primary activity against $CysLTR_1$, and the leukotriene synthesis inhibitors (LTSI), the main member being the 5-lipoxygenase inhibitor, zileuton. A third class, the 5-lipoxygenase activating protein (FLAP) inhibitors have offered promising benefits, but none of these agents are yet FDA-approved. More recent data has suggested that drug-induced inhibition of the leukotriene A_4 hydrolase (LTA_4H) enzyme, which is responsible for the synthesis of LTB_4 from LTA_4, may offer therapeutic benefit against the inflammatory and bronchoconstrictory effects of LTB_4-mediated pathways [17]. Indeed, earlier studies have shown some *LTA4H* gene polymorphisms to be associated with susceptibility to allergy, asthma and Chronic Obstructive Pulmonary Disease (COPD), as well as to contribute to the baseline lung function of smokers and the serum IgE levels of asthmatic patients [31–33]. In addition, specific *LTA4H* polymorphisms have been associated with LTRA treatment outcomes (Table 1). However information on whether *LTA4H* gene polymorphisms may also modify the therapeutic outcome of LTA4 hydrolase inhibitors, remains to be determined.

Functional leukotriene-related pharmacogenetic effects, may be contributed by polymorphic variation in genes regulating various steps within the pathway involved in leukotriene production or receptors that transduce cellular effects. Table 1 lists some examples of functional leukotriene-related genetic variants.

A well-studied Sp1-motif repeat polymorphism in the 5-lipoxygenase (*ALOX5)* gene promoter, is reported to cause a significant reduction in the expression of 5-lipoxygenase, in carriers of variants other than the wild-type penta-repeat. This results in a lower leukotriene-contributed inflammatory component in asthmatic patients carrying non-wild type Sp1-repeats. Consequently, drugs targeting this pathway show a reduced response in patients who are homozygous or heterozygous for Sp1-repeat variants. This was initially demonstrated with the LTSI ABT-761 [34], a 5-lipoxygenase inhibitor which halted development after Phase III clinical trials [45], and later also demonstrated with the LTRA montelukast [36]. Indeed, *ALOX5* promoter Sp1 repeats other than the wild type $(Sp1)_5$, associate with overall worse asthma control [46]. An earlier study contrastingly reported non-wild type Sp1 repeat polymorphisms to be *beneficial*, in terms of an increased reduction in the number of asthma exacerbations in montelukast-treated patients [35], while a

Table 1 A list of some pharmacogenetically relevant genes which influence treatment outcomes of drugs targeting the 5-lipoxygenase pathway

Gene	Locus	Allele	Function	Reference
ALOX5	10q112	Sp1 binding motif repeats other than $[Sp1]_5$	Loss of effect of LTSI, ABT-761	[34]
		Sp1 binding motif repeats other than $[Sp1]_5$	Reduction in asthma exacerbations in patients treated with montelukast for 6 months	[35]
		$[Sp1]_5$	Improved montelukast response in asthmatic patients with at least one $[Sp1]_5$ allele	[36]
		rs2115819	'G' allele associated with improved FEV_1 in patients treated with montelukast for 6 months	[35]
		rs4987105	Improved PEF in asthmatic patients treated with montelukast for 12 weeks	[37]
		rs4986832	Improved PEF in asthmatic patients treated with montelukast for 12 weeks	[37]
		rs2115819	Reduced FEV_1 response to zileuton and montelukast in asthmatic patients	[38]
LTC4S	5q35	−444 A>C rs730012	Reduced risk of asthma exacerbations in patients treated with montelukast for 6 months	[35]
		−444 A>C rs730012	Improved FEV_1 following 1 month treatment with pranlukast	[39]
LTA4H	12q22	rs2660845	'G' allele increases risk of asthma exacerbations in patients treated with montelukast for 6 months	[35]
		rs2540491 (Intron 3, G/A)	'A' allele associated with augmented FEV_1 response to salbutamol in patients who were also treated with one of montelukast, zafirlukast or zileuton	[40]
CYSLTR1	Xq13-q21	−642 A>G	'G' alleles causes lower CysLTR1 expression in transfected THP1 cells and may therefore potentially reduce efficacy of LTRAs	[30]
		−634 C>T	Higher allelic frequency corresponds to higher montelukast dose requirements in aspirin intolerant asthmatics	[41]

Table 1 (continued)

Gene	Locus	Allele	Function	Reference
CYSLTR2	13q14.2	rs912277	Improved PEF in asthmatic patients treated with montelukast for 12 weeks	[37]
		rs912278	Improved PEF in asthmatic patients treated with montelukast for 12 weeks	[37]
		M201V	Loss of LTD_4 binding affinity and LTD_4-mediated signalling *in vitro*	[42]
Haplotypes				
LTC4S	5q35	−1072G/A −444A/C	G-A, G-C and A-C haplotypes (but not A-A) increase dexamethasone-induced LT4CS transcription *in vitro*	[43]
Genes not directly involved in the lipoxygenase pathway				
SLCO2B1	11q13	rs12422149 (Arg312Gln)	Heterozygotes show lower plasma montelukast concentrations than similarly dosed homozygotes	[44]
ABCC1 (MRP1)	16p13.1	rs119774	'T' allele associated with improved FEV_1 in patients treated with montelukast for 6 months	[38]

recent study by Nwokoro and co-workers, reported less unscheduled medical attendences in montelukast-treated children carrying the wild type Sp1-pentarepeat than children carrying other Sp1 repeat variants [47].

ALOX5 SNPs may also carry pharmacogenetic relevance. For example, Tantisira and co-workers associated the *ALOX5* intronic SNP rs2115819 with a reduction in response to both zileuton and montelukast treatment in asthmatic patients, using FEV_1 as a clinical endpoint [38], while *ALOX5* rs4987105 and rs4986832 SNPs have both been associated with improved montelukast treatment outcomes [37]. However, the mechanisms underlying these associations remain to be determined.

The tetra-locus haplotype [−1708G-21C-270G-1728A] in the *ALOX5* gene is associated with aspirin sensitivity to asthma [48]. A potential explanation for aspirin-induced asthma exacerbations, argues for an already potentiated leukotriene production pathway, which, in the presence of aspirin, or another non-steroidal anti-inflammatory drug (NSAID) is loaded with excess arachidonic acid, which would otherwise be shunted down the cyclooxygenase pathway in an NSAID-free environment. The arachidonic acid loading generates a higher production of bronchoconstricting leukotrienes, often resulting in exacerbation of asthma symptoms. The association of the aspirin-sensitive phenotype with genetic variation has valuable clinical applications, as it could potentially discriminate patients in whom NSAID contraindications are warranted from those in whom NSAID use is permissible.

The CysLTR1 and CysLTR2 receptors are the targets through which cysteinyl leukotrienes act with CysLTR1 providing the major binding affinity for LTRAs. The genes for these receptors may therefore be considered to be strong candidates for functional pharmacogenetic variability. However it has been difficult to identify clear associations between clinical therapeutic responses and genetic polymorphisms in these genes. Two particular *CYSLTR1* variants, G300S and I206S, have been reported to show increased LTD_4 potency, and may therefore potentially also exhibit altered potency for other ligands, including LTRAs [49]. In addition, Duroudier et al. (2009) found the upstream *CYSLTR1* −642A>G SNP to be associated with reduced gene expression in transiently transfected THP1 cells, therefore potentially contributing to reduced LTRA response *in vivo* due to reduced expression of the drug target [30]. Kim et al. (2007) similarly found high allelic frequencies of another upstream *CYSLTR1* SNP −634C>T, to correspond to higher montelukast dose requirements in aspirin intolerant asthmatic patients [41]. Some *CYSLTR2* coding sequence SNPs have been associated with the presence of aspirin intolerance in asthma, and haplotype analysis further consolidated this observation [50]. The presence of rs912277 or rs912278 has been reported to result in improved response to montelukast [37], suggesting a potentially modified interaction between the CysLTR2 receptor and LTRAs in patients carrying these genotypes. The CYSLTR2 M201V receptor variant, has been shown to exhibit minimal affinity for LTD_4 in HEK293 stable transfectants, while functional analysis of LTD_4-stimulated CYSLTR2 receptor activation, showed downstream signalling endpoints to be extensively reduced for the expressed variant receptor in this cell line. LTC_4 showed an approximately 50% reduction in binding affinity to the M201V variant, and LTC_4-mediated downstream signalling was reduced at low concentrations, but appeared to be restored at higher LTC_4 levels [42]. Interestingly, this variant had also been associated with atopy [51] and asthma [52] in earlier studies.

Genes not directly involved in cysteinyl leukotriene synthesis or actions may also exert a pharmacogenetic influence (Table 1). For example, the rs2660845 SNP in *LTA4H* which codes for leukotriene A_4 hydrolase, has been reported to exert a negative influence on LTRA treatment outcomes. LTA_4 hydrolase is responsible for the synthesis of LTB_4, a dihydroxy leukotriene, with no affinity for the cysteinyl leukotriene receptors [35]. In addition, genes may also influence LTRA or LTSI treatment outcomes, by directly interfering with pharmacokinetic processes. For example the rs119774 SNP in the multidrug resistance associated protein 1 (*ABCC1*) gene has been associated with an improved FEV_1-based clinical response to both zileuton and montelukast in asthmatic patients [38]. *ABCC1* is involved in the transport of LTC_4 from the cell. In contrast, the membrane transporter gene *SLCO2B1* rs12422149 SNP (Arg312Gln) has been associated with reduced morning plasma montelukast concentrations in asthmatic patients and improved symptom assessment changes [44]. However, not all studies have been able to replicate these findings.

6 Glucocorticoids

Inhaled glucocorticoids (GC) are commonly used in the management of asthma. The molecular actions of glucocorticoids are multifaceted and complex, involving the simultaneous activation and/or repression of many genes. This has often made it difficult to establish clear genotype-phenotype associations. Moreover, although the human genome only contains one known glucocorticoid receptor (GR) gene (*NR3C1* located at cytogenetic band 5q31.3), alternative splicing and alternative translational initiation events give rise to separate receptor isoforms having different functional profiles [53]. There is also recent evidence for a membrane-bound GR isoform, possible expressed by the same gene, and which mediates rapid-onset non-genomic actions of glucocorticoids [54]. Furthermore, glucocorticoid receptor (GR) isoforms may also be subject to a variety of posttranslational modifications which result in altered receptor function.

The major pharmacogenetic issues associated with glucocorticoid therapy relate to patients who fail to respond beneficially to their therapeutic potential. These are often referred to as steroid-resistant individuals. Two forms of glucocorticoid resistance have been described. Type I is an induced type, and usually the result of cytokine-induced modification of glucocorticoid activity on T-lymphocytes, such as IL-2 and IL-4 in T-cells [55], IL-13 in monocytes [56], TNFα+IFNγ in airway smooth muscle [57] and IL-17 in airway epithelial cells [58–60]. This type of resistance may be reversed through cytokine inhibition approaches. Type II, the pharmacogenetic form of glucocorticoid resistance, is irreversible and may be due to genetic variation in the glucocorticoid receptor gene itself, or other genes which contribute to the functional pathways that relate to glucocorticoid receptor activation and function [61, 62].

The *NR3C1* gene is highly functionally conserved, and functional polymorphisms identified to date are rare, with minor allelic frequencies in various populations commonly falling below 3 % [63]. The functional outcomes of these polymorphisms are strongly dependent on the specific receptor isoform in which they are expressed, and the pharmacogenetic phenotype may be the result of the combined effects of the same SNP on different isoforms. Expression of these isoforms is driven by the selective use of at least 5 recognised *NR3C1* promoters [64], and polymorphic variation in these promoters may determine the respective transcript expression level. For example the −22C>A polymorphism located upstream of the *NR3C1* gene significantly decreases the promoter transcriptional activity compared to the wild type C allele as determined by promoter luciferase reporter assays in HepG2 and HEK293 cells, and is likely to be related to lower GRα expression in the clinical setting [65]. Similarly, the G allele of the *Bcl*I GR promoter polymorphism is significantly associated with GC resistance and with the development of severe asthma [66]. Moreover, a second layer of complexity in selection of *NR3C1* isoform expression arises from the epigenetic-driven methylation of the different promoters, occurring in a tissue-dependent fashion [67]. This may partially explain the difficulty of corroborating *in vitro* functional effects of *NR3C1* promoter polymorphisms with *in vivo* clinical outcomes.

The two major GR isoforms which contribute to GC treatment outcomes are the α and β gene products. GRα, is a 777 amino acid protein which is the functional receptor responsible for the majority of glucocorticoid actions. The second isoform, GRβ, is an alternatively spliced product which carries the first identical 727 amino acids as GRα. It has a C-terminal sequence which is both shorter and non-homologous to the C-terminal sequence of GRα, and this renders it unable to bind ligand or be transcriptionally active. Both isoforms are ubiquitously expressed in human cells, with GRβ normally exhibiting lower expression levels [68–70]. Table 2 highlights some polymorphisms which are related to glucocorticoid response.

6.1 Polymorphisms Influencing GRα Function

The dbSNP database (http://www.ncbi.nlm.nih.gov/SNP/) lists more than 18000 human *NR3C1* SNPs, located throughout the 157 kilobases of the whole gene. However, most of these are localised to intronic regions, and exert no known influence on the expressed receptor protein or on its production. About 39 *NR3C1* coding region missense SNPs have been recognised to date. Though several associations with disease severity phenotypes have been reported, the *NR3C1* polymorphisms which are *pharmacogenetically relevant*, only represent a small fraction of the total SNP pool. Moreover, clinically relevant *NR3C1* variants which influence therapeutic outcomes, tend to occur at low allelic frequencies in all populations, though functional *in vitro* work occasionally suggests that more common variants may also contribute to glucocorticoid responses. Niu and co-workers identified 108 polymorphisms within exonic, intronic and untranslated regions of the *NR3C1* gene, in a sample of 240 individuals. *In vitro* functional analysis identified Phe(65)Val and Asp(687)Glu to be associated with higher hGRα protein expression. SNP 746T>C and haplotype 237delC / 238C>T/240G>C were both associated with reduced transcript expression, and Ala(229)Thr and Ile(292)Val was found to be associated with decreased ligand binding. These functional polymorphisms, present in allelic frequencies ranging from 5.8% to 18.3% in different ethnic groups, are potential contributors to reduced glucocorticoid sensitivity in a clinical setting [73].

The presence of variants AG or GG of codon 363 in exon 2 of the *NR3C1* gene has been reported to correlate with better *in vivo* anti-inflammatory effects of glucocorticoid therapy and reduced risk of development of uncontrolled asthma in a study carried out using a sample from the Polish population. The G allele causes a change of asparagine to serine, and it has been suggested that this may modify interactions of the receptor with the transcriptional factors, AP-1 and NF-κB [76]. However, earlier work by Huizenga's workgroup could not identify any alteration in the repressive capacity of the *NR3C1* N363S variant on target genes via negative glucocorticoid-responsive elements or via AP-1 and NF-κB [87]. Though this polymorphism is present at allelic frequencies of 3–7% in Caucasian and African populations, studies in the Swedish population had earlier failed to identify these genotype-phenotype associations [88]. Furthermore, several studies have reported this N363S polymorphism to be either absent, or otherwise occur at very low allelic frequencies (<0.5%) in subjects of various Asian origins [63, 89–91].

Table 2 Pharmacogenetically relevant genes which may be related to glucocorticoid actions

Gene	Locus	Allele	Function	Reference
NR3C1	5q31-q32	Val>Asp641	Threefold lower receptor binding affinity for dexamethasone	[71]
		Val>Ile729	Fourfold decrease in dexamethasone response	[72]
		Phe>Val65	Increased hGRα protein expression *in vitro*	[73]
		Asp>Glu687	Increased hGRα protein expression *in vitro*	[73]
		746T>C	Reduced hGRα transcript expression *in vitro*	[73]
		237delC 238 C>T 240G>C [haplotype]	Reduced hGRα transcript expression *in vitro*	[73]
		Ala>Thr229	Decreased ligand binding affinity	[73]
		Ile>Val292	Decreased ligand binding affinity	[73]
		*Bcl*I RFLP	Higher cortisol levels in GC than CC genotypes, both pre- and post-ICS treatment	[74]
		*Bcl*I RFLP	Increased *in vitro* sensitivity to methylprednisolone	[75]
		*Bcl*I RFLP	'G' allele associated with *in vivo* glucocorticoid resistance and severe asthma	[66]
		Asn>Ser363	Increased sensitivity to GCs	[76]
		−22 C>A	Decreased *in vitro* promoter transcriptional activity	[65]
		G-A-T haplotype (consisting of BclI SNP G, intron B 33389A, intron B 33388T)	Low post-dexamethasone cortisol levels	[77]
CRHR1	17q12-q22	rs1876828, rs242941	Delayed cortisol response, following inhaled glucocorticoid treatment	[74]
TBX21	17q21.2	His>Glu33	Larger improvements in methacholine-induced PC_{20} following treatment with inhaled glucocorticoids	[78]
		rs9910408	Larger improvements in methacholine-induced PD_{20} in asthmatics carrying the AA genotype, following treatment with inhaled glucocorticoids	[79]
ABCB1	7q21.12	G2677T	Reduced glucocorticoid response	[80]
		G2995A	Reduced glucocorticoid response	[80]
		C3435T	Increased glucocorticoid response	[81]

Table 2 (continued)

Gene	Locus	Allele	Function	Refer-ence
TACR2 (*NK2R*)	10q22.1	Gly231Glu	Gly allele associated with improved glucocorticoid response	[82]
STIP1	11q13.1	rs4980524 (G/T) (intron 1)	Altered lung function responses to glucocorticoids	[83]
		rs6591838 (A/G), (intron 1)	Altered lung function responses to glucocorticoids	[83]
		rs2236647 (C/T), (intron 5)	Altered lung function responses to glucocorticoids	[83]
DUSP1	5q35.1	rs881152 (A/G), (5' region)	Associated with bronchodilator response in asthma	[84]
FCER2	19p13.2	rs28364072 (T/C), (intronic)	'C' allele associated with lower FcεRII expression, higher serum IgE and higher risk of asthma exacerbations during budesonide therapy	[85]
GLCCI1	7p21.3	rs37972 (C/T), (5' region)	'C' allele associated with improved glucocorticoid response	[86]

6.2 Polymorphisms Influencing GRβ Function

High expression levels of GRβ are strongly associated with glucocorticoid resistance. The early evidence to support this comes both from clinical observations [92] as well as from *in vitro* overexpression experiments in cell line models [93, 94], but the mechanisms for this action are still widely debated today. In contrast to GRα, the β-isoform interacts poorly with heat shock proteins, does not bind ligands, and is transcriptionally inactive. GRβ-induced glucocorticoid resistance is likely to occur via competition with GRα for binding to glucocorticoid responsive elements (GREs) on gene promoters, via direct GRα inactivation through heterodimerization, and by the inhibition of co-activating proteins which are necessary for GRα activity. Recent evidence suggests that GRβ competes with GRα for binding to glucocorticoid receptor-interacting protein 1 (also called nuclear receptor coactivator 2 and coded for by the *NCOA2* gene), and generates an ineffective co-activator complex. Although individuals having a high GRβ/GRα expression ratio appear to carry this as a lifelong phenotypic trait, the identification of specific gene variants which strongly associate with this aberrant ratio has been elusive. Current evidence suggests that high GRβ/GRα seems to be more related to disease-specific factors, in particular specific cytokine profiles, which may occur in diseases such as rheumatoid arthritis, asthma, and cancer and autoimmune diseases [95, 96]. It has been suggested that high GRβ may be a consequence of a patient-specific disease profile, rather than a contributor to the cause. The associated glucocorticoid insensitivity further complicates the situation, by hindering our therapeutic ability to reverse the causative inflammatory processes. For example recent evidence shows that cytokines IL-17 and IL-23 released by infiltrating T-cells in asthma, contribute to GRβ upregulation and consequently to a degree of glucocorticoid insensitivity [97].

6.3 *Non-NR3C1 Polymorphisms Which Influence Glucocorticoid Response*

The multiplicity of GR-mediated pathways, lend its actions to be potentially influenced by variability in several other genes. However, the functional contribution of such variability is often difficult to predict. For example, allelic variations in *CNTNAP2, LEPR, CRHR1, NTAN1, SLC12A3, ALPL, BGLAP*, and *APOB* have all been associated with prednisolone-induced hypertension in specific patients [98]. Most of these genes are involved in the hypothalamic-pituitary axis pathway. Polymorphic variations in genes that code for lung esterases are interesting candidates for the inhaled prodrug glucocorticoids beclomethasone and ciclesonide, since both are activated in the airways by esterase action. Beclomethasone diproprionate is metabolised to the more active beclomethasone-17-monopropionate, while ciclesonide, an inactive prodrug, is metabolised in airway epithelial cells to the pharmacologically active desisobutyryl-ciclesonide [99]. However, although genetic variability in various esterase genes has been described [100, 101], pharmacogenetic evidence for their clinical contribution to bioactivation-associated pharmacology is lacking. Some pharmacogenetically relevant polymorphism are described below.

6.4 *ATP-Binding Cassette, Sub-Family B Member 1* (ABCB1) *Gene*

The membrane transporter PgP is coded for by *ABCB1* (previously known as MDR1), a 28-exon gene located on 7q21.1. It is a cellular efflux transporter which operates on a large range of substrates including prednisolone, dexamethasone and beclomethasone monopropionate (the active metabolite of the pulmonary administered beclomethasone dipropionate). *ABCB1* overexpression is of concern, since increased Pgp efflux activity contributes to reduced therapeutic outcome of the affected drugs. Examples of high-expression-inducing *ABCB1* variants include G2677T and G2995A [80]. Conversely, low expression variants such as C3435T [81] may contribute to increased glucocorticoid response due to intracellular accumulation of the drug. There is also evidence to show that glucocorticoids upregulate Pgp expression, and may therefore amplify the effects of *ABCB1* allele-specific high activity variants [102].

6.5 *Corticotropin Releasing Hormone Receptor Type 1* (CRHR1) *Gene*

The CRHR type 1 receptor is the major corticotropin receptor, and therefore has a key regulatory role in endogenous corticosteroid synthesis. This gene, located on 17q12–22, exhibits SNPs which have been shown to be associated with cortisol response in humans. For example, Tsartsali's workgroup carried out Synacthen testing on male asthmatic children, treated with inhaled budesonide (400 μg/day) or fluticasone (200 μg/day) for 3 months, and genotyped the patients for *CRHR1* variants. Homozygotes for the *CRHR1* variant rs242941 (TT) demonstrated a delayed cortisol response after treatment when compared to heterozygotes (GT) ($p=0.033$) and wild-type homozygotes (GG) ($p=0.018$). Homozygotes (AA) for a second *CRHR1* SNP, rs1876828, manifested lower baseline cortisol levels before treatment ($p=0.009$) compared to the GG genotype. These patients also showed marginal statistical significance for delayed cortisol response after treatment compared to the GA heterozygotes ($p=0.05$) [74].

6.6 *T-Box Expressed in T Cells* (TBX21) *Gene* and Other Genes

TBX21 encodes the T-bet transcription factor and contributes to an enhanced Th1 lymphocyte population. *TBX21*-knockout mice have been shown to develop airway eosinophilia, bronchial hyperresponsiveness and enhanced airway remodelling, thus suggesting that this gene to have a protective role in asthma [103]. An SNP causing a His>Glu33 substitution has been identified to activate Th1 cytokine synthesis in cell culture models, and this is postulated to protect against allergy and asthma by shifting lymphocyte proliferation away from the Th_2 development. Indeed, both heterozygous and homozygous Glu33 patients have been reported to show a greater reduction in PC_{20}-measured airway hyperresponsiveness, than His33 homozygotes. Unexpectedly, it has been shown that glucocorticoids inhibit T-bet induction, possibly contributing to a shift *towards* a Th_2-based response rather than against [78]. The precise mechanistic contribution of His>Glu33 in this scenario remains to be explained. A second *TBX21* polymorphism, rs9910408, has been associated with increased response to inhaled glucocorticoids in adult asthmatic patients. Work carried out by Lopert's group showed that patients carrying the AA genotype, exhibit better reductions in bronchial hyperresponsiveness, higher improvements in FEV_1, and better quality of life scores, than similarly ICS-treated GG or AG patients [79].

Polymorphisms in the neurokinin-2 receptor gene (*TACR2*–previously known as *NK2R*) influence glucocorticoid-mediated therapeutic outcome, possibly by modifying neurokinin-induced inflammation and bronchoconstriction [82]; intronic genetic variants of the stress-induced phosphoprotein-1 gene (*STIP1*) influence lung function responses to glucocorticoids by mechanisms which are unclear, and which may involve interactions with Hsp70 chaperones, thus influencing GR function [83]. *DUSP1*, a gene that codes for a p38 MAPK (mitogen-activated protein kinase) inhibitor, the expression of which is thought to be influenced by glucocorticoid-mediated regulation, influences glucocorticoid response possibly by polymorphism-dependent alterations in its ability to target the p38 MAPK signalling pathway [84]. Interestingly, polymorphic variation in the low-affinity IgE receptor (FcεRII) has been shown to increase the risk of asthma exacerbations in children on budesonide therapy, possibly due to an associated increase in serum IgE concentrations. These findings have survived replication studies in separate asthma cohorts [85]. The *GLCCI1* (glucocorticoid-induced transcript 1) rs37972 SNP ('C' allele) was found to be associated with improved glucocorticoid response in asthmatic patients, through a genome-wide association study (GWAS) carried out by Tantisira's workgroup [86]. Since rs37972 is localised to the *GLCCI1* promoter, it may be postulated that the mechanism involves polymorphism-induced variations in the transcriptional regulation of the *GLCCI1* gene. However, the precise mechanism by which this influences glucocorticoid response is unclear.

The elucidation of clear glucocorticoid pharmacogenetic genotype-phenotype associations remains a challenge. The complex molecular network which is involved in the outcomes of glucocorticoid receptor activation, suggests that the search for pharmacogenetic variants would benefit from the use of hypothesis-free genome wide approaches. The study of multiple polymorphisms spread over a gene network may be more relevant than attempts to identify single responsible variants. Moreover, age may also influence pharmacogenetically-induced phenotypic traits. Park and co-workers have recently found rs10044254, an *FBXL7* (F-box and leucine-rich repeat protein 7) gene A>G intronic SNP, to be associated with improved response to inhaled glucocorticoids in the paediatric asthmatic population, but not in adult patients [104]. *FBXL7* codes for an F-box protein which plays a role in the phosphorylation-dependent ubiquitination of proteins.

7 Anti-Muscarinic Agents

Anti-muscarinic drugs are commonly used in the management of COPD. However they may also be used in severe asthma as second line bronchodilators, in combination with an adrenergic β_2-agonist, in those patients who do not adequately respond to β_2-agonist-induced bronchodilation alone, or who demonstrate a reduced degree of airway reversibility.

Two main anti-muscarinic agents are used to manage airway disease; ipratropium and tiotropium. Both drugs are administered by inhalation, and are pharmacologically non-selective antagonists of airway muscarinic receptors M1, M2 and M3; however receptor-bound tiotropium exhibits a much slower rate of dissociation from M3 receptors, than it does from M1 and M2, giving it long acting and kinetically M3 selective properties [105].

The coding regions of these receptor genes tend to be highly conserved with only low allelic frequency synonymous SNPs being reported (Table 3). However, a common polymorphism (1696T>A) reported by Fenech and co-workers [106] in the 3'UTR of the M2 gene, was later found by Szczepankiewicz's group to be associated with poor bronchodilator response to ipratropium bromide in paediatric asthmatic patients homozygous for the T allele [107].

The regulation of expression of M2 and M3 receptors is highly promoter-dependent, and polymorphisms have been demonstrated in the promoters of both genes [108, 109]. Indeed, 7 dinucleotide $(CA)_n$ repeat polymorphisms [(insertion $(CA)_1$, wild type $(CA)_{14}$, $\Delta(CA)_1$, $\Delta(CA)_2$, $\Delta(CA)_3$, $\Delta(CA)_4$, $\Delta(CA)_8$] have been identified in a high activity region of the *CHRM2* promoter (Table 3), and *in vitro* luciferase reporter assays carried out in human airway smooth muscle (HASM) and airway epithelial (BEAS-2B) cell transfectants have reported repeat-dependent transcriptional activities. These repeats may therefore potentially alter muscarinic M2 receptor expression *in vivo*. Since airway muscarinic M2 receptors are primarily neurologically expressed and exhibit autoreceptor function, a decrease in expression may contribute to increased airway parasympathetic activity, and thus reduce the response to antimuscarinic drugs.

Maeda reported two *CHRM1* promoter polymorphisms (−9697C>T and −4953A>G) to be associated with asthma susceptibility, with the T-A haplotype showing a significantly lower promoter activity in luciferase reporter assays carried out using IMR32 neuronal transfectants ($p=0.019$) [110]. Such SNPs could be expected to have some degree of pharmacogenetic importance *in vivo,* since the major receptors accessible to anti-muscarinic agents administered by the inhaled route, tend to be M2 and M3 rather than M1.

8 Pharmacogenetic Aspects of Disease Management: *ADRB2* Arg16 and Second Line Therapy in Asthma

While the study of genetic contributions to drug responses have been predominantly in the trial setting and focused to acute responses to medication e.g. change in lung function, a more recent study set out to look in a real life situation where genetic information could influence management of disease over a prolonged period [112]. First line treatment for asthma is a short acting β_2-adrenoceptor agonist e.g. salbutamol as required (step 1); if symptoms persist the addition of an inhaled glucocorticoid e.g. beclomethasone is considered (step 2), and for additional control, the addition of a long acting β_2-adrenoceptor agonist e.g. salmeterol or a leukotriene receptor antagonist (LTRA) e.g. montelukast (step 3) may be used. In this context, Lipworth and colleagues investigated whether genotype information at the *ADRB2* Arg16 locus is useful in order to inform the choice of prescribing add on therapy, i.e. salmeterol or montelukast in a community setting [112]. Children with persistent asthma and homozygous for the Arg16 genotype ($n=62$) were randomised to receive salmeterol (50 μg bd) or montelukast (5 or 10 mg once daily) as an add-on to inhaled fluticasone propionate for 1 year. The aim was to test the hypothesis that carriers of the Arg16 genotype may be more prone to adverse effects e.g. exacerbation associated with prolonged β_2-adrenoceptor agonist as observed in several large scale studies. Using this rationale the montelukast group should have superior control for this preselected population. The study outcomes were school absences (primary outcome), exacerbation score, reliever (salbutamol) use, morning dyspnoea

Table 3 Polymorphisms of potential pharmacogenetic importance, identified in muscarinic receptor genes. Although several polymorphisms have been identified in these genes, most clinical associations have been with disease related rather than with pharmacological response outcomes. The values in brackets provide the reported allelic frequency, where available

Gene	Locus	Allele	Function	Reference
CHRM1	11q12–q13	−9697 C>T	Associated with asthma	[110]
		−4953A>G	Associated with asthma	[110]
CHRM2	7q35–q36	1197T>C *Coding region, synonymous (ACT>ACC, Thr>Thr) (0.5 %)*	No known pharmacogenetic relevance	[106]
		1696T>A 3' *UTR* (63–68 %)	Paediatric TT homozygote asthmatic patients exhibit poor response to ipratropium bromide	[106, 107]
		976 A>C *Coding region, synonymous (AGA>CGA, Arg>Arg) (0.5 %)*	No known pharmacogenetic relevance	[106]
		$[(CA)_{15}]$ (0.6 %) $[(CA)_{14}]$ (22.3 %) $\Delta(CA)_1$ (44.0 %) $\Delta(CA)_2$ (0.6 %) $\Delta(CA)_3$ (14.9 %) $\Delta(CA)_4$ (0.6 %) $\Delta(CA)_8$ (17.1 %) *$(CA)_n$ located ~96 bp downstream of the most 5' transcriptional start site for CHRM2*	$\Delta(CA)_8$ results in significantly lower transcriptional activity of a reporter gene in BEAS-2B and HASM cells. This allele may potentially influence CHRM2 gene expression *in vivo*	[108]
		1050 A>G *Coding region, synonymous (TCA>TCG, Ser>Ser) (14–16 %)*	No known pharmacogenetic relevance	[111]
CHRM3	1q43	261 C>T *Coding region, synonymous (ATC>ATT, Ile>Ile) (0.5 %)*	No known pharmacogenetic relevance	[111]
		−708 A/G *Promoter*	No known pharmacogenetic relevance	[109]
		−627G/C *Promoter*	No known pharmacogenetic relevance	[109]
		−513 C/A *Promoter*	No known pharmacogenetic relevance	[109]
		−492 C/T *Promoter*	No known pharmacogenetic relevance	[109]
		$(CTTT)_{12-20}$ *Promoter*	No known pharmacogenetic relevance	[109]
		$(GT)_{6-19}$ *Promoter*	No known pharmacogenetic relevance	[109]

and quality of life scores which are in keeping with the more real life approach to the study. Montelukast provided superior benefit for all measures with clinically relevant differences being obtained within 3 months. No significant difference in lung function, FEV_1 (% predicted) were observed, adding to the accumulating data that there is a lack of correlation between lung function and symptom based scores [112]. These results generated in a community setting involving disease management decisions are encouraging, and suggest that larger and longer prospective studies are warranted to provide more definitive data regarding the clinical utility of Arg16 stratification. The addition of a Gly16 arm to the study and the potential use of additional markers to define the population, would potentially provide clearer interpretation of the study outcomes.

9 Asthma Pharmacogenetics and Ethnicity

Allelic frequencies are known to often differ substantially across different populations and ethnic groups who do not share a common ancestry. Since pharmacogenetic associations may be haplotype-driven rather than SNP-driven, and inter-ethnic differences may also be observed in gene penetrance, the overall pharmacogenetic genotype-phenotype relationship may vary substantially across ethnic groups. Moreover, even where the genotype-phenotype relationships are unaffected, the mere differences in allelic frequencies across these populations may influence the general effectiveness of a therapeutic agent within a specific ethnic group. Table 5 provides some examples of inter-ethnic group differences in allelic frequencies of known pharmacogenetically relevant loci. This may have implications when applying similar pharmacological treatment guidelines, because while the majority of one ancestry population may adequately respond to a specific drug, the same may not necessarily be said of a second population which happens to harbour higher gene frequencies of a particular drug response-inhibiting allele. Similarly, it may be difficult to replicate clinical pharmacogenetic genotype-phenotype association studies in separate ethnic groups, because the results of such studies may be influenced by substantially dissimilar allelic frequencies of related genes, especially in haplotype-driven phenotypes (Table 4).

Disease phenotypes may also show inter-ethnic differences, and these may in turn influence the management patterns. For example, Puerto Rican and Afro-American asthmatic patients show a higher morbidity and mortality than non-Hispanic white and Mexican American subjects [114]. Though the different health care systems may contribute to this disparity, the underlying cause is more likely to be due to different allelic profiles arising from different ancestral origins [113].

The lack of agreement between studies on the functional role of the two most commonly studied *ADRB2* coding region polymorphisms, Gly16Arg and Gln27Glu (described earlier in this chapter), may be partially attributed to the mixture of patient ethnicities recruited for the different studies, essentially generating outcomes which were difficult to replicate due to patient stratification. This may be further

Table 4 Examples of pharmacogenetic SNPs and their allelic frequencies in different ancestry groups[a] (CEU: Utah residents with ancestry from northern and western Europe; YRI: subjects from Yoruba in Ibadan, Nigeria; ASW: African American subjects from the southwestern United States; MEX: Mexican American subjects from Los Angeles, California; CHB: Han Chinese from Beijing, China; JPT: Japanese from Tokyo, Japan)

Gene	SNP	Major/minor allele	Minor allelic frequency					
			CEU	YRI	ASW	MEX	CHB	JPT
ADRB2	rs1042713 (Gly16Arg)	G/A	0.36	0.53	0.57	0.47	0.58	0.44
ALOX5	rs21158197	A/G	0.37	0.21	0.28	0.53	0.73	0.82
LTC4S	rs730012	A/C	0.32	0.007	0.06	0.22	0.14	0.18
LTA4H	rs266845	T/G	0.17	0.007	0.06	0.22	0.44	0.36
STIP1	rs2236647	C/T	0.43	0.76	0.71	0.6	0.43	0.35
GLCCI1	rs37972	C/T	0.45	0.14	0.29	0.46	0.43	0.38

[a] adapted from Ortega and Meyers (2014) [113]

appreciated, in the light of 49 *ADRB2* polymorphisms identified in a multi-ethnic population, some alleles of which may potentially operate in concert with the Gly16Arg and Gln27Glu variants under study. This generates various haplotypes with frequencies which may be specifically associated with particular groups. For example, the National Heart, Lung, and Blood Institute (NHLBI) BARGE (beta agonist response by genotype) trial, in line with earlier studies, reported that *ADRB2* Gly16 homozygotes demonstrate improved lung function and symptom control improvements during salbutamol administration that were not observable in Arg16 homozygous asthma patients [115]. In constrast, outcomes of the NHLBI's LARGE (long-acting beta agonist response by genotype) trial, showed that Arg16 and Gly16 homozygous patients both experienced similar PEFR improvements following therapy; however Gly16 homozygotes experienced a decrease in methacholine bronchial reactivity, while this was not observable in asthmatic Arg16 homozygotes participating in this study [15]. Deeper analysis of the LARGE study data, however, shows that within the subgroup of African American subjects, Gly16 homozygotes experienced significantly greater improvements in morning and evening peak flow rates with salmeterol than Arg16 homozygotes, although the numbers of subjects in these analyses were low [113].

The inclusion of ethnic groups as part of a wider sample population in clinical pharmacogenetic studies, without statistically addressing genetic diversities, may confound the study outcomes. Ortega and Meyers (2014) argue that in order for prospective pharmacogenetic studies to be carried out in ethnic minorities, the following issues should be addressed; (a) ethnic groups tend to represent subject minorities, thus making the design of suitable statistically powered studies difficult; (b) besides genetic diversity, ethnic grops may also differ in factors related to cultural or habitual issues, such as access to health care, compliance with medications, and different environmental exposures, such as cigarette smoke, (c) inter-ethnic altered allele frequencies may also contribute to ethnic-group specific gene-gene interactions, and genotype-phenotype relationships, and (d) the accuracy of self-reported ethnicity, especially in admixed populations, may have to be considered [113]. For example, a gene-gene interaction has been reported between rs11547772 in the *GSNOR* gene and Gly16Arg in *ADRB2*, which influences salbutamol bronchodilator response in Puerto Rican asthmatic families, while this interaction could not be observed in Mexican American patients [26]. In addition, there is often an increased frequency of rare gene variants in populations with African ancestry compared to non-Hispanic white populations [116].

Addressing pharmacogenetic issues in ethnic groups remains a challenge, especially within the context of large genome wide association studies (GWAS). The sample size requirements needed to achieve the necessary statistical power, imposes restrictions on study feasibility due to difficulties in being able to enrol sufficient patients from one genetically conserved group. On the other hand, incorporating these groups into larger sample cohorts may confound outcomes for reasons previously described. Ortega and Meyers (2014) suggest that one approach would be the use of admixture mapping techniques rather than direct GWAS, since the former method uses estimates of ancestry at each SNP to test for associations with a phenotype. In contrast, GWA studies directly compare allele frequencies with phenotype. In addition, admixture mapping requires a smaller number of genetic markers and therefore offers greater statistical power, while maintaining coverage for genomic regions that contain rare variants. Though this approach offers less resolving power than standard GWAS, the data generated is a more accurate depiction of the underlying pharmacogenetic associations within these groups [117, 118].

10 The Role of Patient Stratification Based on Clinical Data and Genetics in New Drug Development for Asthma

Drug development has been slow to generate new classes of compounds for the management of asthma. The major of recent advances in patient care have come from new indications for existing drug classes, e.g. improved duration of action for LABAs. Much effort has been placed on the inhibition of cytokines in asthma including anti-IL-5, IL-13, IL-4 and TNFα, however initial results have been disappointing. One explanation for this limited progress is the design and recruitment to Phase II trials for the evaluation of new compounds. For example, initial trials of anti-IL5 (mepolizumab) in asthma demonstrated disappointing effects on lung function and asthma symptoms when patient recruitment was not stratified. Only after careful selection of patients, based on refractory asthma and sputum eosinophilia ($>3\,\%$), were clinically relevant improvements in asthma observed, e.g. reduced exacerbation rates [119]. Similarly, a recent study examining anti-IL-13 antibody (lebrikizumab) efficacy in asthma used preselected individuals with an IL-13/Th_2 dominated disease (defined by elevated total serum IgE, blood eosinophilia and serum periostin levels) [120]. Periostin is an extracellular matrix protein induced by IL-13 in airway epithelial cells and lung fibroblasts, and an important biomarker of

Th_2-associated airway inflammation [121]. The preselected asthma patients showed greatest improvements in the primary outcome, being a mean change in baseline FEV_1 (% predicted) of 8.2% compared to the low periostin group which only showed a mean improvement of 1.6%, following 12 weeks treatment [120]. These Phase II studies demonstrate that careful select asthma subjects is critical in asthma drug development and evaluation. It is important to note that serum eosinophilia has a recognised genetic component involving multiple polymorphism within *IL5, IL33, IL1RL1, WDR36* and *MYB* genes [122] and that polymorphisms spanning the *IL13* locus show robust association with asthma [123]. It is therefore highly likely that these drug responsive sub-phenotypes are genetically determined.

Therefore overall, recent approaches to treat asthma include; (a) cytokine inhibition *e.g.* anti IL-5 (mepolizumab), anti-TNFα (etanercept, soluble TNFR), anti-IL-13 (lebrikizumab, tralokinumab), (b) chemokine receptor antagonists *e.g.* chemokine (CXC motif) receptor 2 (CXCR2) (SCH527123) and (c) signalling pathway inhibitors with anti-inflammatory effects *e.g.* targeting p38 MAPK (SD-282), PDE4 (roflumilast), PI3K inhibitors (IPI-145) [124–126].

The specific targeting of single mediators is unlikely to provide a therapeutic option for asthma using the broad definition of the term. However accumulating evidence suggests that targeting of specific subpopulations leads to clinical efficacy *e.g.* anti-IL-5 and anti-IL-13 (Table 5). Very recently, a study investigating a human IL-13-neutralising monoclonal antibody (tralokinumab) used a carefully balanced population of atopic and non-atopic asthmatics with exclusions for additional respiratory pathology, cigarette smoking≥10 pack-years, recent infection or treat-

Table 5 Selected asthma therapies in Phase II development and potential for patient stratification approaches

Anti-cytokine therapies	Evidence for stratification potential	Reference
IL-4Rα antagonist (pitrikinra)	Yes, based on *IL4RA* genotype	[14, 129]
IL-4Rα antagonist (dupliumab)	Yes, based on a blood eosinophil count of at least 300 cells per microliter or a sputum eosinophil level of at least 3% and using inhaled glucocorticoids plus long-acting β-adrenoceptor agonists	[128]
Anti-IL-5 (mepolizumab)	Yes, severe asthma and elevated serum and sputum eosinophils	[119]
Anti-IL-9 (MEDI-528)	Unknown	[130]
Anti-IL-13 (tralokinumab)	Yes, based on exclusions for additional respiratory pathology, cigarette smoking≥10 pack-years, recent infection or treatment with immunosuppressive medication	[127]
Anti-IL-13 (lebrikizumab)	Yes, based on elevated total IgE, blood eosinophilia and serum periostin levels	[120]
Soluble TNFα receptor (etanercept)	Unknown	[131]

ment with immunosuppressive medication in a randomised, double-blind study [127]. Here, tralokinumab was associated with improving lung function based on an increase in FEV_1 and a decrease in daily β_2-adrenoceptor agonist use [127]. Similarly, a monoclonal antibody that blocks IL-4Rα function (dupilumab) has provided some encouraging results, such as a reduction in asthma exacerbations in carefully selected patients that have a blood eosinophil count of at least 300 cells per microliter or a sputum eosinophil level of at least 3 % and who are administered inhaled glucocorticoids plus long-acting beta-agonists [128]. As pointed out, features such as eosinophilia have a genetic basis. It is thus anticipated that the phenotype may be genetically driven, although definitive pharmacogenetic studies remain to be completed for these biologics.

As novel asthma genes are identified, it is important for the evaluation of these potential drug targets, particularly in the context of Phase II trials, to take into account genetic factors. Preliminary data supports genetic testing in Phase II trials of newer compounds, as shown in a recent report of the IL-4/IL-13 dual antagonist pitrikinra. Polymorphic variation in the target receptor for this antagonist i.e. IL-4Rα significantly influenced outcomes in allergic asthma subjects [14]. This antagonist is a recombinant form of IL-4, differing at two amino acid residues, i.e. a mutein, R121D/Y124D. This has been shown to reduce late phase antigen responses (LAR) to inhaled antigen as defined by changes in lung function (FEV_1) over a period of 4–10 h post antigen exposure. The subjects under study had been following a four week period of twice daily active treatment of nebulised pitrikinra or placebo, with an increased LAR ratio correlating with a reduced FEV_1 [129]. Stratification of subjects (pitrikinra $n=15$, placebo $n=14$) into genotype groups for the non-synonomous IL-4Rα SNP rs1801275 (Gln576Arg) identified that Arg/Arg carriers had an attenuated LAR compared to Gln/Gln or Gln/Arg genotypes. Similarly, stratification based on rs1805011 (Glu400Ala) showed an attenuated LAR in the Glu/Ala group but not in the Glu/Glu subjects [14]. The Arg576 variant (when in combination with Val75) has been shown to be a risk factor for allergic asthma and lead to enhanced IL-4Rα signalling post IL-4 stimulation, potentially providing a putative mechanism of action [132].

Therefore, although preliminary, these data suggest that selecting a subgroup of patients based on genotype, where it is anticipated that the receptor/pathway may have a more dominant role in that individual's asthma, may be critical to interpreting Phase II clinical trials.

11 Conclusion

The field of pharmacogenetics has made great strides forward since the original coining of the word by Frederick Vogel more than half a century ago [133]. In recent years, there has been a development of high throughput technologies, both in genetic discovery and drug development, as well as the establishment of various pharmacogenetic working groups operating under both the scientific and regulatory arms of international agencies.

The future of therapeutic management lies in the individual patient tailoring of "real life" data emerging on the usefulness of genotype-driven asthma management. Pharmacogenetics is currently moving through the translational phase from bench to bedside. The establishment of curated knowledgebases such as PharmGKB (https://www.pharmgkb.org), and the setting up of international professional committees such as the Clinical Pharmacogenetics Implementation Consortium (CPIC), established in 2009, contributes to the creation of platforms which facilitate translational pharmacogenetics. The advisory and regulatory roles of major agencies such as the Food and Drug Administration, and the European Medicines Agency, in the field of pharmacogenetics, provide a link between academia and industry, for the optimisation of the discovery process.

The incorporation of pharmacogenetic studies within Phase II clinical trials of new compounds is important for the identification of genetically defined cohorts who may respond differently to the drug. Subject stratification at this stage, provides information which paves the way for more informed decision taking, relating to drug development, and optimisation of the compound's targeted use. Compounds that may fail to show adequate benefit/risk ratios in a genetically undefined study population, may be found to be ideal for genetically defined subsets. Alternatively, subject stratification during early clinical trials, may identify compounds that show unwarranted toxicities in a general population, to be detrimental only to subjects with a particular genetic profile. This information may essentially influence the availability of new drugs on the market.

These initiatives are especially relevant in diseases such as asthma for which therapies directed at new drug targets are under development. In the future, therapeutics will see an increase in the use of biologics as well as targeted therapies, and respiratory disease is no exception. The consolidation of data, together with the assessment of robustness and clinical relevance of pharmacogenotypes, are critical components for the development and establishment of personalised genotype-guided prescribing protocols in bronchial asthma. Rapidly advancing technologies, especially in the areas of GWAS and whole genome sequencing, coupled with improvements in high-throughput techniques and refined bioinformatics analyses, all contribute to accelerating the attainment of these aims.

References

1. GINA (Global Initiative for Asthma) (2014) Global strategy for asthma management and prevention. www.ginasthma.org
2. Chung KF, Wenzel SE, Brozek JL, Bush A, Castro M, Sterk PJ, Adcock IM, Bateman ED, Bel EH, Bleecker ER, Boulet LP, Brightling C, Chanez P, Dahlen SE, Djukanovic R, Frey U, Gaga M, Gibson P, Hamid Q, Jajour NN, Mauad T, Sorkness RL, Teague WG (2014) International ERS/ATS guidelines on definition, evaluation and treatment of severe asthma. Eur Respir J 43(2):343–373. doi:10.1183/09031936.00202013
3. Hawkins GA, Tantisira K, Meyers DA, Ampleford EJ, Moore WC, Klanderman B, Liggett SB, Peters SP, Weiss ST, Bleecker ER (2006) Sequence, haplotype, and association analysis of ADRbeta2 in a multiethnic asthma case-control study. Am J Respir Critical Car Med 174(10):1101–1109. doi:10.1164/rccm.200509-1405OC

4. Reihsaus E, Innis M, MacIntyre N, Liggett SB (1993) Mutations in the gene encoding for the beta 2-adrenergic receptor in normal and asthmatic subjects. Am J Respir Cell Mol Biol 8(3):334–339. doi:10.1165/ajrcmb/8.3.334
5. Drysdale CM, McGraw DW, Stack CB, Stephens JC, Judson RS, Nandabalan K, Arnold K, Ruano G, Liggett SB (2000) Complex promoter and coding region beta 2-adrenergic receptor haplotypes alter receptor expression and predict in vivo responsiveness. Proc Natl Acad Sci USA 97(19):10483–10488
6. Sayers I, Hall IP (2005) Pharmacogenetic approaches in the treatment of asthma. Curr Allergy Asthma Rep 5(2):101–108
7. Hall IP, Sayers I (2007) Pharmacogenetics and asthma: false hope or new dawn? Eur Respir J 29(6):1239–1245. doi:10.1183/09031936.00088206
8. Pascual RM, Bleecker ER (2010) Pharmacogenetics of asthma. Curr Opin Pharmacol 10(3):226–235. doi:10.1016/j.coph.2010.05.013
9. Tse SM, Tantisira K, Weiss ST (2011) The pharmacogenetics and pharmacogenomics of asthma therapy. Pharmacogenomics J 11(6):383–392. doi:10.1038/tpj.2011.46
10. Sayers I, Hawley J, Stewart CE, Billington CK, Henry A, Leighton-Davies JR, Charlton SJ, Hall IP (2009) Pharmacogenetic characterization of indacaterol, a novel beta 2-adrenoceptor agonist. Br J Pharmacol 158(1):277–286. doi:10.1111/j.1476-5381.2009.00224.x
11. Lee MY, Cheng SN, Chen SJ, Huang HL, Wang CC, Fan HC (2011) Polymorphisms of the beta2-adrenergic receptor correlated to nocturnal asthma and the response of terbutaline nebulizer. Pediatr Neonatol 52(1):18–23. doi:10.1016/j.pedneo.2010.12.011
12. Basu K, Palmer CN, Tavendale R, Lipworth BJ, Mukhopadhyay S (2009) Adrenergic beta(2)-receptor genotype predisposes to exacerbations in steroid-treated asthmatic patients taking frequent albuterol or salmeterol. J Allergy Clin Immunol 124(6):1188–1194e1183. doi:10.1016/j.jaci.2009.07.043
13. Bleecker ER, Postma DS, Lawrance RM, Meyers DA, Ambrose HJ, Goldman M (2007) Effect of ADRB2 polymorphisms on response to longacting beta2-agonist therapy: a pharmacogenetic analysis of two randomised studies. Lancet 370(9605):2118–2125. doi:10.1016/S0140-6736(07)61906-0
14. Slager RE, Hawkins GA, Ampleford EJ, Bowden A, Stevens LE, Morton MT, Tomkinson A, Wenzel SE, Longphre M, Bleecker ER, Meyers DA (2010) IL-4 receptor alpha polymorphisms are predictors of a pharmacogenetic response to a novel IL-4/IL-13 antagonist. J Allerg Clin Immunol 126(4):875–878. doi:10.1016/j.jaci.2010.08.001
15. Wechsler ME, Kunselman SJ, Chinchilli VM, Bleecker E, Boushey HA, Calhoun WJ, Ameredes BT, Castro M, Craig TJ, Denlinger L, Fahy JV, Jarjour N, Kazani S, Kim S, Kraft M, Lazarus SC, Lemanske RF, Jr, Markezich A, Martin RJ, Permaul P, Peters SP, Ramsdell J, Sorkness CA, Sutherland ER, Szefler SJ, Walter MJ, Wasserman SI, Israel E (2009) National Heart L, Blood Institute's Asthma Clinical Research N Effect of beta2-adrenergic receptor polymorphism on response to longacting beta2 agonist in asthma (LARGE trial): a genotype-stratified, randomised, placebo-controlled, crossover trial. Lancet 374(9703):1754–1764. doi:10.1016/S0140-6736(09)61492-6
16. Panebra A, Wang WC, Malone MM, Pitter DR, Weiss ST, Hawkins GA, Liggett SB (2010) Common ADRB2 haplotypes derived from 26 polymorphic sites direct beta2-adrenergic receptor expression and regulation phenotypes. PloS One 5(7):e11819. doi:10.1371/journal.pone.0011819
17. Thangapandian S, John S, Son M, Arulalapperumal V, Lee KW (2013) Development of predictive quantitative structure-activity relationship model and its application in the discovery of human leukotriene A4 hydrolase inhibitors. Future Med Chem 5(1):27–40. doi:10.4155/fmc.12.184
18. Litonjua AA, Lasky-Su J, Schneiter K, Tantisira KG, Lazarus R, Klanderman B, Lima JJ, Irvin CG, Peters SP, Hanrahan JP, Liggett SB, Hawkins GA, Meyers DA, Bleecker ER, Lange C, Weiss ST (2008) ARG1 is a novel bronchodilator response gene: screening and replication in four asthma cohorts. Am J Respir Crit Care Med 178(7):688–694. doi:10.1164/rccm.200709-1363OC

19. Meurs H, McKay S, Maarsingh H, Hamer MA, Macic L, Molendijk N, Zaagsma J (2002) Increased arginase activity underlies allergen-induced deficiency of cNOS-derived nitric oxide and airway hyperresponsiveness. Br J Pharmacol 136(3):391–398. doi:10.1038/sj.bjp.0704725
20. Maarsingh H, Zuidhof AB, Bos IS, van Duin M, Boucher JL, Zaagsma J, Meurs H (2008) Arginase inhibition protects against allergen-induced airway obstruction, hyperresponsiveness, and inflammation. Am J Respir Crit Care Med 178(6):565–573. doi:10.1164/rccm.200710-1588OC
21. Vonk JM, Postma DS, Maarsingh H, Bruinenberg M, Koppelman GH, Meurs H (2010) Arginase 1 and arginase 2 variations associate with asthma, asthma severity and beta2 agonist and steroid response. Pharmacogenet Genomics 20(3):179–186. doi:10.1097/FPC.0b013e328336c7fd
22. Duan QL, Gaume BR, Hawkins GA, Himes BE, Bleecker ER, Klanderman B, Irvin CG, Peters SP, Meyers DA, Hanrahan JP, Lima JJ, Litonjua AA, Tantisira KG, Liggett SB (2011) Regulatory haplotypes in ARG1 are associated with altered bronchodilator response. Am J Respir Crit Care Med 183(4):449–454. doi:10.1164/rccm.201005-0758OC
23. Moore PE, Ryckman KK, Williams SM, Patel N, Summar ML, Sheller JR (2009) Genetic variants of GSNOR and ADRB2 influence response to albuterol in African-American children with severe asthma. Pediatr Pulmonol 44(7):649–654. doi:10.1002/ppul.21033
24. Gaston B, Reilly J, Drazen JM, Fackler J, Ramdev P, Arnelle D, Mullins ME, Sugarbaker DJ, Chee C, Singel DJ et al. (1993) Endogenous nitrogen oxides and bronchodilator S-nitrosothiols in human airways. Proc Natl Acad Sci USA 90(23):10957–10961
25. Whalen EJ, Foster MW, Matsumoto A, Ozawa K, Violin JD, Que LG, Nelson CD, Benhar M, Keys JR, Rockman HA, Koch WJ, Daaka Y, Lefkowitz RJ, Stamler JS (2007) Regulation of beta-adrenergic receptor signaling by S-nitrosylation of G-protein-coupled receptor kinase 2. Cell 129(3):511–522. doi:10.1016/j.cell.2007.02.046
26. Choudhry S, Que LG, Yang Z, Liu L, Eng C, Kim SO, Kumar G, Thyne S, Chapela R, Rodriguez-Santana JR, Rodriguez-Cintron W, Avila PC, Stamler JS, Burchard EG (2010) GSNO reductase and beta2-adrenergic receptor gene-gene interaction: bronchodilator responsiveness to albuterol. Pharmacogenet Genomics 20(6):351–358. doi:10.1097/FPC.0b013e328337f992
27. Himes BE, Jiang X, Wagner P, Hu R, Wang Q, Klanderman B, Whitaker RM, Duan Q, Lasky-Su J, Nikolos C, Jester W, Johnson M, Panettieri RA, Jr, Tantisira KG, Weiss ST, Lu Q (2014) RNA-Seq transcriptome profiling identifies CRISPLD2 as a glucocorticoid responsive gene that modulates cytokine function in airway smooth muscle cells. PloS One 9(6):e99625. doi:10.1371/journal.pone.0099625
28. Kellaway CH, Trethewie ER (1940) The liberation of a slow-reacting smooth muscle-stimulating substance in anaphylaxis. Exp Physiol 30(2):121–145
29. Brocklehurst WE (1960) The release of histamine and formation of a slow-reacting substance (SRS-A) during anaphylactic shock. J Physiol 151:416–435
30. Duroudier NP, Tulah AS, Sayers I (2009) Leukotriene pathway genetics and pharmacogenetics in allergy. Allergy 64(6):823–839. doi:10.1111/j.1398-9995.2009.02015.x
31. Via M, De Giacomo A, Corvol H, Eng C, Seibold MA, Gillett C, Galanter J, Sen S, Tcheurekdjian H, Chapela R, Rodriguez-Santana JR, Rodriguez-Cintron W, Thyne S, Avila PC, Choudhry S, Gonzalez Burchard E, Genetics of Asthma in Latino Americans S (2010) The role of LTA4H and ALOX5AP genes in the risk for asthma in Latinos. Clin Exp Allergy: J Br Soc Allergy Clin Immunol 40(4):582–589. doi:10.1111/j.1365-2222.2009.03438.x
32. Tulah AS, Parker SG, Moffatt MF, Wardlaw AJ, Connolly MJ, Sayers I (2011) The role of ALOX5AP, LTA4H and LTB4R polymorphisms in determining baseline lung function and COPD susceptibility in UK smokers. BMC Med Genet 12:173. doi:10.1186/1471-2350-12-173
33. Holloway JW, Barton SJ, Holgate ST, Rose-Zerilli MJ, Sayers I (2008) The role of LTA4H and ALOX5AP polymorphism in asthma and allergy susceptibility. Allergy 63(8):1046–1053. doi:10.1111/j.1398-9995.2008.01667.x
34. Drazen JM, Yandava CN, Dube L, Szczerback N, Hippensteel R, Pillari A, Israel E, Schork N, Silverman ES, Katz DA, Drajesk J (1999) Pharmacogenetic association between ALOX5

promoter genotype and the response to anti-asthma treatment. Nat Genet 22(2):168–170. doi:10.1038/9680
35. Lima JJ, Zhang S, Grant A, Shao L, Tantisira KG, Allayee H, Wang J, Sylvester J, Holbrook J, Wise R, Weiss ST, Barnes K (2006) Influence of leukotriene pathway polymorphisms on response to montelukast in asthma. Am J Respir Crit Care Med 173(4):379–385. doi:10.1164/rccm.200509-1412OC
36. Telleria JJ, Blanco-Quiros A, Varillas D, Armentia A, Fernandez-Carvajal I, Jesus Alonso M, Diez I (2008) ALOX5 promoter genotype and response to montelukast in moderate persistent asthma. Respir Med 102(6):857–861. doi:10.1016/j.rmed.2008.01.011
37. Klotsman M, York TP, Pillai SG, Vargas-Irwin C, Sharma SS, van den Oord EJ, Anderson WH (2007) Pharmacogenetics of the 5-lipoxygenase biosynthetic pathway and variable clinical response to montelukast. Pharmacogenet Genomics 17(3):189–196. doi:10.1097/FPC.0b013e3280120043
38. Tantisira KG, Lima J, Sylvia J, Klanderman B, Weiss ST (2009) 5-lipoxygenase pharmacogenetics in asthma: overlap with Cys-leukotriene receptor antagonist loci. Pharmacogenet Genomics 19(3):244–247. doi:10.1097/FPC.0b013e328326e0b1
39. Asano K, Shiomi T, Hasegawa N, Nakamura H, Kudo H, Matsuzaki T, Hakuno H, Fukunaga K, Suzuki Y, Kanazawa M, Yamaguchi K (2002) Leukotriene C4 synthase gene A(-444)C polymorphism and clinical response to a CYS-LT(1) antagonist, pranlukast, in Japanese patients with moderate asthma. Pharmacogenetics 12(7):565–570
40. Tcheurekdjian H, Via M, De Giacomo A, Corvol H, Eng C, Thyne S, Chapela R, Rodriguez-Cintron W, Rodriguez-Santana JR, Avila PC, Burchard EG, Genetics of Asthma in Latino Americans S (2010) ALOX5AP and LTA4H polymorphisms modify augmentation of bronchodilator responsiveness by leukotriene modifiers in Latinos. J Allergy Clin Immunol 126(4):853–858. doi:10.1016/j.jaci.2010.06.048
41. Kim SH, Ye YM, Hur GY, Lee SK, Sampson AP, Lee HY, Park HS (2007) CysLTR1 promoter polymorphism and requirement for leukotriene receptor antagonist in aspirin-intolerant asthma patients. Pharmacogenomics 8(9):1143–1150. doi:10.2217/14622416.8.9.1143
42. Brochu-Bourque A, Veronneau S, Rola-Pleszczynski M, Stankova J (2011) Differential signaling defects associated with the M201V polymorphism in the cysteinyl leukotriene type 2 receptor. J Pharmacol Exp Ther 336(2):431–439. doi:10.1124/jpet.110.172411
43. Sayers I, Sampson AP, Ye S, Holgate ST (2003) Promoter polymorphism influences the effect of dexamethasone on transcriptional activation of the LTC4 synthase gene. Eur J Hum Genet 11(8):619–622. doi:10.1038/sj.ejhg.5201015
44. Mougey EB, Feng H, Castro M, Irvin CG, Lima JJ (2009) Absorption of montelukast is transporter mediated: a common variant of OATP2B1 is associated with reduced plasma concentrations and poor response. Pharmacogenet Genomics 19(2):129–138. doi:10.1097/FPC.0b013e32831bd98c
45. Reid JJ (2001) ABT-761 (Abbott). Curr Opin Investig Drugs 2(1):68–71
46. Mougey E, Lang JE, Allayee H, Teague WG, Dozor AJ, Wise RA, Lima JJ (2013) ALOX5 polymorphism associates with increased leukotriene production and reduced lung function and asthma control in children with poorly controlled asthma. Clin Exp Allergy: J Br Soc Allergy Clin Immunol 43(5):512–520. doi:10.1111/cea.12076
47. Nwokoro C, Pandya H, Turner S, Eldridge S, Griffiths CJ, Vulliamy T, Price D, Sanak M, Holloway JW, Brugha R, Koh L, Dickson I, Rutterford C, Grigg J (2014) Intermittent montelukast in children aged 10 months to 5 years with wheeze (WAIT trial): a multicentre, randomised, placebo-controlled trial. Lancet Respir Med 2(10):796–803. doi:10.1016/S2213-2600(14)70186-9
48. Choi JH, Park HS, Oh HB, Lee JH, Suh YJ, Park CS, Shin HD (2004) Leukotriene-related gene polymorphisms in ASA-intolerant asthma: an association with a haplotype of 5-lipoxygenase. Hum Genet 114(4):337–344. doi:10.1007/s00439-004-1082-1
49. Thompson MD, Capra V, Takasaki J, Maresca G, Rovati GE, Slutsky AS, Lilly C, Zamel N, McIntyre Burnham W, Cole DE, Siminovitch KA (2007) A functional G300S variant of the cysteinyl leukotriene 1 receptor is associated with atopy in a Tristan da Cunha isolate. Pharmacogenet Genomics 17(7):539–549. doi:10.1097/FPC.0b013e328012d0bf

50. Park JS, Chang HS, Park CS, Lee JH, Lee YM, Choi JH, Park HS, Kim LH, Park BL, Choi YH, Shin HD (2005) Association analysis of cysteinyl-leukotriene receptor 2 (CYSLTR2) polymorphisms with aspirin intolerance in asthmatics. Pharmacogenet Genomics 15(7):483–492
51. Thompson MD, Storm van's Gravesande K, Galczenski H, Burnham WM, Siminovitch KA, Zamel N, Slutsky A, Drazen JM, George SR, Evans JF, O'Dowd BF (2003) A cysteinyl leukotriene 2 receptor variant is associated with atopy in the population of Tristan da Cunha. Pharmacogenetics 13(10):641–649. doi:10.1097/01.fpc.0000054127.14659.42
52. Pillai SG, Cousens DJ, Barnes AA, Buckley PT, Chiano MN, Hosking LK, Cameron LA, Fling ME, Foley JJ, Green A, Sarau HM, Schmidt DB, Sprankle CS, Blumenthal MN, Vestbo J, Kennedy-Wilson K, Wixted WE, Wagner MJ, Anderson WH, Ignar DM, Investigators of the GN (2004) A coding polymorphism in the CYSLT2 receptor with reduced affinity to LTD4 is associated with asthma. Pharmacogenetics 14(9):627–633
53. Nicolaides NC, Galata Z, Kino T, Chrousos GP, Charmandari E (2010) The human glucocorticoid receptor: molecular basis of biologic function. Steroids 75(1):1–12. doi:10.1016/j.steroids.2009.09.002
54. Strehl C, Gaber T, Lowenberg M, Hommes DW, Verhaar AP, Schellmann S, Hahne M, Fangradt M, Wagegg M, Hoff P, Scheffold A, Spies CM, Burmester GR, Buttgereit F (2011) Origin and functional activity of the membrane-bound glucocorticoid receptor. Arthritis Rheum 63(12):3779–3788. doi:10.1002/art.30637
55. Kam JC, Szefler SJ, Surs W, Sher ER, Leung DY (1993) Combination IL-2 and IL-4 reduces glucocorticoid receptor-binding affinity and T cell response to glucocorticoids. J Immunol 151(7):3460–3466
56. Spahn JD, Szefler SJ, Surs W, Doherty DE, Nimmagadda SR, Leung DY (1996) A novel action of IL-13: induction of diminished monocyte glucocorticoid receptor-binding affinity. J Immunol 157(6):2654–2659
57. Tliba O, Damera G, Banerjee A, Gu S, Baidouri H, Keslacy S, Amrani Y (2008) Cytokines induce an early steroid resistance in airway smooth muscle cells: novel role of interferon regulatory factor-1. Am J Respir Cell Mol Biol 38(4):463–472. doi:10.1165/rcmb.2007-0226OC
58. Zijlstra GJ, Ten Hacken NH, Hoffmann RF, van Oosterhout AJ, Heijink IH (2012) Interleukin-17 A induces glucocorticoid insensitivity in human bronchial epithelial cells. Eur Respir J 39(2):439–445. doi:10.1183/09031936.00017911
59. Adcock IM, Barnes PJ (2008) Molecular mechanisms of corticosteroid resistance. Chest 134(2):394–401. doi:10.1378/chest.08-0440
60. Dejager L, Vandevyver S, Petta I, Libert C (2014) Dominance of the strongest: inflammatory cytokines versus glucocorticoids. Cytokine Growth Factor Rev 25(1):21–33. doi:10.1016/j.cytogfr.2013.12.006
61. Panek M, Pietras T, Antczak A, Gorski P, Kuna P, Szemraj J (2012) The role of functional single nucleotide polymorphisms of the human glucocorticoid receptor gene NR3C1 in Polish patients with bronchial asthma. Mol Biol Rep 39(4):4749–4757. doi:10.1007/s11033-011-1267-3
62. Panek M, Pietras T, Fabijan A, Milanowski M, Wieteska L, Gorski P, Kuna P, Szemraj J (2013) Effect of glucocorticoid receptor gene polymorphisms on asthma phenotypes. Exp Ther Med 5(2):572–580. doi:10.3892/etm.2012.809
63. Souza MC, Martins CS, Silva-Junior IM, Chriguer RS, Bueno AC, Antonini SR, Silva WA Jr, Zago MA, Moreira AC, Castro M (2014) NR3C1 polymorphisms in Brazilians of Caucasian, African, and Asian ancestry: glucocorticoid sensitivity and genotype association. Arq Bras Endocrinol Metabol 58(1):53–61
64. Russcher H, Dalm VA, de Jong FH, Brinkmann AO, Hofland LJ, Lamberts SW, Koper JW (2007) Associations between promoter usage and alternative splicing of the glucocorticoid receptor gene. J Mol Endocrinol 38(1–2):91–98. doi:10.1677/jme.1.02117
65. Ikeda Y, Suehiro T, Tsuzura S, Shiinoki T, Kaneda T, Kumon Y, Hashimoto K (2001) A polymorphism in the promoter region of the glucocorticoid receptor gene is associated with its transcriptional activity. Endocr J 48(6):723–726

66. Pietras T, Panek M, Tworek D, Oszajca K, Wujcik R, Gorski P, Kuna P, Szemraj J (2011) The Bcl I single nucleotide polymorphism of the human glucocorticoid receptor gene h-GR/NR3C1 promoter in patients with bronchial asthma: pilot study. Mol Biol Rep 38(6):3953–3958. doi:10.1007/s11033-010-0512-5
67. Turner JD, Alt SR, Cao L, Vernocchi S, Trifonova S, Battello N, Muller CP (2010) Transcriptional control of the glucocorticoid receptor: CpG islands, epigenetics and more. Biochem Pharmacol 80(12):1860–1868. doi:10.1016/j.bcp.2010.06.037
68. Hollenberg SM, Weinberger C, Ong ES, Cerelli G, Oro A, Lebo R, Thompson EB, Rosenfeld MG, Evans RM (1985) Primary structure and expression of a functional human glucocorticoid receptor cDNA. Nature 318(6047):635–641
69. Encio IJ, Detera-Wadleigh SD (1991) The genomic structure of the human glucocorticoid receptor. J Biol Chem 266(11):7182–7188
70. Kino T, Su YA, Chrousos GP (2009) Human glucocorticoid receptor isoform beta: recent understanding of its potential implications in physiology and pathophysiology. Cell Mol Life Sci 66(21):3435–3448. doi:10.1007/s00018-009-0098-z
71. Hurley DM, Accili D, Stratakis CA, Karl M, Vamvakopoulos N, Rorer E, Constantine K, Taylor SI, Chrousos GP (1991) Point mutation causing a single amino acid substitution in the hormone binding domain of the glucocorticoid receptor in familial glucocorticoid resistance. J Clin Invest 87(2):680–686. doi:10.1172/JCI115046
72. Malchoff DM, Brufsky A, Reardon G, McDermott P, Javier EC, Bergh CH, Rowe D, Malchoff CD (1993) A mutation of the glucocorticoid receptor in primary cortisol resistance. J Clin Invest 91(5):1918–1925. doi:10.1172/JCI116410
73. Niu N, Manickam V, Kalari KR, Moon I, Pelleymounter LL, Eckloff BW, Wieben ED, Schaid DJ, Wang L (2009) Human glucocorticoid receptor alpha gene (NR3C1) pharmacogenomics: gene resequencing and functional genomics. J Clin Endocrinol Metab 94(8):3072–3084. doi:10.1210/jc.2008-2109
74. Tsartsali L, Papadopoulos M, Lagona E, Papadimitriou A, Kanaka-Gantenbein C, Louizou E, Kastania A, Priftis KN, Chrousos G (2012) Association of hypothalamic-pituitary-adrenal axis-related polymorphisms with stress in asthmatic children on inhaled corticosteroids. Neuroimmunomodulation 19(2):88–95. doi:10.1159/000329592
75. Cuzzoni E, De Iudicibus S, Bartoli F, Ventura A, Decorti G (2012) Association between BclI polymorphism in the NR3C1 gene and in vitro individual variations in lymphocyte responses to methylprednisolone. Br J Clin Pharmacol 73(4):651–655. doi:10.1111/j.1365-2125.2011.04130.x
76. Panek M, Pietras T, Antczak A, Fabijan A, Przemecka M, Gorski P, Kuna P, Szemraj J (2012) The N363S and I559N single nucleotide polymorphisms of the h-GR/NR3C1 gene in patients with bronchial asthma. Int J Mol Med 30(1):142–150. doi:10.3892/ijmm.2012.956
77. Stevens A, Ray DW, Zeggini E, John S, Richards HL, Griffiths CE, Donn R (2004) Glucocorticoid sensitivity is determined by a specific glucocorticoid receptor haplotype. J Clin Endocrinol Metab 8(2):892–897
78. Tantisira KG, Hwang ES, Raby BA, Silverman ES, Lake SL, Richter BG, Peng SL, Drazen JM, Glimcher LH, Weiss ST (2004) TBX21: a functional variant predicts improvement in asthma with the use of inhaled corticosteroids. Proc Natl Acad Sci USA 101(52):18099–18104. doi:10.1073/pnas.0408532102
79. Lopert A, Rijavec M, Zavbi M, Korosec P, Flezar M (2013) Asthma treatment outcome in adults is associated with rs9910408 in TBX21 gene. Sci Rep 3:2915. doi:10.1038/srep02915
80. Sakaeda T, Nakamura T, Okumura K (2002) MDR1 genotype-related pharmacokinetics and pharmacodynamics. Biol Pharm Bull 25(11):1391–1400
81. Siegsmund M, Brinkmann U, Schaffeler E, Weirich G, Schwab M, Eichelbaum M, Fritz P, Burk O, Decker J, Alken P, Rothenpieler U, Kerb R, Hoffmeyer S, Brauch H (2002) Association of the P-glycoprotein transporter MDR1(C3435T) polymorphism with the susceptibility to renal epithelial tumors. J Am Soc Nephrol 13(7):1847–1854
82. Ye YM, Lee HY, Kim SH, Jee YK, Lee SK, Lee SH, Park HS (2009) Pharmacogenetic study of the effects of NK2R G231E G > A and TBX21 H33Q C > G polymorphisms on asthma control with inhaled corticosteroid treatment. J Clin Pharm Ther 34(6):693–701. doi:10.1111/j.1365-2710.2009.01054.x

83. Hawkins GA, Lazarus R, Smith RS, Tantisira KG, Meyers DA, Peters SP, Weiss ST, Bleecker ER (2009) The glucocorticoid receptor heterocomplex gene STIP1 is associated with improved lung function in asthmatic subjects treated with inhaled corticosteroids. J Allergy Clin Immunol 123(6):1376–1383, e1377. doi:10.1016/j.jaci.2009.01.049
84. Jin Y, Hu D, Peterson EL, Eng C, Levin AM, Wells K, Beckman K, Kumar R, Seibold MA, Karungi G, Zoratti A, Gaggin J, Campbell J, Galanter J, Chapela R, Rodriguez-Santana JR, Watson HG, Meade K, Lenoir M, Rodriguez-Cintron W, Avila PC, Lanfear DE, Burchard EG, Williams LK (2010) Dual-specificity phosphatase 1 as a pharmacogenetic modifier of inhaled steroid response among asthmatic patients. J Allergy Clin Immunol 126(3):618–625 e611–612. doi:10.1016/j.jaci.2010.06.007
85. Tantisira KG, Silverman ES, Mariani TJ, Xu J, Richter BG, Klanderman BJ, Litonjua AA, Lazarus R, Rosenwasser LJ, Fuhlbrigge AL, Weiss ST (2007) FCER2: a pharmacogenetic basis for severe exacerbations in children with asthma. J Allergy Clin Immunol 120(6):1285–1291. doi:10.1016/j.jaci.2007.09.005
86. Tantisira KG, Lasky-Su J, Harada M, Murphy A, Litonjua AA, Himes BE, Lange C, Lazarus R, Sylvia J, Klanderman B, Duan QL, Qiu W, Hirota T, Martinez FD, Mauger D, Sorkness C, Szefler S, Lazarus SC, Lemanske RF Jr, Peters SP, Lima JJ, Nakamura Y, Tamari M, Weiss ST (2011) Genomewide association between GLCCI1 and response to glucocorticoid therapy in asthma. New Engl J Med 365(13):1173–1183. doi:10.1056/NEJMoa0911353
87. Huizenga NA, Koper JW, De Lange P, Pols HA, Stolk RP, Burger H, Grobbee DE, Brinkmann AO, De Jong FH, Lamberts SW (1998) A polymorphism in the glucocorticoid receptor gene may be associated with and increased sensitivity to glucocorticoids in vivo. J Clin Endocrinol Metab 83(1):144–151. doi:10.1210/jcem.83.1.4490
88. Rosmond R, Bouchard C, Bjorntorp P (2001) Tsp509I polymorphism in exon 2 of the glucocorticoid receptor gene in relation to obesity and cortisol secretion: cohort study. BMJ 322(7287):652–653
89. Syed AA, Irving JA, Redfern CP, Hall AG, Unwin NC, White M, Bhopal RS, Alberti KG, Weaver JU (2004) Low prevalence of the N363S polymorphism of the glucocorticoid receptor in South Asians living in the United Kingdom. J Clin Endocrinol Metab 89(1):232–235. doi:10.1210/jc.2003-030995
90. Koyano S, Saito Y, Sai K, Kurose K, Ozawa S, Nakajima T, Matsumoto K, Saito H, Shirao K, Yoshida T, Minami H, Ohtsu A, Saijo N, Sawada J (2005) Novel genetic polymorphisms in the NR3C1 (glucocorticoid receptor) gene in a Japanese population. Drug Metab Pharmacokinet 20(1):79–84
91. Duan ZX, Gu W, Du DY, Hu P, Jiang DP, Zhu PF, Wang ZG, Jiang JX (2009) Distributions of glucocorticoid receptor gene polymorphisms in a Chinese Han population and associations with outcome after major trauma. Injury 40(5):479–483. doi:10.1016/j.injury.2008.09.025
92. Goecke A, Guerrero J (2006) Glucocorticoid receptor beta in acute and chronic inflammatory conditions: clinical implications. Immunobiology 211(1–2):85–96. doi:10.1016/j.imbio.2005.11.002
93. Bamberger CM, Bamberger AM, de Castro M, Chrousos GP (1995) Glucocorticoid receptor beta, a potential endogenous inhibitor of glucocorticoid action in humans. J Clin Invest 95(6):2435–2441. doi:10.1172/JCI117943
94. Pujols L, Mullol J, Roca-Ferrer J, Torrego A, Xaubet A, Cidlowski JA, Picado C (2002) Expression of glucocorticoid receptor alpha- and beta-isoforms in human cells and tissues. Am J Physiol Cell Physiol 283(4):C1324–1331. doi:10.1152/ajpcell.00363.2001
95. Lewis-Tuffin LJ, Cidlowski JA (2006) The physiology of human glucocorticoid receptor beta (hGRbeta) and glucocorticoid resistance. Ann N Y Acad Sci 1069:1–9. doi:10.1196/annals.1351.001
96. Kadmiel M, Cidlowski JA (2013) Glucocorticoid receptor signaling in health and disease. Trends Pharmacol Sci 34(9):518–530. doi:10.1016/j.tips.2013.07.003
97. Vazquez-Tello A, Halwani R, Hamid Q, Al-Muhsen S (2013) Glucocorticoid receptor-beta up-regulation and steroid resistance induction by IL-17 and IL-23 cytokine stimulation in peripheral mononuclear cells. J Clin Immunol 33(2):466–478. doi:10.1007/s10875-012-9828-3

98. Kamdem LK, Hamilton L, Cheng C, Liu W, Yang W, Johnson JA, Pui CH, Relling MV (2008) Genetic predictors of glucocorticoid-induced hypertension in children with acute lymphoblastic leukemia. Pharmacogenet genomics 18(6):507–514. doi:10.1097/FPC.0b013e3282fc5801
99. Nave R, McCracken N (2008) Metabolism of ciclesonide in the upper and lower airways: review of available data. J Asthma Allergy 1:11–18
100. Wu MH, Chen P, Wu X, Liu W, Strom S, Das S, Cook EH Jr, Rosner GL, Dolan ME (2004) Determination and analysis of single nucleotide polymorphisms and haplotype structure of the human carboxylesterase 2 gene. Pharmacogenetics 14(9):595–605
101. Charasson V, Bellott R, Meynard D, Longy M, Gorry P, Robert J (2004) Pharmacogenetics of human carboxylesterase 2, an enzyme involved in the activation of irinotecan into SN-38. Clin Pharmacol Ther 76(6):528–535. doi:10.1016/j.clpt.2004.08.007
102. Crowe A, Tan AM (2012) Oral and inhaled corticosteroids: differences in P-glycoprotein (ABCB1) mediated efflux. Toxicol Appl Pharmacol 260(3):294–302. doi:10.1016/j.taap.2012.03.008
103. Finotto S, Neurath MF, Glickman JN, Qin S, Lehr HA, Green FH, Ackerman K, Haley K, Galle PR, Szabo SJ, Drazen JM, De Sanctis GT, Glimcher LH (2002) Development of spontaneous airway changes consistent with human asthma in mice lacking T-bet. Science 295(5553):336–338. doi:10.1126/science.1065544
104. Park HW, Dahlin A, Tse S, Duan QL, Schuemann B, Martinez FD, Peters SP, Szefler SJ, Lima JJ, Kubo M, Tamari M, Tantisira KG (2014) Genetic predictors associated with improvement of asthma symptoms in response to inhaled corticosteroids. J Allergy Clin Immunol 133(3):664–669.e5. doi:10.1016/j.jaci.2013.12.1042
105. Tautermann CS, Kiechle T, Seeliger D, Diehl S, Wex E, Banholzer R, Gantner F, Pieper MP, Casarosa P (2013) Molecular basis for the long duration of action and kinetic selectivity of tiotropium for the muscarinic M3 receptor. J Med Chem 56(21):8746–8756. doi:10.1021/jm401219y
106. Fenech AG, Ebejer MJ, Felice AE, Ellul-Micallef R, Hall IP (2001) Mutation screening of the muscarinic M(2) and M(3) receptor genes in normal and asthmatic subjects. Br J Pharmacol 133(1):43–48. doi:10.1038/sj.bjp.0704039
107. Szczepankiewicz A, Breborowicz A, Sobkowiak P, Kramer L, Popiel A (2009) Association of A/T polymorphism of the CHRM2 gene with bronchodilator response to ipratropium bromide in asthmatic children. Pneumonol Alergol Pol 77(1):5–10
108. Fenech AG, Billington CK, Swan C, Richards S, Hunter T, Ebejer MJ, Felice AE, Ellul-Micallef R, Hall IP (2004) Novel polymorphisms influencing transcription of the human CHRM2 gene in airway smooth muscle. Am J Respir Cell Mol Biol 30(5):678–686. doi:10.1165/rcmb.2003-0011OC
109. Donfack J, Kogut P, Forsythe S, Solway J, Ober C (2003) Sequence variation in the promoter region of the cholinergic receptor muscarinic 3 gene and asthma and atopy. J Allergy Clin Immunol 111(3):527–532
110. Maeda Y, Hizawa N, Jinushi E, Honda A, Takahashi D, Fukui Y, Konno S, Shimizu T, Shimizu H, Yamaguchi E (2006) Polymorphisms in the muscarinic receptor 1 gene confer susceptibility to asthma in Japanese subjects. Am J Respir Crit Care Med 174(10):1119–1124
111. Yamamoto T, Yamashita N, Kuwabara M, Nakano J, Sugimoto H, Akiyama K, Hirai K, Ishii A, Uehara Y, Ohta K (2002) Mutation screening of the muscarinic m2 and m3 receptor genes in asthmatics, outgrow subjects, and normal controls. Ann Genet 45(3):109–113
112. Lipworth BJ, Basu K, Donald HP, Tavendale R, Macgregor DF, Ogston SA, Palmer CN, Mukhopadhyay S (2013) Tailored second-line therapy in asthmatic children with the Arg(16) genotype. Clin Sci (Lond) 124(8):521–528. doi:10.1042/CS20120528
113. Ortega VE, Meyers DA (2014) Pharmacogenetics: implications of race and ethnicity on defining genetic profiles for personalized medicine. J Allergy Clin Immunol 133(1):16–26. doi:10.1016/j.jaci.2013.10.040
114. Carter-Pokras OD, Gergen PJ (1993) Reported asthma among Puerto Rican, Mexican-American, and Cuban children, 1982 through 1984. Am J Public Health 83(4):580–582

115. Israel E, Chinchilli VM, Ford JG, Boushey HA, Cherniack R, Craig TJ, Deykin A, Fagan JK, Fahy JV, Fish J, Kraft M, Kunselman SJ, Lazarus SC, Lemanske RF Jr, Liggett SB, Martin RJ, Mitra N, Peters SP, Silverman E, Sorkness CA, Szefler SJ, Wechsler ME, Weiss ST, Drazen JM, National Heart L, Blood Institute's Asthma Clinical Research N (2004) Use of regularly scheduled albuterol treatment in asthma: genotype-stratified, randomised, placebo-controlled cross-over trial. Lancet 364(9444):1505–1512. doi:10.1016/S0140-6736(04)17273-5
116. Genomes Project C, Abecasis GR, Auton A, Brooks LD, DePristo MA, Durbin RM, Handsaker RE, Kang HM, Marth GT, McVean GA (2012) An integrated map of genetic variation from 1,092 human genomes. Nature 491(7422):56–65. doi:10.1038/nature11632
117. Shriner D (2013) Overview of admixture mapping. Curr Protoc Hum Genet Chapter 1:Unit 1.23. doi:10.1002/0471142905.hg0123s76
118. Winkler CA, Nelson GW, Smith MW (2010) Admixture mapping comes of age. Annu Rev Genomics Hum Genet 11:65–89. doi:10.1146/annurev-genom-082509-141523
119. Haldar P, Brightling CE, Hargadon B, Gupta S, Monteiro W, Sousa A, Marshall RP, Bradding P, Green RH, Wardlaw AJ, Pavord ID (2009) Mepolizumab and exacerbations of refractory eosinophilic asthma. New Engl J Med 360(10):973–984. doi:10.1056/NEJMoa0808991
120. Corren J, Lemanske RF, Hanania NA, Korenblat PE, Parsey MV, Arron JR, Harris JM, Scheerens H, Wu LC, Su Z, Mosesova S, Eisner MD, Bohen SP, Matthews JG (2011) Lebrikizumab treatment in adults with asthma. New Engl J Med 365(12):1088–1098. doi:10.1056/NEJMoa1106469
121. Parulekar AD, Atik MA, Hanania NA (2014) Periostin, a novel biomarker of TH2-driven asthma. Curr Opin Pulm Med 20(1):60–65. doi:10.1097/MCP.0000000000000005
122. Gudbjartsson DF, Bjornsdottir US, Halapi E, Helgadottir A, Sulem P, Jonsdottir GM, Thorleifsson G, Helgadottir H, Steinthorsdottir V, Stefansson H, Williams C, Hui J, Beilby J, Warrington NM, James A, Palmer LJ, Koppelman GH, Heinzmann A, Krueger M, Boezen HM, Wheatley A, Altmuller J, Shin HD, Uh ST, Cheong HS, Jonsdottir B, Gislason D, Park CS, Rasmussen LM, Porsbjerg C, Hansen JW, Backer V, Werge T, Janson C, Jonsson UB, Ng MC, Chan J, So WY, Ma R, Shah SH, Granger CB, Quyyumi AA, Levey AI, Vaccarino V, Reilly MP, Rader DJ, Williams MJ, van Rij AM, Jones GT, Trabetti E, Malerba G, Pignatti PF, Boner A, Pescollderungg L, Girelli D, Olivieri O, Martinelli N, Ludviksson BR, Ludviksdottir D, Eyjolfsson GI, Arnar D, Thorgeirsson G, Deichmann K, Thompson PJ, Wjst M, Hall IP, Postma DS, Gislason T, Gulcher J, Kong A, Jonsdottir I, Thorsteinsdottir U, Stefansson K (2009) Sequence variants affecting eosinophil numbers associate with asthma and myocardial infarction. Nat Genet 41(3):342–347. doi:10.1038/ng.323
123. Moffatt MF, Gut IG, Demenais F, Strachan DP, Bouzigon E, Heath S, von Mutius E, Farrall M, Lathrop M, Cookson WO (2010) A large-scale, consortium-based genomewide association study of asthma. New Engl J Med 363(13):1211–1221. doi:10.1056/NEJMoa0906312
124. Barnes KC (2010) Ancestry, ancestry-informative markers, asthma, and the quest for personalized medicine. J Allergy Clin Immunol 126(6):1139–1140. doi:10.1016/j.jaci.2010.10.032
125. Marandi Y, Farahi N, Hashjin GS (2013) Asthma: beyond corticosteroid treatment. Arch Med Sci 9(3):521–526. doi:10.5114/aoms.2013.33179
126. Barnes PJ (2012) New drugs for asthma. Semin Respir Crit Care Med 33(6):685–694. doi:10.1055/s-0032-1326965
127. Piper E, Brightling C, Niven R, Oh C, Faggioni R, Poon K, She D, Kell C, May RD, Geba GP, Molfino NA (2013) A phase II placebo-controlled study of tralokinumab in moderate-to-severe asthma. Eur Respir J 41(2):330–338. doi:10.1183/09031936.00223411
128. Wenzel S, Ford L, Pearlman D, Spector S, Sher L, Skobieranda F, Wang L, Kirkesseli S, Rocklin R, Bock B, Hamilton J, Ming JE, Radin A, Stahl N, Yancopoulos GD, Graham N, Pirozzi G (2013) Dupilumab in persistent asthma with elevated eosinophil levels. New Engl J Med 368(26):2455–2466. doi:10.1056/NEJMoa1304048

129. Wenzel S, Wilbraham D, Fuller R, Getz EB, Longphre M (2007) Effect of an interleukin-4 variant on late phase asthmatic response to allergen challenge in asthmatic patients: results of two phase 2a studies. Lancet 370(9596):1422–1431. doi:10.1016/s0140-6736(07)61600-6
130. Parker JM, Oh CK, LaForce C, Miller SD, Pearlman DS, Le C, Robbie GJ, White WI, White B, Molfino NA, Group M-CT (2011) Safety profile and clinical activity of multiple subcutaneous doses of MEDI-528, a humanized anti-interleukin-9 monoclonal antibody, in two randomized phase 2a studies in subjects with asthma. BMC Pulm Med 11:14. doi:10.1186/1471-2466-11-14
131. Holgate ST, Noonan M, Chanez P, Busse W, Dupont L, Pavord I, Hakulinen A, Paolozzi L, Wajdula J, Zang C, Nelson H, Raible D (2011) Efficacy and safety of etanercept in moderate-to-severe asthma: a randomised, controlled trial. Eur Respir J 37(6):1352–1359. doi:10.1183/09031936.00063510
132. Risma KA, Wang N, Andrews RP, Cunningham CM, Ericksen MB, Bernstein JA, Chakraborty R, Hershey GK (2002) V75R576 IL-4 receptor alpha is associated with allergic asthma and enhanced IL-4 receptor function. J Immunol 169(3):1604–1610
133. Vogel F (1959) Moderne problem der humangenetik. Ergeb Inn Med Kinderheilkd 12:52–125

Pharmacogenetics and Antineoplastic Therapies

Jai N. Patel, Christine M. Walko and Federico Innocenti

Abstract The genome of cancer cells differs from that of the host cell from which it arises. These changes in molecular pathways drive cellular proliferation and ultimately tumour growth and progression. As a result, there has been a shift from categorising tumours solely based on their tissue of origin and histology to consideration of their molecular profiles. This transformation has led to the current breadth of treatments available, including largely targeted therapies. When individualising cancer therapy, it is essential to evaluate the expected individual drug exposure, risk for toxicity, and expected drug efficacy; however, a large heterogeneity in response to antineoplastic therapies exists across the human population. Dose-limiting toxicities often lead to dose reductions and delays in therapy, even in a potentially curative setting. It is therefore imperative that clinicians be able to identify the patients most likely to benefit from treatment and those at an increased risk of toxicity. This chapter will focus on the pharmacogenetic associations of antineoplastic therapies, and the prospective identification of clinically validated and/or utilised germ-line and somatic biomarkers, maximising clinical efficacy and minimising toxicity.

Keywords Pharmacogenetics · Germ-line · Somatic · Predictive · Biomarker · Antineoplastics · Cancer

---

F. Innocenti (✉)
UNC Institute for Pharmacogenomics and Individualized Therapy, University of North Carolina Eshelman School of Pharmacy, CB 7361, 120 Mason Farm Road, Chapel Hill, NC 27599-7361, USA
e-mail: innocent@unc.edu

J. N. Patel
Levine Cancer Institute, Carolinas HealthCare System, 1021 Morehead Medical Drive, Charlotte, NC 28204, USA

C. M. Walko
The DeBartolo Family Personalized Medicine Institute, Moffitt Cancer Center, 12902 Magnolia Drive, Tampa, FL 33612, USA

G. Grech, I. Grossman (eds.), *Preventive and Predictive Genetics: Towards Personalised Medicine,* Advances in Predictive, Preventive and Personalised Medicine 9,
DOI 10.1007/978-3-319-15344-5_10

1 Introduction

Administration of an equivalent dose of an antineoplastic drug to a randomly selected population of cancer patients has been shown to result in a range of toxicities, from mild to lethal events [1]. Dose-limiting toxicities often lead to dose reductions and delays in therapy, even in the potentially curative setting. Additionally, poor prediction of driver tumour abnormalities, compensatory cellular pathways leading to resistance, and inter-individual variability of exposure to active drug may result in marginal efficacy of current regimens in the adjuvant and metastatic setting [2]. Drug exposure above a therapeutic threshold often results in severe toxicity without additional clinical benefit, while insufficient drug concentration at the site of action may result in a sub-therapeutic response. Secondary to the inherent narrow therapeutic nature of antineoplastics, the identification of optimal methods for individualising cancer therapy is in itself imperative [3]. Pharmacogenomics, or the study of how genetic inheritance influences one's response to medications, has the potential to change the way medicine is practiced. This shift in cancer treatment can potentially be achieved through optimising drug and dose selection to minimise toxicity and maximise clinical benefit [4].

Since the completion of the Human Genome Project in 2003 and the unveiling of the architecture of somatic alterations in many tumour types through The Cancer Genome Atlas Network, genomics has become a mainstay of cancer research [5]. As a result, there has been a shift from categorising tumours solely based on their tissue of origin and histology to consideration of their molecular profiles. Ultimately, this shift has introduced a transition from the drug development of broadly acting cytotoxic agents to specific targeted therapies, allowing for the current breadth of treatments available. By utilising both germ-line and somatic DNA, pharmacogenomics provides a practical method to optimise systemic drug exposures, to decrease adverse drug reactions, and to maximise clinical efficacy, thereby improving the survival and quality of life of patients. This chapter will highlight the major advances made in pharmacogenetics of antineoplastics, including many clinically validated and/or utilised genetic predictors of drug efficacy and/or toxicity.

2 Approaches to Genomic Cancer Medicine: Germ-Line versus Somatic DNA, Prognostic versus Predictive Biomarkers

A pharmacogenetic or pharmacogenomic marker is defined as any genetic or genomic marker (or set of markers) that is associated with drug response. Genomic cancer research provides a unique opportunity to analyse two sets of DNA: the patient and the tumour, also referred to as germ-line DNA and somatic DNA, respectively. These two sets of DNA differ in many ways, including, but not limited to, location, accessibility, and clinical utility. While germ-line DNA mutations are found within germ cells (sperm or egg), and are therefore inherited and transmitted to the offspring, somatic DNA mutations occur within any other cell (i.e. tumour cell) after conception and subsequently are not passed on to progeny (see review [6, 7]).

Clinically relevant germ-line mutations are often useful for determining both pharmacokinetic and pharmacodynamic responses, including variations in the enzymes thiopurine-S-methyltransferase (TPMT), cytochrome P450 2D6 (CYP2D6), uridine-diphosphate glucuronosyltransferase 1A1 (UGT1A1), and dihydropyrimidine dehydrogenase (DPD). Determining the activity of these enzymes can ultimately help to predict drug or active metabolite exposure and the resulting implication on clinical response and toxicity. Because of their acquired nature, somatic mutations are useful in evaluating pharmacodynamic effects, and ultimately tumour response [6]. A few well known examples include presence of BCR-ABL in chronic myeloid leukaemia (CML), *KRAS* mutations in metastatic colorectal cancer, the epidermal growth factor receptor (*EGFR*) mutations in lung cancer, the overexpression of human EGFR-2 (HER2) in breast and gastric cancer, BRAF in metastatic melanoma, and anaplastic lymphoma kinase (ALK) in lung cancer. These mutated proteins drive affected cells toward transformation and addict cells to their abnormal function, where the continued presence of the abnormal function becomes required for the tumour cell to survive. This forms the basis of targeted therapies, where blockage of these mutations ultimately deprives the cancer cell from stimulation, growth, and survival (see review [8]).

Biomarkers in general can be broadly categorised into two main types: prognostic and predictive. A prognostic biomarker is a single measurable trait, or combination of measurable traits, that has been shown to be associated with clinical outcome, regardless of treatment (see review [9]). Notable examples include the 70-gene profile, MammaPrint™, for oestrogen/progesterone receptor (ER/PR) positive, lymph node-negative breast cancer [10] and microsatellite instability (MSI) in colorectal cancer patients [11]. Gene expression arrays such as MammaPrint™ help to determine the risk of recurrence in women with early stage breast cancer and help to guide treatment recommendations based on recurrence risk [10]. High-frequency MSI (MSI-H), often found in patients with Lynch Syndrome, may indicate a favourable prognosis compared to microsatellite stable/low-frequency MSI (MSS/MSI-L), independently of chemotherapy, in local and advanced colorectal cancer [11, 12]. Additionally, biomarkers such as *BRAF* mutations have been found to have stronger prognostic significance than predictive in certain tumours such as colorectal, as opposed to being a highly predictive biomarker in tumours such as melanoma [13, 14].

A predictive biomarker is a single measurable trait, or combination of measurable traits, that predicts response to specific therapies and can be used to prospectively identify patients who are most likely to benefit from treatment and/or those predisposed to toxicity (see review [15]). Notable examples of predictive biomarkers are listed in Table 1 and are discussed in further detail in the remainder of the chapter. Additionally, some biomarkers are classified as both prognostic and predictive, such as overexpression of HER2 and presence of ER/PR.

Table 1 Summary of biomarker classifications, antineoplastic drugs, and FDA package insert information

Biomarker	Germ-line (G) or Somatic (S)	Predictive or prognostic or both	Antineoplastic drug(s)	Excerpt from FDA package insert regarding the biomarker[a]
CYP2D6	G	Predictive	Tamoxifen	None
DPD	G	Predictive	Capecitabine, 5-fluorouracil	"Capecitabine is contraindicated in patients with known DPD deficiency" "Rarely, unexpected, severe toxicity (e.g., stomatitis, diarrhea, neutropenia and neurotoxicity) associated with 5-fluorouracil has been attributed to a deficiency of DPD activity. A link between decreased levels of DPD and increased, potentially fatal toxic effects of 5-fluorouracil therefore cannot be excluded"
TPMT	G	Predictive	6MP; azathioprine	"Homozygous-deficient patients, if given usual doses of 6-MP, accumulate excessive cellular concentrations of active thioguanine nucleotides predisposing them to 6MP toxicity" "Patients with deficient or null TPMT activity are at increased risk for severe 6MP toxicity from conventional doses of 6MP and generally require substantial dose reduction"
UGT1A1	G	Predictive	Irinotecan; nilotinib	"Individuals who are homozygous for the UGT1A1*28 allele are at increased risk for neutropenia following initiation of irinotecan treatment" "…a reduction in the starting dose by at least one level of irinotecan should be considered for patients known to be homozygous for the UGT1A1*28 allele"
ALK	S	Predictive	Crizotinib	"…is indicated for the treatment of patients with locally advanced or metastatic NSCLC that is anaplastic lymphoma kinase (ALK)-positive as detected by an FDA- approved test" "Detection of ALK-positive NSCLC is necessary for selection of patients for treatment with crizotinib"

Table 1 (continued)

Biomarker	Germ-line (G) or Somatic (S)	Predictive or prognostic or both	Antineoplastic drug(s)	Excerpt from FDA package insert regarding the biomarker[a]
BCR-ABL	S	Both	Imatinib, dasatinib, nilotinib, bosutinib, ponatinib	"…indicated for the treatment of newly diagnosed adult and pediatric patients with Ph+CML and adult patients with relapsed or refractory Ph+ALL." (Imatinib, dasatinib) "…indicated for the treatment of Ph+chronic CML in adult patients resistant to or intolerant to prior therapy that included imatinib." (Nilotinib) "…indicated for the treatment of adult patients with chronic, accelerated, or blast phase Ph+chronic myelogenous leukemia (CML) with resistance or intolerance to prior therapy." (Bosutinib) "…for the treatment of adult patients with Ph+ALL that is resistant or intolerant to prior tyrosine kinase inhibitor therapy." (Ponatinib)
BRAF	S	Prognostic (colorectal) Both (melanoma)	Vemurafenib	"Before taking vemurafenib, patients must have BRAF V600 mutation –positive tumor status confirmed…" "The efficacy and safety of vemurafenib in patients with tumors expressing BRAF V600 non-E mutations have not been convincingly established"
C-KIT	S	Both	Imatinib	"…indicated for the treatment with KIT (CD117) positive unresectable, resectable, and/or metastatic malignant GIST"
EGFR	S	Predictive	Erlotinib	"Erlotinib inhibits the intracellular phosphorylation of tyrosine kinase associated with the epidermal growth factor receptor (EGFR)" "…indicated for first-line use in patients with metastasized NSCLC that tests positive for EGFR mutations"

Table 1 (continued)

Biomarker	Germ-line (G) or Somatic (S)	Predictive or prognostic or both	Antineoplastic drug(s)	Excerpt from FDA package insert regarding the biomarker[a]
ER/PR	S	Both	Tamoxifen, exemestane, letrozole, anastrozole, fulvestrant	"Available evidence indicates that patients whose tumors are estrogen receptor positive are more likely to benefit from tamoxifen therapy" "The estrogen and progesterone receptor values may help to predict whether adjuvant tamoxifen therapy is likely to be beneficial" "Exemestane is indicated for adjuvant treatment of postmenopausal women with estrogen-receptor positive early breast cancer…interferes with estrogen-induced stimulation or maintenance of growth of hormonally responsive (ER/PR+) breast cancers." (Same for all anti-hormonal therapies)
HER2	S	Both	Trastuzumab, pertuzumab, lapatinib, T-DM1	"…should only be used in patients whose tumors have HER2 protein over-expression." (Same for all therapies targeting HER2)
KRAS	S	Both (NSCLC) Predictive (colorectal)	Cetuximab, panitumumab	"Retrospective subset analyses of trials in patients with colorectal cancers having *KRAS* mutations noted a lack of benefit associated with these monoclonal antibodies" "…indicated for first-line treatment in combination with chemotherapy for patients with metastatic colorectal cancer who have EGFR-expressing, and KRAS wild-type tumors"
PML-RAR alpha	S	Both	Arsenic trioxide ATRA	"…indicated for those where APL is characterized by the presence of the t(15;17) translocation or PML/RAR-alpha gene expression"
CD20 antigen	S	Predictive	Rituximab, ofatumumab, tositumumab, ibritumomab	"Rituximab is indicated for the treatment of patients with relapsed or refractory, low-grade or follicular, CD20-positive, B-cell non-Hodgkin's lymphoma." (Same for tositumumab and ibritumomab; however, for patients refractory to rituximab) "Ofatumumab is a CD20-directed cytolytic monoclonal antibody indicated for the treatment of patients with CLL refractory to fludarabine and alemtuzumab"

Table 1 (continued)

Biomarker	Germ-line (G) or Somatic (S)	Predictive or prognostic or both	Antineoplastic drug(s)	Excerpt from FDA package insert regarding the biomarker[a]
CD30 antigen	S	Predictive	Brentuximab	"Brentuximab is a CD30-directed antibody-drug conjugate indicated for the treatment of patients with Hodgkin lymphoma…"
CD52 antigen	S	Predictive	Alemtuzumab	"Alemtuzumab is a recombinant DNA-derived humanized monoclonal antibody that is directed against CD52"

CYP2D6 cytochrome P450 2D6, *DPD* dihydropyrimidine dehydrogenase, *TPMT* thiopurine-S-methyltransferase, *6MP 6* mercaptopurine, *UGT1A1* uridine-diphosphateglucuronosyltransferase, *ALK* anaplastic lymphoma kinase, *Ph* Philadelphia chromosome, *EGFR* epidermal growth factor receptor, *ER/PR* estrogen receptor/progesterone receptor, HER2 human epidermal growth factor receptor 2, *T-DM1* trastuzumab emtansine, *ATRA* all trans retinoic acid, *NSCLC* non-small-cell lung cancer; *CD* cluster of differentiation

[a] FDA website for table of drugs with pharmacogenomic biomarkers on drug label: http://www.fda.gov/drugs/scienceresearch/researchareas/pharmacogenetics/ucm083378.htm

3 Germ-Line Predictive Markers of Anticancer Efficacy and/or Toxicity

3.1 *Thiopurine S-Methyltransferase (TPMT)*

Of the germ-line predictive markers identified, *TPMT* pharmacogenetics is one of the original hallmark examples of how a germ-line genetic mutation can impact clinical outcomes and drug toxicity, particularly in children with acute lymphoid leukemia (ALL). 6-mercaptopurine (6-MP) is one of the backbone agents used to treat ALL. It inhibits the formation of purine ribonucleotides through the formation of thioguanine nucleotide analogues, ultimately leading to cell death. TPMT is a cytosolic enzyme ubiquitously expressed in the human body and catalyses the S-methylation of thiopurines into inactive compounds. Approximately 6–11 % of the population is heterozygous for the defective variants of TPMT and 0.2–0.6 % is homozygous resulting in null enzyme activity (see review [16]). TPMT *2 and *3 account for 95 % of defective TPMT activity. The risk of myelosuppression is significantly elevated in patients who, secondary to a TPMT deficiency, have increased thioguanine nucleotides [16, 17].

In a study of 180 children with ALL treated with a 6-MP-based regimen, patients were genotyped and had TPMT enzyme activity measured. A significant inverse relationship was observed between concentration of thioguanine nucleotides and TPMT enzyme activity. The percentage of wild-type, heterozygous, and homozygous-deficient patients who were able to tolerate the full dose throughout treatment was 84, 65, and 7 %, respectively ($P < 0.001$) [18]. The Clinical Pharmacogenetics Implementation Consortium (CPIC) recommends an initial 90 % dose reduction in homozygous variant patients and a 30–70 % dose reduction in heterozygous patients [19]. The Food and Drug Administration (FDA) recommends, but does not require, genetic testing prior to treatment with 6-MP (Table 1) [20]. As a result, a number of institutions, including St. Jude Children's Hospital, routinely test all patients receiving 6-MP for *TPMT* genetic variants. As noted in Table 2, several key regulatory bodies support genetic testing of *TPMT* to individualise 6-MP dosing.

3.2 *UDP Glucuronosyltransferase 1 Family, Polypeptide A1, UGT1A1*

Irinotecan is effective in a variety of cancers including those of the gastrointestinal tract and lung. It is metabolised by carboxylesterases to the active metabolite, SN-38. UGT1A1 conjugates SN-38 to an inactive glucuronide metabolite, SN-38G, which can undergo enterohepatic recirculation, and is eventually excreted in the bile and urine (see reviews [21, 22]). Polymorphic *UGT1A1* variants may lead to a reduction in enzyme activity or expression, resulting in an increase in SN-38, and ultimately an increased risk of severe neutropenia. The wild type allele is referred

Table 2 Summary of biomarker implication on treatment, clinical assay availability, and incorporation into guidelines and practice

Biomarker	Clinical assay(s) available[a]	Clinical implication	Federal/Regulatory/ National Guidelines supporting association/use	Year of discovery[b, c]	Year of implementation of first targeted therapy or genetic test
CYP2D6	AmpliChip Cytochrome P450 Genotyping test[a], xTAG CYP2D6, P450-Glo	Altered tamoxifen metabolism (potential for increased rates of breast cancer recurrence) [30, 106]	DPWG	2005 [30, 106]	Genetic test not yet frequently utilised
DPD	DPD Enzyme Assay	Increased risk of 5-fluorouracil or capecitabine induced toxicity [39]	CAP, CPIC, DPWG, FDA, NICE	1998 [39]	Genetic test not yet frequently utilised
TPMT	TPMT Enzyme diagnostic test, Prometheus TPMT genetic test	Increased risk of 6-mercaptopurine induced neutropenia [18]	ASCO, CAP, CPIC, DPWG, EMA, ESMO, FDA, NCCN, NICE	1999 [18]	2010 [25]
UGT1A1	Invader UGT1A1 Molecular Assay[a], INFINITI UGT1A1 Assay	Increased risk of irinotecan or nilotinib induced toxicity [23]	CAP, DPWG, FDA, NCCN	2004 [23]	Genetic test not yet frequently utilised
ALK	Vysis ALK Break Apart FISH probe kit[a]	Sensitivity to crizotinib [43]	ASCO, CAP, FDA, EMA, ESMO, NCCN, NICE	2007 [107]	2010 [43]
BCR-ABL (Ph chromosome)	BCR-ABL Mutation test	Sensitivity to imatinib, dasatinib, nilotinib, bosutinib, ponatinib [108]	ASCO, CAP, FDA, EMA, ESMO, NCCN, NICE	1973 [109]	2001 [110]
BRAF	Cobas 4800 V600E[a]	Sensitivity to vemurafenib [13]	ASCO, CAP, FDA, EMA, ESMO, NCCN, NICE	2001 [111]	2011 [13]
C-KIT	c-Kit pharmDx[a]	Sensitivity to imatinib [59]	ASCO, CAP, FDA, EMA, ESMO, NCCN, NICE	2001 [112]	2001 [112]

Table 2 (continued)

Biomarker	Clinical assay(s) available[a]	Clinical implication	Federal/Regulatory/ National Guidelines supporting association/use	Year of discovery[b, c]	Year of implementation of first targeted therapy or genetic test
EGFR	EGFR pharmDx[a], EGFR Mutation Assay[a] Cobas EGFR Mutation Test[a]	Sensitivity to erlotinib [65]	ASCO, CAP, FDA, EMA, ESMO, NCCN, NICE	2004 [64]	2003 [113]
ER/PR	ER/PR pharmDx[a], TargetPrint	Sensitivity to hormonal therapy (tamoxifen, exemestane, letrozole, anastrozole) [71]	ASCO, CAP, FDA, EMA, ESMO, NCCN, NICE	1896 [114]	1896[d] [114]
HER2	HercepTest[a], Pathway[a], Insite[a], PathVysion[a], SPOT-Light[a], HER2 CISH[a]	Sensitivity to trastuzumab, pertuzumab, lapatinib, T-DM1 [115–117]	ASCO, CAP, FDA, EMA, ESMO, NCCN, NICE	1985 [118]	2001 [119]
KRAS	Therascreen KRAS RGQ PCR Kit[a], DxS KRAS Mutation Test Kit, Genzyme's KRAS Mutation Analysis	Resistance to cetuximab or panitumumab [14]	ASCO, CAP, FDA, EMA, ESMO, NCCN, NICE	1987 [120]	2006 [121]
PML-RAR alpha	PML/RARA t(15;17) Quantitative Real-time PCR; FISH AML M3,	Sensitivity to all-trans-retinoic acid and arsenic trioxide [55]	ASCO, CAP, FDA, EMA, ESMO, NCCN, NICE	1990 [122]	1987 [123]

NSCLC non-small-cell lung cancer, *ASCO* American Society of Clinical Oncology, *FDA* Food and Drug Administration, *EMA* European Medicines Agency, *ESMO* European Society for Medical Oncology, *NCCN* National Comprehensive Cancer Network, *CPIC* Clinical Pharmacogenetics Implementation Consortium, *DPWG* Dutch Pharmacogenetics Working Group, *NICE* National Institute for Health and Care Excellence, *CAP* College of American Pathologist

[a] FDA-cleared (and CLIA-approved); Clinical assay(s) available adapted from Patel et al. [124]

[b, c] For germ-line variants (CYP2D6, DPD, TPMT, UGT1A1), the year of first discovery of the association with clinical phenotype was used. For somatic variants (all others), the year of the discovery of the actual aberration was used

[d] 1896—oophorectomy; date of approval of first modern day anti-hormonal therapy: 1998 (letrozole)

to as *UGT1A1**1, while the most common and studied variants include *28, *93, *60, and *6; *UGT1A1**28 leads to a 70 % reduction in expression of the gene. The frequency of the *28 allele is 39 % in patients of European, 16 % of Asian, and 43 % of African origins [22].

The *UGT1A1* genotype and total bilirubin levels were found to be strongly associated with severe neutropenia in 66 patients with advanced cancers receiving irinotecan 350 mg/m^2 every 3 weeks. The relative risk of grade 4 neutropenia was 9.3 (95 % CI, 2.4–36.4) for patients with the *28/*28 genotype versus wild-type and heterozygous patients [23]. One meta-analysis demonstrated a significant increase in toxicity for *28/*28 patients at medium (150–250 mg/m^2) and high doses (>250 mg/m^2) (odds ratio [OR] 3.2 and 27.8, respectively) [24]. The FDA label recommends an initial dose reduction be considered for patients who are homozygous for the *UGT1A1* *28 allele [20]. The Clinical Pharmacogenetics Implementation Consortium (CPIC) and Dutch Pharmacogenetics Working Group (DPWG) guidelines suggest an initial dose reduction by 30 % in *28/*28 patients if the starting recommended dose is >250 mg/m^2 [2, 25]. Additionally, new studies define individualised dosing of irinotecan. A genotype-guided phase I dose escalation study defined the irinotecan maximum tolerated dose (MTD) to be 370 and 310 mg/m^2 for *1/*1 and *1/*28 patients, respectively, suggesting higher tolerable doses in wild-type and heterozygous patients [26]. The observed toxicities were mainly gastrointestinal and neutropenia. A subsequent and similar genotype-guided phase I study determined the MTD in *1/*1, *1/*28, and *28/*28 patients to be 390, 340, and 150 mg/m^2, respectively [27].

3.3 *Cytochrome P450, Family 2, Subfamily D, Polypeptide 6 (CYP2D6)*

The selective Estrogen Receptor (ER) modulator, tamoxifen, is widely used in locally advanced and metastatic ER positive breast cancer patients. Tamoxifen metabolism via CYP2D6 produces a more potent metabolite, endoxifen. *CYP2D6**1 is the wild-type allele (resulting in normal enzyme activity), while *3, *4, and *5 result in null enzyme activity, and *9, *10, and *17 result in decreased enzyme activity. Patients with two *CYP2D6* alleles with normal enzyme activity are considered to be 'extensive metabolisers (EM).' Patients with one or more *CYP2D6* alleles with reduced enzyme activity or heterozygous for an inactive allele are considered to be 'intermediate metabolisers (IM),' while patients with two inactive alleles are considered to be 'poor metabolisers (PM)' [28]. The frequency of the most common variant allele, *4, is approximately 18 % in patients of European origin, while rarely (<1 %) found in patients of Asian origin and moderately (7–9 %) in African origin [29]. Additionally, the frequency of the *10 allele is approximately 40 % in patients of Japanese origin, while the *17 allele frequency is about 25 % in patients of African origin [30, 29].

A retrospective review in 225 breast cancer patients who received adjuvant tamoxifen identified a significantly shorter time to recurrence (hazard ratio [HR] 3.20; $P=0.007$) and worse 2-year relapse-free survival (RFS) (HR 2.69; $P=0.009$) in PMs (*4/*4) as compared to EM patients [31]. A subsequent retrospective study demonstrated higher recurrence rates for PMs and IMs when compared to EMs (29 and 20.9% vs. 14.9%, respectively); compared with EMs, there was a significantly increased risk of recurrence for IMs (HR 1.40; $p<0.05$) and PMs (HR 1.90; $p<0.05$) [32]. An ongoing debate exists on the clinical utility of *CYP2D6* testing as two large studies published in 2011 demonstrated no association between genotype and recurrence rates [33, 34]; however, major flaws in the genomic study design, such as use of tumour DNA as opposed to germ-line DNA, may help explain these opposing results (see commentary [35]). Advocates of the uptake of *CYP2D6* genetic testing indicate that until prospective, adjuvant trial data is available, the current evidence is sufficient to accept the *CYP2D6*-tamoxifen pharmacogenetic relationship in postmenopausal women [36]; however, the appropriate dose adjustment and ultimate impact on breast cancer recurrence appear to be unclear. The DPWG recommends consideration of aromatase inhibitor (AI) therapy in postmenopausal women who are PMs and IMs (along with avoidance of concomitant CYP2D6 inhibitors) [37]. Regardless of expert recommendations, very few regulatory bodies support the use of *CYP2D6* genetic testing (Table 2). Of the validated germ-line biomarkers, *CYP2D6* is the only one not included on the FDA label (Table 1).

Dosing of tamoxifen may be dependent upon the *CYP2D6* genotype. An interventional genotype-guided study in patients of largely Caucasian origin demonstrated a significant increase in endoxifen concentrations in PM and IM patients when doubling the dose from 20 to 40 mg daily [28]. A subsequent study in patients of Japanese origin identified a significant 1.4 and 1.7-fold higher concentration of plasma endoxifen in *CYP2D6* *1/*10 and *10/*10 patients, when dosed at 30 and 40 mg/day, respectively, compared to concentrations prior to dosage increase (all patients received 20 mg/day of tamoxifen before the intervention) [38]. Additionally, these plasma concentrations of endoxifen achieved similar levels of those in *CYP2D6* *1/*1 patients receiving 20 mg/day of tamoxifen [38].

3.4 Dihydropyrimidine Dehydrogenase, DPD

Fluorouracil has been the backbone therapy of several cancers, including but not limited to colon, rectal, and gastric. DPD is the rate-limiting enzyme involved in the metabolism of fluorouracil to the inactive 5,6-dihydrofluorouracil (see review [39]). Genetic variations in the gene coding for DPD, *DPYD*, resulting in deficient DPD activity is estimated to occur in approximately 1% of patients, which has the potential to prolong the half-life of fluorouracil by up to 100-fold, ultimately increasing the risk of fluorouracil-induced toxicity, namely neutropenia and/or diarrhea (see review [40]). To date, among the many proposed defective variants of *DPYD*, IV14+1G>A is the one that has shown the most predictive value. This variant leads to splicing of mature DPD mRNA. Only 1.8% of the overall population carries at least one variant allele [41].

In one study, sixty patients who experienced severe fluorouracil-induced toxicity were genotyped for the splice site mutation [41]. Results indicated that 28% of all patients were heterozygous or homozygous for the mutation compared to 3% who were wild-type ($P<0.001$). An overall decreased DPD activity was noted in roughly 60% of all cases, independently of genotype [41]. In a multicenter trial of 683 patients with various cancers treated with fluorouracil monotherapy, patients were genotyped for *DPYD*. Investigators noted a positive predictive value of 0.46 for overall toxicity in patients with the splice site mutation; however, of interest was a gene/sex-interaction resulting in an OR for toxicity of 41.8 ($P<0.0001$) for male patients but only 1.33 ($P>0.05$) in female patients [42]. Establishing DPD deficiency measuring catalytic DPD activity in peripheral blood mononuclear cells might be challenging, as standardised cut-offs for deficient DPD activity are not available and different laboratories use different thresholds. The FDA label includes a warning for increased risk of toxicity in patients with deficient DPD activity; however, the definition of what constitutes "deficient" is not defined, nor does the FDA mandate genetic testing for any SNPs in *DPYD* (Table 1) [20].

4 Somatic Predictive Markers of Anticancer Efficacy

4.1 ALK

EML4-ALK, a fusion gene found in several types of cancers including anaplastic large-cell lymphoma, neuroblastoma, inflammatory myofibroblastic tumours, and non-small-cell lung cancer (NSCLC), results in constitutive kinase activity allowing activation of the RAS-MEK-ERK, JAK3-STAT3, and PI3K-AKT pathways [43]. This translocation is relatively uncommon occurring in approximately 2–7% of all NSCLC cases [43]. Crizotinib, an oral ATP-competitive selective inhibitor of ALK and MET tyrosine kinases, was granted accelerated approval in 2011 by the FDA, along with a companion diagnostic test for the *ALK* genetic abberation. The FDA label indicates genetic testing is necessary to confirm an *ALK* positive tumour prior to a patient receiving crizotinib (Table 1). Additionally, several key regulatory bodies support genetic testing for *ALK*, along with a number of other somatic biomarkers included in this chapter (Table 2).

In an open-label, multicenter, two-step phase I trial, 82 NSCLC patients harboring the fusion *ALK* gene rearrangements were treated with crizotinib. An overall response rate of 57% was observed along with a disease control rate of 87% at 8 weeks. Historical data indicates a 15–20% response rate with standard chemotherapy [43]. The probability of progression free survival (PFS) at 6 months was estimated at 72% [43]. The impressive activity and clinical benefit from phase I and II clinical trials led to the rapid FDA approval of crizotinib for treatment in

ALK positive NSCLC, prior to validation in phase III clinical trials (see review [44]). As with many targeted therapies, the initial therapeutic benefit is short-lived secondary to acquired resistance, which may present in the form of novel *EGFR*, *KIT*, or *ALK* mutations not previously identified [45].

4.2 BCR-ABL

The development of targeted treatment was made promising by landmark discoveries that identified the chromosomal abnormality responsible for CML: the Philadelphia chromosome (Ph) formed by a translocation between chromosomes 9 and 22, resulting in a fusion between the breakpoint cluster region (*BCR*) and the *c-ABL* oncogene [46]. This chimeric BCR-ABL tyrosine kinase (TK) occurs in >95 % of all CML cases and causes constitutive activation, triggering numerous signal transduction pathways associated with cell survival, proliferation, and resistance to apoptosis [47]. The original tyrosine kinase inhibitor (TKI), imatinib, was discovered in a high-throughput screening assay and was subsequently approved in 2001. As indicated in Table 1, the FDA label recommends therapy with TKIs in hematological malignancies which are Ph positive.

At a 5-year follow up, imatinib demonstrated significantly higher rates of complete hematologic response compared to interferon-alfa plus Ara-C (97 vs. 56 %, $P<0.001$), and significantly higher major and complete cytogenetic response (85 and 74 % vs. 22 and 8 %, respectively, $P<0.001$) [48]. The substantial superiority of imatinib over standard chemotherapy resulted in the study results being disclosed early and most patients being crossed over to the imatinib arm [48]. Secondary to emerging resistance, subsequent TKIs with slightly different mechanisms of action were developed, including nilotinib, dasatinib, bosutinib, and ponatinib. Approximately 10 % of patients will present initially with resistance to TKIs and close to 20 % will develop resistance over time (1–5 years) [49]. The primary mechanisms of resistance are point mutations occurring in the kinase domain resulting in steric hindrance; however, newer agents such as ponatinib feature a carbon-carbon triple bond, which allows the molecule to take up a position with no steric hindrance, ultimately inhibiting a variety of TKs and overcoming the initial resistance (see review [50]).

4.3 PML-RAR Alpha

Similarly to BCR-ABL in CML, the routine detection of the *PML-RAR alpha* translocation has provided a significant and widely incorporated predictive biomarker for diagnosis and treatment of acute promyelocytic leukaemia (APL). Nearly all APL cases, which account for approximately 10 % of all acute myeloid leukaemia cases, express a translocation between chromosomes 15 and 17 (see review [51]). The resulting fusion between the PML gene on chromosome 15 and the retinoic acid receptor alpha gene on chromosome 17, t(15;17), allows increased differentiation of leukaemic cells [51]. This disease has traditionally been the most malignant form of acute leukemia with a fatality within weeks, without treatment. Chemotherapy (daunorubicin, idarubicin, and cytosine arabinoside) became the initial front-line therapy resulting in a complete remission (CR) rate of 75%; however, nearly all patients rapidly progressed within 11 to 25 months [52].

Treatment with all-trans-retinoic acid (ATRA) directly down-regulates PML-RAR alpha, subsequently inducing disease remission in 90–95% of newly diagnosed patients with a 5-year disease free survival (DFS) of 75% (see review [52]). In a phase III study comparing ATRA to daunorubicin, the rates of overall survival (OS) at one, two, and 3 years after entry into the study were 75, 57, and 50%, respectively, among patients assigned to chemotherapy, and 82, 72, and 67% among those assigned to ATRA ($P=0.003$) [53]. Unfortunately, relapse frequently ensues within months following treatment, which is routinely followed by combination chemotherapy [54]. Another agent, arsenic trioxide, triggers the rapid degradation of PML-RAR alpha through the targeting of the promyelocytic leukemic moieties of the fusion protein, demonstrating a high CR rate (close to 60%) in relapsed APL (see review [55]). As a result, current therapeutic approaches for APL in patients with PML-RAR alpha often require a combination approach involving ATRA plus arsenic trioxide plus or minus chemotherapy (anthracycline or gemtuzumab ozogamicin). As noted in Table 1, treatment of APL with these agents requires confirmatory genetic testing to identify PML-RAR alpha gene expression.

4.4 *BRAF*

Approximately 40–60% of cutaneous melanomas on non-sun damaged skin carry mutations in *BRAF* (frequency is approximately 15% on sun damaged skin), which led to constitutive activation of downstream signalling through the MAPK pathway. Roughly 90% of these mutations result in the substitution of glutamic acid for valine at codon 600 (V600E). Vemurafenib, a potent inhibitor of *BRAF* V600E, has marked antitumour activity against metastatic melanoma tumours harboring the *BRAF* mutation and was subsequently approved in 2011 [13]. As noted in Table 1, verification of positivity for *BRAF* V600E is required prior to therapy with vemurafenib.

In a large randomised phase III trial, 675 untreated metastatic melanoma patients expressing the *BRAF* mutation were randomly assigned to receive standard of care (dacarbazine) or vemurafenib. Results demonstrated a HR for death in the vemurafenib arm of 0.37 when compared to dacarbazine ($P<0.001$). At 6 months, OS was 84 and 64% in the vemurafenib and dacarbazine arm, respectively (further follow up is ongoing). The HR for tumour progression in the vemurafenib group was 0.26 ($P<0.001$) [13]. Despite dramatic initial responses, resistance eventually occurs in all patients with a mean time to diagnosis at first lesion at roughly 10 weeks.

Likely resistance mechanisms include the paradoxical activation of the MAPK pathway through *RAS* mutations and signalling of downstream pathway proteins such as MAPK kinase (MEK). Combining a BRAF inhibitor with a MEK inhibitor may thus help overcome this resistance mechanism. In fact, a randomised study demonstrated significantly improved PFS (4.8 vs. 1.5 months; HR 0.47; $P<0.0001$) in patients receiving a MEK inhibitor, trametinib, in combination with a selective BRAF inhibitor, dabrafenib [56]. Both of these drugs ultimately received FDA approved indications in 2013 for the treatment of patients with unresectable or metastatic melanoma with *BRAF* V600E or V600K mutation who have not already received a BRAF inhibitor.

4.5 C-KIT

The *C-KIT* proto-oncogene (also known as CD117) encodes a receptor TK, for which the stem cell factor is the ligand. This interaction results in the development of melanocytes, erythrocytes, germ cells, and mast cells, owing to dimerization, autophosphorylation and continuous signal transduction [57, 58]. Roughly 85 % of gastrointestinal stromal tumours (GIST) express gain-of-function C-KIT mutations in exon 9 or 11 [59]. These tumours are often larger in size and more frequently invade adjacent tissues compared to mutation-negative tumours [60]. Before the identification and targeting of C-KIT with imatinib, tumours with gain-of-function *C-KIT* mutations were more aggressive and resulted in a poorer prognosis. However, the addition of imatinib resulted in a partial response rate of 83.5 % for GISTs harboring the exon 11 *C-KIT* mutations. This was compared to patients with tumours containing an exon 9 *C-KIT* mutation or no detectable mutation, having partial response rates of 47.8 % ($P=0.0006$) and 0 % ($P<.0001$), respectively [59]. While imatinib primarily induces tumour shrinkage in GIST via C-KIT inhibition, further inhibition of other TKs, such as platelet derived growth factor receptor-A (PDGFR-A), may play an additional role in the observed antitumour activity of imatinib [61]. As a result, imatinib is only FDA indicated once a diagnosis of C-KIT (CD117) positive GIST has been made (identified by qualitative immunohistochemistry) (Table 1). Although molecular genetic analysis of *C-KIT* may also help identify patients with *C-KIT* positive GIST, the FDA only recommends genetic testing to identify genes associated with imatinib resistance, not for initial diagnosis.

4.6 EGFR

Mutations resulting in the activation of EGFR are located in the TK domain and allow constitutive signalling through activation of the PI3K-AKT and RAS-MEK-ERK pathways (see review [62]). Deletions in exon 19 and a missense mutation at exon 21, resulting in an arginine to leucine substitution (L858R), account for 90 % of all *EGFR* activating mutations found in patients with NSCLC. Subgroup analyses from initial clinical trials revealed that patients with certain clinical and histologic characteristics, including female patients of East Asian descent, those with adenocarcinomas, and never-smokers, are more likely to harbor *EGFR* mutations

[63]. Erlotinib, an oral small molecule TKI, binds to the ATP binding site of EGFR, inhibiting the downstream cascade of events, resulting in greater efficacy over traditional chemotherapy. Approximately 20% of NSCLC patients have activating *EGFR* mutations and are candidates for first-line therapy with erlotinib [64]. Erlotinib, an oral small molecule TKI, binds to the ATP binding site of EGFR, inhibiting the downstream cascade of events, resulting in greater efficacy over traditional chemotherapy. Approximately 20% of NSCLC patients have activating *EGFR* mutations and are candidates for first-line therapy with erlotinib (Table 1).

In a randomised phase III trial, investigators evaluated erlotinib versus standard chemotherapy (gemcitabine + carboplatin) as first-line treatment in 165 advanced NSCLC patients harboring activating *EGFR* mutations. The median PFS was significantly longer in the erlotinib arm compared to the chemotherapy arm (13.1 vs. 4.6 months; HR 0.16; $P<0.0001$). The objective response rate (ORR) was 83 and 36% for erlotinib and chemotherapy, respectively ($P<0.0001$). Interestingly, there was a significant association between reduced PFS and the presence of the L858R mutation as compared with a deletion at exon 19 in patients receiving erlotinib (HR 1.92; $P=0.01$) [65]. Two mechanisms of resistance to EGFR inhibitors include a secondary point mutation in *EGFR* (T790M), which blocks the ability for erlotinib to bind to the ATP binding pocket due to steric hindrance, and the amplification of MET, which activates similar downstream signalling pathways [43]. However, newer agents such as afatinib, an irreversible ErbB-family blocker, has shown activity in patients with the L858R activating mutation, exon 19 deletion, and mutations that confer resistance to erlotinib, including the T790M mutation [66].

4.7 *ER/PR*

As mentioned, ER/PR status can be used to measure disease prognosis (both distant and locoregional recurrence) utilising the 21-gene recurrence score, Oncotype Dx®, in lymph-node negative breast cancer [67], and also can be used as a predictive marker to identify patients most likely to respond to hormonal therapies. Close to 75% of all breast cancers are ER positive and grow in response to oestrogen while close to 65% of these are also PR positive. Studies indicate that ER/PR positive breast tumours are 60% likely to respond to hormonal therapy, whereas the response to hormonal therapy in ER/PR negative tumours is decreased to 5–10% [68]. Frequently administered hormonal therapies include tamoxifen, a selective oestrogen receptor modulator, and AIs (anastrozole, exemestane, and letrozole). As indicated in Table 1, only patients whose tumours are ER/PR positive are likely to benefit from these therapies.

A meta-analysis of women with early stage ER/PR positive breast cancer demonstrated that AIs produced significantly lower rates of recurrence compared with tamoxifen in postmenopausal women in the adjuvant setting; however the contribution of *CYP2D6* genetic variants on tamoxifen treatment was unknown in this study and may have contributed to the observed difference [69]. AIs are not an option for premenopausal women, as these drugs may stimulate the ovaries to produce oestrogen through a negative feedback loop, making tamoxifen the preferred therapy for premenopausal women (see review [70]). Additionally, one study in women with early stage ER/PR positive breast cancer demonstrated that continuing adjuvant tamoxifen to 10 years rather than stopping at 5 years (which has been traditionally done in clinical practice) produces a further reduction in recurrence and mortality, particularly after year 10 (recurrence rate ratio 0.90 [95 % confidence interval (CI) 0.79–1.02] during years 5–9 and 0.75 [95 % CI 0.62–0.90] after year 10). In fact, these data speculate that 10 years of tamoxifen treatment can roughly halve breast cancer mortality during the second decade after diagnosis [71].

4.8 HER2

HER2 is a transmembrane TK receptor that helps control cellular signalling and proliferation at normal expression levels; however, in approximately 20–25 % of breast cancers, the *HER2* gene is amplified resulting in HER2 protein overexpression, allowing the breast cancer cell to proliferate uncontrollably, resulting in aggressive tumour growth and progression [72, 73]. As mentioned previously, HER2 represents both a prognostic and predictive biomarker (Table 1), as overexpression is associated with a higher rate of recurrence without therapy (see review [74]) but is also predictive of a beneficial therapeutic response and enhanced survival from several HER2-targeted therapies including trastuzumab, lapatinib, pertuzumab, and T-DM1, over traditional cytotoxic chemotherapy. As indicated in the FDA label, these targeted agents should only be used in tumours that have HER2 protein overexpression and/or gene amplification (Table 1). Standard practice involves determining HER2 protein over-expression by immunohistochemistry (0 or +1 being negative, +2 being borderline, and +3 being positive), as this method is less costly. If the test result is IHC +2, fluorescence in-situ hybridisation (FISH) must be used to confirm gene amplification.

In one of the original hallmark trials evaluating the clinical utility of trastuzumab, 1694 HER2 positive locally advanced breast cancer patients were randomly assigned to 2 years of treatment with trastuzumab, 1694 randomly assigned to 1 year of trastuzumab, and 1693 assigned to observation [75]. The unadjusted HR for an event (recurrence of breast cancer, contralateral breast cancer, second non-breast malignant disease, or death) in the trastuzumab group, as compared with the observation group, was 0.54 ($P<0.0001$) [75]. Subsequently in a large randomised phase III trial, investigators randomised 3222 women with HER2-positive early-stage breast cancer to receive doxorubicin and cyclophosphamide followed by docetaxel (AC-T), the same regimen plus 52 weeks of trastuzumab, or docetaxel and carboplatin plus trastuzumab (TCH). The estimated DFS at 5 years were 75 % among patients receiving AC-T, 84 % among those receiving AC-T plus trastuzumab (HR 0.64 compared to AC-T; $P<0.001$), and 81 % among those receiving TCH (HR 0.75 compared to AC-T; $P=0.04$) [76]. Lapatinib, a small molecule TKI targeting the intracellular domain of HER2, led to significant improvements in time to progression in advanced or metastatic HER2 positive breast cancer in combination with capecitabine in patients who have progressed on prior therapy (including an anthracycline, a taxane, and trastuzumab) compared to capecitabine alone (HR 0.49; $P<0.001$) [77]. Pertuzumab, an anti-HER2 humanised monoclonal antibody that inhibits receptor dimerisation, has been shown to prolong PFS in combination with trastuzumab and docetaxel compared to placebo, trastuzumab, and docetaxel in HER2-positive metastatic breast cancer patients (HR 0.62; $P<0.001$) [78]. T-DM1, an antibody-drug conjugate incorporating the targeted therapy of trastuzumab with the cytotoxic properties of emtansine (DM1), significantly prolonged PFS and OS with less toxicity than lapatinib plus capecitabine in patients with HER2 positive advanced breast cancer (previously treated with trastuzumab and a taxane) (HR for PFS and OS 0.65 and 0.68, respectively; $P<0.001$ for both) [79].

4.9 KRAS

KRAS, a member of the Ras family of small G proteins, functions as a mediator between the extracellular ligand binding and intracellular signal transduction from the EGFR to the nucleus where transcription occurs (see reviews [80, 81]). The autophosphorylation of the intracellular TK domains at codons 12 and 13 confers constitutive activity of downstream signalling pathways, including RAS-RAF-MAPK and PI3K-AKT pathways (see review [82]). Activating mutations of the *KRAS* oncogene (codon 12 G12V, G12S, G12R, G12D, G12C, G12A, and codon 13 G13D) have emerged as a negative predictive biomarker to identify patients with metastatic colorectal cancer who will not benefit from EGFR inhibitor therapies, panitumumab and cetuximab. Approximately 35–40 % of colorectal cancer patients will harbor *KRAS* mutations [82]. The FDA (Table 1), along with several other regulatory bodies (Table 2), support and recommend testing for *KRAS* status prior to initiating therapy with EGFR inhibitors.

Investigators prospectively evaluated FOLFIRI (fluorouracil, leucovorin, and irinotecan) alone versus FOLFIRI plus cetuximab in patients with evaluable *KRAS* status [14]. The addition of cetuximab to FOLFIRI in patients with *KRAS* wild-type disease resulted in significant improvements in OS (23.5 vs. 20.0 months; HR 0.796) and PFS (57.3 % vs. 39.7 %; OR 2.069) compared with FOLFIRI alone. In patients whose tumours carried mutations in *KRAS*, there was no evidence of a benefit associated with the addition of cetuximab [14]. In a similar trial, investigators prospectively analysed FOLFOX alone versus FOLFOX plus panitumumab by tumour KRAS status in metastatic colorectal cancer patients. In the wild-type *KRAS* patients, the addition of panitumumab significantly improved PFS (HR 0.80; $P=0.02$), while in mutant *KRAS* patients, PFS was significantly reduced in the panitumumab arm (HR 1.29; $P=0.02$) [83].

Additionally, *KRAS* mutations have been identified as a poor prognostic indicator of NSCLC disease and a predictor of resistance to EGFR-TKIs in NSCLC [84,85]. *KRAS* is mutated in approximately 20 % of NSCLC and is associated with adenocarcinoma histology and a positive smoking history [86]. NSCLC patients with *KRAS* mutations treated with erlotinib experienced a shorter median time to progression than patients without *KRAS* mutations (HR 2.14; $P=0.01$); however, *KRAS* mutation status did not appear to affect OS [87]. Although *EGFR* and *KRAS* mutations appear to be mutually exclusive, *KRAS* mutations may represent an important biomarker of non-response to EGFR-TKIs in NSCLC.

5 Chemoimmunotherapy

Chemoimmunotherapy, or chemotherapy combined with immunotherapy, utilises a cytotoxic or cytostatic compound plus immunotherapy which stimulates the body's immune response against the cancer cells (see review [88]). Monoclonal antibodies exert their cytotoxicity on malignant cells via three major mechanisms: antibody-dependent cellular cytotoxicity (ADCC), complement-dependent cytotoxicity (CDC), or directly induced apoptosis [89, 90]. Flow cytometry and cytogenetics can be used to identify antigen markers present in hematologic malignancies, which can then be targeted with various monoclonal antibodies to increase response rates and survival. For example, chemoimmunotherapy involving rituximab targeting the CD20 antigen on B-cell malignancies, such as Non-Hodgkin's Lymphoma (NHL) and chronic lymphoid leukaemia (CLL), has proved superior to chemotherapy alone in terms of response rates, PFS, and OS [91, 92]. In a large phase III trial, 817 patients with CD20 positive CLL were randomly assigned to standard fludarabine plus cyclophosphamide versus fludarabine, cyclophosphamide, and rituximab. At 3 years after randomisation, 65 % of patients in the rituximab arm were free of progression compared to 45 % in the other arm (HR 0.56; $P<0.0001$) [91]. In another study, rituximab administered as induction and/or maintenance therapy with CHOP chemotherapy significantly prolonged failure-free survival (FFS) in diffuse large B-cell lymphoma patients compared to CHOP alone (2-year FFS was 76 % vs. 61 %; $P=0.009$) [93]. Other malignancies rituximab is commonly used in include lymphoma, ALL, and other B-cell CD20 expressing tumours and autoimmune diseases, such as rheumatoid arthritis.

Although rituximab has proven to increase survival in combination with chemotherapy in CLL patients, it has minimal activity as a single agent for induction purposes. This prompted the development of novel anti-CD20 monoclonal antibodies that may be more effective than rituximab, particularly as monotherapy [89]. Ofatumumab, for example, is a novel fully humanised monoclonal antibody that binds to the CD20 molecule [94]. A phase II study demonstrated encouraging ORR, PFS, and OS in patients who were refractory to fludarabine +/− alemtuzumab (an anti-CD52 monoclonal antibody) [95], and also in refractory patients irrespective of previous rituximab treatment (5.6 and 5.3 months PFS in rituximab-naïve patients and those who had received prior rituximab, respectively [$P=0.04$]; however, no difference in ORR or OS) [96]. The results of another phase II trial demonstrated even higher ORRs (77 %) than the previous study in refractory patients (43–53 %)

when previously untreated patients were treated with a combination of fludarabine, cyclophosphamide, and ofatumumab [97]. Other anti-CD20 immunoconjugates include the older tositumumab and ibritumomab, which take advantage of radiolabeled compounds. The use of radiolabeled anti-CD20 antibodies is associated with substantial delayed hematologic toxicity related to nonspecific targeting of the bone marrow [98]. Consequently, these antibodies should be used as single agents or in sequence with conventional chemotherapy. Phase II and III trials of these agents have demonstrated high CR rates (>70%) and ORRs (>90%) [99, 100].

A novel approach to chemoimmunotherapy are antibody-drug conjugates, such as T-DM1 (discussed in the HER2 section) and brentuximab. Brentuximab was granted FDA approval in 2011 for the treatment of relapsed Hodgkin's lymphoma and anaplastic large cell lymphoma. This agent is a chimeric antibody-drug conjugate directed towards the CD30 antigen. Its composition involves a highly stable valine-citrulline linker and a potent chemotherapeutic agent, monomethyl auristatin E, which inhibits microtubule polymerisation [101]. The stable linker conjugating a highly potent compound with a highly specific targeted antibody allows for effective delivery of the cytotoxic agent to the tumour cells of interest. A study of 45 CD30-positive Hodgkin's lymphoma patients treated with brentuximab as a single agent (heavily pre-treated) demonstrated an ORR of 60% and a CR rate of 22%. The median duration of response was 8 months [102]. Another phase II study assessed the efficacy of brentuximab in relapsed or refractory CD30-positive Hodgkin's lymphoma. Results demonstrated an ORR of 75% with CR rate of 34% [103]. Brentuximab efficacy has also been evaluated in 58 patients with relapsed anaplastic large cell lymphoma with an overall response rate of 86% with a median duration of response of 12.6 months [104].

As indicated in Table 1, the FDA recommends routine flow cytometry or cytogenetics to confirm the CD positivity of various haematological malignancies prior to receiving any targeted chemoimmunotherapy.

6 Conclusion

Over 100 drugs currently approved by the FDA require pharmacogenetic information on the drug label, resulting in approximately 25% of all outpatients who receive one or more drugs which are vulnerable to pharmacogenetic variation [105]. Secondary to their unique mechanisms of action and narrow therapeutic indices, antineoplastic therapies are one class of medications greatly impacted by pharmacogenetics. Cancer biomarkers can be used for the early detection of cancers during the diagnostic phase, and subsequently used to determine disease prognosis. Predictive biomarkers are particularly useful in determining a patient's response to therapy, including efficacy and/or toxicity. Elucidating the molecular profile of tumours will help to identify driver mutations, provide a greater number of treatment options and allow for better patient selection in biomarker-driven clinical trials. Because tumour samples also contain both acquired and inherited alterations, cancer sequencing efforts will also capture germline information, which may inform drug selection and/or dose optimisation, as well as genetic susceptibility to disease. These efforts will generate a tremendous amount of data and clinicians must be prepared to interpret and utilise this information to optimise cancer therapeutics.

7 Expert Recommendation

A concerted effort must be made by researchers, clinicians, and cancer centres to adopt genomic-based cancer medicine as the current standard of care. As cost and turnaround time of genomic testing decreases, more and more patients will have their tumours sequenced, allowing for more effective assignment of targeted therapies for each patient. There is likely to be a shift from single-gene testing to multiplex genomic sequencing. It can be argued that the threshold required to warrant single-gene tests differs greatly from the threshold to consider when sequencing information is readily available. Arguable, it is unethical to ignore this information given the validated phenotypes that exist to predict efficacy and/or toxicity. Generation of actionable clinical recommendations, however, is often burdened by the level of evidence needed to warrant implementation.

Although prospective, randomised controlled clinical trials are the gold standard for an intervention to be accepted into standard of care and enforce clinical implementation, this may not be the most effective method for pharmacogenetic studies. The inherent excessive costs, prolonged time, and large sample sizes associated with these trials may deprive many patients of safer and more effective treatments and dosing. The future of genomic-based cancer medicine should focus on adequate sampling of both tumour and germ-line DNA from large phase III clinical trials with prospectively collected data. This information will be critical in both discovery and validation of drug-gene pairs. Well-validated genes that have passed replication should be assessed for clinical implementation through adaptive, biomarker-driven clinical trials. A shift towards large retrospective validation and replication, randomised phase II clinical trials and adaptive-biomarker designs may allow for a more efficient means of translation.

As our knowledge of cancer at the molecular level continues to expand, clinicians must understand these intricate pathways, the therapeutic implication of mutations within these pathways, and the availability and clinical application of such genetic tests in clinical practice.

Acknowledgements Jessie Bishop for her assistance in editing the final chapter.

References

1. Crona D, Innocenti F (2012) Can knowledge of germline markers of toxicity optimize dosing and efficacy of cancer therapy? Biomark Med 6(3):349–362. doi:10.2217/bmm.12.19
2. Evans WE, Relling MV (1999) Pharmacogenomics: translating functional genomics into rational therapeutics. Science 286 (5439):487–491
3. Soh TI, Yong WP, Innocenti F (2011) Recent progress and clinical importance on pharmacogenetics in cancer therapy. Clin Chem Lab Med 49(10):1621–1632. doi:10.1515/CCLM.2011.715
4. Innocenti F, Cox NJ, Dolan ME (2011) The use of genomic information to optimize cancer chemotherapy. Semin Oncol 38(2):186–195. doi:10.1053/j.seminoncol.2011.01.005
5. Green ED, Guyer MS, National Human Genome Research I (2011) Charting a course for genomic medicine from base pairs to bedside. Nature 470(7333):204–213. doi:10.1038/nature09764
6. Deenen MJ, Cats A, Beijnen JH, Schellens JH (2011) Part 1: background, methodology, and clinical adoption of pharmacogenetics. Oncologist 16(6):811–819. doi:10.1634/theoncologist.2010-0258
7. Gerlinger M, Rowan AJ, Horswell S, Larkin J, Endesfelder D, Gronroos E, Martinez P, Matthews N, Stewart A, Tarpey P, Varela I, Phillimore B, Begum S, McDonald NQ, Butler A, Jones D, Raine K, Latimer C, Santos CR, Nohadani M, Eklund AC, Spencer-Dene B, Clark G, Pickering L, Stamp G, Gore M, Szallasi Z, Downward J, Futreal PA, Swanton C (2012) Intratumor heterogeneity and branched evolution revealed by multiregion sequencing. N Engl J Med 366(10):883–892. doi:10.1056/NEJMoa1113205
8. Weinstein IB, Joe A (2008) Oncogene addiction. Cancer Res 68(9):3077–3080; discussion 3080. doi:10.1158/0008-5472.CAN-07-3293
9. Oldenhuis CN, Oosting SF, Gietema JA, de Vries EG (2008) Prognostic versus predictive value of biomarkers in oncology. Eur J Cancer 44(7):946–953. doi:10.1016/j.ejca.2008.03.006
10. van 't Veer LJ Dai H, van de Vijver MJ, He YD, Hart AA, Mao M, Peterse HL, van der Kooy K, Marton MJ, Witteveen AT, Schreiber GJ, Kerkhoven RM, Roberts C, Linsley PS, Bernards R, Friend SH (2002) Gene expression profiling predicts clinical outcome of breast cancer. Nature 415(6871):530–536. doi:10.1038/415530a
11. Benatti P, Gafa R, Barana D, Marino M, Scarselli A, Pedroni M, Maestri I, Guerzoni L, Roncucci L, Menigatti M, Roncari B, Maffei S, Rossi G, Ponti G, Santini A, Losi L, Di Gregorio C, Oliani C, Ponz de Leon M, Lanza G (2005) Microsatellite instability and colorectal cancer prognosis. Clin Cancer Res 11(23):8332–8340. doi:10.1158/1078-0432.CCR-05-1030
12. Popat S, Hubner R, Houlston RS (2005) Systematic review of microsatellite instability and colorectal cancer prognosis. J Clin Oncol 23(3):609–618. doi:10.1200/JCO.2005.01.086
13. Chapman PB, Hauschild A, Robert C, Haanen JB, Ascierto P, Larkin J, Dummer R, Garbe C, Testori A, Maio M, Hogg D, Lorigan P, Lebbe C, Jouary T, Schadendorf D, Ribas A, O'Day SJ, Sosman JA, Kirkwood JM, Eggermont AM, Dreno B, Nolop K, Li J, Nelson B, Hou J, Lee RJ, Flaherty KT, McArthur GA, Group B-S (2011) Improved survival with vemurafenib in melanoma with BRAF V600E mutation. N Engl J Med 364(26):2507–2516. doi:10.1056/NEJMoa1103782
14. Van Cutsem E, Kohne CH, Lang I, Folprecht G, Nowacki MP, Cascinu S, Shchepotin I, Maurel J, Cunningham D, Tejpar S, Schlichting M, Zubel A, Celik I, Rougier P, Ciardiello F (2011) Cetuximab plus irinotecan, fluorouracil, and leucovorin as first-line treatment for metastatic colorectal cancer: updated analysis of overall survival according to tumor KRAS and BRAF mutation status. J Clin Oncol 29(15):2011–2019. doi:10.1200/JCO.2010.33.5091
15. Mandrekar SJ, Sargent DJ (2009) Clinical trial designs for predictive biomarker validation: theoretical considerations and practical challenges. J Clin Oncol 27(24):4027–4034. doi:10.1200/JCO.2009.22.3701
16. McLeod HL, Krynetski EY, Relling MV, Evans WE (2000) Genetic polymorphism of thiopurine methyltransferase and its clinical relevance for childhood acute lymphoblastic leukemia. Leukemia 14(4):567–572

17. Stanulla M, Schaeffeler E, Flohr T, Cario G, Schrauder A, Zimmermann M, Welte K, Ludwig WD, Bartram CR, Zanger UM, Eichelbaum M, Schrappe M, Schwab M (2005) Thiopurine methyltransferase (TPMT) genotype and early treatment response to mercaptopurine in childhood acute lymphoblastic leukemia. JAMA 293(12):1485–1489. doi:10.1001/jama.293.12.1485
18. Relling MV, Hancock ML, Rivera GK, Sandlund JT, Ribeiro RC, Krynetski EY, Pui CH, Evans WE (1999) Mercaptopurine therapy intolerance and heterozygosity at the thiopurine S-methyltransferase gene locus. J Nat Cancer Inst 91(23):2001–2008
19. Relling MV, Gardner EE, Sandborn WJ, Schmiegelow K, Pui CH, Yee SW, Stein CM, Carrillo M, Evans WE, Klein TE, Clinical Pharmacogenetics Implementation C (2011) Clinical Pharmacogenetics Implementation Consortium guidelines for thiopurine methyltransferase genotype and thiopurine dosing. Clin Pharm Ther 89(3):387–391. doi:10.1038/clpt.2010.320
20. Administration UFaD (2010) Table of Pharmacogenomic Biomarkers in Drug Labels
21. Mathijssen RH, van Alphen RJ, Verweij J, Loos WJ, Nooter K, Stoter G, Sparreboom A (2001) Clinical pharmacokinetics and metabolism of irinotecan (CPT-11). Clin Cancer Res 7(8):2182–2194
22. Perera MA, Innocenti F, Ratain MJ (2008) Pharmacogenetic testing for uridine diphosphate glucuronosyltransferase 1A1 polymorphisms: are we there yet? Pharmacotherapy 28(6):755–768. doi:10.1592/phco.28.6.755
23. Innocenti F, Undevia SD, Iyer L, Chen PX, Das S, Kocherginsky M, Karrison T, Janisch L, Ramirez J, Rudin CM, Vokes EE, Ratain MJ (2004) Genetic variants in the UDP-glucuronosyltransferase 1A1 gene predict the risk of severe neutropenia of irinotecan. J Clin Oncol 22(8):1382–1388. doi:10.1200/JCO.2004.07.173
24. Hoskins JM, Goldberg RM, Qu P, Ibrahim JG, McLeod HL (2007) UGT1A1*28 genotype and irinotecan-induced neutropenia: dose matters. J Nat Cancer Inst 99(17):1290–1295. doi:10.1093/jnci/djm115
25. Relling MV, Klein TE (2011) CPIC: Clinical Pharmacogenetics Implementation Consortium of the Pharmacogenomics Research Network. Clin Pharm Ther 89(3):464–467. doi:10.1038/clpt.2010.279
26. Toffoli G, Cecchin E, Gasparini G, D'Andrea M, Azzarello G, Basso U, Mini E, Pessa S, De Mattia E, Lo Re G, Buonadonna A, Nobili S, De Paoli P, Innocenti F (2010) Genotype-driven phase I study of irinotecan administered in combination with fluorouracil/leucovorin in patients with metastatic colorectal cancer. J Clin Oncol28(5):866–871. doi:10.1200/JCO.2009.23.6125
27. Marcuello E, Paez D, Pare L, Salazar J, Sebio A, del Rio E, Baiget M (2011) A genotype-directed phase I-IV dose-finding study of irinotecan in combination with fluorouracil/leucovorin as first-line treatment in advanced colorectal cancer. Br J Cancer 105(1):53–57. doi:10.1038/bjc.2011.206
28. Irvin WJ Jr, Walko CM, Weck KE, Ibrahim JG, Chiu WK, Dees EC, Moore SG, Olajide OA, Graham ML, Canale ST, Raab RE, Corso SW, Peppercorn JM, Anderson SM, Friedman KJ, Ogburn ET, Desta Z, Flockhart DA, McLeod HL, Evans JP, Carey LA (2011) Genotype-guided tamoxifen dosing increases active metabolite exposure in women with reduced CYP2D6 metabolism: a multicenter study. J Clin Oncol 29(24):3232–3239. doi:10.1200/JCO.2010.31.4427
29. Bradford LD (2002) CYP2D6 allele frequency in European Caucasians, Asians, Africans and their descendants. Pharmacogenomics 3(2):229–243. doi:10.1517/14622416.3.2.229
30. Jin Y, Desta Z, Stearns V, Ward B, Ho H, Lee KH, Skaar T, Storniolo AM, Li L, Araba A, Blanchard R, Nguyen A, Ullmer L, Hayden J, Lemler S, Weinshilboum RM, Rae JM, Hayes DF, Flockhart DA (2005) CYP2D6 genotype, antidepressant use, and tamoxifen metabolism during adjuvant breast cancer treatment. J Nat Cancer Inst 97(1):30–39. doi:10.1093/jnci/dji005
31. Goetz MP, Knox SK, Suman VJ, Rae JM, Safgren SL, Ames MM, Visscher DW, Reynolds C, Couch FJ, Lingle WL, Weinshilboum RM, Fritcher EG, Nibbe AM, Desta Z, Nguyen A, Flockhart DA, Perez EA, Ingle JN (2007) The impact of cytochrome P450 2D6 metabolism in women receiving adjuvant tamoxifen. Breast Cancer Res Treat 101(1):113–121. doi:10.1007/s10549-006-9428-0

32. Schroth W, Goetz MP, Hamann U, Fasching PA, Schmidt M, Winter S, Fritz P, Simon W, Suman VJ, Ames MM, Safgren SL, Kuffel MJ, Ulmer HU, Bolander J, Strick R, Beckmann MW, Koelbl H, Weinshilboum RM, Ingle JN, Eichelbaum M, Schwab M, Brauch H (2009) Association between CYP2D6 polymorphisms and outcomes among women with early stage breast cancer treated with tamoxifen. JAMA 302(13):1429–1436. doi:10.1001/jama.2009.1420
33. Regan MM, Leyland-Jones B, Bouzyk M, Pagani O, Tang W, Kammler R, Dell'orto P, Biasi MO, Thurlimann B, Lyng MB, Ditzel HJ, Neven P, Debled M, Maibach R, Price KN, Gelber RD, Coates AS, Goldhirsch A, Rae JM, Viale G, Breast International Group 1-98 Collaborative G (2012) CYP2D6 genotype and tamoxifen response in postmenopausal women with endocrine-responsive breast cancer: the breast international group 1-98 trial. J Nat Cancer Inst 104(6):441–451. doi:10.1093/jnci/djs125
34. Rae JM, Drury S, Hayes DF, Stearns V, Thibert JN, Haynes BP, Salter J, Sestak I, Cuzick J, Dowsett M, trialists A (2012) CYP2D6 and UGT2B7 genotype and risk of recurrence in tamoxifen-treated breast cancer patients. J Nat Cancer Inst 104(6):452–460. doi:10.1093/jnci/djs126
35. Nakamura Y, Ratain MJ, Cox NJ, McLeod HL, Kroetz DL, Flockhart DA (2012) Re: CYP2D6 genotype and tamoxifen response in postmenopausal women with endocrine-responsive breast cancer: the Breast International Group 1–98 trial. J Nat Cancer Inst 104(16):1264; author reply 1266–1268. doi:10.1093/jnci/djs304
36. Brauch H, Schroth W, Goetz MP, Murdter TE, Winter S, Ingle JN, Schwab M, Eichelbaum M (2013) Tamoxifen use in postmenopausal breast cancer: CYP2D6 matters. J Clin Oncol 31(2):176–180. doi:10.1200/JCO.2012.44.6625
37. Swen JJ, Nijenhuis M, de Boer A, Grandia L, Maitland-van der Zee AH, Mulder H, Rongen GA, van Schaik RH, Schalekamp T, Touw DJ, van der Weide J, Wilffert B, Deneer VH, Guchelaar HJ (2011) Pharmacogenetics: from bench to byte–an update of guidelines. Clin Pharm Ther 89(5):662–673. doi:10.1038/clpt.2011.34
38. Kiyotani K, Mushiroda T, Imamura CK, Tanigawara Y, Hosono N, Kubo M, Sasa M, Nakamura Y, Zembutsu H (2012) Dose-adjustment study of tamoxifen based on CYP2D6 genotypes in Japanese breast cancer patients. Breast Cancer Res Treat 131(1):137–145. doi:10.1007/s10549-011-1777-7
39. Diasio RB, Johnson MR (1999) Dihydropyrimidine dehydrogenase: its role in 5-fluorouracil clinical toxicity and tumor resistance. Clin Cancer Res 5(10):2672–2673
40. Toffoli G, Cecchin E (2007) Pharmacogenetics and stomach cancer: an update. Pharmacogenomics 8(5):497–505. doi:10.2217/14622416.8.5.497
41. Van Kuilenburg AB Meinsma R Zoetekouw L Van Gennip AH (2002) High prevalence of the IVS14 + 1G>A mutation in the dihydropyrimidine dehydrogenase gene of patients with severe 5-fluorouracil-associated toxicity. Pharmacogenetics 12(7):555–558
42. Schwab M, Zanger UM, Marx C, Schaeffeler E, Klein K, Dippon J, Kerb R, Blievernicht J, Fischer J, Hofmann U, Bokemeyer C, Eichelbaum M, German FUTSG (2008) Role of genetic and nongenetic factors for fluorouracil treatment-related severe toxicity: a prospective clinical trial by the German 5-FU Toxicity Study Group. J Clin Oncol 26(13):2131–2138. doi:10.1200/JCO.2006.10.4182
43. Kwak EL, Bang YJ, Camidge DR, Shaw AT, Solomon B, Maki RG, Ou SH, Dezube BJ, Janne PA, Costa DB, Varella-Garcia M, Kim WH, Lynch TJ, Fidias P, Stubbs H, Engelman JA, Sequist LV, Tan W, Gandhi L, Mino-Kenudson M, Wei GC, Shreeve SM, Ratain MJ, Settleman J, Christensen JG, Haber DA, Wilner K, Salgia R, Shapiro GI, Clark JW, Iafrate AJ (2010) Anaplastic lymphoma kinase inhibition in non-small-cell lung cancer. N Engl J Med 363(18):1693–1703. doi:10.1056/NEJMoa1006448
44. Shaw AT, Engelman JA (2013) ALK in lung cancer: past, present, and future. J Clin Oncol 31(8):1105–1111. doi:10.1200/JCO.2012.44.5353
45. Katayama R, Shaw AT, Khan TM, Mino-Kenudson M, Solomon BJ, Halmos B, Jessop NA, Wain JC, Yeo AT, Benes C, Drew L, Saeh JC, Crosby K, Sequist LV, Iafrate AJ, Engelman JA (2012) Mechanisms of acquired crizotinib resistance in ALK-rearranged lung Cancers. Sci Transl Med 4(120):120ra17. doi:10.1126/scitranslmed.3003316

46. Nowell PC, Hungerford DA (1960) Chromosome studies on normal and leukemic human leukocytes. J Nat Cancer Inst 25:85–109
47. de Klein A, van Kessel AG, Grosveld G, Bartram CR, Hagemeijer A, Bootsma D, Spurr NK, Heisterkamp N, Groffen J, Stephenson JR (1982) A cellular oncogene is translocated to the Philadelphia chromosome in chronic myelocytic leukaemia. Nature 300(5894):765–767
48. Druker BJ, Guilhot F, O'Brien SG, Gathmann I, Kantarjian H, Gattermann N, Deininger MW, Silver RT, Goldman JM, Stone RM, Cervantes F, Hochhaus A, Powell BL, Gabrilove JL, Rousselot P, Reiffers J, Cornelissen JJ, Hughes T, Agis H, Fischer T, Verhoef G, Shepherd J, Saglio G, Gratwohl A, Nielsen JL, Radich JP, Simonsson B, Taylor K, Baccarani M, So C, Letvak L, Larson RA, Investigators I (2006) Five-year follow-up of patients receiving imatinib for chronic myeloid leukemia. N Engl J Med 355(23):2408–2417. doi:10.1056/NEJMoa062867
49. Zhang WW, Cortes JE, Yao H, Zhang L, Reddy NG, Jabbour E, Kantarjian HM, Jones D (2009) Predictors of primary imatinib resistance in chronic myelogenous leukemia are distinct from those in secondary imatinib resistance. J Clin Oncol 27(22):3642–3649. doi:10.1200/JCO.2008.19.4076
50. Lierman E, Smits S, Cools J, Dewaele B, Debiec-Rychter M, Vandenberghe P (2012) Ponatinib is active against imatinib-resistant mutants of FIP1L1-PDGFRA and KIT, and against FGFR1-derived fusion kinases. Leukemia 26(7):1693–1695. doi:10.1038/leu.2012.8
51. La Thangue NB Kerr DJ (2011) Predictive biomarkers: a paradigm shift towards personalized cancer medicine. Nat Rev Clin Oncol 8(10):587–596. doi:10.1038/nrclinonc.2011.121
52. Wang ZY, Chen Z (2008) Acute promyelocytic leukemia: from highly fatal to highly curable. Blood 111(5):2505–2515. doi:10.1182/blood-2007-07-102798
53. Tallman MS, Andersen JW, Schiffer CA, Appelbaum FR, Feusner JH, Ogden A, Shepherd L, Willman C, Bloomfield CD, Rowe JM, Wiernik PH (1997) All-trans-retinoic acid in acute promyelocytic leukemia. N Engl J Med 337(15):1021–1028. doi:10.1056/NEJM199710093371501
54. Raelson JV, Nervi C, Rosenauer A, Benedetti L, Monczak Y, Pearson M, Pelicci PG, Miller WH Jr (1996) The PML/RAR alpha oncoprotein is a direct molecular target of retinoic acid in acute promyelocytic leukemia cells. Blood 88(8):2826–2832
55. Ohnishi K (2007) PML-RARalpha inhibitors (ATRA, tamibaroten, arsenic troxide) for acute promyelocytic leukemia. Int J Clin Oncol 12(5):313–317. doi:10.1007/s10147-007-0694-6
56. Flaherty KT, Infante JR, Daud A, Gonzalez R, Kefford RF, Sosman J, Hamid O, Schuchter L, Cebon J, Ibrahim N, Kudchadkar R, Burris HA, 3rd, Falchook G, Algazi A, Lewis K, Long GV, Puzanov I, Lebowitz P, Singh A, Little S, Sun P, Allred A, Ouellet D, Kim KB, Patel K, Weber J (2012) Combined BRAF and MEK Inhibition in Melanoma with BRAF V600 Mutations. N Engl J Med 367(18):1694-1703. doi:10.1056/NEJMoa1210093
57. Fletcher CDM, Berman JJ, Corless C, Gorstein F, Lasota J, Longley BJ, Miettinen M, O'Leary TJ, Remotti H, Rubin BP, Shmookler B, Sobin LH, Weiss SW (2002) Diagnosis of gastrointestinal stromal tumors: a consensus approach. Hum Pathol 33(5):459–465. doi:10.1053/hupa.2002.123545
58. Yasuda A, Sawai H, Takahashi H, Ochi N, Matsuo Y, Funahashi H, Sato M, Okada Y, Takeyama H, Manabe T (2007) Stem cell factor/c-kit receptor signaling enhances the proliferation and invasion of colorectal cancer cells through the PI3K/Akt pathway. Dig Dis Sci 52(9):2292–2300. doi:10.1007/s10620-007-9759-7
59. Heinrich MC, Corless CL, Demetri GD, Blanke CD, von Mehren M, Joensuu H, McGreevey LS, Chen CJ, Van den Abbeele AD, Druker BJ, Kiese B, Eisenberg B, Roberts PJ, Singer S, Fletcher CD, Silberman S, Dimitrijevic S, Fletcher JA (2003) Kinase mutations and imatinib response in patients with metastatic gastrointestinal stromal tumor. J Clin Oncol 21(23):4342–4349. doi:10.1200/JCO.2003.04.190
60. Taniguchi M, Nishida T, Hirota S, Isozaki K, Ito T, Nomura T, Matsuda H, Kitamura Y (1999) Effect of c-kit mutation on prognosis of gastrointestinal stromal tumors. Cancer Res 59(17):4297–4300
61. McLean SR, Gana-Weisz M, Hartzoulakis B, Frow R, Whelan J, Selwood D, Boshoff C (2005) Imatinib binding and cKIT inhibition is abrogated by the cKIT kinase domain I missense mutation Val654Ala. Mol Cancer Ther 4(12):2008–2015. doi:10.1158/1535-7163.MCT-05-0070

62. Yarden Y, Shilo BZ (2007) SnapShot: EGFR signaling pathway. Cell 131(5):1018. doi:10.1016/j.cell.2007.11.013
63. Shepherd FA, Rodrigues Pereira J, Ciuleanu T, Tan EH, Hirsh V, Thongprasert S, Campos D, Maoleekoonpiroj S, Smylie M, Martins R, van Kooten M, Dediu M, Findlay B, Tu D, Johnston D, Bezjak A, Clark G, Santabarbara P, Seymour L, National Cancer Institute of Canada Clinical Trials G (2005) Erlotinib in previously treated non-small-cell lung cancer. N Engl J Med 353(2):123–132. doi:10.1056/NEJMoa050753
64. Lynch TJ, Bell DW, Sordella R, Gurubhagavatula S, Okimoto RA, Brannigan BW, Harris PL, Haserlat SM, Supko JG, Haluska FG, Louis DN, Christiani DC, Settleman J, Haber DA (2004) Activating mutations in the epidermal growth factor receptor underlying responsiveness of non-small-cell lung cancer to gefitinib. N Engl J Med 350(21):2129–2139. doi:10.1056/NEJMoa040938
65. Zhou C, Wu YL, Chen G, Feng J, Liu XQ, Wang C, Zhang S, Wang J, Zhou S, Ren S, Lu S, Zhang L, Hu C, Hu C, Luo Y, Chen L, Ye M, Huang J, Zhi X, Zhang Y, Xiu Q, Ma J, Zhang L, You C (2011) Erlotinib versus chemotherapy as first-line treatment for patients with advanced EGFR mutation-positive non-small-cell lung cancer (OPTIMAL, CTONG-0802): a multicentre, open-label, randomised, phase 3 study. Lancet Oncol 12(8):735–742. doi:10.1016/S1470-2045(11)70184-X
66. Miller VA, Hirsh V, Cadranel J, Chen YM, Park K, Kim SW, Zhou C, Su WC, Wang M, Sun Y, Heo DS, Crino L, Tan EH, Chao TY, Shahidi M, Cong XJ, Lorence RM, Yang JC (2012) Afatinib versus placebo for patients with advanced, metastatic non-small-cell lung cancer after failure of erlotinib, gefitinib, or both, and one or two lines of chemotherapy (LUX-Lung 1): a phase 2b/3 randomised trial. Lancet Oncol 13(5):528–538. doi:10.1016/S1470-2045(12)70087-6
67. Mamounas EP, Tang G, Fisher B, Paik S, Shak S, Costantino JP, Watson D, Geyer CE Jr, Wickerham DL, Wolmark N (2010) Association between the 21-gene recurrence score assay and risk of locoregional recurrence in node-negative, estrogen receptor-positive breast cancer: results from NSABP B-14 and NSABP B-20. J Clin Oncol 28(10):1677–1683. doi:10.1200/JCO.2009.23.7610
68. Allegra JC, Lippman ME, Thompson EB, Simon R, Barlock A, Green L, Huff KK, Do HM, Aitken SC (1979) Distribution, frequency, and quantitative analysis of estrogen, progesterone, androgen, and glucocorticoid receptors in human breast cancer. Cancer Res 39(5):1447–1454
69. Dowsett M, Cuzick J, Ingle J, Coates A, Forbes J, Bliss J, Buyse M, Baum M, Buzdar A, Colleoni M, Coombes C, Snowdon C, Gnant M, Jakesz R, Kaufmann M, Boccardo F, Godwin J, Davies C, Peto R (2010) Meta-analysis of breast cancer outcomes in adjuvant trials of aromatase inhibitors versus tamoxifen. J Clin Oncol 28(3):509–518. doi:10.1200/JCO.2009.23.1274
70. Freedman OC, Verma S, Clemons MJ (2006) Pre-menopausal breast cancer and aromatase inhibitors: treating a new generation of women. Breast Cancer Res Treat 99(3):241–247. doi:10.1007/s10549-006-9208-x
71. Davies C, Pan H, Godwin J, Gray R, Arriagada R, Raina V, Abraham M, Alencar VH, Badran A, Bonfill X, Bradbury J, Clarke M, Collins R, Davis SR, Delmestri A, Forbes JF, Haddad P, Hou MF, Inbar M, Khaled H, Kielanowska J, Kwan WH, Mathew BS, Muller B, Nicolucci A, Peralta O, Pernas F, Petruzelka L, Pienkowski T, Rajan B, Rubach MT, Tort S, Urrutia G, Valentini M, Wang Y, Peto R, for the Adjuvant Tamoxifen: Longer Against Shorter Collaborative G (2012) Long-term effects of continuing adjuvant tamoxifen to 10 years versus stopping at 5 years after diagnosis of oestrogen receptor-positive breast cancer: ATLAS, a randomised trial. Lancet 381(9869):805–816. doi:10.1016/S0140-6736(12)61963-1
72. Owens MA, Horten BC, Da Silva MM (2004) HER2 amplification ratios by fluorescence in situ hybridization and correlation with immunohistochemistry in a cohort of 6556 breast cancer tissues. Clin Breast Cancer 5(1):63–69
73. Yang XR, Chang-Claude J, Goode EL, Couch FJ, Nevanlinna H, Milne RL, Gaudet M, Schmidt MK, Broeks A, Cox A, Fasching PA, Hein R, Spurdle AB, Blows F, Driver K, Flesch-Janys D, Heinz J, Sinn P, Vrieling A, Heikkinen T, Aittomaki K, Heikkila P, Blomqvist C, Lissowska J, Peplonska B, Chanock S, Figueroa J, Brinton L, Hall P, Czene K, Humphreys K, Darabi H, Liu J, Van 't Veer LJ, van Leeuwen FE, Andrulis IL, Glendon G, Knight JA,

Mulligan AM, O'Malley FP, Weerasooriya N, John EM, Beckmann MW, Hartmann A, Weihbrecht SB, Wachter DL, Jud SM, Loehberg CR, Baglietto L, English DR, Giles GG, McLean CA, Severi G, Lambrechts D, Vandorpe T, Weltens C, Paridaens R, Smeets A, Neven P, Wildiers H, Wang X, Olson JE, Cafourek V, Fredericksen Z, Kosel M, Vachon C, Cramp HE, Connley D, Cross SS, Balasubramanian SP, Reed MW, Dork T, Bremer M, Meyer A, Karstens JH, Ay A, Park-Simon TW, Hillemanns P, Arias Perez JI, Menendez Rodriguez P, Zamora P, Benitez J, Ko YD, Fischer HP, Hamann U, Pesch B, Bruning T, Justenhoven C, Brauch H, Eccles DM, Tapper WJ, Gerty SM, Sawyer EJ, Tomlinson IP, Jones A, Kerin M, Miller N, McInerney N, Anton-Culver H, Ziogas A, Shen CY, Hsiung CN, Wu PE, Yang SL, Yu JC, Chen ST, Hsu GC, Haiman CA, Henderson BE, Le Marchand L, Kolonel LN, Lindblom A, Margolin S, Jakubowska A, Lubinski J, Huzarski T, Byrski T, Gorski B, Gronwald J, Hooning MJ, Hollestelle A, van den Ouweland AM, Jager A, Kriege M, Tilanus-Linthorst MM, Collee M, Wang-Gohrke S, Pylkas K, Jukkola-Vuorinen A, Mononen K, Grip M, Hirvikoski P, Winqvist R, Mannermaa A, Kosma VM, Kauppinen J, Kataja V, Auvinen P, Soini Y, Sironen R, Bojesen SE, Orsted DD, Kaur-Knudsen D, Flyger H, Nordestgaard BG, Holland H, Chenevix-Trench G, Manoukian S, Barile M, Radice P, Hankinson SE, Hunter DJ, Tamimi R, Sangrajrang S, Brennan P, McKay J, Odefrey F, Gaborieau V, Devilee P, Huijts PE, Tollenaar RA, Seynaeve C, Dite GS, Apicella C, Hopper JL, Hammet F, Tsimiklis H, Smith LD, Southey MC, Humphreys MK, Easton D, Pharoah P, Sherman ME, Garcia-Closas M (2011) Associations of breast cancer risk factors with tumor subtypes: a pooled analysis from the Breast Cancer Association Consortium studies. J Nat Cancer Inst 103(3):250–263. doi:10.1093/jnci/djq526

74. Press MF, Bernstein L, Thomas PA, Meisner LF, Zhou JY, Ma Y, Hung G, Robinson RA, Harris C, El-Naggar A, Slamon DJ, Phillips RN, Ross JS, Wolman SR, Flom KJ (1997) HER-2/neu gene amplification characterized by fluorescence in situ hybridization: poor prognosis in node-negative breast carcinomas. J Clin Oncol 15(8):2894–2904
75. Piccart-Gebhart MJ, Procter M, Leyland-Jones B, Goldhirsch A, Untch M, Smith I, Gianni L, Baselga J, Bell R, Jackisch C, Cameron D, Dowsett M, Barrios CH, Steger G, Huang CS, Andersson M, Inbar M, Lichinitser M, Lang I, Nitz U, Iwata H, Thomssen C, Lohrisch C, Suter TM, Ruschoff J, Suto T, Greatorex V, Ward C, Straehle C, McFadden E, Dolci MS, Gelber RD, Herceptin Adjuvant Trial Study T (2005) Trastuzumab after adjuvant chemotherapy in HER2-positive breast cancer. N Engl J Med 353(16):1659–1672. doi:10.1056/NEJMoa052306
76. Slamon D, Eiermann W, Robert N, Pienkowski T, Martin M, Press M, Mackey J, Glaspy J, Chan A, Pawlicki M, Pinter T, Valero V, Liu MC, Sauter G, von Minckwitz G, Visco F, Bee V, Buyse M, Bendahmane B, Tabah-Fisch I, Lindsay MA, Riva A, Crown J, Breast Cancer International Research G (2011) Adjuvant trastuzumab in HER2-positive breast cancer. T N Engl J Med 365(14):1273–1283. doi:10.1056/NEJMoa0910383
77. Geyer CE, Forster J, Lindquist D, Chan S, Romieu CG, Pienkowski T, Jagiello-Gruszfeld A, Crown J, Chan A, Kaufman B, Skarlos D, Campone M, Davidson N, Berger M, Oliva C, Rubin SD, Stein S, Cameron D (2006) Lapatinib plus capecitabine for HER2-positive advanced breast cancer. T N Engl J Med 355(26):2733–2743. doi:10.1056/NEJMoa064320
78. Baselga J, Cortes J, Kim SB, Im SA, Hegg R, Im YH, Roman L, Pedrini JL, Pienkowski T, Knott A, Clark E, Benyunes MC, Ross G, Swain SM, Group CS (2012) Pertuzumab plus trastuzumab plus docetaxel for metastatic breast cancer. N Engl J Med 366(2):109–119. doi:10.1056/NEJMoa1113216
79. Verma S, Miles D, Gianni L, Krop IE, Welslau M, Baselga J, Pegram M, Oh DY, Dieras V, Guardino E, Fang L, Lu MW, Olsen S, Blackwell K, Group ES (2012) Trastuzumab emtansine for HER2-positive advanced breast cancer. N Engl J Med 367(19):1783–1791. doi:10.1056/NEJMoa1209124
80. Ono M, Kuwano M (2006) Molecular mechanisms of epidermal growth factor receptor (EGFR) activation and response to gefitinib and other EGFR-targeting drugs. Clin Cancer Res 12(24):7242–7251. doi:10.1158/1078-0432.CCR-06-0646
81. Heinemann V, Stintzing S, Kirchner T, Boeck S, Jung A (2009) Clinical relevance of EGFR- and KRAS-status in colorectal cancer patients treated with monoclonal antibodies directed against the EGFR. Cancer Treat Rev 35(3):262–271. doi:10.1016/j.ctrv.2008.11.005

82. Siddiqui AD, Piperdi B (2010) KRAS mutation in colon cancer: a marker of resistance to EGFR-I therapy. Ann Surg Oncol 17(4):1168–1176. doi:10.1245/s10434-009-0811-z
83. Douillard JY, Siena S, Cassidy J, Tabernero J, Burkes R, Barugel M, Humblet Y, Bodoky G, Cunningham D, Jassem J, Rivera F, Kocakova I, Ruff P, Blasinska-Morawiec M, Smakal M, Canon JL, Rother M, Oliner KS, Wolf M, Gansert J (2010) Randomized, phase III trial of panitumumab with infusional fluorouracil, leucovorin, and oxaliplatin (FOLFOX4) versus FOLFOX4 alone as first-line treatment in patients with previously untreated metastatic colorectal cancer: the PRIME study. J Clin Oncol 28(31):4697–4705. doi:10.1200/JCO.2009.27.4860
84. Martin P, Leighl NB, Tsao MS, Shepherd FA (2013) KRAS mutations as prognostic and predictive markers in non-small cell lung cancer. J Thorac Oncol 8(5):530–542. doi:10.1097/JTO.0b013e318283d958
85. Rosell R, Monzo M, Molina F, Martinez E, Pifarre A, Moreno I, Mate JL, de Anta JM, Sanchez M, Font A (1995) K-ras genotypes and prognosis in non-small-cell lung cancer. Ann Oncol 6(Suppl 3):S15–20
86. Linardou H, Dahabreh IJ, Bafaloukos D, Kosmidis P, Murray S (2009) Somatic EGFR mutations and efficacy of tyrosine kinase inhibitors in NSCLC. Nat Rev Clin Oncol 6(6):352–366. doi:10.1038/nrclinonc.2009.62
87. Massarelli E, Varella-Garcia M, Tang X, Xavier AC, Ozburn NC, Liu DD, Bekele BN, Herbst RS, Wistuba, II (2007) KRAS mutation is an important predictor of resistance to therapy with epidermal growth factor receptor tyrosine kinase inhibitors in non-small-cell lung cancer. Clinical Cancer Res 13(10):2890–2896. doi:10.1158/1078-0432.CCR-06-3043
88. Emens LA (2010) Chemoimmunotherapy. Cancer J 16(4):295–303. doi:10.1097/PPO.0b013e3181eb5066
89. Delgado J, Baumann T, Ghita G, Montserrat E (2012) Chronic lymphocytic leukemia therapy: beyond chemoimmunotherapy. Curr Pharm Des 18(23):3356–3362
90. Oldham RK, Dillman RO (2008) Monoclonal antibodies in cancer therapy: 25 years of progress. J Clin Oncol 26(11):1774–1777. doi:10.1200/JCO.2007.15.7438
91. Hallek M, Fischer K, Fingerle-Rowson G, Fink AM, Busch R, Mayer J, Hensel M, Hopfinger G, Hess G, von Grunhagen U, Bergmann M, Catalano J, Zinzani PL, Caligaris-Cappio F, Seymour JF, Berrebi A, Jager U, Cazin B, Trneny M, Westermann A, Wendtner CM, Eichhorst BF, Staib P, Buhler A, Winkler D, Zenz T, Bottcher S, Ritgen M, Mendila M, Kneba M, Dohner H, Stilgenbauer S, International Group of I, German Chronic Lymphocytic Leukaemia Study G (2010) Addition of rituximab to fludarabine and cyclophosphamide in patients with chronic lymphocytic leukaemia: a randomised, open-label, phase 3 trial. Lancet 376(9747):1164–1174. doi:10.1016/S0140-6736(10)61381-5
92. Robak T, Dmoszynska A, Solal-Celigny P, Warzocha K, Loscertales J, Catalano J, Afanasiev BV, Larratt L, Geisler CH, Montillo M, Zyuzgin I, Ganly PS, Dartigeas C, Rosta A, Maurer J, Mendila M, Saville MW, Valente N, Wenger MK, Moiseev SI (2010) Rituximab plus fludarabine and cyclophosphamide prolongs progression-free survival compared with fludarabine and cyclophosphamide alone in previously treated chronic lymphocytic leukemia. J Clin Oncol 28(10):1756–1765. doi:10.1200/JCO.2009.26.4556
93. Habermann TM, Weller EA, Morrison VA, Gascoyne RD, Cassileth PA, Cohn JB, Dakhil SR, Woda B, Fisher RI, Peterson BA, Horning SJ (2006) Rituximab-CHOP versus CHOP alone or with maintenance rituximab in older patients with diffuse large B-cell lymphoma. J Clin Oncol 24(19):3121–3127. doi:10.1200/JCO.2005.05.1003
94. Teeling JL, Mackus WJ, Wiegman LJ, van den Brakel JH, Beers SA, French RR, van Meerten T, Ebeling S, Vink T, Slootstra JW, Parren PW, Glennie MJ, van de Winkel JG (2006) The biological activity of human CD20 monoclonal antibodies is linked to unique epitopes on CD20. J Immunol 177(1):362–371
95. Wierda WG, Kipps TJ, Mayer J, Stilgenbauer S, Williams CD, Hellmann A, Robak T, Furman RR, Hillmen P, Trneny M, Dyer MJ, Padmanabhan S, Piotrowska M, Kozak T, Chan G, Davis R, Losic N, Wilms J, Russell CA, Osterborg A, Hx CDSI (2010) Ofatumumab as single-agent CD20 immunotherapy in fludarabine-refractory chronic lymphocytic leukemia. J Clin Oncol 28(10):1749–1755. doi:10.1200/JCO.2009.25.3187

96. Wierda WG, Padmanabhan S, Chan GW, Gupta IV, Lisby S, Osterborg A, Hx CDSI (2011) Ofatumumab is active in patients with fludarabine-refractory CLL irrespective of prior rituximab: results from the phase 2 international study. Blood 118(19):5126–5129. doi:10.1182/blood-2011-04-348656
97. Wierda WG, Kipps TJ, Durig J, Griskevicius L, Stilgenbauer S, Mayer J, Smolej L, Hess G, Griniute R, Hernandez-Ilizaliturri FJ, Padmanabhan S, Gorczyca M, Chang CN, Chan G, Gupta I, Nielsen TG, Russell CA, Study I (2011) Chemoimmunotherapy with O-FC in previously untreated patients with chronic lymphocytic leukemia. Blood 117(24):6450–6458. doi:10.1182/blood-2010-12-323980
98. Maloney DG (2012) Anti-CD20 antibody therapy for B-cell lymphomas. N Engl J Med 366(21):2008–2016. doi:10.1056/NEJMct1114348
99. Press OW, Unger JM, Braziel RM, Maloney DG, Miller TP, Leblanc M, Fisher RI, Southwest Oncology G (2006) Phase II trial of CHOP chemotherapy followed by tositumomab/iodine I-131 tositumomab for previously untreated follicular non-Hodgkin's lymphoma: five-year follow-up of Southwest Oncology Group Protocol S9911. J Clin Oncol 24(25):4143–4149. doi:10.1200/JCO.2006.05.8198
100. Morschhauser F, Radford J, Van Hoof A, Vitolo U, Soubeyran P, Tilly H, Huijgens PC, Kolstad A, d'Amore F, Gonzalez Diaz M, Petrini M, Sebban C, Zinzani PL, van Oers MH, van Putten W, Bischof-Delaloye A, Rohatiner A, Salles G, Kuhlmann J, Hagenbeek A (2008) Phase III trial of consolidation therapy with yttrium-90-ibritumomab tiuxetan compared with no additional therapy after first remission in advanced follicular lymphoma. J Clin Oncol 26(32):5156–5164. doi:10.1200/JCO.2008.17.2015
101. Bradley AM, Devine M, Deremer D (2013) Brentuximab vedotin: an anti-CD30 antibody-drug conjugate. Am J Health Syst Pharm 70(7):589–597. doi:10.2146/ajhp110608
102. Rothe A, Sasse S, Goergen H, Eichenauer DA, Lohri A, Jager U, Bangard C, Boll B, von Bergwelt Baildon M, Theurich S, Borchmann P, Engert A (2012) Brentuximab vedotin for relapsed or refractory CD30 + hematologic malignancies: the German Hodgkin Study Group experience. Blood 120(7):1470–1472. doi:10.1182/blood-2012-05-430918
103. Younes A, Gopal AK, Smith SE, Ansell SM, Rosenblatt JD, Savage KJ, Ramchandren R, Bartlett NL, Cheson BD, de Vos S, Forero-Torres A, Moskowitz CH, Connors JM, Engert A, Larsen EK, Kennedy DA, Sievers EL, Chen R (2012) Results of a pivotal phase II study of brentuximab vedotin for patients with relapsed or refractory Hodgkin's lymphoma. J Clin Oncol 30(18):2183–2189. doi:10.1200/JCO.2011.38.0410
104. Pro B, Advani R, Brice P, Bartlett NL, Rosenblatt JD, Illidge T, Matous J, Ramchandren R, Fanale M, Connors JM, Yang Y, Sievers EL, Kennedy DA, Shustov A (2012) Brentuximab vedotin (SGN-35) in patients with relapsed or refractory systemic anaplastic large-cell lymphoma: results of a phase II study. J Clin Oncol 30(18):2190–2196. doi:10.1200/JCO.2011.38.0402
105. Frueh FW, Amur S, Mummaneni P, Epstein RS, Aubert RE, DeLuca TM, Verbrugge RR, Burckart GJ, Lesko LJ (2008) Pharmacogenomic biomarker information in drug labels approved by the United States food and drug administration: prevalence of related drug use. Pharmacotherapy 28(8):992–998. doi:10.1592/phco.28.8.992
106. Nowell SA, Ahn J, Rae JM, Scheys JO, Trovato A, Sweeney C, MacLeod SL, Kadlubar FF, Ambrosone CB (2005) Association of genetic variation in tamoxifen-metabolizing enzymes with overall survival and recurrence of disease in breast cancer patients. Breast Cancer Res Treat 91(3):249–258. doi:10.1007/s10549-004-7751-x
107. Soda M, Choi YL, Enomoto M, Takada S, Yamashita Y, Ishikawa S, Fujiwara S, Watanabe H, Kurashina K, Hatanaka H, Bando M, Ohno S, Ishikawa Y, Aburatani H, Niki T, Sohara Y, Sugiyama Y, Mano H (2007) Identification of the transforming EML4-ALK fusion gene in non-small-cell lung cancer. Nature 448(7153):561–566. doi:10.1038/nature05945
108. Druker BJ (2008) Translation of the Philadelphia chromosome into therapy for CML. Blood 112(13):4808–4817. doi:10.1182/blood-2008-07-077958
109. Rowley JD (1973) Letter: a new consistent chromosomal abnormality in chronic myelogenous leukaemia identified by quinacrine fluorescence and Giemsa staining. Nature 243(5405):290–293

110. Druker BJ, Talpaz M, Resta DJ, Peng B, Buchdunger E, Ford JM, Lydon NB, Kantarjian H, Capdeville R, Ohno-Jones S, Sawyers CL (2001) Efficacy and safety of a specific inhibitor of the BCR-ABL tyrosine kinase in chronic myeloid leukemia. N Engl J Med 344(14):1031–1037. doi:10.1056/NEJM200104053441401
111. Davies H, Bignell GR, Cox C, Stephens P, Edkins S, Clegg S, Teague J, Woffendin H, Garnett MJ, Bottomley W, Davis N, Dicks E, Ewing R, Floyd Y, Gray K, Hall S, Hawes R, Hughes J, Kosmidou V, Menzies A, Mould C, Parker A, Stevens C, Watt S, Hooper S, Wilson R, Jayatilake H, Gusterson BA, Cooper C, Shipley J, Hargrave D, Pritchard-Jones K, Maitland N, Chenevix-Trench G, Riggins GJ, Bigner DD, Palmieri G, Cossu A, Flanagan A, Nicholson A, Ho JW, Leung SY, Yuen ST, Weber BL, Seigler HF, Darrow TL, Paterson H, Marais R, Marshall CJ, Wooster R, Stratton MR, Futreal PA (2002) Mutations of the BRAF gene in human cancer. Nature 417(6892):949–954. doi:10.1038/nature00766
112. Joensuu H, Roberts PJ, Sarlomo-Rikala M, Andersson LC, Tervahartiala P, Tuveson D, Silberman S, Capdeville R, Dimitrijevic S, Druker B, Demetri GD (2001) Effect of the tyrosine kinase inhibitor STI571 in a patient with a metastatic gastrointestinal stromal tumor. N Engl J Med 344(14):1052–1056. doi:10.1056/NEJM200104053441404
113. Kris MG, Natale RB, Herbst RS, Lynch TJ Jr, Prager D, Belani CP, Schiller JH, Kelly K, Spiridonidis H, Sandler A, Albain KS, Cella D, Wolf MK, Averbuch SD, Ochs JJ, Kay AC (2003) Efficacy of gefitinib, an inhibitor of the epidermal growth factor receptor tyrosine kinase, in symptomatic patients with non-small cell lung cancer: a randomized trial. JAMA 290(16):2149–2158. doi:10.1001/jama.290.16.2149
114. Beatson G (1896) On the treatment of inoperable cases of carcinoma of the mamma: suggestions for a new method of treatment, with illustrative cases. Lancet 148:162–165
115. Ahn ER, Vogel CL (2012) Dual HER2-targeted approaches in HER2-positive breast cancer. Breast Cancer Res Treat 131(2):371–383. doi:10.1007/s10549-011-1781-y
116. Daniele L, Sapino A (2009) Anti-HER2 treatment and breast cancer: state of the art, recent patents, and new strategies. Recent Pat Anticancer Drug Discov 4(1):9–18
117. Hicks DG, Kulkarni S (2008) Trastuzumab as adjuvant therapy for early breast cancer: the importance of accurate human epidermal growth factor receptor 2 testing. Arch Pathol Lab Med 132(6):1008–1015
118. King CR, Kraus MH, Aaronson SA (1985) Amplification of a novel v-erbB-related gene in a human mammary carcinoma. Science 229(4717):974–976
119. Slamon DJ, Leyland-Jones B, Shak S, Fuchs H, Paton V, Bajamonde A, Fleming T, Eiermann W, Wolter J, Pegram M, Baselga J, Norton L (2001) Use of chemotherapy plus a monoclonal antibody against HER2 for metastatic breast cancer that overexpresses HER2. N Engl J Med 344(11):783–792. doi:10.1056/NEJM200103153441101
120. Bos JL, Fearon ER, Hamilton SR, Verlaan-de Vries M, van Boom JH, van der Eb AJ, Vogelstein B (1987) Prevalence of ras gene mutations in human colorectal cancers. Nature 327(6120):293–297. doi:10.1038/327293a0
121. Lievre A, Bachet JB, Le Corre D, Boige V, Landi B, Emile JF, Cote JF, Tomasic G, Penna C, Ducreux M, Rougier P, Penault-Llorca F, Laurent-Puig P (2006) KRAS mutation status is predictive of response to cetuximab therapy in colorectal cancer. Cancer Res 66(8):3992–3995. doi:10.1158/0008-5472.CAN-06-0191
122. de The H, Chomienne C, Lanotte M, Degos L, Dejean A (1990) The t(15;17) translocation of acute promyelocytic leukaemia fuses the retinoic acid receptor alpha gene to a novel transcribed locus. Nature 347(6293):558–561. doi:10.1038/347558a0
123. Huang ME, Ye YC, Chen SR, Zhao JC, Gu LJ, Cai JR, Zhao L, Xie JX, Shen ZX, Wang ZY (1987) All-trans retinoic acid with or without low dose cytosine arabinoside in acute promyelocytic leukemia. Report of 6 cases. Chin Med J 100(12):949–953
124. Patel JN, Mandock K, McLeod HL (2014) Clinically relevant cancer biomarkers and pharmacogenetic assays. J Oncol Pharm Pract 20(1):65-72. doi:10.1177/1078155212473862

Pharmacogenetics of Coumarin Anticoagulant Therapy

Rianne M.F. van Schie, Talitha I. Verhoef, Anthonius de Boer, Felix J.M. van der Meer, William K. Redekop, Tom Schalekamp and Anke-Hilse Maitland-van der Zee

Abstract Coumarins are effective drugs for treatment and prevention of thromboembolic events. However, their use requires a balancing act between the chance of underdosing which increases the risk of thromboembolic events and the chance of overdosing which increases the risk of haemorrhages. It has been shown that polymorphisms in *VKORC1* and *CYP2C9* explain 35–50 % of the dose variability, although patient characteristics and environmental factors also play a role. In this book chapter we discuss the pharmacogenetics of coumarin derivatives, clinical trials investigating the effectiveness of pre-treatment genotyping and the cost-effectiveness of pharmacogenetic-guided dosing.

Keywords Adverse drug reaction · Pharmacogenetics · Predictive genotyping · Translation · Abacavir · Hypersensitivity · Malignant

---

Rianne M.F. van Schie and Talitha I. Verhoef authors contributed equally

---

A. H. Maitland-van der Zee (✉) · R. M. F. van Schie · T. I. Verhoef · A. de Boer · T. Schalekamp
Division of Pharmacoepidemiology and Clinical Pharmacology, Utrecht Institute for Pharmaceutical Sciences, Utrecht University, Utrecht, The Netherlands
e-mail: a.h.maitland@uu.nl

R. M. F. van Schie
e-mail: RiannevanSchie@gmail.com

T. I. Verhoef
e-mail: t.verhoef@ucl.ac.uk

A. de Boer
e-mail: a.deboer@uu.nl

T. Schalekamp
e-mail: t.schalekamp@uu.nl

F. J. M. van der Meer
Department of Thrombosis and Hemostasis, Leiden University Medical Center, Leiden, The Netherlands
e-mail: F.J.M.van_der_Meer@lumc.nl

W. K. Redekop
Institute for Medical Technology Assessment, Erasmus University, Rotterdam, The Netherlands

G. Grech, I. Grossman (eds.), *Preventive and Predictive Genetics: Towards Personalised Medicine,* Advances in Predictive, Preventive and Personalised Medicine 9,
DOI 10.1007/978-3-319-15344-5_11

1 Introduction

Coumarin derivatives, such as warfarin, phenprocoumon and acenocoumarol, are very effective in the prevention and treatment of thromboembolic diseases, for example in patients with atrial fibrillation or venous thromboembolism [1–5]. Patients with atrial fibrillation have an annual stroke risk of 4.5 %, which decreases to 1.4 % during treatment with warfarin [1]. Warfarin is the most prescribed coumarin in the world while phenprocoumon and acenocoumarol are the coumarins of first choice in continental Europe [6–8]. Although these drugs have already been on the market for decades, finding the right dose for each patient is still challenging. Coumarins have a narrow therapeutic index, often resulting in an unacceptably low anticoagulant effect with an increased risk of thromboembolism or unacceptably high anticoagulant effect with an increased risk of haemorrhages [9–13]. Furthermore, they are subject to inter- and intra-individual variability in dose requirements [14, 15]. Also, the use of coumarins frequently results in drug-related hospitalisation [16–19]. It has been established that anticoagulation response is affected by environmental, clinical, and genetic factors such as age, height, weight, concurrent drug therapy, morbidities, dietary vitamin K intake, and genetic variation in Cytochrome P450 2C9 (CYP2C9) and vitamin K epoxide reductase complex subunit 1 (VKORC1) [20–25]. This chapter elaborates on the inter- and intra-patient variability in the response to coumarin derivatives, mainly focusing on the pharmacogenetics of these drugs.

2 Mechanism of Action

Inactive coagulation factors II, VII, IX and X require γ-carboxylation of the glutamic acid (Glu) residues into γ-carboxyglutamic (Gla) residues for their coagulation activity (see Fig. 1) [26–28]. In this process, the γ-carboxylase cofactor vitamin K-hydroquinone is oxidised to vitamin K-epoxide. Vitamin K- epoxide is recycled for the carboxylation of new coagulation factors in a 2-step reduction to vitamin K-hydroquinone [27, 28]. Vitamin K epoxide reductase (VKOR) is the catalyser of the first step in the reduction of vitamin K-epoxide into vitamin K-quinone and also contributes to the second reduction step, in which vitamin K-quinone is further reduced to vitamin K-hydroquinone [27, 28]. Cytochrome P450 4F2 (CYP4F2) is a vitamin K-oxidase and metabolises vitamin K-quinone to hydroxyvitamin K [29]. Coumarins, also called vitamin K antagonists, inhibit the reduction of oxidised vitamin K by binding to a small trans membrane protein in the endoplasmic reticulum called vitamin K epoxide reductase complex 1 (VKORC1), which is part of the VKOR complex [30, 31]. As a result, vitamin K-hydroquinone will not become available for the γ-carboxylation of coagulation factors. Coumarins thus act indirectly on the coagulation factors. The half-lives of the coagulation factors range from approximately 6 h for factor VII to 2.5 days for

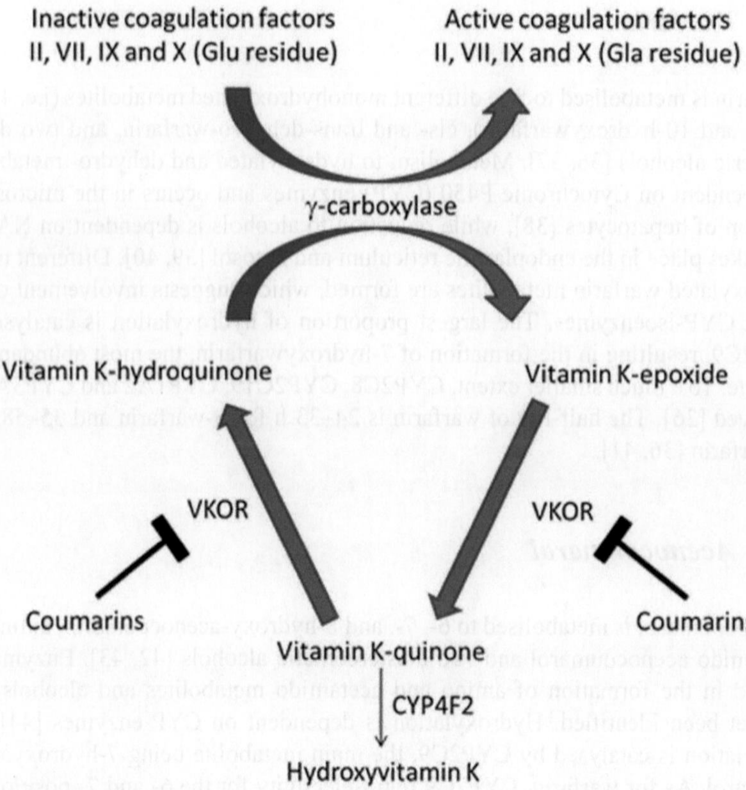

Fig. 1 The mechanism of action

factor II (prothrombin) [32]. This means that the effect of the coumarins in inducing an anticoagulant effect starts 15 h after administration [33] and ends 36–72 h after start of coumarin use [34, 35].

3 Pharmacokinetics

All three coumarin derivatives have a similar chemical structure and belong to the group of 4-hydroxycoumarins. Each coumarin has a single, chiral centre with a R-enantiomeric form or a S enantiomer, which is approximately 2- to 5-fold more potent [36]. Even though the mechanism of action is identical for the three coumarins, there are clear differences in their pharmacokinetic properties and therefore we discuss the pharmacokinetics of the coumarins separately. After administration, all coumarins (except S-acenocoumarol) are absorbed from the gastrointestinal tract with almost complete bioavailability [36].

3.1 Warfarin

Warfarin is metabolised to five different monohydroxylated metabolites (i.e. 4'-, 6-, 7-, 8- and 10-hydroxywarfarin), cis- and trans-dehydro-warfarin, and two diastereomeric alcohols [36, 37]. Metabolism to hydroxylated and dehydro- metabolites is dependent on Cytochrome P450 (CYP) enzymes and occurs in the microsomal fraction of hepatocytes [38], while reduction to alcohols is dependent on NADPH and takes place in the endoplasmic reticulum and cytosol [39, 40]. Different monohydroxylated warfarin metabolites are formed, which suggests involvement of different CYP-isoenzymes. The largest proportion of hydroxylation is catalysed by CYP2C9, resulting in the formation of 7-hydroxywarfarin, the most abundant metabolite. To a much smaller extent, CYP2C8, CYP2C19, CYP1A2 and CYP3A4 are involved [36]. The half-life of warfarin is 24–33 h for S-warfarin and 35–58 h for R-warfarin [36, 41].

3.2 Acenocoumarol

Acenocoumarol is metabolised to 6-, 7-, and 8-hydroxy-acenocoumarol, amino and acetamido acenocoumarol and two diastereometic alcohols [42, 43]. Enzymes involved in the formation of amino and acetamido metabolites and alcohols have not yet been identified. Hydroxylation is dependent on CYP-enzymes [44]. Hydroxylation is catalysed by CYP2C9, the main metabolite being 7-hydroxyacenocoumarol. As for warfarin, CYP2C9 regioselectivity for the 6- and 7- position and stereoselectivity for the S-enantiomer seem to play a role [36]. In contrast, the role of CYP2C19 and CYP1A2 is much smaller [36]. The half-life of acenocoumarol is 1.8 h for S-acenocoumarol—the most potent form—and 6.6 h for R-acenocoumarol [43].

3.3 Phenprocoumon

The metabolites of phenprocoumon are 4'-, 6-, 7- and 8-hydroxy-phenprocoumon and in contrast to warfarin and acenocoumarol all metabolites are hydroxyl-metabolites [36]. The hydroxyl-metabolites are all formed by CYP-enzymes [45, 46]. The 6- and 7-hydroxy phenprocoumon are the most abundant metabolites, 45 and 52 %, respectively [36]. The main metabolising enzymes involved are CYP2C9 for approximately 60–65 % and CYP3A4 for approximately 35–40 % of 6- and 7-hydroxy-phenprocoumon. These CYP-enzymes and CYP2C8 are also involved in the formation of the other metabolites [36]. The half-life of phenprocoumon is much longer compared with the two other coumarins: 110–130 h for S-phenprocoumon (the most potent form) and 110–125 h for R-phenprocoumon [47]. The contribution of CYP2C9 to the metabolism of the different enantiomers of the three coumarins varies [36] and is shown in Fig. 2.

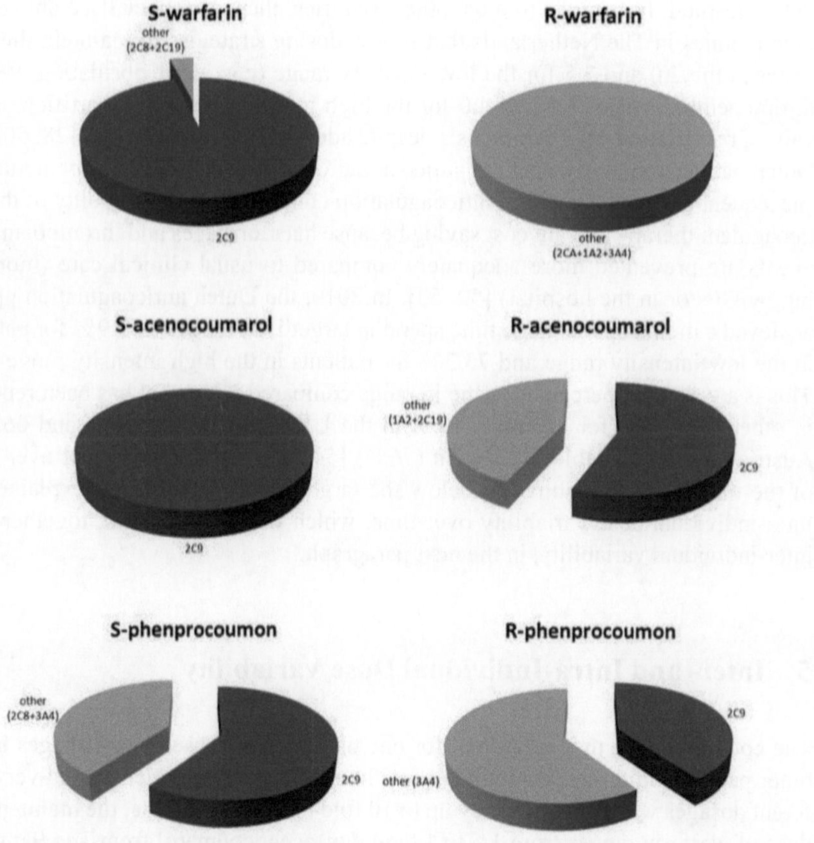

Fig. 2 The contribution of CYP2C9 and other CYP enzymes to the metabolism of the different enantiomers

4 Anticoagulant Therapy

In order to find the most effective and safe balance between underanticoagulation (with a risk of thromboembolic events) and overanticoagulation (with a risk of haemorrhage), a recommendation was made during the first American College of Chest Physicians (ACCP) conference in 1986 that therapy with coumarins should be monitored using the International Normalised Ratio (INR) established by the World Health Organisation [48, 49]. A dose that prolongs the INR to two to three times control (i.e. INR of 2.0–3.0) was recommended for indications such as prophylaxis and treatment of venous thromboembolism, and atrial fibrillation [49]. Higher ranges (i.e. INR of 3.0–4.5) were recommended for other indications including, for example, recurrent venous thrombosis despite adequate anticoagulation [49]. These recommendations are widely accepted and have increased the safety of coumarins [48]. The treatment is often managed by the general practitioner (GP) or a physician in the hospital. In contrast to most other countries, there are specialised anticoagulation clinics in The Netherlands that follow dosing strategies to maintain the INR between the 2.0 and 3.5 for the low intensity range (e.g. atrial fibrillation, venous thromboembolism) or 2.5 and 4.0 for the high intensity range (e.g. artificial heart valves, recurrent venous thrombosis despite adequate anticoagulation) [28, 50, 51]. Dutch patients regularly visit the anticoagulation clinic for INR measurements and subsequent dose adjustments. Anticoagulation clinics improve the quality of the anticoagulant therapy and are cost saving because haemorrhages and thromboembolic events are prevented more adequately compared to usual clinical care (monitoring by GPs or in the hospital) [52, 53]. In 2010, the Dutch anticoagulation clinics achieved a median percentage time spend in target INR range of 77.9% for patients in the low intensity range and 73.2% for patients in the high intensity range [50]. This is a very high percentage time in range compared with what has been reported in other countries (for example, 63% in the UK, 56% in Germany, and 66% in Austria) and comparable to Sweden (76%) [54], but it still means that over 20% of the time, INRs are above or below the target range. This can be explained by intra-individual dose variability over time, which will be discussed, together with inter-individual variability, in the next paragraph.

5 Inter- and Intra-Individual Dose Variability

The coumarin dose that is optimal for one patient may cause haemorrhages in another patient and thromboembolic events in a third patient. Patients need very different dosages which can differ by up to 10 fold [14]. For example, the maintenance dose of warfarin ranges from 1.5 to 12 mg/day, acenocoumarol from 1 to 9 mg/day and phenprocoumon from 0.75 to 9 mg/day [36]. In addition, the required dose may also change over time in an individual patient. There are several factors that cause inter- and intra-individual variability.

5.1 Patient Characteristics and Environmental Factors

Effects of patient characteristics and environmental factors can roughly be divided into 3 categories: effects on the coumarin dose, effects on the stability of the anticoagulant therapy, and effects on clinical outcomes.

5.1.1 Effects on Coumarin Dose

Coumarin dose requirements decrease with increasing age, but increase with increasing weight and height [25, 55]. Many diseases affect the coumarin dosages as well. Patients with hepatic disorders need lower dosages because the synthesis of coagulation factors is reduced in these patients because of Vitamin K deficiency, decreased metabolism due to reduction in hepatocyte mass or hypo-albuminaemia [56, 57]. Hyperthyroidism leads to decreased coumarin dosages compared to euthyroidism, while hypothyroidism is associated with a decreased catabolism of vitamin K-dependent coagulation factors, attenuating the response to oral anticoagulant therapy and resulting in increased dose requirements [56]. Heart failure may cause hepatic congestion, resulting in a decreased synthesis of coagulation factors and therefore lower coumarin maintenance dose requirements [56, 58]. Malignancies might affect the coumarin dose by metastatic liver disease, malnutrition, or use of chemotherapy [56]. Fever decreases coumarin dose requirements, probably by increasing degradation of coagulation factors [9]. Dehydration might affect the INR and therefore the coumarin dose by changing the volume of distribution of the coumarins [57]. Hypo-albuminaemia affects the concentration of unbound coumarins and therefore the coumarin dose requirements [57]. Kidney disorders might also affect the albumin concentration and therefore coumarin dose requirements [57]. Comedication use is also of importance and there are many drugs that can increase or decrease the anticoagulation effect and thereby influence the coumarin dose requirements [22, 23, 25, 59–62]. In the Netherlands, clinically relevant drug interactions with coumarins have been described and regulated in the guidelines for anticoagulation clinics [63, 64]. There are two main categories of drug interactions: first, the pharmacokinetic interactions affecting the absorption, distribution or elimination and second, the pharmacodynamic interactions affecting production or metabolism of coagulation factors, or directly affecting coagulation [57]. Besides affecting the coumarin maintenance dose, comedication might also increase the risk of haemorrhages.

5.1.2 Effects on Stability of the Anticoagulant Therapy

Dietary vitamin K intake interferes with the stability of the oral anticoagulant therapy [65]. Daily supplementation of vitamin K intake possibly contributes to a more stable anticoagulant therapy [66–68]. Other nutrition factors can also be of influence [57]. Because vitamin K is a fat-soluble vitamin, the absorption of vitamin K through the intestines is influenced by fat intake and absorption disorders which might result in instability of the anticoagulant therapy. Gavage feeding might cause fluctuating INRs [57, 69]. This could be due to different concentrations of vitamin K in the gavage in comparison to normal diet. Also, vitamin K might bind to proteins in the gavage feeding, or vitamin K might get lost in the preparation of the gavage or due to adsorption to the tube wall. Disorders of the gastrointestinal tract (e.g. vomiting, diarrhea, malabsorption of fat, or antibiotic use which may affect bacteria in the intestines that produce vitamin K) might affect the stability of anticoagulant therapy [57]. Increased levels of stress are thought to be associated with increased INRs and varying amounts of physical exercise may cause a fluctuation in INR as well [57]. Travelling (and any resulting changes in diet or alcohol consumption) and poor compliance might cause instability as well [57, 70].

5.1.3 Effects on Clinical Outcomes

Hematological disorders, such as thrombocytopenia, might affect the anticoagulant therapy by increasing the risk of haemorrhage. In addition, local disorders such as polyps increase the risk of haemorrhage. Malignancies may increase the risk of both venous thromboembolism and haemorrhages [57].

5.2 Pharmacogenetics

In 1992, Rettie et al. reported that CYP2C9 is the main metabolising enzyme of warfarin [71]. CYP1A2 and CYP3A4 were also found to contribute to the metabolism of the drug [71]. Furuya et al. hypothesised that polymorphisms in *CYP2C9* (resulting in proteins with different catalytic activities) might have a major effect on the clearance of the most potent enantiomer (S-warfarin) and therefore might affect the warfarin maintenance dose [72]. They recruited almost 100 patients who attended the anticoagulation clinic for routine INR monitoring. Information on body weight, height, age, sex, drug history, INRs history, indication for coumarin use, and comorbidities was collected. A blood sample was used to determine the *CYP2C9*2* genotype. Of the 94 included patients, 58 (62 %) were wild type (*CYP2C9*1/*1*) and 36 (38 %) heterozygous for *CYP2C9*2*. There were no patients homozygous for *CYP2C9*2*. Patients carrying the variant allele required significantly lower warfarin dosages than wild type patients (Mann-Whitney U-test, $p=0.02$). In addition, they found an association between age and warfarin dose requirements. The results suggesting an effect of *CYP2C9* genotypes on the coumarin maintenance dose have since been replicated by many research groups [25, 73–78]. Not only *CYP2C9*2* but also *CYP2C9*3* is a common variant allele in Caucasians that reduces the coumarin maintenance dose significantly [25, 73–78]. The CYP2C9*2 allele frequencies vary from 8 to 19 % and the CYP2C9*3 alleles from 3 to 16 % in Caucasians [79]. East Asian and African or Afro-American populations show an absence of CYP2C9*2 and a reduced frequency of CYP2C9*3 (79). The *CYP2C9* genotype explains approximately 4.5–17.5 % of the coumarin (warfarin, acenocoumarol and phenprocoumon) dose variation [25, 76, 80–85].

Rost et al. and Li et al. identified VKORC1 as a target of the coumarins in 2004 [30, 31]. This introduced a new possibility for explaining the coumarin dose variability. Indeed, many researchers showed decreased coumarin dose requirements if patients carried one or two variant alleles in the *VKORC1* gene [73–75, 82, 86, 87]. Two SNPs in *VKORC1*, the −1639G>A and the 1173 C>T, were found to be associated with decreased warfarin dose requirements [28]. It was demonstrated that promotor SNP −1639G>A causes the variability in VKORC1 activity by suppressing the gene expression, but a role for 1173 C>T could not be excluded because of the complete linkage disequilibrium between the two SNPs [88]. Patients carrying one or two variant alleles have decreased levels of *VKORC1* mRNA in the liver and therefore need lower coumarin dosages compared to wild type patients

[88]. Because the two SNPs are in complete linkage disequilibrium [88, 89], studying either of the two SNPs will give the same results. Allele frequencies for the *VKORC1* variant allele are 37–41 % in Caucasians, 10–12 % in African Americans, and 88–92 % in East-Asians [28].

There are many other genes that could potentially affect the coumarin maintenance dose. The association with the coumarin dose might for example be based on other pharmacokinetic or pharmacodynamic mechanisms, for example by affecting the transport of coumarins or vitamin K or by affecting the vitamin K cycle. In the metabolism of phenprocoumon, other metabolising enzymes, especially CYP3A4, also play an important role [36, 90] and therefore SNPs in the genes encoding for these metabolising enzymes are hypothesised to affect the phenprocoumon dose requirements. However, Teichert et al. did not find an association between *CYP3A4*1B* and the phenprocoumon dose [91]. Another gene that has been associated with coumarin response is *CYP4F2* [91–97], which is a vitamin K oxidase. Patients carrying one or two V433M variant alleles in *CYP4F2* have a reduced capacity to metabolise Vitamin K, resulting in increased vitamin K levels and therefore also resulting in higher coumarin dose requirements when compared to non-carriers [29]. SNPs in *CYP4F2* have a nominal effect on the coumarin maintenance dose; it explains an additional 1–2 % of the coumarin dose requirements [92, 94]. Polymorphisms in the gene encoding γ-glutamylcarboxylase (*GGCX*), which is involved in the carboxylation of coagulation factors, have also been shown to have a minor effect on the coumarin dose [74, 98] however other research groups did not find an association between the coumarin dose and polymorphisms in *GGCX* [99, 100]. Other minor influences on the coumarin maintenance dose might be caused by polymorphisms in the genes encoding for the coagulation factors VII and X [101], epoxide hydrolase (*EPHX1*) [100, 102] which encodes a protein subunit of VKOR, apolipoprotein E (*APOE*) [103–107] which encodes for the protein responsible for the vitamin K uptake, and in protein C (*PROC*) [103] which encodes for protein C, responsible for the inactivation of coagulation factors Va and VIIIa. All these polymorphisms show low or no clinical relevance.

Until now, only *VKORC1*, *CYP2C9* and *CYP4F2* genotypes were found to be associated with the coumarin maintenance dose in genome wide association studies (GWAS) [91, 93, 94, 97]. Ross and co-workers studied the allele frequencies of these genes in different populations and found that there are significant differences between populations worldwide [108]. The allele frequencies of the common and variant alleles of *VKORC1, CYP2C9*2, CYP2C9*3* and *CYP4F2* are shown in Fig. 3. One study also found an association between CYP2C18 and the acenocoumarol dose [97]. Another study of 1496 Swedish patients starting warfarin treatment investigated possible associations between183 polymorphisms in 29 candidate genes and warfarin dose and only found an association for *CYP2C9* and *VKORC1* [83].

CYP2C9 and *VKORC1* genotypes together explain approximately 35–50 % of the coumarin dose requirements [83, 87, 109]. To date, a number of studies have reported the development of pharmacogenetics-guided algorithms for coumarins in order to predict the personalised coumarin dose before start of the anticoagulant

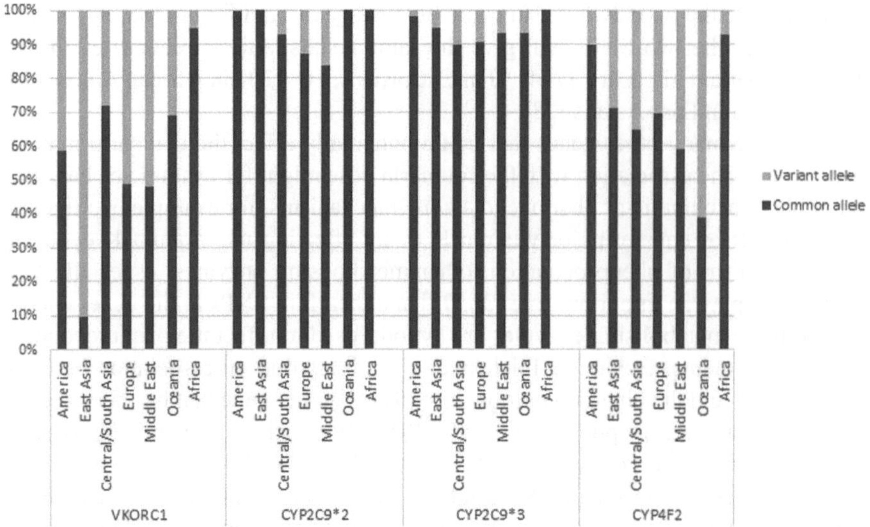

Fig. 3 Allele frequencies of genes associated with coumarin dose requirement among different populations

therapy [25, 76, 80–85]. The predictive value of these algorithms varied from 47 to 60 %. Because of ethnic differences in allele frequencies, it can be expected that pharmacogenetic algorithms have a different predictive value in different populations. Several authors have included race as a parameter in their pharmacogenetic-guided algorithm [76, 80, 81, 83]. The International Warfarin Pharmacogenetics Consortium showed that a model that was adjusted for race performed better than specific models for each ethnicity. However, racial differences were not significantly associated with the required dose when genetic information was added to the model [76].

5.2.1 Clinical Trials

In 2005, the first (pilot) randomised trial on pharmacogenetic-guided warfarin dosing in 38 patients was published [110]. These authors reported no differences in percentage time in INR range or the risk of supratherapeutic INR values. In another randomised trial with 191 patients, the time to stable dose was decreased and the time spent in therapeutic range was increased by pharmacogenetic-guided dosing [111]. In both these studies, only *CYP2C9* genotype was assessed and not *VKORC1* genotype. Anderson et al. [112] investigated the impact of genotyping for both *CYP2C9* and *VKORC1* genotypes in 220 patients. No effect on the number of out-of-range INR values could be demonstrated when looking at all patients, but in wild-type patients and patients carrying multiple variant alleles, genotyping decreased the risk of out-of-range INRs by 10 %. In two small randomised trials in Chinese patients, a stable dose was reached faster in patients receiving a pharmacogenetic-guided dose than in patients receiving a standard dose [113, 114]. Burmester et al., compared dosing using a pharmacogenetic algorithm to a clinical algorithm instead of standard dosing and found no differences in percentage time in therapeutic range between the two arms [115]. The Applying Pharmacogenetic Algorithms to Individualise Dosing of Warfarin (Coumagen-II) trial (NCT00927862) showed that pharmacogenetic dosing was superior to standard dosing for percentage time in and out of therapeutic range [116]. During the first month of the treatment, 31 % of the INR measurements were below or above the therapeutic range in the intervention group vs. 42 % in the control group. The reduction in out-of-range INRs was mainly due to a reduction in INRs below the therapeutic range. The percentage time within the therapeutic range was 69 % in the intervention group and 58 % in the control group. Also, less serious adverse events (including haemorrhagic and thromboembolic events) occurred in the genotype-guided group (4.5 vs. 9.4 %, $p=0.001$). The limitation of this study was the lack of randomised comparison.

The European Pharmacogenetics of Anticoagulant Therapy EU-PACT trial (unique ClinicalTrials.gov Identifiers: NCT01119274, NCT01119261, and NCT01119300) compares a dose algorithm with patient characteristics (or in the case of warfarin standard clinical care) to a dose algorithm with patient characteristics and *VKORC1* and *CYP2C9* genotype [117, 118]. The primary outcome is the time within target INR range. It is the only RCT that investigates all three coumarins (warfarin, phenprocoumon and acenocoumarol). The EU-PACT warfarin arm showed a positive effect of the genotype-guided dosing taking percent time in therapeutic INR range as an outcome. The patients that were genotyped spent 7% more time in range in the first 12 weeks of warfarin therapy compared with the patients in the standard care arm. In the EU-PACT phenprocoumon/acenocoumarol arm there was no statistically significant difference in time in therapeutic range in the first 12 weeks, however there was a statistically significant effect in the first 4 weeks of treatment. Patients in the genotyped arm spend 5% more time within therapeutic range in these first 4 weeks [117]. On the other hand, the Clarification of Optimal Anticoagulation Through Genetics (COAG) (NCT00839657) trial results in no significant difference in the time spent within the therapeutic range in the first 4 weeks of warfarin treatment [119]. These conflicting results are compared in Table 1. One of the reasons for these observed differences might be the comparator, since for warfarin dosage, the genotype guided dose was compared to standard care in the EU-PACT trial, whereas the comparator in the EU-PACT phenprocoumon/acenocoumarol arm and in the COAG trial was a clinical algorithm. Furthermore in the COAG trial it was shown that for African Americans the time in therapeutic range was less in the genotyped arm compared with the clinical algorithm arm. This implies that different algorithms are necessary for different race groups.

Table 1 Overview of randomised clinical trials

	EU-PACT [117]	COAG [119]
Coumarin derivative	Phenprocoumon, acenocoumarol, warfarin	Warfarin
Population	Patients with atrial fibrillation or venous thromboembolism	Patients requiring warfarin therapy with a target INR range of 2–3
Genotypes included	VKORC1, CYP2C9	VKORC1, CYP2C9
Comparator	Clinical algorithm (acenocoumarol, phenprocoumon) Standard care (warfarin)	Clinical algorithm
Number of patients	911	1015
Primary outcome	Percentage time within target INR range	Percentage time within target INR range
Result	Genotype-guided Warfarin Algorithm is superior	No difference

6 Cost-Effectiveness

Clinical trials can provide valuable information about the safety and effectiveness of genotyping before starting coumarin therapy. This information is not only valuable for clinicians but also for policymakers who need to make a decision about whether or not to implement genotype-guided dosing. However, this decision will not only depend on the effectiveness of genotyping, but also on the cost-effectiveness since an important factor for implementation will be reimbursement of the genetic tests. This is the primary reason for performing cost-effectiveness analyses. Some of the cost-effectiveness analyses of genotyping performed in the past have estimated the costs to avoid an adverse event. But for a health insurance company, this way of describing cost-effectiveness makes it difficult to compare with the cost-effectiveness of other drugs for other diseases. Reimbursement authorities therefore often require a so-called *cost-utility analysis* in which the extra costs to gain one quality-adjusted life-year (QALY) are estimated. Since the QALY represents a generic measure of overall health that can be improved by increasing life expectancy and/or quality of life, the cost per QALY gained can therefore be applied for any health technology for any disease area.

One of the first estimates of the cost-effectiveness of genotyping warfarin users was published in 2003. These authors estimated that the cost to avoid one bleeding event were US$5940 [120] if patients were given a dose based on their *CYP2C9* genotype, compared with standard care. Very similar results were obtained by You et al., who calculated a cost-effectiveness ratio of US$5778 per bleeding event avoided [121]. Schalekamp et al. reported that the cost-effectiveness of genotyping acenocoumarol users for their *CYP2C9* genotype was US$5151 per bleeding event avoided [122]. This study focused on the Netherlands, while the other two studies focused on the

US. After the relevance of the *VKORC1* genotype was demonstrated, it was assumed that genotyping the patient for both *CYP2C9* and *VKORC1* genotypes would lead to better dose prediction and therefore a larger effect of pharmacogenetic-guided dosing than genotyping for *CYP2C9* alone. More recent cost-effectiveness analyses therefore also included *VKORC1* genotyping in their assessment. Several authors estimated the cost-utility ratio of genotyping for these two genes compared with standard care in the US [123–127] and reported results that vary from US$60,750 to US$347,000 per QALY gained. Eckman and co-workers performed a meta-analysis of the three trials that were available in 2008 [110–112] and found that pharmacogenetic-guided dosing could reduce the risk of bleeding by 32 % [124]. When they used this data in their economic model, they found that genotyping would cost US$170,000 per QALY gained, a value much higher than the willingness-to-pay thresholds of US$50,000–US$100,000 that are often applied in the US to conclude whether or not an intervention is cost-effective [128]. Sensitivity analyses by Eckman et al. showed that the costs per QALY gained would be less than US$50,000 only if the test would be restricted to patients with a high bleeding risk or if all of the following criteria were met: more bleeding events could be avoided, the test would cost less than US$200 and the results would be available within 1 day. Patrick and co-workers also found that genotyping only patients with a high bleeding risk would increase its chance of being cost-effective [126]. Meckley and co-workers used data from the CoumaGen trial [112] and found a cost-effectiveness ratio of US$60,740 per QALY gained [127]. You et al. reported a much higher cost per QALY gained than previous studies (US$347,000) as well as high costs per life saved (US$1,106,000 per life saved) and high cost per adverse event averted (US$170,000), which combined bleeding events with thromboembolic events [123]. The chance that genotyping would cost less than the US$50,000 threshold was low (38 %) and increased with lower genotyping costs, greater reduction in out-of-range INRs and in specific settings where poor INR control was seen. Using data from the CoumagenII trial [116], in which the time in therapeutic range in the first month was increased by 11 % in the first month, Verhoef et al. reported that pharmacogenetic-guided phenprocoumon dosing would be cost-effective [129] given a cost per QALY gained of 2700 euro.

Recently, novel oral anticoagulant drugs such as dabigatran, rivaroxaban and apixaban have been developed, which appear to be good alternatives to coumarin anticoagulants [130]. You et al. studied the cost-effectiveness of dabigatran and genotype-guided warfarin treatment and showed that dabigatran seems to be a cost-effective treatment [131]. However, they reported that pharmacogenetic-guided warfarin dosing had a higher chance of being cost-effective if it was able to increase the percentage time in target INR range to >77 %.

The main limitation of the cost-effectiveness studies published up to now has been the lack of robust data from appropriately powered clinical trials [132]. Also, the costs of genotyping *VKORC1* and *CYP2C9* polymorphisms are not clear yet. Previous studies have used costs that vary from US$175 to US$575 when the genotype is determined in the lab and US$50 for a point-of-care test [127, 132, 133]. These costs are expected to decrease over time and with increased usage, which will influence the cost-effectiveness as well. In the analysis by Verhoef and co-workers, the use of a point-of-care test was assumed, which provides the results within 2 h and costs less than US$50 [133]. In sum, most of the studies found that, pharmacogenetic-guided dosing did not seem to be cost-effective and their results underline the large influence of effectiveness of genotyping and the costs of the test. Genotype-guided dosing will only be cost-effective if the costs of the test can be kept low or if it has a large effect on INR control and related incidences of adverse events. The results also show that genotyping could be cost-effective if it would be used only with specific patients (with a high bleeding risk) or in specific settings (with a low quality of INR control).

A more reliable estimate of the cost-effectiveness or cost-utility of pharmacogenetic-guided coumarin dosing can be calculated after the results of the large RCTs become available. Because of many differences between countries in costs and organisation of anticoagulation services, the cost-effectiveness of genotyping coumarin users probably varies between countries [54]. Therefore it will also be necessary to carry out country-specific analyses in the future.

7 Conclusion

Coumarins are effective drugs for treatment and prevention of thromboembolic events. However, their use requires a delicate balancing act between the chance of underdosing (which increases the risk of thromboembolic events) and the chance of overdosing (which increases the risk of haemorrhages). It has been shown that polymorphisms in *VKORC1* and *CYP2C9* explain a large part (35–50 %) of the dose variability but patient characteristics and environmental factors also play a role. Clinical trials have researched the added value and cost effectiveness of pre-treatment genotyping. The results from the trials were not convincing, and at this moment there is not enough evidence to recommend genotyping for CYP2C9 and VKORC1 in routine clinical practice. Recent cost-effectiveness studies have shown that the small improvement of time in therapeutic range does not weigh against the costs of genotyping all patients. However, the cost-effectiveness of the intervention will depend on the costs of genotyping and on the availibility of other anticoagulation therapy such as the Direct Oral Anticoagulants (DOACs) [118].

References

1. Albers GW, Sherman DG, Gress DR, Paulseth JE, Petersen P (1991) Stroke prevention in nonvalvular atrial fibrillation: a review of prospective randomized trials. Ann Neurol 30(4): 511–518. doi:10.1002/ana.410300402
2. Petersen P, Boysen G, Godtfredsen J, Andersen ED, Andersen B (1989) Placebo-controlled, randomised trial of warfarin and aspirin for prevention of thromboembolic complications in chronic atrial fibrillation. The Copenhagen AFASAK study. Lancet 1(8631):175–179. doi:10.1016/S0140-6736(89)91200-2

3. The effect of low-dose warfarin on the risk of stroke in patients with nonrheumatic atrial fibrillation. The Boston Area Anticoagulation Trial for Atrial Fibrillation Investigators (1990). N Engl J Med 323(22):1505–1511. doi:10.1056/NEJM199011293232201
4. Stroke Prevention in Atrial Fibrillation Study. Final results (1991). Circulation 84(2):527–539
5. Fareed J, Hoppensteadt DA, Fareed D, Demir M, Wahi R, Clarke M, Adiguzel C, Bick R (2008) Survival of heparins, oral anticoagulants, and aspirin after the year 2010. Semin Thromb Hemost 34(1):58–73. doi:10.1055/s-2008-1066025
6. Pirmohamed M (2006) Warfarin: almost 60 years old and still causing problems. Br J Clin Pharmacol 62(5):509–511. doi:10.1111/j.1365-2125.2006.02806.x
7. Pengo V, Pegoraro C, Cucchini U, Iliceto S (2006) Worldwide management of oral anticoagulant therapy: the ISAM study. J Thromb Thrombolysis 21(1):73–77. doi:10.1007/s11239-006-5580-y
8. Ansell J, Hollowell J, Pengo V, Martinez-Brotons F, Caro J, Drouet L (2007) Descriptive analysis of the process and quality of oral anticoagulation management in real-life practice in patients with chronic non-valvular atrial fibrillation: the international study of anticoagulation management (ISAM). J Thromb Thrombolysis 23(2):83–91. doi:10.1007/s11239-006-9022-7
9. Penning-van Beest FJ, van Meegen E, Rosendaal FR, Stricker BH (2001) Characteristics of anticoagulant therapy and comorbidity related to overanticoagulation. Thromb Haemost 86(2):569–574
10. Hylek EM, Skates SJ, Sheehan MA, Singer DE (1996) An analysis of the lowest effective intensity of prophylactic anticoagulation for patients with nonrheumatic atrial fibrillation. N Engl J Med 335(8):540–546. doi:10.1056/NEJM199608223350802
11. Hylek EM, Singer DE (1994) Risk factors for intracranial hemorrhage in outpatients taking warfarin. Ann Intern Med 120(11):897–902
12. Oden A, Fahlen M, Hart RG (2006) Optimal INR for prevention of stroke and death in atrial fibrillation: a critical appraisal. Thromb Res 117(5):493–499
13. Fuster V, Ryden LE, Cannom DS, Crijns HJ, Curtis AB, Ellenbogen KA, Halperin JL, Heuzey JY L, Kay GN, Lowe JE, Olsson SB, Prystowsky EN, Tamargo JL, Wann S, Smith SC Jr, Jacobs AK, Adams CD, Anderson JL, Antman EM, Hunt SA, Nishimura R, Ornato JP, Page RL, Riegel B, Priori SG, Blanc JJ, Budaj A, Camm AJ, Dean V, Deckers JW, Despres C, Dickstein K, Lekakis J, McGregor K, Metra M, Morais J, Osterspey A, Zamorano JL (2006) ACC/AHA/ESC 2006 guidelines for the management of patients with atrial fibrillation: a report of the American college of cardiology/American heart association task force on practice guidelines and the European society of cardiology committee for practice guidelines (writing committee to revise the 2001 guidelines for the management of patients with atrial fibrillation): developed in collaboration with the European heart rhythm association and the heart rhythm society. Circulation 114(7):e257–e354
14. Rosendaal FR (1996) The Scylla and Charybdis of oral anticoagulant treatment. N Engl J Med 335(8):587–589. doi:10.1056/NEJM199608223350810
15. James AH, Britt RP, Raskino CL, Thompson SG (1992) Factors affecting the maintenance dose of warfarin. J Clin Pathol 45(8):704–706
16. van der Hooft CS, Sturkenboom MC, van Grootheest K, Kingma HJ, Stricker BH (2006) Adverse drug reaction-related hospitalisations: a nationwide study in The Netherlands. Drug Saf 29(2):161–168
17. Schneeweiss S, Hasford J, Gottler M, Hoffmann A, Riethling AK, Avorn J (2002) Admissions caused by adverse drug events to internal medicine and emergency departments in hospitals: a longitudinal population-based study. Eur J Clin Pharmacol 58(4):285–291. doi:10.1007/s00228-002-0467-0
18. Leendertse AJ, Egberts AC, Stoker LJ, van den Bemt PM (2008) Frequency of and risk factors for preventable medication-related hospital admissions in the Netherlands. Arch Intern Med 168(17):1890–1896. doi:10.1001/archinternmed.2008.3
19. Budnitz DS, Shehab N, Kegler SR, Richards CL (2007) Medication use leading to emergency department visits for adverse drug events in older adults. Ann Intern Med 147(11):755–765

20. Penning-van Beest FJ, Geleijnse JM, van Meegen E, Vermeer C, Rosendaal FR, Stricker BH (2002) Lifestyle and diet as risk factors for overanticoagulation. J Clin Epidemiol 55(4): 411–417
21. Carlquist JF, Horne BD, Muhlestein JB, Lappe DL, Whiting BM, Kolek MJ, Clarke JL, James BC, Anderson JL (2006) Genotypes of the cytochrome p450 isoform, CYP2C9, and the vitamin K epoxide reductase complex subunit 1 conjointly determine stable warfarin dose: a prospective study. J Thromb Thrombolysis 22(3):191–197. doi:10.1007/s11239-006-9030-7
22. Schalekamp T, van Geest-Daalderop JH, Kramer MH, van Holten-Verzantvoort AT, de Boer A (2007) Coumarin anticoagulants and co-trimoxazole: avoid the combination rather than manage the interaction. Eur J Clin Pharmacol 63(4):335–343. doi:10.1007/s00228-007-0268-6
23. Schalekamp T, Klungel OH, Souverein PC, de Boer A (2008) Increased bleeding risk with concurrent use of selective serotonin reuptake inhibitors and coumarins. Arch Intern Med 168(2):180–185. doi:10.1001/archinternmed.2007.32
24. Gage BF, Eby C, Milligan PE, Banet GA, Duncan JR, McLeod HL (2004) Use of pharmacogenetics and clinical factors to predict the maintenance dose of warfarin. Thromb Haemost 91(1):87–94. doi:10.1267/THRO04010087
25. van Schie RM, Wessels JA, le Cessie S, de Boer A, Schalekamp T, van der Meer FJ, Verhoef TI, van Meegen E, Rosendaal FR, Maitland-van der Zee AH (2011) Loading and maintenance dose algorithms for phenprocoumon and acenocoumarol using patient characteristics and pharmacogenetic data. Eur Heart J 32(15):1909–1917. doi:10.1093/eurheartj/ehr116
26. Ansell J, Hirsh J, Hylek E, Jacobson A, Crowther M, Palareti G (2008) Pharmacology and management of the vitamin K antagonists: American college of chest physicians evidence-based clinical practice guidelines (8th Ed). Chest 133(6 Suppl):160S–198S. doi:10.1378/chest.08-0670
27. Stafford DW (2005) The vitamin K cycle. J Thromb Haemost 3(8):1873–1878
28. Schalekamp T, de Boer A (2010) Pharmacogenetics of oral anticoagulant therapy. Curr Pharm Des 16(2):187–203. doi:10.2174/138161210790112737
29. McDonald MG, Rieder MJ, Nakano M, Hsia CK, Rettie AE (2009) CYP4F2 is a vitamin K1 oxidase: an explanation for altered warfarin dose in carriers of the V433M variant. Mol Pharmacol 75(6):1337–1346. doi:10.1124/mol.109.054833
30. Rost S, Fregin A, Ivaskevicius V, Conzelmann E, Hortnagel K, Pelz HJ, Lappegard K, Seifried E, Scharrer I, Tuddenham EG, Muller CR, Strom TM, Oldenburg J (2004) Mutations in VKORC1 cause warfarin resistance and multiple coagulation factor deficiency type 2. Nature 427(6974):537–541. doi:10.1038/nature02214
31. Li T, Chang CY, Jin DY, Lin PJ, Khvorova A, Stafford DW (2004) Identification of the gene for vitamin K epoxide reductase. Nature 427(6974):541–544. doi:10.1038/nature02254
32. Mann KG (2005) The challenge of regulating anticoagulant drugs: focus on warfarin. Am Heart J 149(1 Suppl):36–42
33. Kroon C, de Boer A, Hoogkamer JF, Schoemaker HC, van der Meer EJ, Edelbroek PM, Cohen AF (1990) Detection of drug interactions with single dose acenocoumarol: new screening method? Int J Clin Pharmacol Ther Toxicol 28(8):355–360
34. Summary of the Product Characteristics (SPC) Phenprocoumon (2011). http://db.cbg-meb.nl/IB-teksten/h03819.pdf.
35. Summary of the Product Characteristics (SPC) Acenocoumarol (2012). http://db.cbg-meb.nl/IB-teksten/h04464.pdf.
36. Ufer M (2005) Comparative pharmacokinetics of vitamin K antagonists: warfarin, phenprocoumon and acenocoumarol. Clin Pharmacokinet 44(12):1227–1246
37. Trager WF, Lewis RJ, Garland WA (1970) Mass spectral analysis in the identification of human metabolites of warfarin. J Med Chem 13(6):1196–1204
38. Kaminsky LS, Dunbar DA, Wang PP, Beaune P, Larrey D, Guengerich FP, Schnellmann RG, Sipes IG (1984) Human hepatic cytochrome P-450 composition as probed by in vitro microsomal metabolism of warfarin. Drug Metab Dispos 12(4):470–477

39. Hermans JJ, Thijssen HH (1989) The in vitro ketone reduction of warfarin and analogues. Substrate stereoselectivity, product stereoselectivity and species differences. Biochem Pharmacol 38(19):3365–3370
40. Moreland TA, Hewick DS (1975) Studies on a ketone reductase in human and rat liver and kidney soluble fraction using warfarin as a substrate. Biochem Pharmacol 24(21):1953–1957
41. Kelly JG, O'Malley K (1979) Clinical pharmacokinetics of oral anticoagulants. Clin Pharmacokinet 4(1):1–15
42. Dieterle W, Faigle JW, Montigel C, Sulc M, Theobald W (1977) Biotransformation and pharmacokinetics of acenocoumarol (Sintrom) in man. Eur J Clin Pharmacol 11(5):367–375
43. Thijssen HH, Flinois JP, Beaune PH (2000) Cytochrome P4502C9 is the principal catalyst of racemic acenocoumarol hydroxylation reactions in human liver microsomes. Drug Metab Dispos 28(11):1284–1290
44. Hermans JJ, Thijssen HH (1991) Comparison of the rat liver microsomal metabolism of the enantiomers of warfarin and 4'-nitrowarfarin (acenocoumarol). Xenobiotica 21(3): 295–307
45. Toon S, Heimark LD, Trager WF, O'Reilly RA (1985) Metabolic fate of phenprocoumon in humans. J Pharm Sci 74(10):1037–1040
46. He M, Korzekwa KR, Jones JP, Rettie AE, Trager WF (1999) Structural forms of phenprocoumon and warfarin that are metabolized at the active site of CYP2C9. Arch Biochem Biophys 372(1):16–28. doi:10.1006/abbi.1999.1468
47. Jahnchen E, Meinertz T, Gilfrich HJ, Groth U, Martini A (1976) The enantiomers of phenprocoumon: pharmacodynamic and pharmacokinetic studies. Clin Pharmacol Ther 20(3):342–349
48. Dalen JE (2012) Prevention of embolic strokes: the role of the American college of chest physicians. Chest 141(2):294–299. doi:10.1378/chest.11-2641
49. ACCP-NHLBI National Conference on Antithrombotic Therapy. American College of Chest Physicians and the National Heart, Lung and Blood Institute (1986). Chest Feb;89(2 Suppl):1S–106S
50. Federation of Dutch anticoagulation clinics. Samenvatting medische jaarverslagen 2010. http://www.trombosestichting.nl/media/pagecontent/documents/jaarverslagen/Jaarverslag_Trombosestichting_2010.pdf
51. Cannegieter SC, Rosendaal FR, Wintzen AR, van der Meer FJ, Vandenbroucke JP, Briet E (1995) Optimal oral anticoagulant therapy in patients with mechanical heart valves. N Engl J Med 333(1):11–17. doi:10.1056/NEJM199507063330103
52. Chiquette E, Amato MG, Bussey HI (1998) Comparison of an anticoagulation clinic with usual medical care: anticoagulation control, patient outcomes, and health care costs. Arch Intern Med 158(15):1641–1647
53. van Walraven C, Jennings A, Oake N, Fergusson D, Forster AJ (2006) Effect of study setting on anticoagulation control: a systematic review and metaregression. Chest 129(5):1155–1166
54. Verhoef TI, Redekop WK, van Schie RM, Bayat S, Daly AK, Geitona M, Haschke-Becher E, Hughes DA, Kamali F, Levin LA, Manolopoulos VG, Pirmohamed M, Siebert U, Stingl JC, Wadelius M, de Boer A, Maitland-van der Zee AH (2012) Cost-effectiveness of pharmacogenetics in anticoagulation: international differences in healthcare systems and costs. Pharmacogenomics 13(12):1405–1417. doi:10.2217/pgs.12.124
55. Gurwitz JH, Avorn J, Ross-Degnan D, Choodnovskiy I, Ansell J (1992) Aging and the anticoagulant response to warfarin therapy. Ann Intern Med 116(11):901–904
56. Demirkan K, Stephens MA, Newman KP, Self TH (2000) Response to warfarin and other oral anticoagulants: effects of disease states. South Med J 93(5):448–454; quiz 455
57. Commissie SMedischHvandeFvanNTrombosediensten (2010) De kunst van het doseren. Richtlijn, leidraad en informatie voor het doseren van vitamine K-antagonisten. Voorschoten: Federatie van Nederlandse Trombosediensten
58. Visser LE, Bleumink GS, Trienekens PH, Vulto AG, Hofman A, Stricker BH (2004) The risk of overanticoagulation in patients with heart failure on coumarin anticoagulants. Br J Haematol 127(1):85–89. doi:10.1111/j.1365-2141.2004.05162.x

59. Schalekamp T, Klungel OH, Souverein PC, de Boer A (2008) Effect of oral antiplatelet agents on major bleeding in users of coumarins. Thromb Haemost 100(6):1076–1083
60. Verhoef TI, Zuurhout MJ, van Schie RM, Redekop WK, van der Meer FJ, le Cessie S, Schalekamp T, de Boer A, Maitland-van der Zee AH (2012) The effect of omeprazole and esomeprazole on the maintenance dose of phenprocoumon. Br J Clin Pharmacol 74(6):1068–1069. doi:10.1111/j.1365-2125.2012.04295.x
61. Teichert M, van Noord C, Uitterlinden AG, Hofman A, Buhre PN, De Smet PA, Straus S, Stricker BH, Visser LE (2011) Proton pump inhibitors and the risk of overanticoagulation during acenocoumarol maintenance treatment. Br J Haematol 153(3):379–385. doi:10.1111/j.1365-2141.2011.08633.x
62. Howard PA, Ellerbeck EF, Engelman KK, Patterson KL (2002) The nature and frequency of potential warfarin drug interactions that increase the risk of bleeding in patients with atrial fibrillation. Pharmacoepidemiol Drug Saf 11(7):569–576. doi:10.1002/pds.748
63. Federatie van Nederlandse Trombosediensten, Wetenschappelijk Instituut Nederlandse Apothekers. http://www.fnt.nl/behandeling/cumarine-interacties.html
64. Wittkowsky AK. Warfarin and other coumarin derivatives: pharmacokinetics, pharmacodynamics, and drug interactions. Semin Vasc Med 2003(3):221–230
65. Franco V, Polanczyk CA, Clausell N, Rohde LE (2004) Role of dietary vitamin K intake in chronic oral anticoagulation: prospective evidence from observational and randomized protocols. Am J Med 116(10):651–656. doi:10.1016/j.amjmed.2003.12.036
66. Sconce E, Khan T, Mason J, Noble F, Wynne H, Kamali F (2005) Patients with unstable control have a poorer dietary intake of vitamin K compared to patients with stable control of anticoagulation. Thromb Haemost 93(5):872–875
67. Reese AM, Farnett LE, Lyons RM, Patel B, Morgan L, Bussey HI (2005) Low-dose vitamin K to augment anticoagulation control. Pharmacotherapy 25(12):1746–1751. doi:10.1592/phco.2005.25.12.1746
68. Rombouts EK, Rosendaal FR, Van Der Meer FJ (2007) Daily vitamin K supplementation improves anticoagulant stability. J Thromb Haemost 5(10):2043–2048
69. Dickerson RN (2008) Warfarin resistance and enteral tube feeding: a vitamin K-independent interaction. Nutrition 24(10):1048–1052. doi:10.1016/j.nut.2008.05.015
70. van der Meer FJ, Briet E, Vandenbroucke JP, Sramek DI, Versluijs MH, Rosendaal FR (1997) The role of compliance as a cause of instability in oral anticoagulant therapy. Br J Haematol 98(4):893–900
71. Rettie AE, Korzekwa KR, Kunze KL, Lawrence RF, Eddy AC, Aoyama T, Gelboin HV, Gonzalez FJ, Trager WF (1992) Hydroxylation of warfarin by human cDNA-expressed cytochrome P-450: a role for P-4502C9 in the etiology of (S)-warfarin-drug interactions. Chem Res Toxicol 5(1):54–59
72. Furuya H, Fernandez-Salguero P, Gregory W, Taber H, Steward A, Gonzalez FJ, Idle JR (1995) Genetic polymorphism of CYP2C9 and its effect on warfarin maintenance dose requirement in patients undergoing anticoagulation therapy. Pharmacogenetics 5(6): 389–392
73. Schalekamp T, Brasse BP, Roijers JF, van Meegen E, van der Meer FJ, van Wijk EM, Egberts AC, de Boer A (2007) VKORC1 and CYP2C9 genotypes and phenprocoumon anticoagulation status: interaction between both genotypes affects dose requirement. Clin Pharmacol Ther 81(2):185–193
74. Wadelius M, Chen LY, Downes K, Ghori J, Hunt S, Eriksson N, Wallerman O, Melhus H, Wadelius C, Bentley D, Deloukas P (2005) Common VKORC1 and GGCX polymorphisms associated with warfarin dose. Pharmacogenomics J 5(4):262–270
75. Schalekamp T, Brasse BP, Roijers JF, Chahid Y, van Geest-Daalderop JH, de Vries-Goldschmeding H, van Wijk EM, Egberts AC, de Boer A (2006) VKORC1 and CYP2C9 genotypes and acenocoumarol anticoagulation status: interaction between both genotypes affects overanticoagulation. Clin Pharmacol Ther 80(1):13–22

76. Klein TE, Altman RB, Eriksson N, Gage BF, Kimmel SE, Lee MT, Limdi NA, Page D, Roden DM, Wagner MJ, Caldwell MD, Johnson JA (2009) Estimation of the warfarin dose with clinical and pharmacogenetic data. N Engl J Med 360(8):753–764. doi:10.1056/NEJMoa0809329
77. Limdi NA, McGwin G, Goldstein JA, Beasley TM, Arnett DK, Adler BK, Baird MF, Acton RT (2008) Influence of CYP2C9 and VKORC1 1173C/T genotype on the risk of hemorrhagic complications in African-American and European-American patients on warfarin. Clin Pharmacol Ther 83(2):312–321
78. Herman D, Locatelli I, Grabnar I, Peternel P, Stegnar M, Mrhar A, Breskvar K, Dolzan V (2005) Influence of CYP2C9 polymorphisms, demographic factors and concomitant drug therapy on warfarin metabolism and maintenance dose. Pharmacogenomics J 5(3):193–202
79. Xie HG, Prasad HC, Kim RB, Stein CM (2002) CYP2C9 allelic variants: ethnic distribution and functional significance. Adv Drug Deliv Rev 54(10):1257–1270
80. Gage BF, Eby C, Johnson JA, Deych E, Rieder MJ, Ridker PM, Milligan PE, Grice G, Lenzini P, Rettie AE, Aquilante CL, Grosso L, Marsh S, Langaee T, Farnett LE, Voora D, Veenstra DL, Glynn RJ, Barrett A, McLeod HL (2008) Use of pharmacogenetic and clinical factors to predict the therapeutic dose of warfarin. Clin Pharmacol Ther 84(3):326–331. doi:10.1038/clpt.2008.10
81. Lenzini P, Wadelius M, Kimmel S, Anderson JL, Jorgensen AL, Pirmohamed M, Caldwell MD, Limdi N, Burmester JK, Dowd MB, Angchaisuksiri P, Bass AR, Chen J, Eriksson N, Rane A, Lindh JD, Carlquist JF, Horne BD, Grice G, Milligan PE, Eby C, Shin J, Kim H, Kurnik D, Stein CM, McMillin G, Pendleton RC, Berg RL, Deloukas P, Gage BF (2010) Integration of genetic, clinical, and INR data to refine warfarin dosing. Clin Pharmacol Ther 87(5):572–578. doi:10.1038/clpt.2010.13
82. Sconce EA, Khan TI, Wynne HA, Avery P, Monkhouse L, King BP, Wood P, Kesteven P, Daly AK, Kamali F (2005) The impact of CYP2C9 and VKORC1 genetic polymorphism and patient characteristics upon warfarin dose requirements: proposal for a new dosing regimen. Blood 106(7):2329–2333
83. Wadelius M, Chen LY, Lindh JD, Eriksson N, Ghori MJ, Bumpstead S, Holm L, McGinnis R, Rane A, Deloukas P (2009) The largest prospective warfarin-treated cohort supports genetic forecasting. Blood 113(4):784–792. doi:10.1182/blood-2008-04-149070
84. Markatos CN, Grouzi E, Politou M, Gialeraki A, Merkouri E, Panagou I, Spiliotopoulou I, Travlou A (2008) VKORC1 and CYP2C9 allelic variants influence acenocoumarol dose requirements in Greek patients. Pharmacogenomics 9(11):1631–1638. doi:10.2217/14622416.9.11.1631
85. Geisen C, Luxembourg B, Watzka M, Toennes SW, Sittinger K, Marinova M, von Ahsen N, Lindhoff-Last E, Seifried E, Oldenburg J (2011) Prediction of phenprocoumon maintenance dose and phenprocoumon plasma concentration by genetic and non-genetic parameters. Eur J Clin Pharmacol 67(4):371–381. doi:10.1007/s00228-010-0950-y
86. D'Andrea G, D'Ambrosio RL, Perna P D, Chetta M, Santacroce R, Brancaccio V, Grandone E, Margaglione M (2005) A polymorphism in the VKORC1 gene is associated with an interindividual variability in the dose-anticoagulant effect of warfarin. Blood 105(2):645–649. doi:10.1182/blood-2004-06-2111
87. Bodin L, Verstuyft C, Tregouet DA, Robert A, Dubert L, Funck-Brentano C, Jaillon P, Beaune P, Laurent-Puig P, Becquemont L, Loriot MA (2005) Cytochrome P450 2C9 (CYP2C9) and vitamin K epoxide reductase (VKORC1) genotypes as determinants of acenocoumarol sensitivity. Blood 106(1):135–140
88. Wang D, Chen H, Momary KM, Cavallari LH, Johnson JA, Sadee W (2008) Regulatory polymorphism in vitamin K epoxide reductase complex subunit 1 (VKORC1) affects gene expression and warfarin dose requirement. Blood 112(4):1013–1021. doi:10.1182/blood-2008-03-144899
89. Limdi NA, Wadelius M, Cavallari L, Eriksson N, Crawford DC, Lee MT, Chen CH, Motsinger-Reif A, Sagreiya H, Liu N, Wu AH, Gage BF, Jorgensen A, Pirmohamed M, Shin JG, Suarez-Kurtz G, Kimmel SE, Johnson JA, Klein TE, Wagner MJ (2010) Warfarin pharma-

cogenetics: a single VKORC1 polymorphism is predictive of dose across 3 racial groups. Blood 115(18):3827–3834. doi:10.1182/blood-2009-12-255992

90. Ufer M, Svensson JO, Krausz KW, Gelboin HV, Rane A, Tybring G (2004) Identification of cytochromes P450 2C9 and 3A4 as the major catalysts of phenprocoumon hydroxylation in vitro. Eur J Clin Pharmacol 60(3):173–182. doi:10.1007/s00228-004-0740-5
91. Teichert M, Eijgelsheim M, Uitterlinden AG, Buhre PN, Hofman A, De Smet PA, Visser LE, Stricker BH (2011) Dependency of phenprocoumon dosage on polymorphisms in the VKORC1, CYP2C9, and CYP4F2 genes. Pharmacogenet Genomics 21(1):26–34. doi:10.1097/FPC.0b013e32834154fb
92. Caldwell MD, Awad T, Johnson JA, Gage BF, Falkowski M, Gardina P, Hubbard J, Turpaz Y, Langaee TY, Eby C, King CR, Brower A, Schmelzer JR, Glurich I, Vidaillet HJ, Yale SH, Zhang K Q, Berg RL, Burmester JK (2008) CYP4F2 genetic variant alters required warfarin dose. Blood 111(8):4106–4112. doi:10.1182/blood-2007-11-122010
93. Cooper GM, Johnson JA, Langaee TY, Feng H, Stanaway IB, Schwarz UI, Ritchie MD, Stein CM, Roden DM, Smith JD, Veenstra DL, Rettie AE, Rieder MJ (2008) A genome-wide scan for common genetic variants with a large influence on warfarin maintenance dose. Blood 112(4):1022–1027. doi:10.1182/blood-2008-01-134247
94. Takeuchi F, McGinnis R, Bourgeois S, Barnes C, Eriksson N, Soranzo N, Whittaker P, Ranganath V, Kumanduri V, McLaren W, Holm L, Lindh J, Rane A, Wadelius M, Deloukas P (2009) A genome-wide association study confirms VKORC1, CYP2C9, and CYP4F2 as principal genetic determinants of warfarin dose. PLoS Genet 5(3):e1000433. doi:10.1371/journal.pgen.1000433
95. Pautas E, Moreau C, Gouin-Thibault I, Golmard JL, Mahe I, Legendre C, Taillandier-Heriche E, Durand-Gasselin B, Houllier AM, Verrier P, Beaune P, Loriot MA, Siguret V (2010) Genetic factors (VKORC1, CYP2C9, EPHX1, and CYP4F2) are predictor variables for warfarin response in very elderly, frail inpatients. Clin Pharmacol Ther 87(1):57–64. doi:10.1038/clpt.2009.178
96. Perez-Andreu V, Roldan V, Anton AI, Garcia-Barbera N, Corral J, Vicente V, Gonzalez-Conejero R (2009) Pharmacogenetic relevance of CYP4F2 V433M polymorphism on acenocoumarol therapy. Blood 113(20):4977–4979. doi:10.1182/blood-2008-09-176222
97. Teichert M, Eijgelsheim M, Rivadeneira F, Uitterlinden AG, van Schaik RH, Hofman A, De Smet PA, van Gelder T, Visser LE, Stricker BH (2009) A genome-wide association study of acenocoumarol maintenance dosage. Hum Mol Genet 18(19):3758–3768. doi:10.1093/hmg/ddp309
98. Kimura R, Miyashita K, Kokubo Y, Akaiwa Y, Otsubo R, Nagatsuka K, Otsuki T, Okayama A, Minematsu K, Naritomi H, Honda S, Tomoike H, Miyata T (2007) Genotypes of vitamin K epoxide reductase, gamma-glutamyl carboxylase, and cytochrome P450 2C9 as determinants of daily warfarin dose in Japanese patients. Thromb Res 120(2):181–186
99. Herman D, Peternel P, Stegnar M, Breskvar K, Dolzan V (2006) The influence of sequence variations in factor VII, gamma-glutamyl carboxylase and vitamin K epoxide reductase complex genes on warfarin dose requirement. Thromb Haemost 95(5):782–787
100. Loebstein R, Vecsler M, Kurnik D, Austerweil N, Gak E, Halkin H, Almog S (2005) Common genetic variants of microsomal epoxide hydrolase affect warfarin dose requirements beyond the effect of cytochrome P450 2C9. Clin Pharmacol Ther 77(5):365–372
101. Aquilante CL, Langaee TY, Lopez LM, Yarandi HN, Tromberg JS, Mohuczy D, Gaston KL, Waddell CD, Chirico MJ, Johnson JA (2006) Influence of coagulation factor, vitamin K epoxide reductase complex subunit 1, and cytochrome P450 2C9 gene polymorphisms on warfarin dose requirements. Clin Pharmacol Ther 79(4):291–302
102. Luxembourg B, Schneider K, Sittinger K, Toennes SW, Seifried E, Lindhoff-Last E, Oldenburg J, Geisen C (2011) Impact of pharmacokinetic (CYP2C9) and pharmacodynamic (VKORC1, F7, GGCX, CALU, EPHX1) gene variants on the initiation and maintenance phases of phenprocoumon therapy. Thromb Haemost 105(1):169–180. doi:10.1160/TH10-03-0194

103. Wadelius M, Chen LY, Eriksson N, Bumpstead S, Ghori J, Wadelius C, Bentley D, McGinnis R, Deloukas P (2007) Association of warfarin dose with genes involved in its action and metabolism. Hum Genet 121(1):23–34. doi:10.1007/s00439-006-0260-8
104. Visser LE, Trienekens PH, De Smet PA, Vulto AG, Hofman A, van Duijn CM, Stricker BH (2005) Patients with an ApoE epsilon4 allele require lower doses of coumarin anticoagulants. Pharmacogenet Genomics 15(2):69–74
105. Kimmel SE, Christie J, Kealey C, Chen Z, Price M, Thorn CF, Brensinger CM, Newcomb CW, Whitehead AS (2008) Apolipoprotein E genotype and warfarin dosing among Caucasians and African Americans. Pharmacogenomics J 8(1):53–60
106. Kohnke H, Scordo MG, Pengo V, Padrini R, Wadelius M (2005) Apolipoprotein E (APOE) and warfarin dosing in an Italian population. Eur J Clin Pharmacol 61(10):781–783. doi:10.1007/s00228-005-0982-x
107. Sconce EA, Daly AK, Khan TI, Wynne HA, Kamali F (2006) APOE genotype makes a small contribution to warfarin dose requirements. Pharmacogenet Genomics 16(8):609–611. doi:10.1097/01.fpc.0000220567.98089.b5
108. Ross KA, Bigham AW, Edwards M, Gozdzik A, Suarez-Kurtz G, Parra EJ (2010) Worldwide allele frequency distribution of four polymorphisms associated with warfarin dose requirements. J Hum Genet 55(9):582–589. doi:10.1038/jhg.2010.73
109. Becquemont L (2008) Evidence for a pharmacogenetic adapted dose of oral anticoagulant in routine medical practice. Eur J Clin Pharmacol 64(10):953–960. doi:10.1007/s00228-008-0542-2
110. Hillman MA, Wilke RA, Yale SH, Vidaillet HJ, Caldwell MD, Glurich I, Berg RL, Schmelzer J, Burmester JK (2005) A prospective, randomized pilot trial of model-based warfarin dose initiation using CYP2C9 genotype and clinical data. Clin Med Res 3(3):137–145
111. Caraco Y, Blotnick S, Muszkat M (2008) CYP2C9 genotype-guided warfarin prescribing enhances the efficacy and safety of anticoagulation: a prospective randomized controlled study. Clin Pharmacol Ther 83(3):460–470
112. Anderson JL, Horne BD, Stevens SM, Grove AS, Barton S, Nicholas ZP, Kahn SF, May HT, Samuelson KM, Muhlestein JB, Carlquist JF (2007) Randomized trial of genotype-guided versus standard warfarin dosing in patients initiating oral anticoagulation. Circulation 116(22):2563–2570
113. Huang SW, Chen HS, Wang XQ, Huang L, Xu DL, Hu XJ, Huang ZH, He Y, Chen KM, Xiang DK, Zou XM, Li Q, Ma LQ, Wang HF, Chen BL, Li L, Jia YK, Xu XM (2009) Validation of VKORC1 and CYP2C9 genotypes on interindividual warfarin maintenance dose: a prospective study in Chinese patients. Pharmacogenet Genomics 19(3):226–234. doi:10.1097/FPC.0b013e328326e0c7
114. Wang M, Lang X, Cui S, Fei K, Zou L, Cao J, Wang L, Zhang S, Wu X, Wang Y, Ji Q (2012) Clinical application of pharmacogenetic-based warfarin-dosing algorithm in patients of Han nationality after rheumatic valve replacement: a randomized and controlled trial. Int J Med Sci 9(6):472–479. doi:10.7150/ijms.4637
115. Burmester JK, Berg RL, Yale SH, Rottscheit CM, Glurich IE, Schmelzer JR, Caldwell MD (2011) A randomized controlled trial of genotype-based Coumadin initiation. Genet Med 13(6):509–518. doi:10.1097/GIM.0b013e31820ad77d
116. Anderson JL, Horne BD, Stevens SM, Woller SC, Samuelson KM, Mansfield JW, Robinson M, Barton S, Brunisholz K, Mower CP, Huntinghouse JA, Rollo JS, Siler D, Bair TL, Knight S, Muhlestein JB, Carlquist JF (2012) A randomized and clinical effectiveness trial comparing two pharmacogenetic algorithms and standard care for individualizing warfarin dosing (CoumaGen-II). Circulation 125(16):1997–2005. doi:10.1161/CIRCULATIONAHA.111.070920
117. van Schie RM, Wadelius MI, Kamali F, Daly AK, Manolopoulos VG, de Boer A, Barallon R, Verhoef TI, Kirchheiner J, Haschke-Becher E, Briz M, Rosendaal FR, Redekop WK, Pirmohamed M, Maitland van der Zee AH (2009) Genotype-guided dosing of coumarin derivatives: the European pharmacogenetics of anticoagulant therapy (EU-PACT) trial design. Pharmacogenomics 10(10):1687–1695. doi:10.2217/pgs.09.125

118. Baranova EV, Verhoef TI, Asselbergs FW, de Boer A, Maitland-van der Zee AH. Genotype-guided coumarin dosing: where are we now and where do we need to go next? Expert Opin Drug Metab Toxicol 2015(11):509–522
119. French B, Joo J, Geller NL, Kimmel SE, Rosenberg Y, Anderson JL, Gage BF, Johnson JA, Ellenberg JH (2010) Statistical design of personalized medicine interventions: the clarification of optimal anticoagulation through genetics (COAG) trial. Trials 11:108. doi:10.1186/1745-6215-11-108
120. Higashi MK, Veenstra DL (2003) Managed care in the genomics era: assessing the cost effectiveness of genetic tests. Am J Manag Care 9(7):493–500
121. You JH, Chan FW, Wong RS, Cheng G (2004) The potential clinical and economic outcomes of pharmacogenetics-oriented management of warfarin therapy - a decision analysis. Thromb Haemost 92(3):590–597. doi:10.1267/THRO04090000
122. Schalekamp T, Boink GJ, Visser LE, Stricker BH, de Boer A, Klungel OH (2006) CYP2C9 genotyping in acenocoumarol treatment: is it a cost-effective addition to international normalized ratio monitoring? Clin Pharmacol Ther 79(6):511–520
123. You JH, Tsui KK, Wong RS, Cheng G (2009) Potential clinical and economic outcomes of CYP2C9 and VKORC1 genotype-guided dosing in patients starting warfarin therapy. Clin Pharmacol Ther 86(5):540–547. doi:10.1038/clpt.2009.104
124. Eckman MH, Rosand J, Greenberg SM, Gage BF (2009) Cost-effectiveness of using pharmacogenetic information in warfarin dosing for patients with nonvalvular atrial fibrillation. Ann Intern Med 150(2):73–83
125. Leey JA, McCabe S, Koch JA, Miles TP (2009) Cost-effectiveness of genotype-guided warfarin therapy for anticoagulation in elderly patients with atrial fibrillation. Am J Geriatr Pharmacother 7(4):197–203. doi:10.1016/j.amjopharm.2009.07.002
126. Patrick AR, Avorn J, Choudhry NK (2009) Cost-effectiveness of genotype-guided warfarin dosing for patients with atrial fibrillation. Circ Cardiovasc Qual Outcomes 2(5):429–436. doi:10.1161/CIRCOUTCOMES.108.808592
127. Meckley LM, Gudgeon JM, Anderson JL, Williams MS, Veenstra DL (2010) A policy model to evaluate the benefits, risks and costs of warfarin pharmacogenomic testing. Pharmacoeconomics 28(1):61–74. doi:10.2165/11318240-000000000-00000
128. Shiroiwa T, Sung YK, Fukuda T, Lang HC, Bae SC, Tsutani K (2010) International survey on willingness-to-pay (WTP) for one additional QALY gained: what is the threshold of cost effectiveness? Health Econ 19(4):422–437. doi:10.1002/hec.1481
129. Verhoef TI, Redekop WK, Veenstra DL, Thariani R, Beltman PA, van Schie RM, de Boer A, Maitland-van der Zee AH (2013) Cost-effectiveness of pharmacogenetic-guided dosing of phenprocoumon in atrial fibrillation. Pharmacogenomics 14(8):869–883. doi:10.2217/pgs.13.74
130. Dogliotti A, Paolasso E, Giugliano RP (2013) Novel oral anticoagulants in atrial fibrillation: a meta-analysis of large, randomized, controlled trials vs warfarin. Clin Cardiol 36(2):61–67. doi:10.1002/clc.22081
131. You JH, Tsui KK, Wong RS, Cheng G (2012) Cost-effectiveness of dabigatran versus genotype-guided management of warfarin therapy for stroke prevention in patients with atrial fibrillation. PLoS ONE 7(6):e39640. doi:10.1371/journal.pone.0039640
132. Verhoef TI, Redekop WK, Darba J, Geitona M, Hughes DA, Siebert U, de Boer A, Maitland-van der Zee AH, Barallon R, Briz M, Daly A, Haschke-Becher E, Kamali F, Kirchheiner J, Manolopoulos VG, Pirmohamed M, Rosendaal FR, van Schie RM, Wadelius M (2010) A systematic review of cost-effectiveness analyses of pharmacogenetic-guided dosing in treatment with coumarin derivatives. Pharmacogenomics 11(7):989–1002. doi:10.2217/pgs.10.74
133. Howard R, Leathart JB, French DJ, Krishan E, Kohnke H, Wadelius M, van Schie R, Verhoef T, Maitland-van der Zee AH, Daly AK, Barallon R (2011) Genotyping for CYP2C9 and VKORC1 alleles by a novel point of care assay with HyBeacon(R) probes. Clin Chim Acta 412(23–24):2063–2069. doi:10.1016/j.cca.2011.07.013

Implementation of Genomic Medicine: Tools and Challenges

Godfrey Grech, Ron van Schaik and Joseph Borg

Abstract The challenge of personalised medicine have been the centre of constructive debate in the last years, and have activated multidisciplinary approaches to solve the issues for translation to the clinic. In the preface, the editors summarise the challenges to translate genetic-based knowledge to the clinical setting. In this chapter, we will be reviewing the pre-requisites that will lessen the struggles associated with the implementation of pharmacogenetic testing protocols. To achieve this we will focus on the needs and recommendations to promote patient molecular classification to enhance risk assessment and drug response research; stratification of well-defined therapeutic groups, using genetic analysis, into subgroups of responders to specific therapies; the development of technologies and integrative information systems to provide the healthcare system with optimised and sustainable genetic testing protocols; the need of harmonised guidelines for the proper selection of patient groups for clinical trials; and advances in research to generate evidence based knowledge that can be smoothly translated for healthcare use.

Translation from research to the clinic is mainly associated with (1) the co-evolution of biomarkers and targeted therapy and (2) genotypes associated with drug metabolising enzymes. Although significant pharmacogenetic analyses are currently in use, the main obstacles associated with the efficient implementation include the uptake by the health care systems, incorporation into clinical guidelines as well as economic, educational, legal and commercial aspects. Understanding the approach for successful translation to clinical practice and resolving, in-part, the confounding

G. Grech (✉)
Department of Pathology, Faculty of Medicine and Surgery, University of Malta,
Msida MSD2080, Malta
e-mail: godfrey.grech@um.edu.mt

R. van Schaik
Expert Center Pharmacogenetics, IFCC Reference Center for Pharmacogenetics, Dept. Clinical Chemistry, Erasmus University Medical Center, Rotterdam, The Netherlands

J. Borg
Department of Applied Biomedical Science, Faculty of Health Sciences, Mater Dei Hospital,
Block A Level 1 Rm 58, Msida MSD2090, Malta
e-mail: joseph.borg@biotech.um.edu.mt

G. Grech, I. Grossman (eds.), *Preventive and Predictive Genetics: Towards Personalised Medicine,* Advances in Predictive, Preventive and Personalised Medicine 9,
DOI 10.1007/978-3-319-15344-5_12

aspects that negatively impact efficient implementation, shall support the transition from an empirical approach to a personalised health care system.

Keywords Genomic medicine · Pharmacogenetics implementation · Biomarker-driven prescribing · Genomic medicine in health care · Evidence-based medicine · Predictive genetics · Ethnic variations · Genetic-guided therapy · Pharmacogenetic technologies · Next generation sequencing

1 Current Knowledge and Expectations

Understanding the cause of disease at a molecular level permitted the development of drugs, following the discovery of potential therapeutic targets. This biomarker-driven prescribing deals mainly with somatic mutation that are causative of the disease. Somatic DNA, derived from the tissue (biopsies or resections), provides information on mutations that represent the variations associated with the tumour initialisation, progression of disease or secondary mutations due to the tumour itself [1]. The presence of the fusion protein BCR/ABL in chronic myeloid leukaemia (CML), overexpression of human Epidermal Growth Factor Receptor 2 (Her2) in breast cancer (Fig. 1), Epidermal Growth Factor Receptor (*EGFR*) mutants in lung cancer, are genetic markers currently used in the clinic to select therapy. The translation of these genetic tests into the clinic was smooth, mainly due to the clear patient benefits, offering patients access to improved healthcare management, avoiding unnecessary treatment and reducing adverse drug reactions. In addition, the simple algorithm used to apply the results in the health care system, provides the health care provider assurance for correct clinical decisions. A major challenge in targeted therapy is resistance due to acquired mutations of the drug target protein, exemplified by the development of resistance to imatinib, due to acquired mutation in BCR/ABL fusion protein in CML patients or cKit mutants in gastrointestinal stromal tumours (GIST) [2, 3]. Second-generation drugs were developed to treat secondary resistance to imatinib. Hence, genetic analysis to assess development of disease is required for patient management. In addition, to patient-specific adjusted therapy and dosing, classification of the tumour type in individual patients is important to derive information to predict re-occurrence of the disease. This is achieved by using advanced technologies (discussed below) to measure panels of gene expression utilising well established gene signatures associated with disease re-occurrence, such as the FDA approved tests, Mammaprint (Agendia BV) and Oncotype DX (Genomic Health) designed to predict event free survival in breast cancer patients.

After identification/classification of the disease, drug treatment specific for this disease can be started. But also for the same disease, patients may benefit from altered dosing based on their inherited capacity to metabolise these drugs in the liver, as detailed in Chap. 5. The clinical therapeutic approach is evolving from the

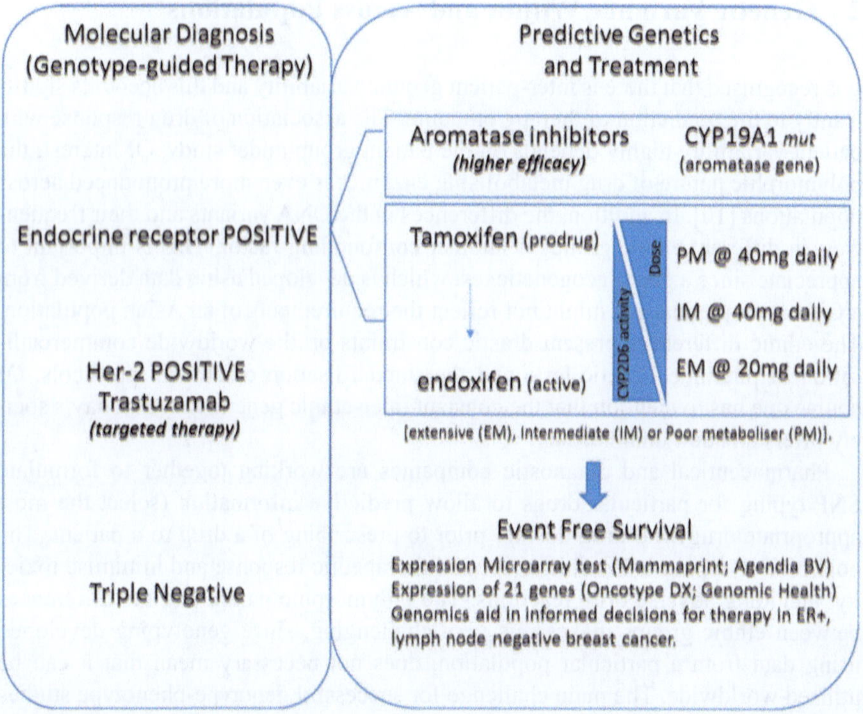

Fig. 1 Genotype-Phenotype evidence. Molecular diagnostics provides genotype-guided therapy by classifying patients into therapeutic groups (*left* panel). Taking endocrine receptor positive cases, the prediction of therapy outcome is summarised in the *right* panel. The efficacy of aromatase inhibitors depends on mutations in the aromatase gene, and the dosage of tamoxifen is guided by *CYP2D6* genotypes associated with the enzyme activity

knowledge of the effect of a drug in the average patient, towards the understanding of the response of drugs in individual patients. Today, it is recognised that personalised medicine is the future. Anticipation of a drug response, adverse drug reaction and selection of a specific therapy contributes significantly to a better quality of life due to evidence-based prescribing and personalised management of a therapy. The potential use of genotype-drug response, well established for *HLA-B*5701* screening for Abacavir therapy, and currently debated for tamoxifen dosing used for estrogen receptor positive breast cancer patients [4–6], warfarin dosing [7], *DPYD* screening for colon cancer treatment with 5-FU or capecitabine [8] and *CYP19A1* SNPs associated with enhanced efficacy of aromatase inhibitors [9], may increase patient safety and therapy efficacy (Fig. 1). Genotype-based prescribing is gaining importance in the clinic, and interrogation of genomic data is expected to overcome the complexity of multifactorial drug-genotype associations. The algorithms used to improve therapy based on genotype alone may not always be successful, depending on the non-genetic factors contribution to drug efficacy and toxicity.

2 Genetic Variance Within and Across Populations

It is recognised that there is inter-patient genetic variability and this accounts significantly to the prediction of therapy outcome. The association of drug response with patient variability highly depends on the patient group under study. Of interest, the polymorphic nature of drug metabolising enzymes is even more pronounced across populations [10]. In addition, the differences in the DNA variants and their frequencies, in different ethnic groups is another confounding factor. This is important to appreciate since a pharmacogenetic test which is developed using data derived from a Caucasian population, might not reflect the requirements of an Asian population. The ethnic differences present drastic constraints on the worldwide commercialisation of pharmacogenetic tests and the standardisation of clinical protocols. Of course one has to mention that the constant inter-ethnic genetic mix in today's society offers an additional hurdle.

Pharmaceutical and diagnostic companies are working together to formulate SNP typing for particular drugs to allow predictive information (select the most appropriate drug) and drug dosing, prior to prescribing of a drug to a patient. The goal of SNP typing is to achieve optimal therapeutic response and minimise toxicity, including adverse drug reactions. The polymorphic nature and the differences between ethnic groups makes this very challenging, since genotyping developed using data from a particular population, does not necessary mean that it can be utilised worldwide. The main challenge for successful genotype-phenotype studies is the selection of patients, design of the study and a thorough and uniform registration of phenotypes. There are various variables that needs to be harmonised to allow population studies and meta-analysis. First, phenotype definition for proper selection of patients should follow specific guidelines. Questions on how are the efficacy and toxicity measured, and which thresholds are being used, if standardised therapy protocols are used, and at which dose is being administered, should be addressed before initiating any study. Secondly, outcome measures used for genetic associations, should be defined properly [11]. There are various outcome measures including (a) genotypic eg viral load; standardised qPCR kits, (b) biochemical assays (c) pharmacokinetics with also various measurable variables such as maintenance dose; half-life; (d) pharmacodynamics and drug response, and (e) clinical outcome. In addition, clinical assessment of a phenotype can be reported as a presence/absence value; grading and scoring of toxicity grade levels. In addition, there are various methodologies used to correlate genotypic differences with drug responses. Randomised controlled trials are commonly used to establish genetic determinants of drug efficacy and toxicity, although also this approach may not always give the clear outcome expected, as was recently demonstrated with two randomised controlled trials on pharmacogenetics for warfarin, where two randomised controlled trials reached opposing conclusions [7, 12], due to slightly different set up of these studies. One approach is to compare outcomes of genetic-guided therapy with clinical-guided care [7, 13]. This approach provides the required evidence, but the method will delay treatment. To overcome this limitation, the use of case-control studies is preferred. This study design depends on defining cases with adverse reaction or reduced drug efficacy. Differential allele frequencies between cases and controls are measured. Case-control studies provide genotype-phenotype association, but require prospective studies to validate the effect of the genotype on the measured outcome. This variability in patient selection, outcome measures and study design, hampers the analysis of data shared from different studies, and results in failure to replicate pharmacogenetic findings [14, 15]. Hence, the lack of reproducibility of results slows down the implementation of genetic testing in the clinic, as was demonstrated with the *CYP2D6*/tamoxifen controversy [16, 17]. Although knowledge is increasing, the clinical response to drugs is still unexpectedly low, and this may be improved by implementing genetic testing in the clinical setting. The challenges to implement genomic medicine are discussed in the following section.

3 Uptake of Genomic Medicine in the Health Care System

The intended use of targeted drugs are clearly indicated on the drug labels. The health care provider is provided with clinically validated companion diagnostic tests to ensure compliance with the recommendations. To date the FDA labelled more than 120 drugs with label notifications that address the use of genetic tests to predict efficacy and toxicity of specific drugs, prior to prescription (http://www.fda.gov/drugs/scienceresearch/researchareas/pharmacogenetics/ucm083378.htm). The challenge is to encourage the adoption of companion diagnostics in the day to day clinical practice.

The expression and activity of various drug metabolising enzymes, determine the pharmacokinetic properties of the drugs, as detailed in Chap. 5. In addition, accumulation of active metabolites is associated with the occurrence of serious adverse drug reactions (ADRs), which is of major concern in therapeutics as explained in Chap. 6. A group of enzymes, the CYPs are the major contributors in drug metabolism and activation. Genetic variations in drug metabolising enzymes, such as thiopurine-S-methyltransferase (TPMT), cytochrome P450 2D6 (CYP2D6) and uridine-diphosphate glucuronosyltransferase 1A1 (UGT1A1) are germ-line variants that are useful to determine both pharmacokinetic and pharmacodynamics responses, as detailed in Chap. 10. Algorithms using information on genetic variants are used to determine the dose of specific pharmaceuticals. This genotype driven drug dosing has been extensively used in dosing of warfarin, which is described in Chap. 11.

3.1 Education of the Healthcare Professionals

Clinical implementation of new technologies does not necessary depend on the in-depth understanding of the technological process by the healthcare professionals,

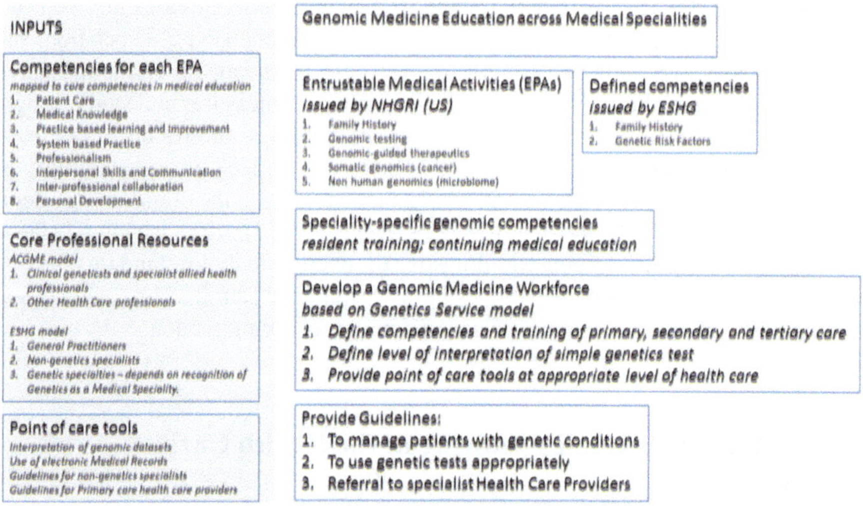

Fig. 2 Genomic Medicine Education across Medical Specialities. Composition of the proposed core professional resources, issued by the National Human Genome Research Institute (NHGRI) as a model for the American Council of Graduate Medical Education (ACGME) and competencies defined by the European Society of Human Genetics (ESHG). The *grey* box lists the competencies, target professional resources and tools required (inputs) and the *right* panel summarises the framework of the models including the deliverables of the proposed education programme

but is highly dependent on the presentation of results in a clear and user friendly manner, specifically addressing the clinical aspect that raised the need of the test. Hence, education of healthcare professionals, should include basic understanding of the technology used, but focus on knowledge to gain competency in the clinical utility of genomics.

The first challenge in providing the proper outcome-based education is to stratify the health care providers into core professional resources and define the competencies to be achieved for each group of providers (Fig. 2; [18]). Of interest the composition of the core professional resources, differ in the models proposed by the American Council of Graduate Medical Education (ACGME) and the European Society of Human Genetics (ESHG) [19]. The competencies are defined based on the desired outcomes [20], using an educational programme designed to establish competence in genetics throughout medical education, residency and speciality training [21]. Speciality-specific genomic competencies are defined to ensure proper residency training and continuous medical education. The proposed Genomic Medicine workforce includes training of primary, secondary and tertiary health care professionals [19, 22]. The educational programmes are supported by guidelines, providing details to use genetic tests appropriately, treatment and management of patients with genetic conditions, tools for clinical decisions, including proper referrals to specialist healthcare providers, use of electronic medical records and guidelines to support primary care health care providers [22]. Various models are current used to implement genetic services in the healthcare system, and different sectors, such as rare and common disorders require different means to deliver proper services. Whereas, specialist genetic clinics drive the service for rare genetic diseases, inter-professional collaboration bringing together geneticists and other specialists is required for common diseases, such as cancer care [23]. The coordination of activities between different health care professionals in primary, secondary and tertiary care require well defined competencies, implementation of training and educational programmes, sharing of data through a regional hub with well-defined outreach procedures, and harmonised guidelines. An overview of the models used to implement genetic services are described in detail in Chap. 2. In addition to screening and diagnosis, the implementation of genetics services within the healthcare system includes the molecular classification of patients into therapeutic groups and predictive genetics/genomics with the ultimate aim to predict therapy response, resistance and relapse. Implementation of pharmacogenetics is purely exercised by specialists in the field and the educational component is mainly covered by continuous professional education and inter-professional communication.

3.2 Evidence Generation

The Clinical Pharmacogenetics Implementation Consortium (CPIC) under the governance of the Pharmacogenomics Research Network (http://www.pgrn.org), and the Pharmacogenomics Knowledge Base (PharmGKB, http://www.pharmgk.org; 24] provide open access, peer-reviewed, updated, evidence-based pharmacogenetic clinical practice guidelines [25]. Justification of pharmacogenetic test implementation require an understanding of the clinical utility and a level of evidence defined by the CPIC framework. Evidence generation is measured against a set of considerations including (a) a significant association between genomic variation/s and the phenotype, (b) the possibility to alter dosages or give alternative drugs for patients with high-risk genotypes, (c) the cohort (size and age group) used to generate the data. Genetic information is periodically updated online on the PharmGKB website. The Pharmacogenetic Clinical Practice Guidelines should be discussed in the various clinical disciplines to promote uptake in the appropriate clinical guidelines. The guidelines can take the form of a drug data sheet or indicate testing of a particular gene for various drug administration. These guidelines are based on studies in specific populations and due to global genetic variation, the guidelines shall be revised to conform to this variance. In addition, the relevance in terms of incidence and phenotype-association, of a particular genetic variant needs more investigation at a global perspective.

4 Current Technologies

With the advent of new technologies in the post-genomic era, a new field called pharmacogenomics has flourished. The aims of pharmacogenomics are to correlate genomic variations and gene expression with drug efficacy, response and/or toxicity [26]. There are several genes, also referred to as pharmacogenes, which are associated with absorption, distribution, metabolism, excretion and toxicity of several drugs [ADMET; www.pharmgkb.org; Denus et al. 2013], that are actively studied using current technologies for gene specific or genome-wide typing. There are a number of genotyping methods, namely Polymerase Chain Reaction- or even microarray-based assays to perform genetic screening of known pharmacogenomic markers in well-documented pharmacogenes.

The most commonly used microarray-based platforms are the AmpliChip P450 platform (Roche Diagnostics, Basel, Switzerland), a CE-IVD certified assay that analyses 33 known *CYP2D6* gene variants, including 7 gene duplications and deletion, as well as 2 frequent *CYP2C19* gene variants (http://molecular.roche.com/assays/Pages/AmplichipCYP450Test.aspx). The Roche AmpliChip Cytochrome P450 Genotyping test is used in connection with the Affymetrix GeneChip Microarray Instrumentation System. This particular platform aims to assist doctors and medical specialists to prescribe personalised treatment options for their patients. Medical professionals can then use their patient's DNA profile to assist them further and better to prescribe appropriate medicine and at correct doses. In turn this will help reduce the unwanted harmful drug reactions and able to prevent patients from being improperly treated with sub-optimal and incorrect doses. The test uses human DNA extracted from either patient's blood or saliva to detect common genetic variants that alter the body's ability to break down (metabolise) specific types of drugs. The enzyme produced from the gene that is tested, called cytochrome P4502D6 (CYP4502D6), is active in metabolising many types of drugs including antidepressants, antipsychotics, β-blockers, and some chemotherapy drugs. Variations in this gene can cause a patient to metabolise these drugs abnormally fast, abnormally slow, or not at all as described in detail in Chaps. 5 and 6.

Also, the DMET™ Plus assay (Affymetrix, Santa Clara, CA, USA), allows simultaneous analysis of 1936 pharmacogenomic biomarkers in 231 pharmacogenes [27]. The DMET™ Plus Solution consists of a Molecular Inversion Probe (MIP) panel that amplifies the precise target DNA of interest, and the allele-specific oligonucleotide array provides a single color readout on the GeneChip® Scanner 3000 or GeneChip® Scanner 3000Dx v.2. The MIP method employed by the DMET™ Plus assay is an efficient technology for large-scale Single Nucleotide Polymorphisms (SNPs) analysis and hence well suited for SNP discovery and genotyping. This technique produces "inverted" probes in which SNPs are introduced into tag sequences that could be analysed using a universal sequence tag [28].

In addition, a cost-effective genotyping array using *TaqMan* chemistry is also available [29]. This method enjoys the concept of a customised array. The array is developed by utilising the Life Technologies QuantStudio 12 K Flex system with OpenArray technology. This array platform was designed with important clinical diagnostic elements in mind that are deemed important for clinical application of the genotyping information. Even more so, the turnaround time from DNA to genotype is only 5 h. The QuantStudio system requires minimal technical support time. Each SNP genotype is run in a separate reaction; therefore new SNPs can be substituted into the array based on new data, hence providing substantial flexibility.

Chips on the open array can be designed in any combination of 3072 SNPs, and heavily depend on the curated pharmacogenetics data and clinical annotations in PharmGKB database.

The above platforms, like every other genetic screening approach, have the inherent danger of missing novel unique or rare variants in ADMET-related genes, which may either affect ADMET gene expression or enzyme structure and may hence lead to variable response or increased toxicity in commonly prescribed drugs. The advent of next-generation sequencing technologies have created unprecedented opportunities to analyse whole genomes, which, unlike conventional medium or even high throughput genetic screening approaches, allows obtaining a full picture with respect to people's variomes. To date, whole exome and/or whole genome sequencing can be easily performed using several commercially available or proprietary platforms, to comprehensively analyse genome variation with a high degree of accuracy and with reasonable costs, compared to the not so distant past. Despite the fact that whole-exome sequencing is currently more cost-effective than whole genome sequencing, it has a number of limitations [30–32]: (a) our knowledge of all truly protein-coding exons in the genome is still incomplete, so current capture probes can only target known exons, (b) there is a degree of variability between the various commercial target enrichment kits, (c) regulatory and untranslated regions are not sequenced, (d) there is a significant bias in the target enrichment step, since the efficiency of capture probes varies considerably and some sequences fail to be targeted by capture probe design altogether. As such, not all templates are sequenced with equal efficiency, not all sequences can be aligned to the reference genome so as to allow base calling and as a result a significant proportion of variants may go undetected.

4.1 Next Generation Sequencing

The way we think about scientific approaches in clinical and applied research was changed by the introduction of Next Generation Sequencing (NGS) technologies and their ability to produce large volumes of data at a relatively cheaper cost than it used to be [33]. Different NGS platforms such as Roche/454, Illumina/Solexa, SOLiD/Life/APG, Helicos BioSciences and Polantor Instrument are commercially available and these technologies differ from the traditional Sanger sequencing mainly in three ways that include; (1) DNA sequencing libraries are clonally amplified in vitro; (2) DNA is sequenced by synthesis and (3) the amplified DNA templates are sequenced simultaneously.

4.1.1 Sample Preparation for NGS

One similarity between all current NGS technologies is the immobilisation of the DNA sample template to a solid support. This facilitates the sequencing of billions of reactions to be performed in massive parallel runs. Two different methods are employed by the different NGS platforms for DNA sample preparation. These are clonally amplified templates that originate from single DNA molecules [33]. For clonally amplified DNA templates, solid-phase amplification [34] or emulsion Polymerase Chain Reaction (emPCR) are utilised and they are the two most common methods used [35]. The solid-phase amplification employed by the Illumina/Solexa involves two consecutive steps. The primary step involves the initial priming and extending of the single-stranded DNA (ssDNA) followed by bridge amplification of the immobilised template with immediately adjacent primers resulting in the formation of clusters. This type of amplification can give rise between 100 to 200 million spatially separated clusters. A hallmark of this method is the fact that these clusters contain free ends to which universal sequencing primers can be added and hybridised therefore enabling the NGS reaction to commence [33].

In emPCR, the sequencing DNA templates are prepared in a cell-free system, resulting in the formation of fragment targets. The universal priming sites adaptors are ligated to the target ends. This allows relatively complex genomes to be amplified and extended using common PCR primers. The next step involves separation of dsDNA to ssDNA and these are then captured onto beads under conditions that preferentially bind to one DNA molecule per bead. The emPCR is used by different platforms. After successful amplification and enrichment of emPCR beads, in the case of Polantor platform, millions of beads are immobilised in a polyacrylamide gel on a standard microscope slide [36]. In the case of SOLiD/Life/APG platform, the beads are chemically cross-linked to an amino-coated glass surface [37] while in the case of Roche/454 the beads are placed into individual Pico Titre Plate [38].

Single-molecule template preparation requires much less starting DNA template than clonally amplified templates and it is perhaps easier to perform. Between 3 and 20 μg of starting material is required in clonally amplified templates whilst in single-molecule templates only 1 μg or less of starting material is required. This method involves the immobilisation of single molecule templates on a solid support that can be carried out by three different approaches before NGS reaction is performed [39]. The Helicos BioSciences platform uses two different approaches. The first approach involves attachment of spatially distributed individual primer molecules to the solid support. The template, which is then hybridised to the immobilised primer, is prepared by randomly fragmenting the starting DNA template into smaller sizes and common adaptors are added to the fragment ends. The second approach also involves the attachment of spatially distributed single-molecule templates to the solid support. These are covalently attached by priming and extending single-stranded, single- molecule templates from immobilised primers. This is followed by hybridisation of a common primer to the template. In both approaches, DNA polymerase binds to the immobilised primer template configuration to initiate the NGS reaction [39]. Pacific Biosciences platforms uses the third approach in which spatially distributed single polymerase molecules are attached to the solid support that contains a bound primed template molecule. In contrast to the first two approaches, the third approach can be used with larger DNA molecules and it can also be used with real-time methods, which give rise to potentially longer read lengths [40].

4.1.2 Sequencing and Imaging

When coming to sequencing, there are fundamental differences between sequencing clonally amplified templates and sequencing single molecule templates. When carrying our clonally amplification it gives rise to a population of identical templates each of which has undergone the sequencing reaction. When imaging is carried out, the observed signal is in harmony with the probes or nucleotides, which were added to the identical templates for a given cycle. Incomplete extension of the template ensemble can result in lagging-strand dephasing while leading-strand dephasing can occur when there is the addition of multiple nucleotides and probes. Signal dephasing give rise to an increase in fluorescence noise resulting in basecalling errors and shorter reads [41]. With single molecules, dephasing is not a problem but these can give rise to multiple nucleotides or probe addition. When this occurs, it gives rise to deletion errors that in turn result in quenching effects between adjacent dye molecules. In the case where there is the incorporation of a nucleotide or probe that does not contain a fluorescent label, no signal will be detected. For sequencing and imaging the different platforms uses four different strategies which are; (a) cyclic reversible termination (CRT), (b) sequencing by ligation (SBL), (c) single nucleotide addition—pyrosequencing, (d) real-time sequencing [33] and (e) DNA nanoball sequencing.

CRT technique uses reversible terminator nucleotides in a cyclic method that encompasses incorporation of nucleotide, fluorescence imaging and cleavage [42]. The first step involves the addition of one fluorescently labeled nucleotide by the bounding of the DNA polymerase to the primed DNA template and simultaneously any unincorporated nucleotides are washed away. The terminating group and the fluorescent dye are removed by a cleavage step and before the next incorporation step, an additional washing step is carried out. Reversible termination can be either of two types; 3' blocked or 3' unblocked [33]. The Illumina/Solexa Genome analyser uses a four-colour CRT cycle. All four nucleotides together with DNA polymerase are added simultaneously to the flow cell channels. These result in the incorporation of the nucleotides together with DNA polymerase into the oligo-primered cluster fragments. The nucleotides carrying a base-unique fluorescent label and the 3-OH group are chemically blocked and therefore each incorporation is a unique event. After each nucleotide incorporation, an imaging step follows. Once the imaging step terminates, the 3' blocking group is chemically removed and this prepares each strand for the next incorporation by DNA polymerase [43]. Total internal reflection fluorescence (TIRF) imaging is used and it detects the four colours. The Helicos BioScience platform uses a similar methodology, however it employs a non-colour cycle.

Sequencing by ligation (SBL) is another cyclic method, however uses DNA ligase instead of DNA polymerase as in the CRT. In SBL one-base-encoded probes or two-base encoded probes are used. It involves the hybridisation of a fluorescently labeled probe to its complementary sequence adjacent to the primed template. Addition of DNA ligase follows and this joins the dye-labelled probe to the primer. Washing is carried out to remove any unligated probes. The identity of the ligated probe is determined by fluorescence imaging [44]. SBL is used by SOLiD/Life/APG platform and sequencing of the *Escherichia coli* MG 1655 genome was carried out by the SBL method [36].

Pyrosequencing, which is a non-electrophoretic, bioluminescence technique, measures the release of inorganic pyrophosphate. This is carried out by proportionally converting the release of inorganic pyrophosphate into invisible light by using a series of enzymatic reactions [45, 46]. Instead of using modified nucleotides to terminate DNA synthesis, this method uses single addition of dNTP in limiting amounts to manipulate DNA polymerase. DNA polymerase extends the primer and then pauses upon the incorporation of the complementary dNTP. Upon addition of the next complementary dNTP in the dispensing cycle, DNA synthesis is reinitiated. The underlying DNA sequence is revealed by flow grams, which are recorded by the order and intensity of the light peaks [33]. Pyrosequencing is used by Roche/454 instrument.

Real-Time sequencing, used by Pacific Biosciences, is different from other sequencing methods, it does not halt the process of DNA synthesis instead during the process of DNA synthesis it involves imaging of the continuous incorporation of dye-labelled nucleotides [42]. The single molecule real time sequencer used in the Pacific Biosciences platform, segregates DNA templates and single polymerase molecules onto a plate containing thousands of nanometre-sized wells. Polymerase molecules are bound to the bottom of the wells and the fluorescence emitted from the bottom of the well is measured by a finely tuned optical system. Fluorescently labelled nucleotides are then incorporated into the wells, and once incorporation takes place, the optical system detects it and a fluorescent signal is given off [40].

DNA nanoball sequencing, is a high throughput next generations sequencing technology that is used to determine the entire genomic sequencing of an organism. The method uses rolling circle replication to amplify small fragments of genomic DNA into *DNA nanoballs*. Fluorescent probes bind to complementary DNA and the probes are then ligated to anchor sequences bound to known sequences on the DNA template. The base order is determined *via* the fluorescence of the ligated and bound probes [47]. This platform has been chiefly pioneered by Complete Genomics, that has recently merged with BGI Technologies and offering clinical and basic research at the whole genome level.

4.1.3 Data Analysis

NGS sequencing data can be analysed by a large variety of software tools, whose functions can fit into several categories such (a) alignment of sequence reads to a reference, (b) basecalling and/or DNA variant detection, (c) *de novo* assembly and (d) genome variation browsing and annotation [48]. Once NGS reads have been generated they are either aligned to a known reference sequence or assembled *de novo* [49, 50]. Intended biological application, cost, effort and time considerations are the basis on which one decides to use one strategy or the other [33]. Alignment to a reference genome in terms of computational methods is the simplest to perform but relatively challenging to decide which reference genome to choose, and whether its ethnically correct for your samples of interest since this may obstruct the detection of structural variations and single nucleotide polymorphisms (SNPs) in the patients specimen. Alignment of short-read sequence data to a reference genome is performed by a variety of computational methods. The most common strategies used are either the performance of a Burrows-Wheeler transform to construct a matrix of all possible rotations of a given sequence or converting the sequence data into a servers of unique integer values (Hash tables) [51]. When performing de novo genomic assembly one must create long stretches of DNA sequence from shorter read length data. Algorithms have been incorporated into software programmes to perform successful *de novo* whole genome assembly [52].

4.1.4 Applications of NGS

The production of large numbers of low-cost reads makes the NGS platforms useful for many applications [33], including pharmacogenetic testing as described above. It has helped research laboratories to investigate disease-causing genes, and mechanisms by identifying novel DNA variants, that can lead to perturbation to transcriptional regulation and hence RNA expression. One of the biggest challenges associated with complex diseases is the identification of specific genetic loci underlying the disease. Until now, one approach to this challenge is to catalogue SNPs across the genome and then using genome-wide association studies to associate the variants with a particular phenotype [53]. In large-scale projects such as the International HapMap Consortium, the predominant methodology for SNP genotyping to date was the high-density SNP arrays but this approach has achieved its limited scope due to the density of the array [54, 55]. Single nucleotide resolution can now be achieved by the new NGS platforms and therefore rare variants can now be detected and categorised. Recently unknown causative mutations have been identified in a family with a recessive form of Charcot-Marie-Tooth disease using SOLiD platform while another family with primary ciliary dyskinesia and Miller syndrome [56, 57]. NGS can improve the detection of rare sequence variations in the discovery to novel somatic mutations in cancer such as acute myeloid leukaemia and myeloma [43]. With NGS the predictive power of GWA studies in the comprehensive SNP identification will be improved and this can lead to more understanding of complex disease trait loci and pharmacogenomics.

4.2 Haplotype Analysis

The determination whether two sequence variants are present *in cis* (on the same copy of a chromosome) or *in trans* (on opposite chromosome) is referred to as haplotype analysis. By using the Sanger sequencing protocols it is difficult to assess the *cis/trans* distinction but the new NGS offers a clever solution without the need to carry out bacterial cloning through an *in vitro* clonal amplification step. The 454, SOLiD platform is able to carry out haplotype analysis [58]. Complete Genomics has also optimised accurate whole genome sequencing coupled with haplotyping [59] over the whole genome using a technique called Long Fragment Reads (LFR). LFR is similar to sequencing long single DNA molecules without cloning or separation of metaphase chromosomes. Other applications of NGS involve copy number variation, epigenetics, transcriptome anlaysis and metagenomics and minimal residual disease detection.

5 New Challenges and Recommendations

The main challenge in genomic medicine is to create the proper vehicle to communicate genome information within the healthcare system. Health care systems are heterogeneous and it is imperative to engage institutional leaders to promote electronic health records including clinical data, disease progression and therapy outcome. The Global Alliance for Genomic Health (GA4GH) aims to standardise data creation, access and retrieval in an effort to realise the full potential of genetic and clinical data in research. To achieve this objective the alliance established an international framework for data sharing promoting effective and ethically responsible approaches. Similarly, the International Collaboration for Clinical Genomics (ICCG) was set to acquire and annotate genomic variation with clinical utility information and submit the genotype/phenotype data into a database for public access. The Clinical Variation (ClinVar) Public Database [60] is an example of these databases. Various databases are being annotated and a concerted action to allow proper interface for simultaneous acquisition of data is required [61]. Hence creation of databases shall comply with an international set of guidelines that define the process of data submission, data formatting and sharing and ensure proper quality checks and curation. In addition, proper annotation depends on the clinical data captured and hence the clinical data elements must be clearly defined. In addition to clinical data, the collection of cost-related data elements are useful for pharmacoeconomic studies.

To the same extent, the presentation of the genetic data to the clinic shall ensure a simple and clear understanding of the use of genetic data to allow health care providers to perform genetic-based informed decisions. Adoption of electronic health records (EHRs) to store and interpret genetic results, based on clinically validated algorithms, provides a tool that will allow genetic based informed decisions [62].

Provision of clinical data between countries is challenging and might be one of the major hurdles. The issue of data-sharing at the level of clinical data is not only challenging, but creates concern in protecting patient identity and dignity. Hence, although data sharing at the level of genomic information is of great benefit, the use of clinical data should be restricted within a regional/national database. Algorithms generated following genetic data sharing, can be easily interrogated for clinical utility using annotated data sets within national databases.

In conclusion, a well defined interface between genomic data sets, clinical data, therapeutic options and dosing algorithms is required to create a vehicle to communicate genome data with the healthcare system. Provision of this interface needs to be accompanied with proper education and selection of actionable genetic tests with robust evidence to support the genotypic association with the specific phenotype and subsequently clinical action.

References

1. Gerlinger M, Rowan AJ, Horswell S, Larkin J, Endesfelder D, Gronroos E, Martinez P, Matthews N, Stewart A, Tarpey P, Varela I, Phillimore B, Begum S, McDonald NQ, Butler A, Jones D, Raine K, Latimer C, Santos CR, Nohadani M, Eklund AC, Spencer-Dene B, Clark G, Pickering L, Stamp G, Gore M, Szallasi Z, Downward J, Futreal PA, Swanton C (2012) Intratumor heterogeneity and branched evolution revealed by multiregion sequencing. New Engl J Med 366(10):883–892. doi:10.1056/NEJMoa1113205
2. Druker BJ (2008) Translation of the Philadelphia chromosome into therapy for CML. Blood 112(13):4808–4817. doi:10.1182/blood-2008-07-077958
3. Stegmeier F, Warmuth M, Sellers WR, Dorsch M (2010) Targeted cancer therapies in the twenty-first century: lessons from imatinib. Clin Pharmacol Ther 87(5):543–552. doi:10.1038/clpt.2009.297
4. Lum DW, Perel P, Hingorani AD, Holmes MV (2013) CYP2D6 genotype and tamoxifen response for breast cancer: a systematic review and meta-analysis. PLoS ONE 8(10):e76648. doi:10.1371/journal.pone.0076648
5. Ruddy KJ, Desantis SD, Gelman RS, Wu AH, Punglia RS, Mayer EL, Tolaney SM, Winer EP, Partridge AH, Burstein HJ (2013) Personalized medicine in breast cancer: tamoxifen, endoxifen, and CYP2D6 in clinical practice. Breast Cancer Res Treat 141(3):421–427. doi:10.1007/s10549-013-2700-1
6. Province MA, Goetz MP, Brauch H, Flockhart DA, Hebert JM, Whaley R, Suman VJ, Schroth W, Winter S, Zembutsu H, Mushiroda T, Newman WG, Lee MT, Ambrosone CB, Beckmann MW, Choi JY, Dieudonne AS, Fasching PA, Ferraldeschi R, Gong L, Haschke-Becher E, Howell A, Jordan LB, Hamann U, Kiyotani K, Krippl P, Lambrechts D, Latif A, Langsenlehner U, Lorizio W, Neven P, Nguyen AT, Park BW, Purdie CA, Quinlan P, Renner W, Schmidt M, Schwab M, Shin JG, Stingl JC, Wegman P, Wingren S, Wu AH, Ziv E, Zirpoli G, Thompson AM, Jordan VC, Nakamura Y, Altman RB, Ames MM, Weinshilboum RM, Eichelbaum M, Ingle JN, Klein TE (2014) CYP2D6 genotype and adjuvant tamoxifen: meta-analysis of heterogeneous study populations. Clin Pharmacol Ther 95(2):216–227. doi:10.1038/clpt.2013.186
7. Pirmohamed M, Burnside G, Eriksson N, Jorgensen AL, Toh CH, Nicholson T, Kesteven P, Christersson C, Wahlstrom B, Stafberg C, Zhang JE, Leathart JB, Kohnke H, Maitland-van der Zee AH, Williamson PR, Daly AK, Avery P, Kamali F, Wadelius M (2013) A randomized trial of genotype-guided dosing of warfarin. New Engl J Med 369(24):2294–2303. doi:10.1056/NEJMoa1311386
8. Rosmarin D, Palles C, Church D, Domingo E, Jones A, Johnstone E, Wang H, Love S, Julier P, Scudder C, Nicholson G, Gonzalez-Neira A, Martin M, Sargent D, Green E, McLeod H, Zanger UM, Schwab M, Braun M, Seymour M, Thompson L, Lacas B, Boige V, Ribelles N, Afzal S, Enghusen H, Jensen SA, Etienne-Grimaldi MC, Milano G, Wadelius M, Glimelius B, Garmo H, Gusella M, Lecomte T, Laurent-Puig P, Martinez-Balibrea E, Sharma R, Garcia-Foncillas J, Kleibl Z, Morel A, Pignon JP, Midgley R, Kerr D, Tomlinson I (2014) Genetic markers of toxicity from capecitabine and other fluorouracil-based regimens: investigation in

the QUASAR2 study, systematic review, and meta-analysis. J Clin Oncol 32(10):1031–1039. doi:10.1200/JCO.2013.51.1857

9. Colomer R, Monzo M, Tusquets I, Rifa J, Baena JM, Barnadas A, Calvo L, Carabantes F, Crespo C, Munoz M, Llombart A, Plazaola A, Artells R, Gilabert M, Lloveras B, Alba E (2008) A single-nucleotide polymorphism in the aromatase gene is associated with the efficacy of the aromatase inhibitor letrozole in advanced breast carcinoma. Clin Cancer Res 14(3):811–816. doi:10.1158/1078-0432.CCR-07-1923
10. Makeeva O, Stepanov V, Puzyrev V, Goldstein DB, Grossman I (2008) Global pharmacogenetics: genetic substructure of Eurasian populations and its effect on variants of drug-metabolizing enzymes. Pharmacogenomics 9(7):847–868. doi:10.2217/14622416.9.7.847
11. Haibe-Kains B, El-Hachem N, Birkbak NJ, Jin AC, Beck AH, Aerts HJ, Quackenbush J (2013) Inconsistency in large pharmacogenomic studies. Nature 504(7480):389–393. doi:10.1038/nature12831
12. Kimmel SE, French B, Kasner SE, Johnson JA, Anderson JL, Gage BF, Rosenberg YD, Eby CS, Madigan RA, McBane RB, Abdel-Rahman SZ, Stevens SM, Yale S, Mohler ER, 3rd, Fang MC, Shah V, Horenstein RB, Limdi NA, Muldowney JA, 3rd, Gujral J, Delafontaine P, Desnick RJ, Ortel TL, Billett HH, Pendleton RC, Geller NL, Halperin JL, Goldhaber SZ, Caldwell MD, Califf RM, Ellenberg JH (2013) A pharmacogenetic versus a clinical algorithm for warfarin dosing. New Engl J Med 369(24):2283–2293. doi:10.1056/NEJMoa1310669
13. Gage BF, Eby C, Johnson JA, Deych E, Rieder MJ, Ridker PM, Milligan PE, Grice G, Lenzini P, Rettie AE, Aquilante CL, Grosso L, Marsh S, Langaee T, Farnett LE, Voora D, Veenstra DL, Glynn RJ, Barrett A, McLeod HL (2008) Use of pharmacogenetic and clinical factors to predict the therapeutic dose of warfarin. Clin Pharmacol Ther 84(3):326–331. doi:10.1038/clpt.2008.10
14. Ross S, Anand SS, Joseph P, Pare G (2012) Promises and challenges of pharmacogenetics: an overview of study design, methodological and statistical issues. JRSM Cardiovasc Dis 1(1):2
15. Holmes MV, Shah T, Vickery C, Smeeth L, Hingorani AD, Casas JP (2009) Fulfilling the promise of personalized medicine? Systematic review and field synopsis of pharmacogenetic studies. PLoS ONE 4(12):e7960. doi:10.1371/journal.pone.0007960
16. Rae JM (2013) CYP2D6 genotype should not be used to determine endocrine therapy in postmenopausal breast cancer patients. Clin Pharmacol Ther 94(2):183–185. doi:10.1038/clpt.2013.102
17. Ratain MJ, Nakamura Y, Cox NJ (2013) CYP2D6 genotype and tamoxifen activity: understanding interstudy variability in methodological quality. Clin Pharmacol Ther 94(2):185–187. doi:10.1038/clpt.2013.66
18. Kaye C, Korf B (2013) Genetic Literacy and Competency. Pediatrics 132(Suppl):S224–S230. doi:10.1542/peds.2013-1032G
19. Skirton H, Lewis C, Kent A, Coviello DA (2010) Genetic education and the challenge of genomic medicine: development of core competences to support preparation of health professionals in Europe. Eur J Hum Genet 18(9):972–977. doi:10.1038/ejhg.2010.64
20. Englander R, Cameron T, Ballard AJ, Dodge J, Bull J, Aschenbrener CA (2013) Toward a common taxonomy of competency domains for the health professions and competencies for physicians. Acad Med 88(8):1088–1094. doi:10.1097/ACM.0b013e31829a3b2b
21. Korf BR (2013) Genomic medicine: educational challenges. Mol Genet Genomic Med 1(3):119–122. doi:10.1002/mgg3.38
22. Gurwitz D, Lunshof JE, Dedoussis G, Flordellis CS, Fuhr U, Kirchheiner J, Licinio J, Llerena A, Manolopoulos VG, Sheffield LJ, Siest G, Torricelli F, Vasiliou V, Wong S (2005) Pharmacogenomics education: international society of pharmacogenomics recommendations for medical, pharmaceutical, and health schools deans of education. Pharmacogenomics J 5(4):221–225. doi:10.1038/sj.tpj.6500312
23. Eeles R, Purland G, Maher J, Evans DG (2007) Delivering cancer genetics services–new ways of working. Fam Cancer 6(2):163–167. doi:10.1007/s10689-007-9137-9

24. Whirl-Carrillo M, McDonagh EM, Hebert JM, Gong L, Sangkuhl K, Thorn CF, Altman RB, Klein TE (2012) Pharmacogenomics knowledge for personalized medicine. Clin Pharmacol Ther 92(4):414–417. doi:10.1038/clpt.2012.96
25. Relling M, Klein T (2011) CPIC: clinical pharmacogenetics implementation consortium of the pharmacogenomics research network. Clin Pharmacol Ther 89(3):464–467
26. Evans WE, Relling MV (1999) Pharmacogenomics: translating functional genomics into rational therapeutics. Science 286(5439):487–491
27. Burmester JK, Sedova M, Shapero MH, Mansfield E (2010) DMET microarray technology for pharmacogenomics-based personalized medicine. Methods Mol Biol 632:99–124. doi:10.1007/978-1-60761-663-47
28. Absalan F, Ronaghi M (2007) Molecular inversion probe assay. Methods Mol Biol 396:315–330
29. Johnson JA, Burkley BM, Langaee TY, Clare-Salzler MJ, Klein TE, Altman RB (2012) Implementing personalized medicine: development of a cost-effective customized pharmacogenetics genotyping array. Clin Pharmacol Ther 92(4):437–439. doi:10.1038/clpt.2012.125
30. Altman RB, Whirl-Carrillo M, Klein TE (2013) Challenges in the pharmacogenomic annotation of whole genomes. Clin Pharmacol Ther 94(2):211–213. doi:10.1038/clpt.2013.111
31. Bamshad MJ, Ng SB, Bigham AW, Tabor HK, Emond MJ, Nickerson DA, Shendure J (2011) Exome sequencing as a tool for Mendelian disease gene discovery. Nat Rev Genet 12(11):745–755. doi:10.1038/nrg3031
32. Gamazon ER, Perera M (2012) Genome-wide approaches in pharmacogenomics: heritability estimation and pharmacoethnicity as primary challenges. Pharmacogenomics 13(10):1101–1104. doi:10.2217/pgs.12.88
33. Metzker ML (2010) Sequencing technologies—the next generation. Nat Rev Genet 11(1):31–46. doi:10.1038/nrg2626
34. Fedurco M, Romieu A, Williams S, Lawrence I, Turcatti G (2006) BTA, a novel reagent for DNA attachment on glass and efficient generation of solid-phase amplified DNA colonies. Nucleic Acids Res 34(3):e22
35. Dressman D, Yan H, Traverso G, Kinzler KW, Vogelstein B (2003) Transforming single DNA molecules into fluorescent magnetic particles for detection and enumeration of genetic variations. Proc Natl Acad Sci USA 100(15):8817–8822. doi:10.1073/pnas.1133470100
36. Shendure J, Porreca GJ, Reppas NB, Lin X, McCutcheon JP, Rosenbaum AM, Wang MD, Zhang K, Mitra RD, Church GM (2005) Accurate multiplex polony sequencing of an evolved bacterial genome. Science 309(5741):1728–1732
37. Kim JB, Porreca GJ, Song L, Greenway SC, Gorham JM, Church GM, Seidman CE, Seidman JG (2007) Polony multiplex analysis of gene expression (PMAGE) in mouse hypertrophic cardiomyopathy. Science 316(5830):1481–1484
38. Leamon JH, Lee WL, Tartaro KR, Lanza JR, Sarkis GJ, deWinter AD, Berka J, Weiner M, Rothberg JM, Lohman KL (2003) A massively parallel PicoTiterPlate based platform for discrete picoliter-scale polymerase chain reactions. Electrophoresis 24(21):3769–3777. doi:10.1002/elps.200305646
39. Harris TD, Buzby PR, Babcock H, Beer E, Bowers J, Braslavsky I, Causey M, Colonell J, Dimeo J, Efcavitch JW, Giladi E, Gill J, Healy J, Jarosz M, Lapen D, Moulton K, Quake SR, Steinmann K, Thayer E, Tyurina A, Ward R, Weiss H, Xie Z (2008) Single-molecule DNA sequencing of a viral genome. Science 320(5872):106–109. doi:10.1126/science.1150427
40. Eid J, Fehr A, Gray J, Luong K, Lyle J, Otto G, Peluso P, Rank D, Baybayan P, Bettman B, Bibillo A, Bjornson K, Chaudhuri B, Christians F, Cicero R, Clark S, Dalal R, Dewinter A, Dixon J, Foquet M, Gaertner A, Hardenbol P, Heiner C, Hester K, Holden D, Kearns G, Kong X, Kuse R, Lacroix Y, Lin S, Lundquist P, Ma C, Marks P, Maxham M, Murphy D, Park I, Pham T, Phillips M, Roy J, Sebra R, Shen G, Sorenson J, Tomaney A, Travers K, Trulson M, Vieceli J, Wegener J, Wu D, Yang A, Zaccarin D, Zhao P, Zhong F, Korlach J, Turner S (2009) Real-time DNA sequencing from single polymerase molecules. Science 323(5910):133–138. doi:10.1126/science.1162986

41. Erlich Y, Mitra PP, delaBastide M, McCombie WR, Hannon GJ (2008) Alta-Cyclic: a self-optimizing base caller for next-generation sequencing. Nat Methods 5(8):679–682. doi:10.1038/nmeth.1230
42. Metzker ML (2009) Sequencing in real time. Nat Biotechnol 27(2):150–151. doi:10.1038/nbt0209-150
43. Mardis ER (2008) Next-generation DNA sequencing methods. Annu Rev Genomics Hum Genet 9:387–402. doi:10.1146/annurev.genom.9.081307.164359
44. Landegren U, Kaiser R, Sanders J, Hood L (1988) A ligase-mediated gene detection technique. Science 241(4869):1077–1080
45. Ronaghi M, Karamohamed S, Pettersson B, Uhlen M, Nyren P (1996) Real-time DNA sequencing using detection of pyrophosphate release. Anal Biochem 242 (1):84–89
46. Ronaghi M, Uhlen M, Nyren P (1998) A sequencing method based on real-time pyrophosphate. Science 281(5375):363, 365
47. Drmanac R, Sparks AB, Callow MJ, Halpern AL, Burns NL, Kermani BG, Carnevali P, Nazarenko I, Nilsen GB, Yeung G, Dahl F, Fernandez A, Staker B, Pant KP, Baccash J, Borcherding AP, Brownley A, Cedeno R, Chen L, Chernikoff D, Cheung A, Chirita R, Curson B, Ebert JC, Hacker CR, Hartlage R, Hauser B, Huang S, Jiang Y, Karpinchyk V, Koenig M, Kong C, Landers T, Le C, Liu J, McBride CE, Morenzoni M, Morey RE, Mutch K, Perazich H, Perry K, Peters BA, Peterson J, Pethiyagoda CL, Pothuraju K, Richter C, Rosenbaum AM, Roy S, Shafto J, Sharanhovich U, Shannon KW, Sheppy CG, Sun M, Thakuria JV, Tran A, Vu D, Zaranek AW, Wu X, Drmanac S, Oliphant AR, Banyai WC, Martin B, Ballinger DG, Church GM, Reid CA (2010) Human genome sequencing using unchained base reads on self-assembling DNA nanoarrays. Science 327(5961):78–81. doi:10.1126/science.1181498
48. Shendure J, Ji H (2008) Next-generation DNA sequencing. Nat Biotechnol 26(10):1135–1145. doi:10.1038/nbt1486
49. Chaisson MJ, Brinza D, Pevzner PA (2009) De novo fragment assembly with short mate-paired reads: Does the read length matter? Genome Res 19(2):336–346. doi:10.1101/gr.079053.108
50. Trapnell C, Salzberg SL (2009) How to map billions of short reads onto genomes. Nat Biotechnol 27(5):455–457. doi:10.1038/nbt0509-455n
51. Flicek P, Birney E (2009) Sense from sequence reads: methods for alignment and assembly. Nat Methods 6(11 Suppl):S6–S12. doi:10.1038/nmeth.1376
52. Simpson JT, Wong K, Jackman SD, Schein JE, Jones SJ, Birol I (2009) ABySS: a parallel assembler for short read sequence data. Genome Res 19(6):1117–1123. doi:10.1101/gr.089532.108
53. Manolio TA, Collins FS, Cox NJ, Goldstein DB, Hindorff LA, Hunter DJ, McCarthy MI, Ramos EM, Cardon LR, Chakravarti A, Cho JH, Guttmacher AE, Kong A, Kruglyak L, Mardis E, Rotimi CN, Slatkin M, Valle D, Whittemore AS, Boehnke M, Clark AG, Eichler EE, Gibson G, Haines JL, Mackay TF, McCarroll SA, Visscher PM (2009) Finding the missing heritability of complex diseases. Nature 461(7265):747–753. doi:10.1038/nature08494i]
54. Frazer KA, Ballinger DG, Cox DR, Hinds DA, Stuve LL, Gibbs RA, Belmont JW, Boudreau A, Hardenbol P, Leal SM, Pasternak S, Wheeler DA, Willis TD, Yu F, Yang H, Zeng C, Gao Y, Hu H, Hu W, Li C, Lin W, Liu S, Pan H, Tang X, Wang J, Wang W, Yu J, Zhang B, Zhang Q, Zhao H, Zhou J, Gabriel SB, Barry R, Blumenstiel B, Camargo A, Defelice M, Faggart M, Goyette M, Gupta S, Moore J, Nguyen H, Onofrio RC, Parkin M, Roy J, Stahl E, Winchester E, Ziaugra L, Altshuler D, Shen Y, Yao Z, Huang W, Chu X, He Y, Jin L, Liu Y, Sun W, Wang H, Wang Y, Xiong X, Xu L, Waye MM, Tsui SK, Xue H, Wong JT, Galver LM, Fan JB, Gunderson K, Murray SS, Oliphant AR, Chee MS, Montpetit A, Chagnon F, Ferretti V, Leboeuf M, Olivier JF, Phillips MS, Roumy S, Sallee C, Verner A, Hudson TJ, Kwok PY, Cai D, Koboldt DC, Miller RD, Pawlikowska L, Taillon-Miller P, Xiao M, Tsui LC, Mak W, Song YQ, Tam PK, Nakamura Y, Kawaguchi T, Kitamoto T, Morizono T, Nagashima A, Ohnishi Y, Sekine A, Tanaka T, Tsunoda T, Deloukas P, Bird CP, Delgado M, Dermitzakis ET, Gwilliam R, Hunt S, Morrison J, Powell D, Stranger BE, Whittaker P, Bentley DR, Daly MJ, de Bakker PI, Barrett J, Chretien YR, Maller J, McCarroll S, Patterson N, Pe'er I, Price A, Purcell S, Richter DJ, Sabeti P, Saxena R, Schaffner SF, Sham PC, Varilly P, Stein LD, Krishnan L,

Smith AV, Tello-Ruiz MK, Thorisson GA, Chakravarti A, Chen PE, Cutler DJ, Kashuk CS, Lin S, Abecasis GR, Guan W, Li Y, Munro HM, Qin ZS, Thomas DJ, McVean G, Auton A, Bottolo L, Cardin N, Eyheramendy S, Freeman C, Marchini J, Myers S, Spencer C, Stephens M, Donnelly P, Cardon LR, Clarke G, Evans DM, Morris AP, Weir BS, Mullikin JC, Sherry ST, Feolo M, Skol A, Zhang H, Matsuda I, Fukushima Y, Macer DR, Suda E, Rotimi CN, Adebamowo CA, Ajayi I, Aniagwu T, Marshall PA, Nkwodimmah C, Royal CD, Leppert MF, Dixon M, Peiffer A, Qiu R, Kent A, Kato K, Niikawa N, Adewole IF, Knoppers BM, Foster MW, Clayton EW, Watkin J, Muzny D, Nazareth L, Sodergren E, Weinstock GM, Yakub I, Birren BW, Wilson RK, Fulton LL, Rogers J, Burton J, Carter NP, Clee CM, Griffiths M, Jones MC, McLay K, Plumb RW, Ross MT, Sims SK, Willey DL, Chen Z, Han H, Kang L, Godbout M, Wallenburg JC, L'Archeveque P, Bellemare G, Saeki K, An D, Fu H, Li Q, Wang Z, Wang R, Holden AL, Brooks LD, McEwen JE, Guyer MS, Wang VO, Peterson JL, Shi M, Spiegel J, Sung LM, Zacharia LF, Collins FS, Kennedy K, Jamieson R, Stewart J (2007) A second generation human haplotype map of over 3.1 million SNPs. Nature 449(7164):851–861

55. LaFramboise T (2009) Single nucleotide polymorphism arrays: a decade of biological, computational and technological advances. Nucleic Acids Res 37(13):4181–4193. doi:10.1093/nar/gkp552
56. Lupski JR, Reid JG, Gonzaga-Jauregui C, Rio Deiros D, Chen DC, Nazareth L, Bainbridge M, Dinh H, Jing C, Wheeler DA, McGuire AL, Zhang F, Stankiewicz P, Halperin JJ, Yang C, Gehman C, Guo D, Irikat RK, Tom W, Fantin NJ, Muzny DM, Gibbs RA (2010) Whole-genome sequencing in a patient with Charcot-Marie-Tooth neuropathy. New Engl J Med 362(13):1181–1191. doi:10.1056/NEJMoa0908094
57. Roach JC, Glusman G, Smit AF, Huff CD, Hubley R, Shannon PT, Rowen L, Pant KP, Goodman N, Bamshad M, Shendure J, Drmanac R, Jorde LB, Hood L, Galas DJ (2010) Analysis of genetic inheritance in a family quartet by whole-genome sequencing. Science 328(5978):636–639. doi:10.1126/science.1186802
58. Anderson MW, Schrijver I (2010) Next generation DNA sequencing and the future of genomic medicine. Genes (Basel) 1(1):38–69. doi:10.3390/genes1010038
59. Peters BA, Kermani BG, Sparks AB, Alferov O, Hong P, Alexeev A, Jiang Y, Dahl F, Tang YT, Haas J, Robasky K, Zaranek AW, Lee JH, Ball MP, Peterson JE, Perazich H, Yeung G, Liu J, Chen L, Kennemer MI, Pothuraju K, Konvicka K, Tsoupko-Sitnikov M, Pant KP, Ebert JC, Nilsen GB, Baccash J, Halpern AL, Church GM, Drmanac R (2012) Accurate whole-genome sequencing and haplotyping from 10 to 20 human cells. Nature 487(7406):190–195. doi:10.1038/nature11236
60. Landrum MJ, Lee JM, Riley GR, Jang W, Rubinstein WS, Church DM, Maglott DR (2014) ClinVar: public archive of relationships among sequence variation and human phenotype. Nucleic Acids Res 42 (D1):D980–D985. doi:10.1093/nar/gkt1113
61. Zheng CJ, Han LY, Xie B, Liew CY, Ong S, Cui J, Zhang HL, Tang ZQ, Gan SH, Jiang L, Chen YZ (2007) PharmGED: pharmacogenetic effect database. Nucleic Acids Res 35(Database):D794–D799. doi:10.1093/nar/gkl853
62. Kho AN, Rasmussen LV, Connolly JJ, Peissig PL, Starren J, Hakonarson H, Hayes MG (2013) Practical challenges in integrating genomic data into the electronic health record. Genet Med 15(10):772–778. doi:10.1038/gim.2013.131

Ethical Considerations in the Genomic Era

Bridget Ellul

Abstract Pharmacogenomics is a powerful molecular tool in biomedical research aimed at providing personalised medicine in everyday clinical practice, best described as the provision of 'the right drug for the right patient at the right dose', that is safe, effective therapy, with minimal adverse reactions. The patient, is the main beneficiary but is also the indispensable key player, providing biological material for research.

This chapter focuses primarily on ethical issues as they affect the patient undergoing pharmacogenetic tests for personalised treatment, the subject enrolled in a clinical trial or participating in genomic research or the healthy person donating biological material for biobanking and research. Issues affecting the other stakeholders will also be pointed out, but again mainly from the perspective of the consumer.

Discussion centres on the right to beneficience, explored through benefit to risk ratio and the right to autonomy, exercised through informed consent with safeguards to ensure privacy and confidentiality in the handling of biological samples and data. Elements of justice will be introduced in relation to the target of equitable access to healthcare.

The basic ethical principles must be upheld through regulatory frameworks. States have embraced various instruments, from local and international guidelines to national legislation, but as genomic research increasingly moves into the global non interventional arena, the vision is of facilitation of international cooperation through harmonised regulations.

Keywords Informed consent · Ethical approval · Pharmacogenetic test uptake · Data protection · Clinical trials · Biobank

B. Ellul (✉)
Faculty of Medicine and Surgery, University of Malta Medical School, Block A, Level 0 Mater Dei Hospital, Msida MSD 2090, Malta
e-mail: bridget.ellul@um.edu.mt

G. Grech, I. Grossman (eds.), *Preventive and Predictive Genetics: Towards Personalised Medicine,* Advances in Predictive, Preventive and Personalised Medicine 9,
DOI 10.1007/978-3-319-15344-5_13

1 Introduction

The concept that genes play a part in controlling response to drugs was recognised in the late 1950s [1]. The speciality of pharmacogenetics [2–4] led to the discovery of specific DNA polymorphisms, notably in drug metabolising enzymes. As molecular techniques developed, in the late 1990's, pharmacogenomics emerged as a new discipline [5, 6] with an important role in the field of drug development. The identification of person to person 'variations of DNA and RNA characteristics as related to drug response' [7] can be applied to the quest for new drug targets and for safe drugs, balancing efficacy with minimal adverse reactions.

Though pharmacogenetics and pharmacogenomics are terms that are often used interchangeably, pharmacogenomics is centred on information from the entire genome. In fact it was the advent of new technologies, such as whole-genome sequencing, that drove research swiftly. Next generation sequencing, whole-exome sequencing and the development of bioinformatics, make the possibility of carrying out large population studies and data analysis, more feasible and will prove effective tools to identify biomarkers related to an individual's likely reaction to a particular medicinal product. This shall support the development of drugs and the required predictive genetic tests and companion diagnostic tests. Once such clinical application becomes an everyday reality, the promise of personalised medicine, the ideal healthcare programme, will finally be realised.

Such a goal requires solid interaction between the scientific and medical communities and the public. The index patient, or the healthy person seeking a predictive test, may well be seen as the ultimate beneficiary of personalised medicine but he is also the key player in the quest for the 'right drug for the right patient at the right dose' [8] since he is the one to contribute the biological material. This complex relationship must be fostered and nurtured on a sound foundation of ethical principles that enhance trust between all stakeholders.

Ethical behaviour in medical science, and specifically in genomic research, offers the foundation for the protection of the basic human rights of an individual and of society, but is also relevant to the other parties involved in research practices, the healthcare providers, the scientific investigators, the pharmaceutical and diagnostic companies, the funders and the policy makers. The perspectives of each group will vary but as all have a vested interest in a successful outcome, and since they depend on each another, there is a willingness to harmonise effective practices to move forward. However to guarantee that ethical issues are respected, this is not enough and there is a requirement for good governance, with a variety of regulatory instruments applicable at various stages of research.

This paper will focus primarily on ethical issues as they affect an individual, as a patient undergoing pharmacogenetic tests for personalised treatment, or as a subject enrolled in a clinical trial or participating in basic scientific research or as a healthy person donating biological material for research. Issues affecting the other stakeholders will also be pointed out, but again mainly from the perspective of the donor.

Ethical aspects will centre on the right to beneficience, explored through benefit to risk ratio, the right to autonomy, exercised through informed consent with safeguards of privacy and confidentiality, and on elements of justice, in relation to the target of equitable access to healthcare.

2 Personalised Therapy

2.1 *Benefits v Risks*

The pharmacogenomics target is to change healthcare management, in particular drug therapy, from general to personalised prescribing of evidence based effective and safe medicines. The benefits of taking the right medicine at the right dose, with minimal side effects are obvious. There can be little doubt that this is the ideal situation for the patient but it also embodies a prime objective in the provision of healthcare. Moreover providing optimum treatment is also a lack of maleficience, one of the main tenets of ethical medical practice. However the benefits must be weighed against the risks.

Personalised medicine is most practised in oncology [9]. For some tumours, genetic testing has become essential, and sometimes mandatory for Good Clinical Practice, before therapy is started. A success story is the clinical application of mandatory pharmacogenetic testing for gefitinib by the European Medicines Agency, EMA. Gefitinib, a tyrosine kinase inhibitor, is an epidermal growth factor receptor, EGFR, antagonist, and when used in patients with advanced non small cell lung cancer with EGFR activating mutations, it provides a significant increase in median survival [10].

However the best known example of pharmacogenomics testing is for variants of the enzyme CYP2D6 [11, 12] in the treatment of breast cancer with tamoxifen [13], where response to tamoxifen is reduced. Ethnic differences in genetic polymorphisms are very variable, with decreased activity, more commonly present in non Caucasians, making them less likely to benefit from tamoxifen [14]. Yet so far there is no mandatory regulation for pharmacogenetic testing prior to starting therapy. In fact recent review and meta-analysis concluded that there is 'insufficient evidence to recommend CYP2D6 genotyping to guide tamoxifen treatment' [15]. This exemplifies the problem of translation of genomic research to clinical use, some issues giving rise to risks for the individual, as will be discussed below. It also highlights the need for assessing the benefit to risk ratio for a particular individual.

The prime risks of a medication are the adverse drug events. Side effects exist for all drugs. In fact 'any drug involves some kind of risk-taking on the part of the patient' [16], but particularly burdensome are the serious adverse drug reactions requiring hospitalisation. However they are difficult to quantify, both as to prevalence and severity. Research has focused on the resulting hospital admissions, but often in individual hospitals rather than national [17] studies. The often quoted meta-analysis study by Lazarou [18] way back in 1998 had estimated they were the 4th to 6th commonest cause of mortality in the states, when hospital admissions due to serious events were 6.7%. In a more recent overview of 95 published studies, related to hospitalisation following adverse drug events, with admissions ranging from 0.1 to 54%, there was such great variation in methodology that it was concluded that 'extrapolation based on a meta-analysis of unselected studies may be biased' [19]. Some have also attempted to quantify the economic impact from the burden on healthcare management [20, 21].

Marketing of drugs is based on a balance between efficacy and safety but for the patient, the decision to take a medicine depends on the benefit to risk ratio, which 'must always be compared with existing alternatives' [16], that is one must ensure that the proposed new therapy is better than the current treatment and management available [22] for each specific patient. However patients with disease causing serious morbidity may be prepared to take more risks, such as use a medication with higher adverse reactions than normally accepted by less ill individuals. Also one might be prepared to try a drug with serious adverse reactions if there is no other alternative available.

Availability and access to a specific therapy and its accompanying diagnostic test may be related to area of residence or may be a question of cost of treatment, not necessarily whether a medicine is actually on the market. This of course raises issues of justice and will be discussed later on in this paper.

3 Pharmacogenomic Tests

To benefit from personalised therapy, the patient needs to know whether his genotype puts him at risk for serious side effects or if a drug will be inefficacious or if an adjustment in dose is required. He has therefore to submit to a pharmacogenomic test developed for the relevant predictive biomarkers.

To put this into perspective, at present, pharmacogenomic information in drug labelling, by the Food and Drug Administration, FDA, is only available for about 150 drugs [23], with 16 having information about more than one gene. However mandatory pharmacogenomics testing prior to starting therapy is only required for 31 and recommended for 6 drugs by the FDA and for 17 by EMA [24]. Ideally regulation of diagnostic tests and drugs should occur together [24].

3.1 Informed Consent

The personal choice to consent to take a pharmacogenomics test respects the right of the patient to be personally involved in his own healthcare management. For valid consent, an adult must be competent to understand and evaluate options and so come to a decision. However for autonomy to be entirely respected, consent must be genuinely informed and not be reduced to the legal requirement of validity and signing a form. The patient must be given the tools to reach a decision, specifically sufficient information, in a language they can understand, to effectively be engaged in evaluating the benefits versus the risks.

Consent by vulnerable patients such as the elderly or the very young may prove problematic as the person may not be fit to fully comprehend information regarding the state of health, let alone the relevance of a genomic test or of treatment options. These groups of people are well protected by regulatory mechanisms that insist on a guardian or legal representative to give consent.

Information should be provided about the specific indication for the pharmacogenetic test, the genetic abnormalities being detected and the interpretation of the result in relation to treatment options available. However the patient must also be made aware of issues regarding handling and secure storage of the sample and the data generated, including the long term dispositions and particularly if there are plans for use in future research. This allows the patient to assess the level of privacy and confidentiality afforded.

The patient must also be given enough time to consider all options and time to ask questions and clarify any confusing issues. The patient has the right to be informed as to benefits and risks, common and unusual. It stands to reason that choice must be free of any coercion. Healthcare professionals are ethically bound to offer only tests that are relevant to the medical problem and tests that are clinically valid, in keeping with good medical practice guidelines.

However there is also a fundamental right not to know [25] and such refusal of consent must be accepted, provided it is a genuine autonomous decision based on evaluation of adequate information.

3.2 *Benefits: Uptake of Tests*

For therapeutic purposes, when there is a definite recommended medicinal available, with the promise of beneficial impact on choice of treatment as well as a better patient outcome, most physicians value the tests [26]. Such tests are generally well accepted also by patients [27, 28] since the patient expects to benefit greatly from knowing the genetic variations which will predict the efficacy of the recommended treatment and/or whether there is any significant toxicity or if the dose needs to be adjusted. A 2009 study among a diverse US population revealed that 77% were 'very likely' or 'somewhat likely' to take a pharmacogenetic test [29]. A US patient survey, with just a response rate of 42%, showed that 73, 85, 91 and 92%, respectively were in favour of obtaining a test to identify if they were likely to have mild side effects, suffer serious side effects, to have the appropriate dose prescribed or to choose a specific drug [30].

Moreover it has been argued that a patient whose genetic test identifies an expected good response or at least an absence of serious side effects, is much more likely to comply with medication [31]. Taking a test may also be beneficial as it reduces anxiety [32].

3.3 Risks: Limitations to Uptake of Tests

3.3.1 Understanding the Value of Genomic Tests

For an individual the decision to take a test may actually prove to be a very difficult and painful decision. The primary reason may be that few patients can really comprehend genetic and genomic tests [33, 34] and appreciate the real benefits or risks. Healthcare providers use the term 'genetic tests' loosely for a multitude of different procedures that explore the function of genes and their products. So it is not surprising that there is a 'confused public perception of genetic testing' [35]. Just the mention of a genetic test to a patient can be a highly emotive experience, with the spectre of genetic exceptionalism in the background, leading one to immediately equate all tests with the possibility of establishing identity. There is also the immediate association that any genetic test must necessarily indicate inheritable disease of the monogenic type. The patient must be educated as to the possible value of a pharmacogenomic test, one that only provides estimates of risks for a particular variant, not definitive results [31].

Uptake of these genomic tests by society may depend on the perception of individuals as to the uncertainty of results of predictive testing. Patients require assurance as to the value of tests, that development is in line with the US ACCE framework model [36], which applies to analytic validity, clinical validity, clinical utility and associated ethical, legal and social implications that ensure adequate safeguards to the scientific measures.

Analytical validity is the ability to measure the relevant biomarker, with reproducible accuracy and reliability, which is the first step before any test can be developed further for the market. Estimate of errors in identifying a gene variant in whole genome sequencing is given as less than 0.5 % [37] but this does assume great importance when dealing with rare variants of disease. Different results in genome sequences of the same sample have been quoted as between 4–14 % [38]. Such variances may give rise to imperfect or erroneous deductions in the interpretation of a predictive test.

Many cannot appreciate that the clinical validity lies in the ability of the test to identify the phenotype from the genotype [39]. Clinical validity is a function of the complex relationship between penetrance of the genomic variant, gene heterogeneity and the test sensitivity and specificity. Genomic tests, though less invasive than phenotypic tests, often have lower sensitivity and specificity [40]. Variations of the latter two test characteristics will give rise to false positives and false negatives.

A low predictive value may be one of the reasons why a test is underused [39]. However it may be even harder to explain that clinical validity may be much lower than expected because response to drugs is not limited only to genetic factors. There is interrelationship with the environment, lifestyle, age, race, comorbidities and other drug treatments [6, 39]. Moreover gene variants are sometimes pleiotropic [41], and are associated with more than one disease or drug response. The contribution of the genotype to a particular drug response is reported as very variable, anything between

20–95 % of all variability [42]. For warfarin, over 40 % of variability in dose requirement cannot be attributed to any known genetic or non genetic factor [43].

Moreover genuine laboratory errors may also occur, for example analysis of the wrong sample or a problem with techniques or equipment. Laboratories providing genetic tests should be accredited and there are established practice guidelines and standards for laboratories conducting pharmacogenetic tests [44] and for ensuring quality assurance of genetic tests [45].

To be useful in clinical practice, a predictive test requires to show clinical utility [39]. It must reflect the expected health benefit attributed to the result of the test, which may include adherence to the drug, for which the patient was tested [46]. Realising there is a specific treatment might actually improve compliance to drug taking [31].

A new concept is that of personal utility [47] that is how a test will prove of benefit to the patient in terms of disease outcomes. Obviously for pharmacogenomics tests, this relates to how useful is the drug therapy available or how important it is to avoid a particular drug with serious adverse reactions. This may be a very individual assessment, depending on the seriousness of the disease being managed. Clinical utility may however also spur a healthy person to take a test in the asymptomatic stage, just for relief of uncertainty. This alone may provide psychological benefit. On the other hand lack of sufficient clinical utility may still push a patient with a serious disease to ask for the therapy even though the test result may not be promising [48].

The test result will always have a psychological effect with the risk of anxiety and depression from false positive tests to misplaced relief or a euphoric state from false negative tests. However a negative test, or one that suggests a reduction in drug dose, may not only cause anxiety but may make the patient unduly worried of the inability to take the recommended therapy at the usual dose, and so the patient may not adhere to what is regarded as less than best therapy [31]. The latter scenario may prove most difficult since it gives a false hope, which is doomed to total shattering effects if the disease actually manifests itself, let alone opening the spectre of litigation for the clinicians. An unexpected lack of response or increased risk may thus occur leading to possible litigation [31, 49]. This alone should encourage the doctor to think twice about which information to give the patient and the reliability of the test is one piece of information which should always be imparted.

Understanding pharmacogenetic tests is even more difficult when the patient has a lower educational level [50]. In a survey of oncology services providers, many cancer patients are thought to be unable to 'adequately comprehend the purposes and complexities of pharmacogenomic testing' [9], in particular in appreciating differences between somatic and germline testing and that only the latter have potential for inheritance and an effect on family members.

It is also probable that uptake of tests depends on the effort made by the healthcare provider in obtaining consent. The onus of promoting a test through enhancing the patient's understanding lies with the medical provider, who is morally and professionally expected to aim at maximising benefit to patients. Yet studies indicate that there is little expertise in genomics among clinicians in oncology [9]. This leads to problems in provision of services [26] and in giving the right advice to patients and/or in interpreting genetic results [51].

Clinicians also worry that they do not have adequate guidelines how to use tests [52], although now there are many such guidelines [52, 53] to consult. Canadian cardiologists, oncologists and family physicians identified various difficulties that prevented use of pharmacogenetic tests, mainly lack of clinical guidelines (60 %), lack of personal knowledge (57 %), no evidence based clinical information (53 %) and expense (48 %). 37 % also recognised that they did not have the time and resources to educate patients [26]. US primary care physicians also reported being uncomfortable with the level of knowledge expected to interpret the genomic tests [54]. There is little formal training of healthcarers, with as many as 92 % reporting no formal undergraduate training [26, 55] and there is a good argument for introduction or increase in the teaching of pharmacogenomics in medical curricula [56–58].

Such physician surveys highlight the need for adequate explanation and counselling by well trained individuals. Counselling requires ensuring that the patient understands the implications of testing, whichever result is obtained. This requires commitment by the healthcare providers to explain in lay man's terms and to ensure there is adequate understanding. Such consultation is time consuming. With regard to counselling for hereditary disease, there are recommendations for counselling pre and post test when the disease is severe, with the counsellor giving the patient sufficient time to weigh up the odds [59].

3.3.2 Discrimination

Patients do refuse to take tests because of fear of the test results finding their way into the wrong hands, such as an employer, or an insurer, which exposes them to discrimination. Those with higher levels of education express fewer concerns about possible misuse of genetic information [34]. Despite anti discriminatory laws being enacted in all democratic countries, patients are still worried [9]. In a Canadian study, 40 % of clinicians admitted that their patients had suffered from the fear of discrimination in relation to genetic testing [26].

Even when there is legislation against discriminatory practices, it may not offer comprehensive protection. The US Genetic Information Nondiscrimination Act, GINA, prohibits employers from discriminating against their employees on the basis of genetic information and prohibits health insurers from refusing to provide insurance, or asking for higher premiums, on the basis solely of genetic tests, but it does not offer protection for life insurance or disability insurance.

Legislation usually takes a firm stand against discrimination in terms of employment because the right to work is a fundamental human right but as to insurance, this is often a personal voluntary choice of the consumer to buy certain products. There may be instances where the consumer feels coerced into making a choice, for example in requiring insurance related to certain transactions, like obtaining a bank loan or buying a house.

3.3.3 Ethnicity

Certain gene variants will be predominant in particular ethnic groups. So clinical validity is higher in such groups. In a multicultural society, due to heterogeneity and cross culture, gene variations within a racial group, may differ more than the variations between different races and so there seems to be no need to have gross racial sub classifications [60]. A focus on race might not take into account the environmental factors. However one cannot negate the fact that certain drugs are certainly contraindicated in certain ethnic groups, e.g. ACE inhibitors in African Americans, so there certainly remains scope for more research into racial genetic differences. African Americans are more likely than other groups to believe that genetic test results will be misused [34], that genetic test results lead to racial discrimination [61] or for the racial/ethnic group to be labelled as inferior [33, 62].

3.3.4 Privacy and Confidentiality

Patients also feel threatened by the risk of breach of privacy and confidentiality. In a telephone survey of US adults, 78 % stated that they were unlikely to have a pharmacogenomic test if there was a risk that their DNA sample or test result could be shared without their permission [30].

The family doctor offering companion diagnostic tests may also be put in a dilemma as to whether they should inform the family, of any positive results, especially when they are also the doctor's patients. This disclosure is always to be considered as unethical professional behaviour, without the consent of the index patient. In fact a person may refuse to be tested just because of fear that they may be asked to make their test results available to relatives. Although one can argue, from an altruistic point of view, that such disclosure should occur, most patients are reluctant to show others their disease status. Also, patients may be prepared to tell their family doctor the result of a pharmacogenetic test but they may have some reservations at sharing the results with other healthcarers involved in their health care management [63].

If tests become widely available, a healthy patient might decide to take a test years before he is likely to develop a disease. So the result will end in his medical file [64]. Protection of data from access by third parties must therefore be ensured through regulatory instruments. Patients should be reassured as to storage facilities for samples and data from results, both paper and electronic formats.

What does the physician do if a patient refuses to take a pharmacogenomic test? Should the medication still be provided, even if there may be adverse reactions? The answer lies in ensuring real informed consent has been obtained because an individual retains the fundamental right of refusing treatment. However physicians are concerned as to possible litigation in the future from the patient or the family.

The American College of Medical Genetics recommends that incidental findings obtained in a clinical (not research) setting should be disclosed to the patient and the clinicians [65]. However not everyone agrees with these guidelines [66–68] and it is best to have an agreed policy at the time of the initial consent, with the added safeguard of utilising counselling if there are unexpected results.

3.4 Direct to Consumer Tests

It is increasingly possible now to obtaining genetic information without direct consultation of a physician by purchasing personal genome tests, PGTs, also called direct to consumer tests, bought directly through the company developing and marketing the test, or via retailers. Commercially available tests, based on whole genome sequencing are now widely available for multifactorial diseases and some also for pharmacogenomic tests. The companies may offer a bundle, with tests providing results about several diseases or groups of diseases, or about therapy options, which vary in nature and therefore also have differing ethical implications. Some pharmacogenomic tests for antidepressants are also being combined with susceptibility tests [69] in psychiatric disease and this raises concern in relation to clinical validity of such tests [70].

Common to all there is the central issue of consent and how to ensure that it is really informed. Article 7 of the Additional Protocol of the Council of Europe on Genetic Testing for Health Purposes [71] states that 'a genetic test for health purposes may only be performed under individualised medical supervision' with a view to offering protection to the person tested and also the possibility of informed consent and counselling.

Bunnik et al state that 'because of the complexity and the quantity of the information offered in PGT, informed consent cannot be fully specific' [72]. They propose a model of consent, with three layers, tiered, layered and staged, which can also be intertwined. Tiered consent is based on giving a choice to the individual as to which type of disease the consumer is interested in. The layered consent relates to the amount of information made available at different times, starting with a minimum basic amount of knowledge, labelled as the first layer; so there are options to know more, with the choice left freely to the consumer. Staged consent refers to provision of information over a specific timeframe, when the consumer has to give consent at various stages of the process, in relation to a certain process, for example pre purchase of the test, prior to being sent the results and prior to receiving updates.

The other ethical dilemma is disclosure of information. The companies provide different levels of assurance as to confidentiality and disclosure of information for their clients. Consumers have expressed a preference for tiered consent schemes that allow individuals to specify the level of data sharing permitted with respect to their genome [73]. Of course not everyone is prepared to share their results, not even with their doctor [74].

There has not been any strong evidence of harm to consumers from availability of direct to consumer tests. In fact a study revealed that the type of information received did not result in any psychological harm [75]. Possibly this reflects the personality of the person willing to obtain such tests.

Other concerns relate to criminal abuse, such as the possibility of submitting biological samples of third parties who have not actually consented, as well as lack of transparency as to what type of research is carried out on the samples submitted [76].

4 Research

4.1 *Clinical Trials*

4.1.1 Informed Consent

The ethical issues related to research in general and to clinical trials in particular are well established and safeguards are faithful to the principles in the Declaration of Helsinki, with its latest amendment in 2013 [77]. From a participant's point of view, informed consent, privacy and confidentiality are guaranteed through scrutiny by Research Ethics Committees, RECs, based in universities and in health departments and institutions, sometimes covered by state legislation. For clinical trials there is specific legislation in most countries reflecting the higher stakes. In Phase I research there may be healthy volunteers while in Phase II or III clinical trials, the patient may get the placebo or the least effective drug. So the information prior to obtaining consent has to be comprehensive and transparent and well explained.

In the EU, states have transposed the EU Directive [78] into national legislation. The Directive lays down detailed guidelines as to what the RECs should assess. Again there are guidelines available through the Oviedo Convention [25] and the Additional Protocol [79], which clearly distinguish subjects capable of giving consent from those who for some reason (age, mental infirmity, and emergency situations) are unable to consent. Research on such vulnerable people, including persons with mental impairment and minors, carries the same problems in relation to consent, as that for diagnostic purposes but there are some specific issues. The subject should only participate if the trial is personal of benefit, or to others suffering from the same type of disease, or if there is no other way of obtaining the same information. Although not able to understand all information, vulnerable individuals may be able to decide and consent to simple procedures and to take part in the decision making process. With respect to children, a child should be involved in evaluation of the benefits and risks of participating in research, and in coming to a decision, in accordance with maturity but consent from the parent or legal guardian is also required, though the minor is allowed to object [80, 79]. Moreover if a competent child refuses trial participation, they must not be coerced to participate just because the parents agree to participation. Ethics mandates that their wishes should be respected, even though legislation only requires consideration of their views [80]. Yet again the main problem here is ensuring adequate information and time to enable the potential trial participants to make up their minds.

Obtaining consent in multicentre clinical trials may be hampered by lack of harmonisation of law. Even in the EU states this has this been possible, for example there is no harmonisation of standards for ethical committees.

4.1.2 Consent for Genetic Studies

Consent for genetic tests, either on blood or tissues, in a clinical trial is usually obtained completely separately from the consent for the rest of the protocol. However the content of the legally binding form to be signed and the information supplied, vary from one trial to another.

The main problem is obtaining consent for future studies. The Oviedo Convention allows additional use of biological materials if 'done in conformity with appropriate information and consent procedures' [25]. The pharmaceutical companies claim that often it is impossible at the first instance to outline exactly the future research. So researchers have sought different models of consent, focusing mainly on broad [81, 82] or open consent for any future research, whether of a genetic nature or not, that is, there would be no need to get back to the trial participant to ask for consent for future studies, or for studies not contemplated in the original design of the trial or because new technologies become available. However broad consent does not mean 'vague' [83] but broad in relation to the original idea of consent to a specific protocol. It usually implies consent related to future research, either on the same disease or some new biomarker, but not tied to a specific project. Effectively this is 'consent to governance' [81] by some authority or person to take the decision in future as to whether to use the material or data for research. However sometimes such distinctions are not even mentioned and this actually amounts more to 'blanket' rather than a broad type of consent. The only safeguard is the requirement to have the future project reviewed by a REC, which is always a prerequisite for substantial amendments to a trial.

Other researchers may opt to give the participant the option to choose whether to be recalled in the future for further consent. Yet this is very cumbersome, not to mention that of course it limits anonymisation of material collected, as the participants have to be traced to be recalled, thus compromising privacy.

There is some disagreement as to whether broad consent can ever be equated with genuine informed consent [85, 86] although most ethicists are in favour of its use and agree that this is a decision that fulfils the original intention behind the introduction of informed consent, that of ensuring the participant's autonomy is protected [82]. Similarly the trial participant may affirm that they do not want to be re-contacted in future and are prepared to give authorisation for the researchers to use their material anyway. From an ethical perspective such a position would also be a voluntary decision and thus guarantees the principle of autonomy [87].

Legally, broad or open consent or waiver of consent cannot by its very nature be considered as informed consent as covered by the EU Directive or by the Council of Europe. However the Nuffield Council on Bioethics did approve broad consent for the 'use of samples that are anonymous or anonymised' [88] while it recommended collection of a separate broad consent if the samples where identifiable.

The possibility to opt out of the trial, must be present for the length of the trial, but also for the stored samples and data. It may be possible at the outset to refuse to be part of the genetic research arm, or to refuse to all future research, although often these comprise exclusion criteria from the trial. When opting out occurs after destruction of the biological sample, the data already collected is generally retained for the research - this must be highlighted in the original consent.

Finally, the issue of coercion arises when a patient agrees to being a trial subject if they obtain access to a new medicine, which they hope will work [88]. Is this a valid consent? One can argue that this decision is conditioning a better outcome in the participant, akin to a placebo effect.

4.1.3 Privacy and Confidentiality

Participants may be worried about privacy and confidentiality of stored biological samples, and sharing of data and results, which in the EU are offered protection through the EU Directive [88, 89]. However Data Protection laws are not comparable in non EU states.

The EU Directive on Clinical Trials does emphasise the 'rights of the subject to physical and mental integrity, to privacy' in accordance with the Data Protection Directive. In general privacy issues related to fear of discrimination are not an issue, as occurs in the diagnostic field. However subjects need to know that insurance companies do not get access to data. For this reason, they may be more likely to agree to participate in a trial if there is anonymisation of samples, though they may be satisfied by coding which allows them potential access, particularly if they trust the pharmaceutical company or researcher.

However the problem of confidentiality is paramount for uncommon orphan diseases, particularly in a small community. With rare diseases, it seems pointless to have anonymity when it might be beneficial to contact the participants to impart individual results. Should there be disclosure of results to such participants, and/or to other family members, particularly if the information is beneficial to them or their family? Potential participants should be encouraged to speak with their families regarding genetic trials, so that the subject can share information with family members [90–92].

4.1.4 Data

There is a duty on the scientists to impart sufficient information at the time of recruiting participants to a clinical trial to enable informed consent. The potential trial subjects should be told about the benefits and risks of the research but there is a need to balance the knowledge divulged with what a reasonable person would expect to be told and to express consent forms in a straightforward unambiguous language. The investigator should find out what is important to a specific group of persons or ethnic group or study group in the inclusion criteria.

Subjects must be informed as to the length of time and site of storage of biological samples and data, the security provided and who has access to the data, whether only researchers or local authorities or third countries, in conformity with Data Protection legislation.

In the process of data sharing, data should be anonymised and some countries have regulatory mechanisms in place, particularly in the case of clinical trials, e.g. EMA guidelines. Data sharing is of course enhanced and expedited through publication in open access journals but this needs to be funded. A survey among trialists reported willingness to share data among respondents (albeit there was only a 46 % response) but they were concerned about appropriate interpretation of data, protecting their own interests with respect to publication or academic recognition, as well as some concern about patient confidentiality being maintained [93]. In the interests of the public, negative results should be reported, so trials can be repeated.

There has been a campaign to increase the transparency of clinical trials and to make results available to the public. It is not clear how companies decide which information to make available to the public or third parties. RECs should actually make sure that both trial registration and the publication of results are mandatory prior to ethics approval [94]. EMA's policy of providing clinical trial data to third parties was weakened due to legal action from pharmaceutical companies. Similar incentives are happening in many countries. Once the EU Directive regulating clinical trials is repealed in 2016 and replaced by the Clinical Trials Regulation [95] registration of all trials in the EU will become mandatory as will the publication of trial results. A full study report must be published in line with guidelines by ICH [7] and again this would become available in the public domain.

Trial subjects may worry about the commercial interests of the company overriding their basic rights. They have to be informed as to commercial interests of the company as well as to the fact that intellectual property rights are vested with the sponsors or the pharmaceutical company.

Drugs that are new on the market need to be followed up for a considerable length of time, in fact ideally throughout the drug's lifetime, the so called 'life-cycle approach to risk management' [96]. This will ensure that rare side effects are identified [16, 96]. EMA guidelines [97] aim to strengthen evidence about the effect of genomic labelling and the use of genomic biomarkers in the post marketing stage, to use it in clinical practice. Post marketing surveillance is when certain rare adverse reactions are identified. These may be due to complex interactions of genetic variations with environment and lifestyle. The pharmaceutical companies may proceed to utilise these results to develop personalised medicines.

4.2 Genomics Research and Biobanks

4.2.1 Informed Consent

Genomics research may lead to development of new drugs, which can then be assessed through clinical trials. By its very nature such research requires a large database so as to enable examination of genomes from a large number of individuals, to obtain meaningful results that identify either susceptibility genes for specific diseases or for variability in drug responses.

This has led to the establishment of biobanks, repositories of biological materials and / or data derived from such samples, from specified populations, sometimes healthy individuals, invited to participate in being donors for research purposes. Some material is in private collections but most is in the public domain. Many banks contain more information, in the form of personal data, which may be also linked to medical and lifestyle data. This data is necessary because of the interrelationship between genes and the environment but it is often linked to other information held on a national level, for example social security or tax information.

Such biological collections are now often managed by multinational groups or consortia. This means we have entered the global era of research where international collaboration is the only way to obtain sufficient research material that can be shared globally for meaningful results for practical clinical application.

Recruitment is increasingly coming indirectly through the use of the biological material already stored in national archives or biobanks. Consent is obtained at the time of donation of samples, often long before the research project is fashioned. The commonest type of consent used currently is broad consent [98]. Such research is non interventional and offers a low risk to participants [99], seeing that most results do not apply directly to the individual who has submitted the material; this renders broad consent more acceptable. To donors, 'practical utility' may be more important than knowing the details of each project [84].

As for clinical trials purposes, broad consent is acceptable provided certain safeguards are in place; these would include measures to maintain confidentiality by ensuring all personal information, whether in the form of biological material or data, is stored securely in a suitably coded fashion. Anonymisation is not usual as it would preclude clinical monitoring or adding new data to the bank. Maintaining privacy is important because of the links to other data. Consent should include information related to who has access to data and with whom data is shared. The principal investigator should not be involved in obtaining consent but a contact person needs to be identified.

The autonomy of consent can also be preserved if, within the constraints of the biobank set up, donors are allowed the possibility to have samples and data withdrawn [100]. A crucial safety measure is to ensure adequate governance regulations of the biobank facility and to have RECs approval for any new research project that uses material from the bank.

Because most of the research is carried out by international collaboration, of different bodies, difficulties arise in obtaining consent from the individual national RECs. In an effort to simplify such authorisation and to ensure it is timely, there is a need for setting up a framework based on cooperation between the multiple research centres, be they health institutions or industry based companies, with consideration of the local policies, regulatory mechanisms and legislation.

A recent recommendation is to set up a Safe Harbor Framework for International Ethics Equivalency [101], creating an International Ethics Review body that harmonises procedures based on the same principles and satisfying the varied national legislation [102]. Management will rely on electronic means to expedite matters.

One way to promote donation to biobanks is to induce patients to donate residual biological material after diagnostic procedures, material that may otherwise be discarded after biopsy [103] or even to obtain material during recruitment in clinical trials. This may be quite acceptable for donors as it benefits them by promoting a sense of altruism in helping society. Some ethicists argue that there is no need for consent [104], particularly if the material stored can be considered as having already been discarded by the patient. On the other hand there are advantages to the patient giving consent. This would enhance trust between patients and researchers, auguring well for future participation in research, since it would show respect for patients' views. They may have strong objections to being involved in any genomic research, especially if it has commercial potential but may be prepared to consent to use of their biological material if the research allows clinically relevant results to be passed on to the patient or their family [105]. Current opinion is that national banks should be able to inform clients of any positive results if it is going to be to their benefit.

In fact in a study [106] of the UK public in 2012, 55% of those surveyed believed that it was 'extremely important' and 25% that it was important to be asked to provide consent for residual samples, and the majority agreed to have an opt in type of consent, to speak directly to a healthcare professional and not just fill a form. 27% of those surveyed and 57% of those in focus groups did prefer an opt out type of consent. However this again would be broad consent at the time of collection.

Much biological material is already present as archived biological material, initially retrieved with another purpose in mind, in particular human tissues in pathology departments in all hospitals, previously collected with consent for diagnostic purposes. The question is whether such material should now be available for genomic research and then how to obtain consent to use such material. Recall of patients is very unlikely to be feasible. In some countries, legislation provides for use without consent if there is appropriate protection, such as ethics approval by RECs, for each new project. If samples are securely coded and stored and access to data is limited, and if the donor has not specifically refused consent, then most agree that specific consent is not necessary [107]. The Council of Europe provides for use without consent if reasonable efforts to contact the owner are not possible, provided the research is of important scientific interest and could not be addressed by using other biological material for which consent was available [108].

There should be a policy as to what to do with samples from minors. Should these be initially excluded from inclusion in a biobank until the minor turns 18?; should they then be contacted for consent?

Other models of consent may be used, such as the authorisation model [85], which allows the donor to decide for which type of research they are willing to contribute and it is up to the donor to lay down particular conditions as to the level of involvement they want in the future, especially regarding recall for future consent or not. Staged or stepwise [109] informed consent would allow potential donors to understand what is happening. Dynamic consent [110] takes involvement a step further as the donors are kept informed by the researchers as to what is happening and are asked to re-consent, with various interactions occurring between the donor and the biobank, regarding samples and data use. It is claimed that this would increase recruitment, trust and transparency. However Segerdahl [111] has been critical of such consent, claiming that this places a lifelong burden on the donor while the researchers have the liberty to move on.

Another problem particular to biobanks is what would happen to banked material after death of the donor. The OECD [112] leaves the options to the biobank but the relevant policy must be made known to the participant from the outset. The sample and / or data could be withdrawn or the option given to relatives to decide or they can be anonymised.

Biobanks have variable policies [113] regarding return of unexpected findings to research participants, whether these are incidental findings unrelated to the aims of the research or whether they are 'individual research results' that are part of the study variables, with less likelihood of the latter being passed on to the subjects. Such policies must be clear before consent is obtained.

4.2.2 Data Privacy

Now that technology has really advanced, the limiting factor in genomic research is actually the data analysis. Electronic records facilitate the extraction of results but electronic data is never really completely safe. With today's bioinformatics tools it is possible to analyse large public data bases and use data linkage disequilibrium to identify individuals, even if data was anonymised.

Electronic medical records may be less secure than envisaged but even researchers have claimed that 'efficiency and utility of securing accurate personal genomic information through genomic testing and electronic medical records may outweigh patient privacy concerns if cancer treatment outcomes can be improved' [9]. The US has an electronic Medical Records and Genomics network, eMERGE, with biobanks linked to electronic medical records, specifically aimed at finding genomic markers and genotype-phenotype associations.

The Personal Genome Project [37] hinges on publicly sharing genome date from self referrals, such genomic data being combined with public health data. The participants are 'explicitly not promised anonymity' because the project leaders argue that although protection of data is possible in such research, privacy cannot be always ensured when there is public release of data for sharing among researchers.

Therefore it is crucial to ensure confidentiality. This is possible in EU states where it is covered by the Data Protection Directive [89]. However the collection, storage and sharing of samples is not harmonised. Data may be transferred across the EU in line with the Data Protection Directive but transfer to third parties is allowed if they are deemed to have data protection laws similar to the EU; this excludes the United States of America. In an attempt to solve this impediment to sharing information, in 2000, the European Commission and the US Department of Commerce agreed on a Safe Harbor Framework, agreeable to both sides [114] for ensuring privacy.

It is a much bigger problem if such material is unfairly disclosed to authorities for educational decisions, employment and legal decisions on culpability [87]. This will erode donors' trust in the system but there is no specific legislation as yet, at least on an EU level, to give biobanks the right to protect their data at all costs. It is envisaged that the data may be useful for allocation of health resources. Accessibility to biobanks must therefore be controlled.

5 Responsibilities of Other Stakeholders

Commercial companies in the pharmaceutical industry have a duty to society to invest in research and development of new drugs to provide efficacious and safe drugs through properly conducted clinical trials. But Phase I development of new drugs is expensive and traditionally drugs have been aimed at 'one size fits all'.

It is now accepted that it is unethical to ignore the needs of non responders, when such reaction is the result of recognised pharmacogenomic differences that can be targeted in drug development for the production of personalised drugs for these people.

However if the number of potential patients is low, it will not be economically feasible for a pharmaceutical company to invest in drug development specific to such patient groups since the drug will be too expensive to manufacture, even if altruistically this will be of long term benefit to such individuals, if not a life saver.

The company may need to look into allele distributions for specific populations before deciding to invest in developing drugs for such a group of people. Of course this seems unjust to a minority of people with orphan diseases.

Some have argued that companies have to be given incentives to pursue drug development, such as allowing them exclusive research for rare diseases for which they aim to provide therapy. This may be public funding [115], as happened for orphan diseases, for which research has been adopted by international networks. The US Orphan Drug Act of 1983 allowed tax refunds and 7 year exclusivity for drugs for orphan diseases.

Knowing which individuals are likely to benefit from a new drug based on their genetic variations can lead to clinical trials only in patients with the particular genotype, thus leading to a new personalised product at a quicker rate. This minimises costs.

Maybe this is why the major investment has been in personalised therapy in oncology. The increasing number of elderly people has brought about an increase in the prevalence of cancer and an easily available source of biological material to work on. However, one can argue that either someone looked far ahead and invested in oncology work, or it was fortuitous and initial success led to development of more drugs in oncology [9].

A company may also consider to rescue drugs that have failed to pass a clinical trial because the majority suffered toxic effects. Yet the drug may be developed for the relatively few responders. Such a scenario is more likely for specific ethnic groups [116], as happened with BiDil (isosorbide and hydralazine) [117]. Repurposing of drugs, the use of approved drugs for another indication, may also prove possible.

The Nuffield Council on Bioethics [88] had proposed that licences for marketing of medicines could be tied to a pharmacogenetic test to ensure that it is carried out and not bypassed in a measure to avoid costs.

5.1 *Justice*

Healthcare providers may be willing to allocate resources for pharmacogenetic tests or personalised drugs if there is evidence of the benefit in terms of savings in health expenditure, in the long term, for treating the rare conditions.

More commonly the problem is that the tests are still expensive [49] and for pharmacogenetically designed drugs they are likely to be more expensive tests, leading to inequalities in access [118]. In fact not all genomic tests are covered by insurance [119] and this deters patients from obtaining the test and moreover the availability of insurance cover is not uniform, for example in the US, this varies from state to state [120].

Access to genetic diagnostic testing may be limited for various reasons, including residence, distance from healthcare facility, socio-economic status and insurance coverage. It may not be available at all in certain countries or at least it may only be available in tertiary centres, rendering access difficult and entailing expenses to contact the providers. If the test is positive, there is the problem that the relevant therapy, again likely to be more expensive than for common diseases, may not be available or may not be free to all. Treatment for rare conditions is much more expensive than expected. Sometimes the impetus must come from lobbying by patient groups to put pressure on insurance companies to provide cover and on healthcare providers to make expensive tests available [3, 121].

Pharmacogenomics will benefit society by improving the benefit to risk ratio for a particular drug in a particular population but it can never guarantee improvement for an individual [49]. Putting the onus on a person to take decisions about their health can be interpreted as empowerment but it may also be placing a burden of responsibility on their shoulders, instead of healthcarers. Patients might have to conform to social expectations [122].

6 Outcome and Recommendations

It is the hope of all stakeholders, but primarily of the main beneficiary, the potential patient, that the promise of pharmacogenomics to provide 'the right drug for the right patient at the right dose' is fulfilled. Advances in molecular technology have enabled the application of genomics to the development of specific drugs for personalised therapy. At present the number of drugs with pharmacogenomic labels is limited and the challenge is to identify genetic variability of drug targets and develop biomarker diagnostic tests for common diseases. This requires investment in large scale research and robust clinical trials on an international basis, with multidisciplinary international collaboration. However the starting point for such projects is the pool of biological material from the patient or potential patient.

This chapter has focused on the ethical issues that guarantee standards and promote cooperation by society. The primary emphasis is on informed consent, which is central to the involvement of the patient. It is a crucial step for the uptake of a pharmacogenetic test, for enrolment in a clinical trial or genomic research and for the donation of biological material for a biobank. Continuing cooperation of the patient is enhanced through safeguards to ensure privacy and confidentiality in the handling of biological samples and data. Consent is also promoted through dissemination of the knowledge that there is equitable access to personalised medicine and not solely a reliance on the cost effectiveness of the tests and the therapy.

So society must be kept fully informed about pharmacogenomic progress to encourage open interaction with scientific and medical researchers. There is a definite educational role for healthcare professionals, provided they are themselves also well informed and trained to explain the benefits of personalised medicine, to choose and interpret relevant genomic tests and to encourage research participation. Genomic training must be included in medical and health science curricula at all levels, with emphasis on continuing professional development to keep abreast of advances.

Participation in multidisciplinary international research may be enhanced by opting for broad consent for genetic studies in clinical trials and for collection of biobank samples. The move to increased transparency in the conduct of clinical trials will provide more information to individuals as trial results are made public. Post marketing surveillance of drugs needs to be pursued on a regular basis to identify the effects of genomic labelling, pick up rare side effects and identify potential new niches for specific drugs.

Ethical principles are now embedded in various regulatory frameworks and legal instruments, but states have embraced different norms. As genomic research increasingly moves into the global arena, the vision is to work towards establishing harmonised regulations to facilitate international collaboration and so achieve the goal of making personalised medicine a reality in everyday medical practice.

References

1. Motulsky AG (1957) Drug reactions enzymes, and biochemical genetics. J Am Med Assoc 165(7):835–837
2. Vogel F (1959) Moderne problem der humangenetik. Ergeb Inn Med Kinderheilkd 12:52–125
3. Gurwitz D, Motulsky AG (2007) 'Drug reactions, enzymes, and biochemical genetics': 50 years later. Pharmacogenomics 8(11):1479–1484. doi:10.2217/14622416.8.11.1479
4. Kalow W (ed) (1962) Pharmacogenetics. Heredity and the response to drugs. W. B. Saunders Company, Philadelphia
5. Marshall A (1997) Genset-Abbott deal heralds pharmacogenomics era. Nat Biotechnol 15(9):829–830. doi:10.1038/nbt0997-829b
6. Pirmohamed M (2011) Pharmacogenetics: past, present and future. Drug Discov Today 16(19–20):852–861. doi:10.1016/j.drudis.2011.08.006

7. ICH harmonized tripartite guideline: definitions for genomic biomarkers, pharmacogenomics, pharmacogenetics, genomic data and sample coding categories E 15. (2007) Paper presented at the International Conference on Harmonisation of Technical Requirements for Registration of Pharmaceuticals for Human Use
8. The Five Rights (1999) ISMP medication safety alert. Acute Care Edition, vol 4. Institute for Safe Medication Practices, ISMP
9. McGowan ML, Settersten RA Jr, Juengst ET, Fishman JR (2014) Integrating genomics into clinical oncology: ethical and social challenges from proponents of personalized medicine. Urol Oncol 32(2):187–192. doi:10.1016/j.urolonc.2013.10.009
10. Rosell R, Moran T, Queralt C, Porta R, Cardenal F, Camps C, Majem M, Lopez-Vivanco G, Isla D, Provencio M, Insa A, Massuti B, Gonzalez-Larriba JL, Paz-Ares L, Bover I, Garcia-Campelo R, Moreno MA, Catot S, Rolfo C, Reguart N, Palmero R, Sanchez JM, Bastus R, Mayo C, Bertran-Alamillo J, Molina MA, Sanchez JJ, Taron M (2009) Screening for epidermal growth factor receptor mutations in lung cancer. N Engl J Med 361(10):958–967. doi:10.1056/NEJMoa0904554
11. Hoskins JM, Carey LA, McLeod HL (2009) CYP2D6 and tamoxifen: DNA matters in breast cancer. Nat Rev Cancer 9(8):576–586. doi:10.1038/nrc2683
12. The Human Cytochrome P450 (CYP) Allele Nomenclature Database. http://www.cypalleles.ki.se/. Accessed 04 Oct 2014
13. Early Breast Cancer Trialists' Collaborative Group (EBCTCG) (2005) Effects of chemotherapy and hormonal therapy for early breast cancer on recurrence and 15-year survival: an overview of the randomised trials. Lancet 365(9472):1687–1717
14. Becquemont L (2011) Practical recommendations for pharmacogenomics-based prescription: 2010 ESF–UB Conference on Pharmacogenetics and Pharmacogenomics. Pharmacogenomics 12(1):113–124
15. Lum DW, Perel P, Hingorani AD, Holmes MV (2013) CYP2D6 genotype and tamoxifen response for breast cancer: a systematic review and meta-analysis. PLoS One 8(10):e76648. doi:10.1371/journal.pone.0076648
16. Hansson M (2010) Taking the patient's side: the ethics of pharmacogenetics. Pers Med 7(1):75–85
17. Isacson D, Johansson L, Bingefors K (2008) Nationwide survey of subjectively reported adverse drug reactions in Sweden. Ann Pharmacother 42(3):347–353. doi:10.1345/aph.1K488
18. Lazarou J, Pomeranz BH, Corey PN (1998) Incidence of adverse drug reactions in hospitalized patients: a meta-analysis of prospective studies. JAMA 279(15):1200–1205
19. Leendertse AJ, Visser D, Egberts AC, van den Bemt PM (2010) The relationship between study characteristics and the prevalence of medication-related hospitalizations: a literature review and novel analysis. Drug Saf 33(3):233–244. doi:10.2165/11319030-000000000-00000
20. Gyllensten H, Hakkarainen KM, Hagg S, Carlsten A, Petzold M, Rehnberg C, Jonsson AK (2014) Economic impact of adverse drug events–a retrospective population-based cohort study of 4970 adults. PLoS One 9(3):e92061. doi:10.1371/journal.pone.0092061
21. Gyllensten H, Rehnberg C, Jonsson AK, Petzold M, Carlsten A, Andersson Sundell K (2013) Cost of illness of patient-reported adverse drug events: a population-based cross-sectional survey. BMJ Open 3(6). doi:10.1136/bmjopen-2013-002574
22. Limdi NA, Veenstra DL (2010) Expectations, validity, and reality in pharmacogenetics. J Clin Epidemiol 63(9):960–969. doi:10.1016/j.jclinepi.2009.09.006
23. Table of Pharmacogenomic Biomarkers in Drug Labeling. US Food and Drug Administration. http://www.fda.gov/drugs/scienceresearch/researchareas/pharmacogenetics/ucm083378.htm. Accessed 18 Oct 2014
24. Buchanan A, Califano A, Kahn J, McPherson E, Robertson J, Brody B (2002) Pharmacogenetics: ethical and regulatory issues in research and clinical practice. Report of the consortium on pharmacogenetics. Findings and recommendations. http://www.utexas.edu/law/faculty/jrobertson/finalcop.pdf. Accessed 04 Oct 2014
25. Convention for the protection of human rights and dignity of the human being with regard to the application of biology and medicine: convention on human rights and biomedicine (1996)

Council of Europe. http://conventions.coe.int/Treaty/en/Treaties/Html/164.htm. Accessed 11 Oct 2014
26. Bonter K, Desjardins C, Currier N, Pun J, Ashbury FD (2011) Personalised medicine in Canada: a survey of adoption and practice in oncology, cardiology and family medicine. BMJ Open 1(1):e000110. doi:10.1136/bmjopen-2011-000110. http://bmjopen.bmj.com/content/1/1/e000110.long. Accessed 04 Oct 2014
27. Issa AM, Tufail W, Hutchinson J, Tenorio J, Baliga MP (2009) Assessing patient readiness for the clinical adoption of personalized medicine. Public Health Genomics 12(3):163–169. doi:10.1159/000189629
28. Rogausch A, Prause D, Schallenberg A, Brockmoller J, Himmel W (2006) Patients' and physicians' perspectives on pharmacogenetic testing. Pharmacogenomics 7(1):49–59. doi:10.2217/14622416.7.1.49
29. O'Daniel J, Lucas J, Deverka P, Ermentrout D, Silvey G, Lobach DF, Haga SB (2010) Factors influencing uptake of pharmacogenetic testing in a diverse patient population. Public Health Genomics 13(1):48–54. doi:10.1159/000217795
30. Haga SB, O'Daniel JM, Tindall GM, Lipkus IR, Agans R (2012) Survey of US public attitudes toward pharmacogenetic testing. Pharmacogenomics J 12(3):197–204. doi:10.1038/tpj.2011.1
31. Haga SB, LaPointe NM (2013) The potential impact of pharmacogenetic testing on medication adherence. Pharmacogenomics J 13(6):481–483. doi:10.1038/tpj.2013.33
32. van Bokhoven MA Koch H van der Weijden T Grol RP Kester AD Rinkens PE Bindels PJ Dinant GJ (2009) Influence of watchful waiting on satisfaction and anxiety among patients seeking care for unexplained complaints. Ann Fam Med 7(2):112–120. doi:10.1370/afm.958
33. Peters N, Rose A, Armstrong K (2004) The association between race and attitudes about predictive genetic testing. Cancer Epidemiol Biomarkers Prev 13(3):361–365
34. Suther S, Kiros GE (2009) Barriers to the use of genetic testing: a study of racial and ethnic disparities. Genet Med 11(9):655–662. doi:10.1097/GIM.0b013e3181ab22aa
35. Roses AD (2000) Pharmacogenetics and the practice of medicine. Nature 405(6788):857–865. doi:10.1038/35015728
36. Genomic Testing ACCE Model List of 44 targeted questions aimed at a comprehensive review of genetic testing. Centers for Disease Control and Prevention, CDC. http://www.cdc.gov/genomics/gtesting/ACCE/acce_proj.htm. Accessed 04 Oct 2014
37. Ball M Personal Genome Project Study Material. http://www.personalgenomes.org/static/docs/harvard/v20120430-study-guide.pdf. Accessed 11 Oct 2014
38. Rosenfeld JA, Mason CE, Smith TM (2012) Limitations of the human reference genome for personalized genomics. PLoS One 7(7):e40294. doi:10.1371/journal.pone.0040294
39. Scott SA (2011) Personalizing medicine with clinical pharmacogenetics. Genet Med 13(12):987–995. doi:10.1097/GIM.0b013e318238b38c
40. van Staveren MC Guchelaar HJ van Kuilenburg AB Gelderblom H Maring JG (2013) Evaluation of predictive tests for screening for dihydropyrimidine dehydrogenase deficiency. Pharmacogenomics J 13(5):389–395. doi:10.1038/tpj.2013.25
41. Shields AE (2011) Ethical concerns related to developing pharmacogenomic treatment strategies for addiction. Addict Sci Clin Pract 6(1):32–43
42. Wang L, McLeod HL, Weinshilboum RM (2011) Genomics and drug response. N Engl J Med 364(12):1144–1153. doi:10.1056/NEJMra1010600
43. Au N, Rettie AE (2008) Pharmacogenomics of 4-hydroxycoumarin anticoagulants. Drug Metab Rev 40(2):355–375. doi:10.1080/03602530801952187
44. Valdes R Jr, Payne DA, Linder MW (eds) (2010) Laboratory analysis and application of pharmacogenetics to clinical practice. Laboratory medicine practice guidelines. National Academy of Clinical Biochemistry. https://www.aacc.org/~/media/practice-guidelines/pharmacogenetics/pgx_guidelines.pdf?la=en. Accessed Oct 2014
45. OECD, Organisation for Economic Co-operation and Development, Guidelines for quality assurance in molecular genetic testing (2007). http://www.oecd.org/science/biotech/38839788.pdf. Accessed 18 Oct 2014

46. Jorgensen AL, Hughes DA, Hanson A, van Eker D, Toh CH, Pirmohamed M, Williamson PR (2013) Adherence and variability in warfarin dose requirements: assessment in a prospective cohort. Pharmacogenomics 14(2):151–163. doi:10.2217/pgs.12.199
47. Foster MW, Mulvihill JJ, Sharp RR (2009) Evaluating the utility of personal genomic information. Genet Med 11(8):570–574. doi:10.1097/GIM.0b013e3181a2743e
48. Trosman JR, Van Bebber SL, Phillips KA (2010) Coverage policy development for personalized medicine: private payer perspectives on developing policy for the 21-gene assay. J Oncol Pract 6(5):238–242. doi:10.1200/JOP.000075
49. Shah RR, Shah DR (2012) Personalized medicine: is it a pharmacogenetic mirage? Br J Clin Pharmacol 74(4):698–721. doi:10.1111/j.1365–2125.2012.04328.x
50. Galesic M, Garcia-Retamero R (2010) Statistical numeracy for health: a cross-cultural comparison with probabilistic national samples. Arch Intern Med 170(5):462–468. doi:10.1001/archinternmed.2009.481
51. Scheuner MT, Sieverding P, Shekelle PG (2008) Delivery of genomic medicine for common chronic adult diseases: a systematic review. JAMA 299(11):1320–1334. doi:10.1001/jama.299.11.1320
52. Relling MV, Klein TE (2011) CPIC: clinical pharmacogenetics implementation consortium of the pharmacogenomics research network. Clin Pharmacol Ther 89(3):464–467. doi:10.1038/clpt.2010.279
53. Amstutz U, Carleton BC (2011) Pharmacogenetic testing: time for clinical practice guidelines. Clin Pharmacol Ther 89(6):924–927. doi:10.1038/clpt.2011.18
54. Haga SB, Burke W, Ginsburg GS, Mills R, Agans R (2012) Primary care physicians' knowledge of and experience with pharmacogenetic testing. Clin Genet 82(4):388–394. doi:10.1111/j.1399–0004.2012.01908.x
55. Zika E, Gurwitz D, Ibarreta D (2006) Pharmacogenetics and pharmacogenomics: state-of-the-art and potential socio-economic impact in the EU. Scientific and Technical Research Series. European Commission. EUR22214—DG Joint Research Centre, Institute for Prospective Technological Studies. http://ftp.jrc.es/EURdoc/eur22214en.pdf. Accessed 18 Oct 2014
56. Green JS, O'Brien TJ, Chiappinelli VA, Harralson AF (2010) Pharmacogenomics instruction in US and Canadian medical schools: implications for personalized medicine. Pharmacogenomics 11(9):1331–1340. doi:10.2217/pgs.10.122
57. Higgs JE, Andrews J, Gurwitz D, Payne K, Newman W (2008) Pharmacogenetics education in British medical schools. Genomic Med 2(3–4):101–105. doi:10.1007/s11568-009-9032-6
58. Priorities for Personalised Medicine (2008) Report of the President's Council of Advisors on Science and Technology. https://www.whitehouse.gov/files/documents/ostp/PCAST/pcast_report_v2.pdf. Accessed 04 Oct 2014
59. Kääriäinen H, Hietala M, Kristoffersson U, Nippert I, Rantanen E, Sequeiros J, Schmidtke J, Kerzin-Storrar L (2008) Recommendations for genetic counselling related to genetic testing. Eurogentest. http://www.eurogentest.org/index.php?id=674. Accessed 18 Oct 2014
60. Bryc K, Auton A, Nelson MR, Oksenberg JR, Hauser SL, Williams S, Froment A, Bodo JM, Wambebe C, Tishkoff SA, Bustamante CD (2010) Genome-wide patterns of population structure and admixture in West Africans and African Americans. Proc Natl Acad Sci U S A 107(2):786–791. doi:10.1073/pnas.0909559107
61. Zimmerman RK, Tabbarah M, Nowalk MP, Raymund M, Jewell IK, Wilson SA, Ricci EM (2006) Racial differences in beliefs about genetic screening among patients at inner-city neighborhood health centers. J Natl Med Assoc 98(3):370–377
62. Thompson HS, Valdimarsdottir HB, Jandorf L, Redd W (2003) Perceived disadvantages and concerns about abuses of genetic testing for cancer risk: differences across African American, Latina and Caucasian women. Patient Educ Couns 51(3):217–227
63. Haga SB, Kawamoto K, Agans R, Ginsburg GS (2011) Consideration of patient preferences and challenges in storage and access of pharmacogenetic test results. Genet Med 13(10):887–890. doi:10.1097/GIM.0b013e31822077a5

64. Wertz DC (2003) Ethical, social and legal issues in pharmacogenomics. Pharmacogenomics J 3(4):194–196. doi:10.1038/sj.tpj.6500188
65. Green RC, Berg JS, Grody WW, Kalia SS, Korf BR, Martin CL, McGuire AL, Nussbaum RL, O'Daniel JM, Ormond KE, Rehm HL, Watson MS, Williams MS, Biesecker LG (2013) ACMG recommendations for reporting of incidental findings in clinical exome and genome sequencing. Genet Med 15(7):565–574. doi:10.1038/gim.2013.73
66. Allyse M, Michie M (2013) Not-so-incidental findings: the ACMG recommendations on the reporting of incidental findings in clinical whole genome and whole exome sequencing. Trends Biotechnol 31(8):439–441. doi:10.1016/j.tibtech.2013.04.006
67. McGuire AL, Joffe S, Koenig BA, Biesecker BB, McCullough LB, Blumenthal-Barby JS, Caulfield T, Terry SF, Green RC (2013) Point-counterpoint. Ethics and genomic incidental findings. Science 340(6136):1047–1048. doi:10.1126/science.1240156
68. Wolf SM, Annas GJ, Elias S (2013) Point-counterpoint. Patient autonomy and incidental findings in clinical genomics. Science 340(6136):1049–1050. doi:10.1126/science.1239119
69. Gene Planet. http://www.geneplanet.com. Accessed 04 Oct 2014
70. Bunnik EM, Schermer MH, Janssens AC (2012) The role of disease characteristics in the ethical debate on personal genome testing. BMC Med Genomics 5:4. doi:10.1186/1755-8794-5-4
71. Additional Protocol to the Convention on human rights and biomedicine, concerning genetic testing for health purposes (2008) Council of Europe. http://conventions.coe.int/Treaty/en/Treaties/Html/203.htm. Accessed 11 Oct 2014
72. Bunnik EM, Janssens AC, Schermer MH (2013) A tiered-layered-staged model for informed consent in personal genome testing. Eur J Hum Genet 21(6):596–601. doi:10.1038/ejhg.2012.237
73. McGuire AL, Hamilton JA, Lunstroth R, McCullough LB, Goldman A (2008) DNA data sharing: research participants' perspectives. Genet Med 10(1):46–53. doi:10.1097/GIM.0b013e31815f1e00
74. Darst BF, Madlensky L, Schork NJ, Topol EJ, Bloss CS (2014) Characteristics of genomic test consumers who spontaneously share results with their health care provider. Health Commun 29(1):105–108. doi:10.1080/10410236.2012.717216
75. Bloss CS, Schork NJ, Topol EJ (2011) Effect of direct-to-consumer genomewide profiling to assess disease risk. N Engl J Med 364(6):524–534. doi:10.1056/NEJMoa1011893
76. Ayme S GL, Matthijs G, Waligora J, Aygun SB, Barton D, Borry P, Brdicka R, Brookes AJ, Dequeker E, De la Villa I, Goossens M, Howard HC, Kääriäinen H, Kroese M, Levy P, Macchia F, Macek M, Nourissier C, Owen L, Sanne J-L, Schmidtke J, Sequeiros J, Stenhouse S, Taruscio D, Westwood J (2013) European Workshop on Genetic Testing Offer in Europe, JRC 77944 EUR 25684 EN European Commission. http://www.eurogentest.org/fileadmin/templates/eugt/pdf/News_documents/JRC_Genetic_testing_offer_in_Europe.pdf. Accessed 11 Oct 2014
77. World Medical Association. http://www.wma.net/en/30publications/10policies/b3/. Accessed October 04, 2014
78. Directive 2001/20/EC of the European Parliament and of the Council of 4 April 2001 on the approximation of the laws, regulations and administrative provisions of the Member States relating to the implementation of good clinical practice in the conduct of clinical trials on medicinal products for human use Official Journal of the European Communities 2001;L121:34-44. http://www.eortc.be/services/doc/clinical-eu-directive-04-april-01.pdf. Accessed 04 Oct 2014
79. Additional Protocol to the Convention on human rights and biomedicine concerning biomedical research (2005) Council of Europe. http://conventions.coe.int/Treaty/en/Treaties/html/195.htm. Accessed 11 Oct 2014
80. Directive 2001/20/EC Article 4 Official Journal of the European Communities 2001;L121:34-44. http://www.eortc.be/services/doc/clinical-eu-directive-04-april-01.pdf. Accessed 11 Oct 2014

81. Sheehan M (2011) Can broad consent be informed consent? Public Health Ethics 4(3):226–235. doi:10.1093/phe/phr020
82. Helgesson G (2012) In defense of broad consent. Cambridge Quarterly of Healthcare Ethics 21(01):40–50
83. Segerdahl P (2012) How unspecific is broad consent? http://ethicsblog.crb.uu.se/2012/09/13/how-unspecific-is-broad-consent/. Accessed 04 Oct 2014
84. Caulfield T, Upshur RE, Daar A (2003) DNA databanks and consent: a suggested policy option involving an authorization model. BMC Med Ethics 4:E1
85. Hofmann B (2009) Broadening consent–and diluting ethics? J Med Ethics 35(2):125–129. doi:10.1136/jme.2008.024851
86. Hoppe N (2013) The issue with tissue: why making human biomaterials available for research purposes is still controversial. Diagn Histopathol 19:315–321
87. Pharmacogenetics: ethical issues (2003) London edn. Nuffield Council on Bioethics. http://nuffieldbioethics.org/wp-content/uploads/2014/07/Pharmacogenetics-Report.pdf. Accessed 18 Oct 2014
88. Directive 95/46/EC of the European Parliament and of the Council of 24 October 1995 on the protection of individuals with regard to the processing of personal data and on the free movement of such data Official Journal L 281, 23/11/1995 P. 0031–0050. http://eur-lex.europa.eu/legal-content/EN/TXT/?uri=uriserv:OJ.L_.1995.281.01.0031.01EN. Accessed 11 Oct 2014
89. Forrest LE, Delatycki MB, Skene L, Aitken M (2007) Communicating genetic information in families—a review of guidelines and position papers. Eur J Hum Genet 15(6):612–618
90. Gaff CL, Clarke AJ, Atkinson P, Sivell S, Elwyn G, Iredale R, Thornton H, Dundon J, Shaw C, Edwards A (2007) Process and outcome in communication of genetic information within families: a systematic review. Eur J Hum Genet 15(10):999–1011
91. Forrest LE, Curnow L, Delatycki MB, Skene L, Aitken M (2008) Health first, genetics second: exploring families' experiences of communicating genetic information. Eur J Hum Genet 16:1329–1335
92. Rathi V, Dzara K, Gross CP, Hrynaszkiewicz I, Joffe S, Krumholz HM, Strait KM, Ross JS (2012) Sharing of clinical trial data among trialists: a cross sectional survey. BMJ 345:e7570. doi:10.1136/bmj.e7570
93. Chalmers I (2013) Health Research Authority's great leap forward on UK trial registration. BMJ 347:f5776. doi:10.1136/bmj.f5776
94. Regulation No 536/2014 of the European Parliament and of the Council of 16 April 2014 on clinical trials on medicinal products for human use, and repealing Directive 2001/20/EC Official Journal of the European Union, L 158, 27.5.2014, 1–76. http://eur-lex.europa.eu/legal-content/EN/TXT/?uri=uriserv:OJ.L_.2014.158.01.0001.01.ENG. Accessed 11 Oct 2014
95. Eichler HG, Pignatti F, Flamion B, Leufkens H, Breckenridge A (2008) Balancing early market access to new drugs with the need for benefit/risk data: a mounting dilemma. Nat Rev Drug Discov 7(10):818–826. doi:10.1038/nrd2664
96. Guideline on key aspects for the use of pharmacogenomic methodologies in the pharmacovigilance evaluation of medicinal products (2013) European Medicines Agency. http://www.ema.europa.eu/docs/en_GB/document_library/Scientific_guideline/2014/01/WC500160232.pdf. Accessed 18 Oct 2014
97. Master Z, Nelson E, Murdoch B, Caulfield T (2012) Biobanks, consent and claims of consensus. Nat Methods 9(9):885–888. doi:10.1038/nmeth.2142
98. Caulfield T, Kaye J (2009) Broad consent in biobanking: reflections on seemingly insurmountable dilemmas. Med Law Int 102:85–100
99. Hansson MG, Dillner J, Bartram CR, Carlson JA, Helgesson G (2006) Should donors be allowed to give broad consent to future biobank research? Lancet Oncol 7(3):266–269
100. Dove ES, Knoppers BM, Zawati MH (2014) Towards an ethics safe harbor for global biomedical research. J Law Biosci 1(1):3–51
101. Rial-Sebbag E, Cambon-Thomsen A (2012) The emergence of biobanks in the legal landscape: towards a new model of governance. J Law Soc 39(1):113–130

102. Caenazzo L, Tozzo P, Pegoraro R (2013) Biobanking research on oncological residual material: a framework between the rights of the individual and the interest of society. BMC Med Ethics 14:17. doi:10.1186/1472-6939-14-17
103. van Diest PJ (2002) No consent should be needed for using leftover body material for scientific purposes. For. BMJ 325(7365):648–651
104. Savulescu J (2002) No consent should be needed for using leftover body material for scientific purposes. Against. BMJ 325(7365):648–651
105. Lewis C, Clotworthy M, Hilton S, Magee C, Robertson MJ, Stubbins LJ, Corfield J (2013) Consent for the use of human biological samples for biomedical research: a mixed methods study exploring the UK public's preferences. BMJ Open 3(8):e003022 doi:10.1136/bmjopen-2013-003022
106. Helgesson G, Dillner J, Carlson J, Bartram CR, Hansson MG (2007) Ethical framework for previously collected biobank samples. Nat Biotechnol 25(9):973–976
107. Recommendation Rec (2006) 4 of the Committee of Ministers to member states on research on biological materials of human origin. https://wcd.coe.int/ViewDoc.jsp?id=977859. Accessed 31 Oct 2014
108. Mascalzoni D, Hicks A, Pramstaller P, Wjst M (2008) Informed consent in the genomics era. PLoS Med 5(9):e192. doi:10.1371/journal.pmed.0050192
109. Kanelloupoulou NK, Kaye J, Whitley E, Creese S, Lund D, Hughes K (2011) Dynamic consent—a solution to a perennial problem. BMJ Recent Rapid Responses available at http://www.bmj.com/content/343/bmj.d6900?tab=responses. Accessed 04 Oct 2014
110. Segerdahl P (2013) Dynamic consent in biobank research: better than broad consent? http://ethicsblog.crb.uu.se/2013/10/02/dynamic-consent-in-biobank-research-better-than-broad-consent/. Accessed 04 Oct 2014
111. OECD Guidelines on human biobanks and genetic research databases (2009). http://www.oecd.org/science/biotech/44054609.pdf. Accessed 04 Oct 2014
112. Wolf SM, Crock BN, Van Ness B, Lawrenz F, Kahn JP, Beskow LM, Cho MK, Christman MF, Green RC, Hall R, Illes J, Keane M, Knoppers BM, Koenig BA, Kohane IS, Leroy B, Maschke KJ, McGeveran W, Ossorio P, Parker LS, Petersen GM, Richardson HS, Scott JA, Terry SF, Wilfond BS, Wolf WA (2012) Managing incidental findings and research results in genomic research involving biobanks and archived data sets. Genet Med 14(4):361–384. doi:10.1038/gim.2012.23
113. Commission Decision 2000/520/EC pursuant to Directive 95/46/EC of the European Parliament and of the Council on the adequacy of the protection provided by the safe harbour privacy principles and related frequently asked questions issued by the US Department of Commerce Official Journal of the European Communities L215:25/08/2000 P. 0007–0047. Accessed 31 Oct 2014
114. Frueh FW, Rudman A, Simon K, Gutman S, Reed C, Dorner AJ (2006) Experience with voluntary and required genomic data submissions to the FDA: summary report from track 1 of the third FDA-DIA-PWG-PhRMA-BIO pharmacogenomics workshop. Pharmacogenomics J 6(5):296–300
115. Roberts DE (2008) Is race-based medicine good for us?: African American approaches to race, biomedicine, and equality. J Law Med Ethics 36(3):537–545. doi:10.1111/j.1748-720X.2008.302.x
116. Brody H, Hunt LM (2006) BiDil: assessing a race-based pharmaceutical. Ann Fam Med 4(6):556–560
117. Bevan JL, Lynch JA, Dubriwny TN, Harris TM, Achter PJ, Reeder AL, Condit CM (2003) Informed lay preferences for delivery of racially varied pharmacogenomics. Genet Med 5(5):393–399. doi:10.109701.GIM.0000087989.12317.3F
118. Hresko AHS (2012) Insurance coverage policies for personalised medicine. J Pers Med 2:201–216

119. Weldon CB, Trosman JR, Gradishar WJ, Benson AB, 3rd, Schink JC (2012) Barriers to the use of personalized medicine in breast cancer. J Oncol Pract 8(4):e24–31. doi:10.1200/JOP.2011.000448
120. Phillips KA, Veenstra DL, Oren E, Lee JK, Sadee W (2001) Potential role of pharmacogenomics in reducing adverse drug reactions: a systematic review. JAMA 286(18):2270–2279
121. Juengst ET FM, Settersten RA Jr (2012) Personalised genomic medicine and the rhetoric of empowerment. Hastings Cent Rep 42(5):34–34. doi:10.1002/hast.65

Index

G. Grech, I. Grossman (eds.), *Preventive and Predictive Genetics: Towards Personalised Medicine,* Advances in Predictive, Preventive and Personalised Medicine 9,
DOI 10.1007/978-3-319-15344-5

MIX
Papier aus verantwortungsvollen Quellen
Paper from responsible sources
FSC® C105338
www.fsc.org

If you have any concerns about our products,
you can contact us on
ProductSafety@springernature.com

In case Publisher is established outside the EU,
the EU authorized representative is:
Springer Nature Customer Service Center GmbH
Europaplatz 3, 69115 Heidelberg, Germany

Printed by Libri Plureos GmbH
in Hamburg, Germany